AF854457

Avec le soutien du

CNL

Centre national du livre

Flag Varieties

ACTUALITÉS MATHÉMATIQUES

Collection dirigée par Lê Dung Tráng

V. Lakshmibai, N. Gonciulea

Flag Varieties

ACTUALITÉS MATHÉMATIQUES

HERMANN ÉDITEURS DES SCIENCES ET DES ARTS

ISBN 2 7056 6389 4

© 2001, HERMANN, ÉDITEURS DES SCIENCES ET DES ARTS, 293 RUE LECOURBE, 75015 PARIS

Toute reproduction ou représentation de cet ouvrage, intégrale ou partielle, serait illicite sans l'autorisation de l'éditeur et constituerait une contrefaçon. Les cas strictement limités à usage privé ou de citation sont régis par la loi du 11 mars 1957.

Summary

Preface

This book provides an introduction to flag varieties and their Schubert varieties. Because of their rich geometry and combinatorics, there are very many interesting and important geometric, combinatorial and representation-theoretic results on flag varieties in the literature over the past few decades due to several authors. Most of these results appear only in papers in journals. To our knowledge, except for Fulton's book "Young tableaux", there does not exist a book in the literature that could serve as a source – especially for graduate students – for learning about the geometric, combinatorial and representation-theoretic aspects of flag varieties and their Schubert varieties. The emphasis in Fulton's book "Young tableaux" is on the role Young tableaux and explains the above mentioned three-phased aspects of flag varieties from the Young tableaux viewpoint. Hence we felt the need in the literature for a book on the flag varieties that would serve as a reference for basic results on flag varieties and Schubert varieties. Needless to say that this is not an easy task, given the breadth and depth of this beautiful area of mathematics. Nevertheless, in this book we have attempted to deal at length with Schubert varieties in the Grassmannian as well as the flag variety. Thus the main theme of the book is "presentation of results for a Schubert variety". These results include the normality, Cohen-Macaulayness, ideal theory, singularity theory of Schubert varieties. Alongside, we have also included the discussion on certain affine varieties related to Schubert varieties – the classical determinantal varieties, ladder determinantal varieties, quiver varieties, varieties of complexes and certain affine toric varieties.

The prerequisite for this book is some familiarity with Commutative Algebra, Algebraic Geometry and Algebraic Groups. A basic reference to Commutative Algebra is [**49**], Algebraic Geometry [**67**], and Algebraic Groups [**22**], [**73**]. Most of the basic results from Commutative Algebra, Algebraic Geometry and Algebraic Groups are summarized in Chapters 1, 2 and 3. We have mostly used standard notation and terminology and have tried to keep notation to a minimum.

This book can be used for a year long course on flag and Schubert varieties. The material covered in this book should provide adequate preparation for graduate students and researchers in the areas of Algebraic Geometry and Algebraic Groups to work on open problems on Schubert varieties, and also for reading about "generalized Schubert varieties".

We are grateful to Victor Kreiman for helping us with the editing of this book.

N. Gonciulea
Ann Arbor, December 15, 1999.

V. Lakshmibai
Boston, December 15, 1999.

Introduction

Homogeneous spaces are fundamental objects in Geometry. The flag varieties constitute an important class of homogeneous spaces. Because of their rich geometry and combinatorics, the flag varieties form interesting and important objects of study in the areas of Algebraic Geometry, Algebraic Groups and Combinatorics. The Schubert varieties in the flag varieties provide a powerful inductive machinery for the study of flag varieties.

Let k be the base field, $V = k^n$, and $G = SL(V)$, the group of linear automorphisms of V of determinant 1. Fixing a basis in V, say the standard basis, we shall identify G with the group of $n \times n$-matrices of determinant 1. Let $\mathcal{F}(V)$ denote the set of all flags in V, namely $\mathcal{F}(V) = \{F = (V_0 \subset \cdots \subset V_i \subset \cdots \subset V_n),\ \dim V_i = i\}$. We have a canonical identification of the *flag variety* $\mathcal{F}(V)$ with G/B, where B is the subgroup of G consisting of upper triangular matrices (see Chapter 7 for details).

In the early part of this century, Schur (cf. **[145]**, **[146]**), Young (cf. **[163]**), Weyl (cf. **[162]**) etc., did a combinatorial analysis of G by studying the representations of G via the combinatorics of the "Young tableaux". During the middle part of this century, Borel (cf. **[24]**), Bott (cf. **[26]**), Weil (cf. **[161]**) etc., analyzed the geometric aspects of G through a study of the flag variety $\mathcal{F}(V)$ as explained in the next paragraph.

It is a well-known fact that the (finite-dimensional) irreducible representations of G are indexed by Young diagrams with columns of length at most $n - 1$. In characteristic 0, the representation theory of G is completely known; in view of complete reducibility of G in characteristic 0, all finite-dimensional G-modules will be known by knowing the irreducible ones and the latter are determined by the well-known Weyl characters (see Chapter 5). On the other hand, the finite-dimensional irreducible G-modules (in characteristic 0) can be geometrically realized as spaces of sections of effective line bundles on G/B (see Chapter 4). This should already give the reader a flavour of the interplay among geometry,representation theory and combinatorics (on the flag variety). It should be remarked that the modular theory of G (the representation theory of G in positive characteristics) is still open, though much progress has been made in this area (thanks to the works **[118]**, **[82]**, **[79]**, **[7]**).

In 1934, Ehresmann (cf. **[48]**) showed that the classes of the Schubert subvarieties in the Grassmannian give a $\mathbb{Z}$-basis for the cohomology ring of the Grassmannian, and thus established a key relationship between the geometry of the flag varieties and the theory of characteristic classes. In 1956 Chevalley (**[37]**) further enhanced this relationship by showing that the Schubert classes form a $\mathbb{Z}$-basis for the Chow ring of the generalized flag variety. The work of Ehresmann and Chevalley suggests that one approach to understand $\mathcal{F}(V)$ is through a study of its Schubert varieties (B-orbit closures in $\mathcal{F}(V)$, the B-action being the action induced by the natural G-action on $\mathcal{F}(V)$). At the end of the nineteenth century, Schubert (**[144]**) introduced the "Schubert calculus" which gives recipes to determine the numbers of solutions of geometric problems; for example, the number of lines in $\mathbb{P}^3(\mathbb{C})$ intersecting four lines in general position is two. Hodge (cf. **[71]**) used Schubert varieties in the Grassmannian and constructed a nice basis consisting of "standard monomials" in the *Plücker coordinates* for the homogeneous coordinate ring of the Grassmannian (as well as the homogeneous coordinate rings of its Schubert varieties) for the Plücker embedding. Hodge has also given a similar result for $SL(n)/B$ in terms of "standard Young tableaux". This work of Hodge was first generalized to classical groups in the series Geometry G/P I–V (cf. **[102]**, **[149]**, **[104]**, **[100]**, **[101]**, **[106]**), and then to Kac-Moody groups associated to symmetrizable Kac-Moody algebras (cf. **[115]**). We shall refer to this generalization as "Standard Monomial Theory". The Standard Monomial Theory has led to very many interesting and important geometric and representation-theoretic consequences. The geometric consequences are the vanishing of the higher cohomology groups of effective line bundles on G/B, normality and Cohen-Macaulayness of Schubert varieties, determination of singular loci of Schubert varieties, explicit bases for rings of invariants in the classical invariant theory, degeneration of $\mathcal{F}(V)$ to a toric variety etc (cf. **[106]**, **[77]**, **[105]**, **[104]**, **[61]**). The representation-theoretic consequences are the construction of explicit bases for the space of sections of effective line bundles over Schubert varieties and a positive-termed character formula for these spaces (*Lakshmibai-Seshadri character formula*), good filtration property for these spaces, generalized Littlewood-Richardson rule, branching rules etc (**[114]**, **[115]** **[116]**).

Our main focus in this book is on Schubert varieties in the Grassmannian as well as Schubert varieties in the flag variety $SL(n)/B$. After recalling generalities and some basic results on semisimple algebraic groups and their representations in Chapters 2, 3, 4 and 5, we have devoted Chapters 6–11 to the study of Schubert varieties. The Grassmannian and its Schubert varieties are studied in Chapter 6. After introducing them, we present the Standard Monomial Theory for them, and then deduce the geometric consequences – the vanishing of the higher cohomology groups of effective line bundles on G/B, normality and Cohen-Macaulayness of Schubert varieties. We present similar results for $SL(n)/B$ and its Schubert varieties in Chapter 7. In Chapter

7, we also present the ideal theory of Schubert varieties. In Chapters 8 and 9, we study the singularities of Schubert varieties in $SL(n)/B$. In Chapter 8, we also present the conjecture of [**103**] on the irreducible components of the singular locus of a Schubert variety (in $SL(n)/B$). In Chapters 10 and 11, we study the degenerations of Schubert varieties; in particular, we present the result that the flag variety $SL(n)/B$ degenerates to a toric variety (cf. Chapter 11). In Chapter 11, we also describe Gröbner bases for Schubert varieties in $SL(n)/B$. In Chapter 12, we present some important affine algebraic varieties related to Schubert varieties. Certain affine algebraic varieties – the classical determinantal varieties, ladder determinantal varieties, quiver varieties – are realized as some "nice" open subsets of certain Schubert varieties in $SL(n)/Q$ (for suitable parabolic subgroups Q), and hence are concluded to be normal, Cohen-Macaulay etc. In Chapter 12, we have also included a detailed treatment of the classical determinantal varieties – their normality, Cohen-Macaulayness, singular loci, relationship to classical invariant theory. Our treatment uses their relationship to Schubert varieties. In Chapter 13, we discuss the homology and the cohomology of the flag variety. We describe an algebraic and a geometric basis for the cohomology ring of the flag variety. Further, we present the Schubert polynomials. These results are then spelled out for the Grassmannian and the relationship between the Littlewood-Richardson coefficients and the structure coefficients in the cohomology ring of the Grassmannian is brought out.

It should be remarked that there is yet another important approach to the geometric study of Schubert varieties, namely, using the theory of *Frobenius splitting* (cf. [**122**], [**136**], [**135**], [**85**]). In this book, for the study of Schubert varieties, we have taken exclusively the "standard monomial theoretic approach". It should be remarked that the Frobenius splitting theory (for Schubert varieties) itself can be given a standard monomial theoretic treatment (cf. [**115**], [**88**]).

We have tried to keep the material presented in this book as self-contained as possible, by providing proofs of the results presented in Chapters 6–13, except for Chapter 9. (Chapter 9 is just a summary of results on rational smoothness and Kazdhan-Lusztig theory. Because of the vast nature of the Kazdhan-Lusztig theory, we decided not to include proofs of results presented in this chapter. Nevertheless, we have provided the relevant references for these results). Most of the proofs presented in this book should fall within the reach of the background material provided in Chapters 1–5. Cohomology theory of sheaves (not covered in the background material) is a prerequisite for the material on vanishing theorems discussed in Chapter 6, §5 and Chapter 7, §4. (We have included a brief review of cohomology theory of sheaves in Chapter 6, §5). Familiarity with basic results in Geometric Invariant Theory is required for the material discussed in Chapter 12, §2.10 - especially the proofs of Theorems 2.10.8 and 2.10.10 in Chapter 12; here again, we have included a brief review of the basi c definitions and results from Geometric Invariant Theory in Chapter 12, §2.10.

CHAPTER 1

Preliminaries

This chapter is a brief review of Commutative Algebra and Algebraic Geometry. We have included just those facts which will be required for the discussion of the topics included in the text. For a more in-depth treatment of Commutative Algebra, we refer the reader to [**49**]; and for Algebraic Geometry, we recommend [**67**], [**124**].

1. Commutative Algebra

Throughout this section, A shall denote a commutative ring with 1.

1.1. Noetherian rings

A ring A is said to be *noetherian* if every ideal of A is finitely generated, or, equivalently, A satisfies a.c.c. (*ascending chain condition*), namely every strictly increasing chain of ideals terminates or, equivalently, every nonempty collection of ideals has a maximal element relative to inclusion.

THEOREM 1.1.1. *(Hilbert Basis Theorem) If A is noetherian, then so is the polynomial ring $A[x]$. In particular, if k is a field, then $k[x_1, \ldots, x_n]$ is noetherian.*

1.2. Localization

Let S be a *multiplicative set* in A, i.e.,$0 \notin S$, $1 \in S$, and $a, b \in S \implies ab \in S$. The *ring of quotients* (also known as *ring of fractions*) $S^{-1}A$ is constructed using equivalence classes of pairs $(a, s) \in A \times S$, where $(a_1, s_1) \sim (a_2, s_2)$ if there exists $s \in S$ such that $s(s_2a_1 - s_1a_2) = 0$. Denoting the equivalence class of (a, s) by $\frac{a}{s}$, the multiplication and addition in $S^{-1}A$ are defined by

$$\frac{a_1}{s_1} \cdot \frac{a_2}{s_2} = \frac{a_1a_2}{s_1s_2}, \quad \frac{a_1}{s_1} + \frac{a_2}{s_2} = \frac{s_2a_1 + s_1a_2}{s_1s_2}.$$

We have the natural map $A \to S^{-1}A$, $a \mapsto \frac{a}{1}$, which is universal for homomorphisms from A rendering the elements of S invertible. The (prime) ideals of $S^{-1}A$ correspond bijectively to the (prime) ideals of A not meeting S.

If S is the set of all nonzero divisors in A, then $S^{-1}A$ is called the *full ring of fractions of A*; note that in this case $A \to S^{-1}A$ is injective. If A is an integral domain, then the full ring of fractions of A consists of fractions of the form $\frac{a}{s}$, with $a \in A$, $s \in A \setminus \{0\}$, and is in fact a field, called the *field of fractions* (or also *the quotient field*) of A.

If $S = A \setminus \mathcal{P}$, where $\mathcal{P}$ is a prime ideal of A, then $S^{-1}A$ is denoted $A_{\mathcal{P}}$. It is a local ring (i.e. has a unique maximal ideal $\mathcal{P}A_{\mathcal{P}}$, consisting of the nonunits in $A_{\mathcal{P}}$). The prime ideals of $A_{\mathcal{P}}$ correspond to the prime ideals of A contained in $\mathcal{P}$. If A is noetherian, then so is $A_{\mathcal{P}}$. If $\mathfrak{m}$ is a maximal ideal, then the fields $A/\mathfrak{m}$ and $A_{\mathfrak{m}}/\mathfrak{m}A_{\mathfrak{m}}$ are naturally isomorphic. The canonical map $A \to A_{\mathfrak{m}}$ induces a vector space isomorphism $\mathfrak{m}/\mathfrak{m}^2 \cong \mathfrak{m}A_{\mathfrak{m}}/(\mathfrak{m}A_{\mathfrak{m}})^2$.

For an A-module M, $S^{-1}M$ is defined by using classes of pairs $(m, s) \in M \times S$, where $(m_1, s_1) \sim (m_2, s_2)$ if there exists $s \in S$ such that $s(s_2m_1 - s_1m_2) = 0$. Then $S^{-1}M$ is an $S^{-1}A$ module (in a natural way), and is naturally isomorphic to $S^{-1}A \otimes M$. The functor $M \to S^{-1}M$ from A-modules to $S^{-1}A$-modules is exact, i.e., it takes exact sequences to exact sequences. Further, it preserves tensors and Hom's in the following sense: if M and N are A-modules, the natural map $S^{-1}(M \otimes_A N) \to S^{-1}M \otimes_{S^{-1}A} S^{-1}N$ is an isomorphism, and the natural map

$$S^{-1}(\mathrm{Hom}_A(M, N)) \to \mathrm{Hom}_{S^{-1}A}(S^{-1}M, S^{-1}N)$$

is an isomorphism if M is finitely generated.

1.3. Radical of an ideal and the nilradical of a ring

If I is an ideal in a ring A, then the set $\sqrt{I} = \{a \in A \mid a^n \in I\}$ is an ideal of A, called the *radical of* I. We also have

$$\sqrt{I} = \bigcap_{\mathcal{P} \supset I,\ \mathcal{P} \text{ prime}} \mathcal{P}.$$

Note that $I \subset \sqrt{I}$. The ideal I is called *radical* or *reduced* if $\sqrt{I} = I$. Prime ideals are examples of radical ideals.

The set of nilpotent elements in a ring A is an ideal denoted $\mathrm{nil}(A)$. Note that $\mathrm{nil}(A) = \sqrt{(0)}$, and $\mathrm{nil}(A/I) = \sqrt{I}/I$. The ring A is said to be *reduced* if $\mathrm{nil}(A) = (0)$.

If S is a multiplicative set, then

$$\sqrt{I \cdot S^{-1}A} = \sqrt{I} \cdot S^{-1}A.$$

In particular a ring A is reduced if and only if the full ring of fractions of A is reduced.

1.4. Transcendence degree

Let K be a field extension of k. The elements $a_1, \ldots, a_d \in K$ are said to be *algebraically independent over* k if no nonzero polynomial $f(x_1, \ldots, x_d)$ over k satisfies $f(a_1, \ldots, a_d) = 0$. A maximal subset of algebraically independent elements (over k) in K is called a *transcendence basis of* K/k. If K is a finitely generated extension of k, say $K = k(a_1, \ldots, a_n)$, then a transcendence basis can be chosen from among the a_i's, say $a_1, \ldots, a_d$. The integer d (which is independent of the choice of the transcendence basis) is called the *transcendence degree of* K *over* k, and is denoted $\mathrm{tr.deg}_k K$. We have $k(a_1, \ldots, a_d)$ is purely transcendental over k, and K is algebraic over $k(a_1, \ldots, a_d)$.

1.5. Integral closure

Let $B \supset A$ be an extension of rings. An element $b \in B$ is said to be *integral* over A if there exists a monic polynomial $f(x) \in A[x]$ having b as a zero. The *integral closure of A in B* is the subring of B consisting of elements of B which are integral over A. The ring B is said to be *integral over A* if every element of B is integral over A. An integral domain A is said to be *integrally closed* if A is the integral closure of A in F, F being the field of fractions of A. (Such a domain is also said to be a *normal* domain).

EXAMPLE 1.5.1. $\mathbb{Z}$ is integrally closed.

1.6. $\text{tr.deg}_k A$

Let $A = k[a_1, \ldots, a_n]$ be a finitely generated k-algebra; further, let A be an integral domain with quotient field K. Then $\text{tr.deg}_k A$ is defined as

$$\text{tr.deg}_k A = \text{tr.deg}_k K.$$

PROPOSITION 1.6.1. *(Noether normalization Lemma) Let $A = k[a_1, \ldots, a_n]$ be a finitely generated k-algebra. Further, let A be an integral domain. Let $d = \text{tr.deg}_k A$. Then there exist elements $x_1, \ldots, x_d \in A$ algebraically independent over k such that A is integral over $k[x_1, \ldots, x_d]$.* See [**124**] for a proof.

1.7. Krull dimension of a ring

Let A be a noetherian ring. The Krull dimension of A is defined to be the maximal length k of a chain of prime ideals $(0) \subset \mathcal{P}_1 \subset \mathcal{P}_2 \subset \cdots \subset \mathcal{P}_k \subset A$, and is denoted $\dim A$. We list below some facts on $\dim A$.

1. $\dim A = \text{Sup}_{\mathfrak{m}}(\dim A_{\mathfrak{m}})$, $\mathfrak{m}$ being a maximal ideal in A.
2. Let $A = k[x_1, \ldots, x_n]$. Then $\dim A = n$.
3. Let A be an integral domain which is a finitely generated k-algebra.
   Then $\dim A = \text{tr.deg}_k A$.
4. Let A be an integral domain which is a finitely generated k-algebra. Let $\mathcal{P}$ be a prime ideal in A. Then

$$\dim A/\mathcal{P} + \dim A_{\mathcal{P}} = \dim A.$$

Defining the height (resp. coheight) of a prime ideal $\mathcal{P}$ as $\dim A_{\mathcal{P}}$ (resp. $\dim A/\mathcal{P}$), the above formula is quite often expressed as

$$\text{height}\mathcal{P} + \text{coheight}\mathcal{P} = \dim A.$$

1.8. Cohen-Macaulay rings and modules

Let A be a noetherian ring, and M a finitely generated A-module.

DEFINITION 1.8.1. An element $a \in A$ is said to be a *zero divisor* in M if $am = 0$, for some nonzero element m in M.

Note that a is not a zero divisor if and only if the map $M \to M$, $m \mapsto am$ is a monomorphism.

DEFINITION 1.8.2. A sequence $a_1, \ldots, a_r$ of elements of A is said to be an *M-regular sequence*, or just an *M-sequence* if
(1) $(a_1, \ldots, a_r)M \neq M$
(2) a_i is not a zero divisor in $M_{i-1} := M/(a_1, \ldots, a_{i-1})M$, $1 \leqslant i \leqslant r$ (here, $a_0 = 0$).

Let now A be local with $\mathfrak{m}$ as its maximal ideal.

PROPOSITION 1.8.3. *Any two maximal M-sequences of elements in $\mathfrak{m}$ have the same length.* See **[49]** for a proof.

DEFINITION 1.8.4. Let A, M be as above. The length of a maximal M-sequence of elements in $\mathfrak{m}$ is defined as the *depth of M*, and is denoted depth$M$.

PROPOSITION 1.8.5. *Let A, M as above. Then* depth$M \leqslant \dim M$, *where* $\dim M = \dim(A/\text{ann}M)$, ann$M$ *being the annihilator of M.* (see **[49]** for a proof).

DEFINITION 1.8.6. Let A, M as above. Then M is said to be *Cohen-Macaulay* if either $M = (0)$, or depth$M = \dim M$. The ring A is said to be *Cohen-Macaulay* if A is Cohen-Macaulay as an A-module.

1.9. Regular rings

Let $(A, \mathfrak{m})$ be a local noetherian ring, and let $K = A/\mathfrak{m}$.

PROPOSITION 1.9.1. *(Nakayama's Lemma) Let A, M as above. Suppose that $\mathfrak{m}M = M$. Then $M = (0)$.* See **[49]** for a proof.

1.9.2. Consequences of Nakayama's Lemma

(1) Let A, M as above. Let N be a submodule of M such that $M = \mathfrak{m}M + N$. Then $M = N$. (This follows by applying Nakayama's Lemma to M/N; note that $\mathfrak{m} \cdot M/N = (\mathfrak{m}M + N)/N = M/N$).
(2) Let $V = M/\mathfrak{m}M$; note that V is a finite dimensional vector space over $K(= A/\mathfrak{m})$. Let $\dim_K M/\mathfrak{m}M = n$. Suppose that $x_1, \ldots, x_n$ are elements in M such that $\overline{x}_1, \ldots, \overline{x}_n$ in $M/\mathfrak{m}M$ is a K-basis for V. Then $x_1, \ldots, x_n$ generate M. (To see this, let N be the submodule of M generated by $x_1, \ldots, x_n$. Then under the canonical map $M \to M/\mathfrak{m}M$, N maps onto $M/\mathfrak{m}M$. Hence $N + \mathfrak{m}M = M$, and (1) implies $N = M$).

Let us now take $M = \mathfrak{m}$. Let $x_1, \ldots, x_n$ in $\mathfrak{m}$ be such that $\overline{x}_1, \ldots, \overline{x}_n$ is a K-basis for $\mathfrak{m}/\mathfrak{m}^2$. Then $x_1, \ldots, x_n$ generate $\mathfrak{m}$; further, n is the smallest number of elements in a generating set for $\mathfrak{m}$. Let $\dim A = r$. In general, we have $r \leqslant n$.

DEFINITION 1.9.3. $(A, \mathfrak{m})$ is said to be *regular* if $r = n$, or equivalently, $\mathfrak{m}$ is generated by r elements.

DEFINITION 1.9.4. A noetherian ring is said to be *regular* if $A_{\mathcal{P}}$ is a regular local ring for all prime ideals $\mathcal{P}$.

EXAMPLE 1.9.5. $k[x_1, \dots, x_n]$ is a regular ring.

THEOREM 1.9.6. *Let A be a regular ring. Then*

(1) *A is a normal domain,*
(2) *A is a U.F.D. – unique factorization domain (recall that a U.F.D. is a domain in which every nonzero nonunit element has an expression as a product of prime elements; here, an element $a \in A$ is said to be a prime element if the principal ideal Aa is a prime ideal).*

2. Affine varieties

Let k be the base field, which we suppose to be algebraically closed of arbitrary characteristic.

2.1. The affine space $\mathbb{A}^n$

We shall denote by $\mathbb{A}^n_k$, or just $\mathbb{A}^n$, the affine n-space, consisting of $(a_1, \dots, a_n)$, $a_i \in k$. For $P = (a_1, \dots, a_n) \in \mathbb{A}^n$, the a_i's are called the *affine coordinates of* P.

2.2. Affine varieties

Given an ideal I in $k[x_1, \dots, x_n]$, let

$$V(I) = \{(a_1, \dots, a_n) \in \mathbb{A}^n \mid f(a_1, \dots, a_n) = 0 \text{ for all } f \in I\}.$$

The set $V(I)$ is called an *affine variety*. Clearly $V(I) = V(\sqrt{I})$. Fixing a (finite) set of generators $\{f_1, \dots, f_r\}$ for I, $V(I)$ can be thought of as the set of common zeros of $f_1, \dots, f_r$.

Conversely, given a subset $X \subset \mathbb{A}^n$, let

$$\mathcal{I}(X) = \{f \in k[x_1, \dots, x_n \mid f(x) = 0 \text{ for all } x \in X\}.$$

We have

$$X \subset V(\mathcal{I}(X)), \quad I \subset \mathcal{I}(V(I)).$$

THEOREM 2.2.1. *(Hilbert's Nullstellensatz) Let I be an ideal in $k[x_1, \dots, x_n]$. Then $\sqrt{I} = \mathcal{I}(V(I))$.*

For a proof, refer to [**124**].

As a consequence of Hilbert's Nullstellensatz we obtain an inclusion-reversing bijection

$$\{\text{radical ideals in } k[x_1, \dots, x_n]\} \longleftrightarrow \{\text{affine varieties in } \mathbb{A}^n\},$$
$$I \longmapsto V(I)$$
$$\mathcal{I}(X) \longleftarrow X.$$

The noetherian property of $k[x_1, \dots, x_n]$ implies d.c.c. (descending chain condition) on the set of affine varieties in $\mathbb{A}^n$. Now, if $\mathfrak{m}$ is a maximal ideal, then by Hilbert's Nullstellensatz $V(\mathfrak{m}) \neq \emptyset$. Let $a \in V(\mathfrak{m})$. We have $\mathfrak{m} \subset \mathcal{I}(\{a\}) \subset k[x_1, \dots, x_n]$. Hence we get $\mathfrak{m} = \mathcal{I}(\{a\})$ (since $\mathfrak{m}$ is a maximal ideal). Conversely, if $a \in \mathbb{A}^n$, consider the homomorphism $k[x_1, \ldots, x_n] \to k$, $f \mapsto f(a)$. Its kernel is precisely $\mathfrak{I}(\{a\})$, and is a maximal ideal (since k is field). Thus, under the above bijection, points of $\mathbb{A}^n$ correspond 1-1 to the maximal ideals of $k[x_1, \ldots, x_n]$. (The maximal ideal corresponding to $(a_1, \ldots, a_n)$ is the ideal $(x_1 - a_1, x_2 - a_2, \ldots, x_n - a_n)$.)

2.3. Zariski topology on $\mathbb{A}^n$

Define a topology on $\mathbb{A}^n$ by declaring $\{V(I), I \text{ ideal in } k[x_1, \ldots, x_n]\}$ as the set of closed sets. We now check that this defines a topology on $\mathbb{A}^n$. We have
(1) $\mathbb{A}^n = V((0))$, $\emptyset = V(k[x_1, \ldots, x_n])$.
(2) $V(I) \cup V(J) = V(I \cap J)$.

The inclusion $V(I) \cup V(J) \subset V(I \cap J)$ is clear. To see the reverse inclusion, let $a \in V(I \cap J)$. If possible, let us assume that $a \notin V(I)$, $a \notin V(J)$. Assumption implies that there exist $f \in I$, $g \in J$ such that $f(a) \neq 0$, $g(a) \neq 0$. Since $fg \in I \cap J$, we have $f(a)g(a) = 0$, a contradiction. Hence our assumption is wrong, and the reverse inclusion follows.
(3) $\bigcap_\alpha V(I_\alpha) = V(\sum_\alpha I_\alpha)$ (clear).

REMARK 2.3.1. 1. $\mathbb{A}^n$ is a T_1-space for the Zariski topology, i.e. points are closed subsets.

2. $\mathbb{A}^n$ is *not* Hausdorff for the Zariski topology. For, consider $\mathbb{A}^1$. The closed subsets are precisely the finite sets, and hence no two nonempty open sets can be disjoint.

3. The d.c.c. on closed sets implies the a.c.c on open sets, i.e., any nonempty collection of open sets has a maximal element. Hence $\mathbb{A}^n$ is quasi-compact, i.e., every open cover admits a finite subcover (the term "quasi" is used since $\mathbb{A}^n$ is not Hausdorff).

4. If $k = \mathbb{C}$, then the zero-set of a polynomial $f \in \mathbb{C}[x_1, \ldots, x_n]$ is closed in the usual topology of $\mathbb{C}^n$, being the inverse image of the closed set $\{0\}$ in $\mathbb{C}$ under the continuous map $\mathbb{C}^n \to \mathbb{C}$, $a \mapsto f(a)$. The set of common zeros of a collection of polynomials is also closed in the usual topology, being the intersection of closed sets. Of course, the complex n-space $\mathbb{C}^n$ has plenty of other closed sets which are not obtained this way, as is clear in the case $n = 1$. Thus the usual topology is stronger than the Zariski topology.

2.4. Irreducible components

A topological space X is said to be *irreducible* if X cannot be written as the union of two proper nonempty open sets in X, or equivalently, any two nonempty open sets in X have a nonempty intersection, or equivalently, any nonempty open set is dense. A subspace of X is irreducible if and only if its closure $\overline{Y}$ is irreducible. By Zorn's lemma, every irreducible subspace of X is contained in a maximal one; the maximal irreducible subsets are closed, and are called the *irreducible components* of X.

2.5. Noetherian spaces

A topological space is said to be *noetherian* if every open set in X is quasi-compact, or equivalently if open sets satisfy the maximal condition or equivalently, if each nonempty collection of open sets has a maximal element, or equivalently, if open sets satisfy a.c.c, or equivalently, if closed sets satisfy d.c.c.

PROPOSITION 2.5.1. *Let X be noetherian. Then X has only finitely many irreducible components X_i, and $X = \bigcup X_i$.* For a proof, see **[22]**, or **[73]**.

2.6. Irreducible affine varieties

PROPOSITION 2.6.1. *A closed set X in $\mathbb{A}^n$ is irreducible if and only if $\mathfrak{I}(X)$ is prime. In particular, $\mathbb{A}^n$ is irreducible.*

PROOF. Denote $\mathfrak{I}(X)$ by I. Let X be irreducible. Let $f_1 f_2 \in I$. Then each $a \in X$ is a zero of f_1 or f_2. Thus $X \subset V(I_1) \cup V(I_2)$, where $I_1 = (f_1)$, $I_2 = (f_2)$. The irreducibility of X implies that either $X \subset V(I_1)$ or $X \subset V(I_2)$. Hence we obtain either f_1 or f_2 belongs to I. Thus I is prime.

Let now I be prime. If possible, let us assume that $X = X_1 \cup X_2$, with $X_i \subset X$, $i = 1, 2$ closed in X. Then we can find $f_i \in \mathfrak{I}(X_i)$ such that $f_i \notin I$. Now, for $a \in X$, $a \in X_1$ or $a \in X_2$. hence $f_1(a) = 0$ or $f_2(a) = 0$. This implies $f_1 f_2(a) = 0$ for all $a \in X$, i.e. $f_1 f_2 \in I$, with $f_i \notin I, i = 1, 2$. This contradicts the primality of I. Hence our assumption is wrong and the result follows. □

COROLLARY 2.6.2. *Under the bijection in 2.2, the prime ideals correspond to irreducible affine varieties.*

2.7. The affine algebra $k[X]$

A finitely generated k-algebra is also called an *affine k-algebra* (or simply an affine algebra, the field k being fixed). Let X be an affine variety in $\mathbb{A}^n$. The affine algebra $k[x_1, \ldots, x_n]/\mathfrak{I}(X)$ is called the *affine algebra of X*, and is denoted $k[X]$. Now each $\overline{f} \in k[X]$ defines a function $X \to k$, $a \mapsto f(a)$ (note that this is well-defined). Thus each element of $k[X]$ may be thought of as a polynomial function on X (with values in k). For this reason, $k[X]$ is also called the *algebra of polynomial functions on X*, or also the *algebra of regular functions on X*, or the *coordinate ring of X*. If X is irreducible, then $k[X]$ is an integral domain (since $\mathfrak{I}(X)$ is prime), and the quotient field $k(X)$ of $k[X]$ is called the *function field of X* (or *the field of rational functions on X*).

As in the case of $\mathbb{A}^n$, we see that we have a bijection between closed subsets of X and the radical ideals of $k[X]$, under which the irreducible closed subsets of X correspond to the prime ideals of $k[X]$. In particular, the points of X are in 1-1 correspondence with the maximal ideals of $k[X]$. Further, X is a noetherian topological space, and the *principal open subsets* $X_f = \{x \in X \mid f(x) \neq 0\}$, $f \in k[X]$, give a base for the Zariski topology.

2.8. Morphisms

Let $X \subset \mathbb{A}^n, Y \subset \mathbb{A}^m$ be two affine varieties. A *morphism* $\varphi : X \to Y$ is a mapping of the form $\varphi(a) = (\psi_1(a), \psi_2(a), \ldots, \psi_m(a))$, where $a \in X$, and for $i = 1, \ldots, m$, $\psi_i \in k[x_1, \ldots, x_n]$. A morphism $\varphi : X \to Y$ defines a k-algebras morphism $\varphi^* : k[Y] \to k[X]$ given by $\varphi^*(f) = f \circ \varphi$. We have

THEOREM 2.8.1. *The map $X \mapsto k[X]$ defines a (contravariant) equivalence of the category of affine varieties (with morphisms as defined above) and the category of affine k-algebras, i.e., finitely generated k-algebras (with k-algebra morphisms as morphisms). Further, irreducible affine varieties correspond to affine k-algebras which are integral domains.*

For a proof, see [**67**].

2.9. Products of affine varieties

The product of the Zariski topologies on $\mathbb{A}^n$ and $\mathbb{A}^m$ does not give the Zariski topology on $\mathbb{A}^{n+m}$. For example, in $\mathbb{A}^1 \times \mathbb{A}^1$ the only closed sets in the set product topology are finite unions of horizontal and vertical lines, while $\mathbb{A}^2$ has many more sets that are closed in the Zariski topology.

To arrive at a correct definition (so that we will have $\mathbb{A}^n \times \mathbb{A}^m \cong \mathbb{A}^{n+m}$), one takes the general category-theoretical definition, and defines the product as a triple (Z, p, q), where Z is an affine variety and $p : Z \to X$, $q : Z \to Y$ are morphisms such that given a triple (M, α, β), where M is an affine variety and $\alpha : M \to X$, $\beta : M \to Y$ are morphisms, there exists a unique morphism $\theta : M \to Z$ such that the following diagram is commutative:

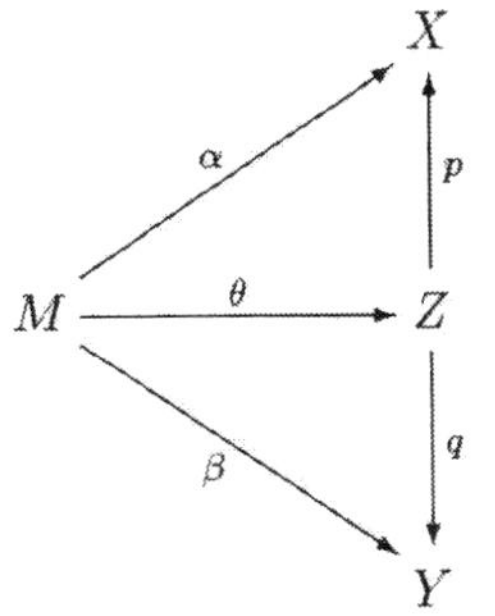

REMARK 2.9.1. If a product (Z, p, q) exists, then the underlying set has to be the cartesian product of the sets X and Y; for, consider $W = \{\text{a single point}\}$. Then by definition we have

$$\operatorname{Mor}(W, Z) = \operatorname{Mor}(W, X) \times \operatorname{Mor}(W, Y).$$

For any affine variety V, the morphisms of W to V clearly correspond to the points of V.

THEOREM 2.9.2. *(Existence of products) Let X, Y be two affine varieties with coordinate rings R, S respectively. Then the affine variety Z with coordinate ring $R \otimes_k S$ together with the canonical maps $p : Z \to X$, $q : Z \to Y$ (induced by $R \to R \otimes S$, $r \mapsto r \otimes 1$, and $S \to R \otimes S$, $s \mapsto 1 \otimes s$ respectively) is a product of X and Y.*

The uniqueness up to isomorphism of a product follows from the universal mapping property of a product. The product of X and Y is denoted by $(X \times Y, p, q)$.

REMARK 2.9.3. Let $X \subset \mathbb{A}^n$, $Y \subset \mathbb{A}^m$ be two affine varieties. Then the product variety $X \times Y$ (as defined above) is nothing but the set $X \times Y \subset \mathbb{A}^{n+m}$, together with the induced topology.

3. Projective varieties

3.1. The projective space $\mathbb{P}^n$

We shall denote by $\mathbb{P}^n_k$, or just $\mathbb{P}^n$, the set $(\mathbb{A}^{n+1} \setminus \{0\})/\sim$, where $\sim$ is the equivalence relation $(a_0, \dots, a_n) \sim (b_0, \dots, b_n)$ if there exists $\lambda \in k^*$ such that $(a_0, \dots, a_n) = \lambda(b_0, \dots, b_n)$. Thus, a point $P \in \mathbb{P}^n$ is determined by an equivalence class $[a_0, \dots, a_n]$, and for any $n+1$-tuple $(b_0, \dots, b_n)$ in this equivalence class, the b_i's will be referred as the *projective* (or *homogeneous*) *coordinates of* P.

Sometimes we write $\mathbb{P}^n$ also as $\mathbb{P}(V)$, where V is an $(n+1)$-dimensional k-vector space, the points of $\mathbb{P}^n$ being identified with 1-dimensional subspaces of V. Let $f(x_0, \dots, x_n) \in k[x_0, \dots, x_n]$. Further, let f be homogeneous of degree d. The homogeneity of f implies $f(\lambda x_o, \dots, \lambda x_n) = \lambda^d f(x_0, \dots, x_n)$, $\lambda \in k^*$. Hence it makes sense to talk about f being zero or nonzero at a point $P \in \mathbb{P}^n$.

3.2. Projective varieties

Let I be a homogeneous ideal in $k[x_0, \dots, x_n]$, i.e. for each $f \in I$, the homogeneous parts of f belong to I, or equivalently, I is generated by some homogeneous polynomials. Let

$$V(I) = \{P \in \mathbb{P}^n \mid f(P) = 0,\ f \text{ a homogeneous element of } I\}.$$

The set $V(I)$ is called a projective variety. Conversely, given a subset of $X \subset \mathbb{A}^n$, let $\mathfrak{I}(X)$ be the ideal generated by

$$\{f \in k[x_0, \dots, x_n],\ f \text{ homogeneous} \mid f(P) = 0 \text{ for all } P \in X\}.$$

As in the affine case, $\mathfrak{I}(X)$ is a radical ideal. We have a similar version (as in the affine case) of the Nullstellensatz with one minor adjustment, namely the ideal I_0 in $k[x_0, \dots, x_n]$ generated by $x_0, \dots, x_n$ is a proper radical ideal, but clearly has no zeros in $\mathbb{P}^n$. Deleting I_0, we have a similar formulation given by the following

THEOREM 3.2.1. *The maps $I \mapsto V(I)$, $X \mapsto \mathfrak{I}(X)$ define an inclusion-reversing bijection between the set of homogeneous radical ideals of $k[x_0, \dots, x_n]$ other than I_0 and the projective varieties in $\mathbb{P}^n$.*

3.3. Zariski topology on $\mathbb{P}^n$

This is defined in exactly the same way as in the affine case, by declaring $\{V(I) \mid I$ homogeneous radical ideal in $k[x_0, \ldots, x_n]$ other than $I_0\}$ as closed sets. As in the affine case, under the above bijection the homogeneous prime ideals (other than I_0) correspond to irreducible projective varieties. Let $U_i = \{[a] \in \mathbb{P}^n \mid a_i \neq 0\}$, $0 \leqslant i \leqslant n$. (These are some special open sets.) The map $U_i \to \mathbb{A}^n$,

$$[a] \mapsto \left(\frac{a_0}{a_i}, \ldots, \frac{a_{i-1}}{a_i}, \frac{a_{i+1}}{a_i}, \ldots, \frac{a_n}{a_i}\right)$$

defines an isomorphism of affine varieties. The quotients

$$\frac{a_0}{a_i}, \ldots, \frac{a_{i-1}}{a_i}, \frac{a_{i+1}}{a_i}, \ldots, \frac{a_n}{a_i}$$

are called the *affine coordinates on* U_i, $0 \leqslant i \leqslant n$. Note that $\{U_i, 0 \leqslant i \leqslant n\}$ is an open cover for $\mathbb{P}^n$.

3.4. Weil divisors

et $X = \mathbb{P}^n$. Let g be a homogeneous polynomial of degree d in $k[x_o, \ldots, x_n]$. We shall denote by H_g the *hypersurface in X determined by* g, namely

$$H_g = \{[a_0, \ldots, a_n] \in \mathbb{P}^n | g(a_0, \ldots, a_n) = 0\}.$$

The *degree* $\deg H_g$ of the hypersurface H_g is defined as $d = \deg g$.

Let Div$X$ be the free abelian group generated by the hypersurfaces in X. An element of Div$X$ is called a *Weil divisor*. Let $D \in \text{Div}X$, with $D = \sum_1^m n_i Y_i$, where $Y_i = H_{g_i}$, $g_i \in k[x_0, \ldots, x_n]$ homogeneous polynomials of degree d_i, $1 \leqslant i \leqslant m$. Then $\deg D$ is defined to be the integer $\sum_1^m n_i d_i$.

Let g be a homogeneous polynomial of degree d. Let $g = g_1^{m_1} \ldots g_r^{m_r}$, where the g_i's are irreducible, homogeneous polynomials of degree d_i. Let $Y_i = H_{g_i}$. Then $\sum_1^r m_i Y_i$ is called *the divisor of* g, and is denoted (g).

Let $f \in k(x_0, \ldots, x_n)$, say $f = g/h$, with g and h homogeneous polynomials of the same degree. Then $(g) - (h)$ is called the *divisor of* f, and is denoted (f). A divisor D in Div$X$ is called *principal* if $D = (f)$, for some f as above.

Two Weil divisors D_1, D_2 in Div$X$ are said to be *linearly equivalent*, and we write $D_1 \sim D_2$, if $D_1 - D_2$ is a principal divisor. The abelian group $\text{Cl}X = \text{Div}X/\sim$ is called the *divisor class group of* X.

LEMMA 3.4.1. *Let* H *be the hyperplane* $x_0 = 0$.

(1) *If* D *is a divisor of degree* d, *then* $D \sim dH$.

(2) *The map* deg : $\text{Cl}\,\mathbb{P}^n \to \mathbb{Z}$, $\deg[D]$ =*degree of* D *is well-defined, and is an isomorphism.*

PROOF. (1) Let us write $D = D_1 - D_2$, where D_1, D_2 are effective divisors. Let degree of D_i be d_i, $i = 1, 2$. Then the degree of D is $d_1 - d_2$. Let $D_i = (g_i)$. Now $D - dH = (f)$, where $f = g_1/x_0^d g_2$. From this (1) follows.

(2) The assertion that the map deg is well-defined follows from the fact that the degree of a principal divisor is 0. The assertion that deg is an isomorphism follows from (1) (since the degree of H is 1). □

4. Schemes – affine and projective

4.1. Presheaves

Let X be a topological space. Let top(X) be the category whose objects are open sets in X, and whose morphisms are inclusions. Let $\mathcal{C}$ be a category. A *$\mathcal{C}$-valued presheaf on X* is a contravariant functor $U \mapsto F(U)$ from top(X) to $\mathcal{C}$. Thus, if $V \subset U$ are open sets in X, then we have a $\mathcal{C}$-morphism

$$\mathrm{res}_V^U : F(U) \longrightarrow F(V)$$

called the *restriction map*.

A *morphism* $\varphi : F \to F'$ of presheaves is a morphism of functors. Suppose $\mathcal{C}$ is a category of "sets with structure", like groups, rings, modules etc., Then we say that F is a presheaf of groups, rings, modules etc., respectively.

If $x \in X$, then the collection $\{F(U), U \text{open neighborhood of } x\}$ is a directed system, and $F_x = \varinjlim_{x\in U} F(U)$ is called the *stalk of F at x*.

4.2. Sheaves

Let F be a $\mathcal{C}$-valued presheaf on X, where $\mathcal{C}$ is some category of "sets with structure". Then F is called a *sheaf* if it satisfies the following "sheaf axioms":
(S1) If $U \subset V$ are two open sets, and $f \in F(V)$, then $f|_U \in F(U)$
(S2) For every collection $\{U_i\}$ of open sets in X with $U = \bigcup U_i$
(a) if $f, g \in F(U)$ are such that $f|_{U_i} = g|_{U_i}$ for all i, then $f = g$.
(b) if $\{f_i \in F(U_i)\}$ is a collection such that $f_i|_{U_i \cap U_j} = f_j|_{U_i \cap U_j}$ for all i and j, then there exists an $f \in F(U)$ such that $f|_{U_i} = f_i$.

REMARK 4.2.1. To define a $\mathcal{C}$-valued sheaf $\mathcal{F}$ on X, where $\mathcal{C}$ is some category of "sets with structure", it suffices to define it on a base. For, let $\mathcal{B} = \{U_\alpha\}$ be a base for the topology. Given $\mathcal{F}(U_\alpha)$, for any open U we define

$$\mathcal{F}(U) = \varprojlim_{U_\alpha \subset U} \mathcal{F}(U_\alpha).$$

Example 1. Let X be a topological space, and for U open in X, let $F(U)$ be the ring of continuous real valued functions on U. The assignment $U \mapsto F(U)$, for U open, defines a sheaf.

Example 2. Let $X \in \mathbb{A}^n$ be an irreducible affine variety, with function field $k(X)$. Let $R = k[X]$. Define

$$\mathcal{O}_{X,x} = \{f \in k(X) \mid f \text{ is regular at } x\}$$

(note that f is regular at x if $f = g/h$, with $g, h \in R$, $h(x) \neq 0$). We have, $\mathcal{O}_{X,x}$ is simply $R_{\mathcal{P}}$, where $\mathcal{P}$ is the prime ideal $\{f \in R \mid f(x) = 0\}$; in particular, $\mathcal{O}_{X,x}$ is a local ring. The assignment $U \mapsto \bigcap_{x \in U} \mathcal{O}_{X,x}$, for U open defines a sheaf called the *structure sheaf*, and denoted $\mathcal{O}_X$.

Example 3. Let now X be an affine variety with irreducible components X_i. Let $U \subset X$ open, and $x \in U$. A function $f : U \to k$ is said to be *regular at* x if there exist $g, h \in k[X]$ and an open set $V \subset U$, $x \in V$, such that for all $y \in V$, $h(y) \neq 0$, $f(y) = \frac{g(y)}{h(y)}$. Set

$$\mathcal{O}_X(U) = \{f : U \to k \mid f \text{ is regular at all } x \in U\}.$$

Then $\mathcal{O}_X$ is a sheaf, again called the *structure sheaf*. Note that $\mathcal{O}_X(X) = k[X]$.

4.3. Sheafification

Let F be a $\mathcal{C}$-valued presheaf on X, where $\mathcal{C}$ is some category of "sets with structure". Then there is a sheaf F', called the *sheafification of* F, or the *sheaf associated with* F, and a morphism $f : F \to F'$ such that the map Mor$(F', G) \to$ Mor(F, G) (induced by f) is bijective whenever G is a sheaf.

4.4. Construction of F'

Let $E = \dot{\bigcup}_{x \in X} F_x$ (disjoint union of sets). Let $p : E \to X$ be the "projection map", namely $p(a) = x$, if $a \in F_x$. For U open in X, and $\sigma \in F(U)$, we have a canonical map (also denoted by σ) $\sigma : U \to E$, $\sigma(x) = \overline{\sigma_x} \in F_x$, where $\overline{\sigma_x}$ is the image of σ under the canonical map $F(U) \to F_x$. Equip E with the strongest topology which makes $\sigma : U \to E$ continuous for all $\sigma \in F(U)$, and all open U. It can be seen easily that a set G in E is open if and only if for every open U in X and $\sigma \in F(U)$, the set $W = \{x \in U \mid \overline{\sigma_x} \in G\}$ is open. The space (E, p) is called the *etale space of* F.

Let F' be the sheaf of continuous sections of p, i.e. for U open in X,

$$F'(U) = \{s : U \to E \text{ continuous such that } p \circ s = \mathrm{id}_U\}.$$

4.5. Given a sheaf $\mathcal{F}$ on X, $\mathcal{F}(X)$ is usually denoted by $\Gamma(X, \mathcal{F})$, and its elements are called *global sections* of $\mathcal{F}$

4.6. Geometric spaces

A *geometric space* is a topological space X together with a sheaf $\mathcal{O}_X$ of rings (commutative with identity element) on X whose stalks $\mathcal{O}_{X,x}$ are local rings. The sheaf $\mathcal{O}_X$ is called the structure sheaf of X. We denote the maximal ideal of $\mathcal{O}_{X,x}$ by $\mathfrak{m}_x$, and the residue class field by $k(x)$.

A morphism $(X, \mathcal{O}_X) \to (Y, \mathcal{O}_Y)$ of geometric spaces consists of a continuous map $f : X \to Y$ together with ring homomorphisms

$$f_U^V : \mathcal{O}_Y(V) \longrightarrow \mathcal{O}_X(U)$$

for $U \subset X$, $V \subset Y$ open sets such that $f(U) \subset V$. These maps are required to be compatible with the respective restriction maps in $\mathcal{O}_X$ and $\mathcal{O}_Y$. Then it is easy to see that if $x \in X$ and $y = f(x)$, then f induces a local homomorphism $f_x : \mathcal{O}_{Y,y} \to \mathcal{O}_{X,x}$, i.e. $f_x(\mathfrak{m}_y) \subset \mathfrak{m}_x$.

5. The scheme Spec(A)

Let A be a ring, and let Spec(A) be the set of all prime ideals in A. Define a topology on Spec(A) (called the *Zariski topology* on Spec(A) by declaring the closed sets as $V(I) = \{\mathfrak{p} \in \text{Spec}(A) \mid \mathfrak{p} \supset I\}$. For $Y \subset X$, let $\mathfrak{I}(Y) = \bigcap_{\mathfrak{p} \in Y} \mathfrak{p}$. Then $V(\mathfrak{I}(Y)) = \overline{Y}$. Further, we have $\mathfrak{I}(V(I)) = \sqrt{I}$. Thus we have an inclusion-reversing bijection between the set of closed sets in Spec(A), and the set of radical ideals in A, under which irreducible closed sets correspond to prime ideals. If A is noetherian, then Spec(A) is a noetherian topological space, and the irreducible components of X correspond to the minimal primes in A.

5.1. The principal open sets

Let $f \in A$, and $\mathfrak{p} \in \text{Spec}(A)$. Let $f(\mathfrak{p})$ be the image of f in the residue class field of $A_\mathfrak{p}$ (which is simply the field of fractions of $A/\mathfrak{p}$). Let $X = \text{Spec}(A)$, and $X_f = X \setminus V((f)) = \{\mathfrak{p} \in X \mid f(\mathfrak{p}) \neq 0\}$. The set X_f is called a *principal open set.* For any ideal I, we have $V(I) = \bigcap_{f \in I} V((f))$. Thus the principal open sets form a base of the Zariski topology on X.

5.2. Closed points of Spec(A)

Let $Y \subset X$. Then we have $V(\mathfrak{I}(Y)) = \overline{Y}$, the closure of Y. In particular, a point $\{y\}$ is closed if and only if the ideal y is maximal.

PROPOSITION 5.2.1. *Let A be an affine k-algebra. Let $x \in$ Spec(A). Let $\mathfrak{m}_x$ be the maximal ideal of the local ring $\mathcal{O}_{X,x}$, and $k(x) = \mathcal{O}_{X,x}/\mathfrak{m}_x$. Then x is a closed point if and only if $k(x) = k$.*

PROOF. Let $\mathfrak{p}$ be the prime ideal corresponding to x. Assume that x is a closed point. Then $\mathfrak{p}$ is a maximal ideal. Hence $k(x) = A/\mathfrak{p}$, and is a finite extension of k (since A is a k-algebra of finite type). Hence $k(x) = k$ (since $k = \overline{k}$).

Let now $k(x) = k$. We have $k \subset A/\mathfrak{p} \subset k(x)$ (note that $k(x)$ is the quotient field of $A/\mathfrak{p}$). Thus we obtain $A/\mathfrak{p} = k$, a field, and hence $\mathfrak{p}$ is a maximal ideal. This implies that x is a closed point. □

5.3. Geometric space structure on Spec(A)

Let $X = \text{Spec}(A)$, and let $\mathcal{O}_X$ be the sheaf associated to the presheaf $U \mapsto S_U^{-1}A$, where S_U (for U open in X) is the set of all $f \in A$ vanishing nowhere on U, i.e. $f(\mathfrak{p}) \neq 0$ for all $\mathfrak{p} \in U$ (here $f(\mathfrak{p})$ is as in 5.1). It is easily seen that the stalk $\mathcal{O}_{X,x}$ is simply A_x. Thus we have,

THEOREM 5.3.1. *(Spec(A), $\mathcal{O}_{\text{Spec}(A)}$) is a geometric space.*

REMARK 5.3.2. If A is an integral domain with quotient field K, then the A_x's are subrings of K, and $\mathcal{O}_X$ can be defined directly by $\mathcal{O}_X(U) = \bigcap_{x\in U} A_x$.

Let $X = \mathrm{Spec}(A)$, $Y = \mathrm{Spec}(B)$. Then it is seen easily that a morphism $\varphi : (X, \mathcal{O}_X) \to (Y, \mathcal{O}_Y)$ induces a ring homomorphism $B \to A$. Conversely, a ring homomorphism $B \to A$ induces a morphism $X \to Y$ (see **[124]** for a proof).

5.4. Affine schemes

DEFINITION 5.4.1. An *affine scheme* is a geometric space $(X, \mathcal{O}_X)$ which is isomorphic to $(\mathrm{Spec}\, A, \mathcal{O}_{\mathrm{Spec}\, A})$.

THEOREM 5.4.2. *The map $A \mapsto (\mathrm{Spec}(A), \mathcal{O}_{\mathrm{Spec}(A)})$ defines a (contravariant) equivalence of the category of rings and the category of affine schemes.*

DEFINITION 5.4.3. A *scheme* is a geometric space $(X, \mathcal{O}_X)$ in which every point has an open neighborhood U such that $(U, \mathcal{O}|_U)$ is an affine scheme.

6. The scheme Proj$S$

Let $S = \bigoplus_{d\geqslant 0} S_d$ be *a graded ring*, i.e., S_d is an abelian group and $S_d S_e \subset S_{d+e}$. Let S_+ be the ideal $S_+ = \bigoplus_{d>0} S_d$. Define Proj$S$ to be the set of all homogeneous prime ideals of S not containing S_+. For a homogeneous ideal $\mathfrak{a}$ of S, set

$$V(\mathfrak{a}) = \{\mathfrak{p} \in \mathrm{Proj}S \mid \mathfrak{p} \supseteq \mathfrak{a}\}.$$

LEMMA 6.0.1. (1) *If $\mathfrak{a}$ and $\mathfrak{b}$ are two homogeneous ideals in S, then $V(\mathfrak{ab}) = V(\mathfrak{a}) \cup V(\mathfrak{b})$.*

(2) *If $\{\mathfrak{a}_i\}$ is any family of homogeneous ideals in S, then $V(\sum \mathfrak{a}_i) = \bigcap V(\mathfrak{a}_i)$.*

See **[67]** for a proof.

In view of the Lemma, we can define a topology on Proj$S$ by taking the closed sets to be the subsets of the form $V(\mathfrak{a})$.

6.0.2. The structure sheaf $\mathcal{O}$ on Proj$S$**.** For $\mathfrak{p} \in \mathrm{Proj}S$, let $S_{(\mathfrak{p})}$ denote the *homogeneous localization* of S at $\mathfrak{p}$ consisting of elements of degree 0 in $S_\mathfrak{p}$, i.e.,

$$S_{(\mathfrak{p})} = \{\frac{f}{g} \in S_\mathfrak{p} \mid f, g \text{ homogeneous of the same degree}\}.$$

For an open set $U \subseteq \mathrm{Proj}S$, define $\mathcal{O}(U)$ as the set of functions $s : U \to \coprod S_{(\mathfrak{p})}$ such that the following hold:

(1) For each $\mathfrak{p} \in U$, $s(\mathfrak{p}) \in S_{(\mathfrak{p})}$.

(2) s is locally a quotient of elements of S, i.e., for each $\mathfrak{p}$ in U, there exists a neighborhood V of $\mathfrak{p}$ in U, and homogeneous elements $g, f \in S$ of the same degree, such that for all $\mathfrak{q} \in V, f \notin \mathfrak{q}$, $s(\mathfrak{q}) = \frac{g}{f}$ in $S_{(\mathfrak{q})}$.

It is clear that $\mathcal{O}$ is a presheaf of rings, with the natural restrictions, and it is also clear from the local nature of the definition that $\mathcal{O}$ is in fact a sheaf.

DEFINITION 6.0.3. We define $(\mathrm{Proj}S, \mathcal{O})$ to be Proj$S$ with the sheaf of rings constructed as above.

PROPOSITION 6.0.4. *$(ProjS, \mathcal{O})$ is a scheme. Further, for $\mathfrak{p} \in ProjS$, the stalk $\mathcal{O}_{\mathfrak{p}}$ is isomorphic to the local ring $S_{(\mathfrak{p})}$.* See **[67]** for a proof.

6.0.5. The special open subsets $D_+(f)$. For a homogeneous element $f \in S_+$, define $D_+(f) = \{\mathfrak{p} \in \text{Proj}S \mid f \notin \mathfrak{p}\}$. Then $D_+(f)$ is open in Proj$S$. Further, these open sets cover Proj$S$, and we have an isomorphism of geometric spaces:

$$(D_+(f), \mathcal{O}|_{D_+(f)}) \cong \text{Spec}S_{(f)},$$

where $S_{(f)}$ is the subring of elements of degree 0 in S_f.

See **[67]** for details.

7. Sheaves of $\mathcal{O}_X$-modules

Let $(X, \mathcal{O}_X)$ be a scheme. A sheaf on X is said to be a sheaf of $\mathcal{O}_X$-modules if for $U \subset X$ open, $F(U)$ is an $\mathcal{O}_X(U)$-module.

Example. Let M be an A-module. Then M defines a presheaf $\overline{M}$, namely for all $f \in A$, $\overline{M}(X_f) = M_f$ (note that in view of Remark 4.2.1, we have for any open U, $\overline{M}(U) = \varprojlim_{X_f \subset U} M_f$). Let $\tilde{M}$ be the sheaf associated to $\overline{M}$. Then $\tilde{M}$ is a sheaf of $\mathcal{O}_X$-modules.

7.1. The sheaf $\tilde{M}$ on Proj$S$

Let S be a graded ring, and M a graded S-module, i.e., M is an S-module together with a decomposition $M = \bigoplus_{d \in \mathbb{Z}} M_d$ such that $S_d \cdot M_r \subseteq M_{d+r}$. The sheaf $\tilde{M}$ on Proj$S$ is defined as follows. For $\mathfrak{p} \in \text{Proj}S$, let $M_{(\mathfrak{p})}$ denote the group of elements of degree 0 in $M_{\mathfrak{p}}$. For an open set $U \subseteq \text{Proj}S$, define $\tilde{M}(U)$ as the set of functions $s : U \to \coprod M_{(\mathfrak{p})}$ such that the following holds:

(1) For each $\mathfrak{p} \in U$, $s(\mathfrak{p}) \in M_{(\mathfrak{p})}$.

(2) s is locally a quotient, i.e., for each $\mathfrak{p}$ in U, there exists a neighborhood V of $\mathfrak{p}$ in U, and homogeneous elements $m \in M, f \in S$ of the same degree, such that for all $\mathfrak{q} \in V, f \notin \mathfrak{q}$, $s(\mathfrak{q}) = \frac{m}{f}$ in $M_{\mathfrak{q}}$.

We make $\tilde{M}$ into a sheaf with the natural restriction maps. We have the following facts:

(1) For $\mathfrak{p} \in \text{Proj}S$, the stalk $\tilde{M}_{\mathfrak{p}} \cong M_{(\mathfrak{p})}$.

(2) For a homogeneous element $f \in S_+$, we have,

$$\tilde{M}|_{D_+(f)} \cong (\tilde{M_{(f)}})$$

via the isomorphism $D_+(f) \cong \text{Spec}S_{(f)}$, where $M_{(f)}$ denotes the group of elements of degree 0 in the localized module M_f.

See **[67]** for details.

7.2. The twisting sheaf

Let $X = \text{Proj}S$. For $n \in \mathbb{Z}$, set $M(n)$ to be the graded S-module with $M(n)_k = M_{n+k}$, for all $k \in \mathbb{Z}$. Define the sheaf $\mathcal{O}_X(n)$ to be $\tilde{S(n)}$. The sheaf $\mathcal{O}_X(1)$ is called the *twisting sheaf of Serre*. For any sheaf F of $\mathcal{O}_X$-modules, we define

$$F(n) = F \otimes_{\mathcal{O}_X} \mathcal{O}_X(n).$$

7.3. The sheaf $f_*\mathcal{F}$

Let $f : X \to Y$ be a morphism between two schemes. Let $\mathcal{F}$ be any sheaf on X. We define the direct image sheaf $f_*\mathcal{F}$ on Y by $(f_*\mathcal{F})(V) = \mathcal{F}(f^{-1}(V))$ for any open subset V of Y.

7.4. The sheaf $f^*\mathcal{G}$

Let $f : X \to Y$ be as above. Let $\mathcal{G}$ be any sheaf on Y. We define the sheaf $f^{-1}\mathcal{G}$ on X to be the sheaf associated to the presheaf $U \mapsto \varinjlim_{V \supseteq f(U)} \mathcal{G}(V)$, for U open in X, where the limit is taken over all open subsets of Y containing $f(U)$.

We define the sheaf $f^*\mathcal{G}$, the inverse image of $\mathcal{G}$ by the morphism f, to be $f^{-1}\mathcal{G} \otimes_{f^{-1}\mathcal{O}_Y} \mathcal{O}_X$.

7.5. The sheaf $\mathcal{O}_X(1)$

For a subscheme X of $\mathbb{P}^n$, we define $\mathcal{O}_X(1)$ to be $i^*(\mathcal{O}_{\mathbb{P}^n}(1))$, where $i : X \to \mathbb{P}^n$ is the inclusion.

7.6. Locally free sheaves

A sheaf F of $\mathcal{O}_X$-modules is said to be *free* if F is isomorphic to a direct sum of copies of $\mathcal{O}_X$, and the number of copies of $\mathcal{O}_X$ is called its *rank*. The sheaf F is said to be *locally free* if X can be covered by open sets U such that $F|_U$ is a free $\mathcal{O}_U$-module (here $\mathcal{O}_U$ is just $\mathcal{O}_X|_U$). If the rank of (the locally free sheaf) F on any such U is the same, say n, then F is said to be a *locally free sheaf of rank n on X*. A locally free sheaf of rank 1 is called an *invertible sheaf*.

7.7. The scheme $V(\mathcal{E})$ associated to a rank n locally free sheaf $\mathcal{E}$

Let $(X, \mathcal{O}_X)$ be a scheme. Let $\mathcal{E}$ be a locally free sheaf of rank n on X. Let $\{X_\alpha = \mathrm{Spec} A_\alpha, \alpha \in I\}$ be an affine open cover of X. Let B_α be the symmetric algebra of the A_α-module $\mathcal{E}(X_\alpha)$. Let $Y_\alpha = \mathrm{Spec} B_\alpha$, and $f_\alpha : Y_\alpha \to X_\alpha$ the canonical morphism. For every pair $\alpha, \beta \in I$, we have canonical isomorphisms $f_\alpha^{-1}(X_\alpha \cap X_\beta) \xrightarrow{\cong} f_\beta^{-1}(X_\alpha \cap X_\beta)$ (by the compatibility conditions on the affine covering $\{X_\alpha, \alpha \in I\}$). Hence the family $\{Y_\alpha, \alpha \in I\}$ can be glued together to define a scheme $(Y, \mathcal{O}_Y)$, and a morphism $f : Y \to X$ such that $f|_{Y_\alpha} = f_\alpha$.

Remark. The scheme Y is unique (up to isomorphism), i.e. if (Y', f') is another such pair, then there exists an isomorphism $\varphi : Y \to Y'$ such that the following diagram is commutative.

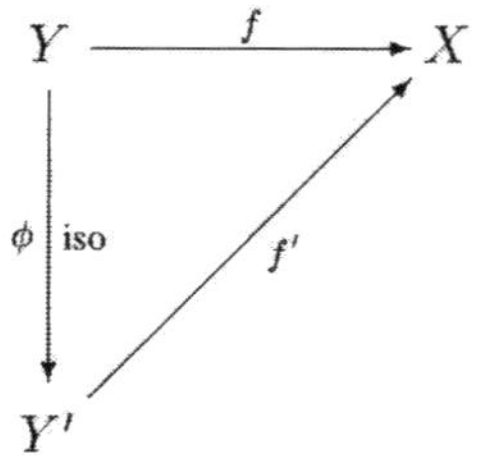

The scheme Y is denoted by $V(\mathcal{E})$.

For $U = \mathrm{Spec}A$ open affine in X such that $\mathcal{E}|_U$ is free, we have $f^{-1}(U) \cong \mathrm{Spec}A[x_1, \dots, x_n]$. We shall denote this isomorphism by ψ_U.

7.8. Vector bundles

Let $(X, \mathcal{O}_X)$ be a scheme. A (geometric) *vector bundle* of rank n over X is a scheme Y, and a morphism $f : Y \to X$, together with additional data consisting of an open covering $\{U_i\}$ of X, and isomorphisms $\psi_i : f^{-1}(U_i) \stackrel{\cong}{\to} \mathrm{Spec}A_i[x_1, \dots, x_n]$ (here $\mathrm{Spec}A_i = U_i$), such that for any i, j and for any open affine subset $V = \mathrm{Spec}A \subset U_i \cap U_j$, the automorphism $\psi_j \circ \psi_i^{-1}$ of $\mathrm{Spec}A[x_1, \dots, x_n]$ is given by a linear automorphism θ of $A[x_1, \dots, x_n]$, i.e. $\theta(a) = a$ for all $a \in A$, and $\theta(x_i) = \sum a_{ij}x_j$ for suitable $a_{ij} \in A$.

An *isomorphism*

$$g : \{Y, f, \{U_i\}, \{\psi_i\}\} \longrightarrow \{Y', f', \{U_i'\}, \{\psi_i'\}\}$$

of two rank n vector bundles over X is an isomorphism $g : Y \to Y'$ such that $f = f' \circ g$ and such that Y, f, together with the covering of X consisting of all the U_i's and U'_j's, and the isomorphisms ψ_i and $\psi'_j \circ g$ is also a vector bundle on X.

PROPOSITION 7.8.1. *Let $\mathcal{E}$ be a locally free sheaf of rank n on X. Let $\{U_i, i \in I\}$ be an affine covering of X such that $\mathcal{E}|_{U_i}$ is free (of rank n). Let $U_i = \mathrm{Spec}A_i$, and $\psi_i : f^{-1}(U_i) \to \mathrm{Spec}A_i[x_1, \dots, x_n]$ the corresponding isomorphism. Then $\{Y, f, \{U_i\}, \{\psi_i\}\}$ is a vector bundle of rank n over X.*

Conversely, let $f : Y \to X$ be a rank n vector bundle over X. Let $\mathcal{S}_f$ be the presheaf given by $\mathcal{S}_f(U) = \{s : U \to Y$ morphism such that $fs = \mathrm{id}_U\}$. Then $\mathcal{S}_f$ is in fact a locally free sheaf of $\mathcal{O}_X$-modules of rank n.

The proof is easy.

As a consequence of Proposition 7.8.1, we have

THEOREM 7.8.2. *There is a one-to one correspondence between isomorphism classes of locally free sheaves of rank n on X, and isomorphism classes of rank n vector bundles over X.*

In view of the above bijection, we use the words "locally free sheaf" and "vector bundles" interchangeably.

7.8.3. The space $H^0(X, \mathcal{E})$. Given a vector bundle $\mathcal{E}$ over X, we shall denote the space of sections of $\mathcal{E}$ over X by $H^0(X, \mathcal{E})$, namely

$$H^0(X, \mathcal{E}) = \{s : X \to V(\mathcal{E}) \mid f \circ s = \mathrm{id}_X\}$$

(here f is as in §7.7).

7.9. Invertible sheaves and line bundles

An *invertible sheaf* on X is a locally free sheaf of rank 1. The associated rank 1 vector bundle is called a *line bundle*.

7.10. The group Pic$X$

Let Pic$X$ be the *Picard group of* X, i.e. the group of isomorphism classes of invertible sheaves on X. Taking $X = \mathbb{P}^n$, we have, Cl$X \cong$ Pic$X$ (ClX being as in §3.4), and under this isomorphism $[H]$ corresponds to $\mathcal{O}_X(1)$, where H is the hypersurface $x_0 = 0$, and $\mathcal{O}_X(1)$ is the *twisting sheaf of Serre*. One refers to $\mathcal{O}_X(1)$ as the *hyperplane bundle*, or also as the *tautological line bundle* on X. We have

$$H^0(\mathbb{P}^n, \mathcal{O}_{\mathbb{P}^n}(r)) = \mathrm{Sym}^r(V^*), \quad r \in \mathbb{Z}^+, \quad V = k^{n+1}.$$

8. k-schemes and k-varieties

8.1. The affine k-scheme Spec$_k A$

Let A be an affine k-algebra. Let Spec$_k A = \{$ all maximal ideals in $A\}$. As seen in §2.2, there is a canonical bijection

$$\mathrm{Spec}_k A \longrightarrow \mathrm{Hom}_{k\text{-alg}}(A, k)$$

(namely $\mathfrak{m}$ corresponds to the homomorphism $e_{\mathfrak{m}} : A \longrightarrow k$, $e_{\mathfrak{m}}(f) = f(\mathfrak{m})$). The function $f : X \to k$, $f \in A$, $\mathfrak{m} \mapsto f(\mathfrak{m})$ determines f modulo the nilradical of A. We define the sheaf $\mathcal{O}_X$ exactly as in §4.2, Example 3.

8.2. k-spaces

A *k-space* is a geometric space $(X, \mathcal{O}_X)$ such that for U open in X, $\mathcal{O}_X(U)$ is a k-algebra. A *morphism* $f : X \to Y$ of k-spaces is a morphism of geometric spaces such that the ring homomorphisms F_U^V are k-algebra homomorphisms.

DEFINITION 8.2.1. A k-space isomorphic to $(X, \mathcal{O}_X)$, $X = \mathrm{Spec}_k A$ is called an *affine k-scheme*.

As in Theorem 5.4.2, we have

THEOREM 8.2.2. *The map $A \mapsto \mathrm{Spec}_k(A)$ is a contravariant equivalence from the category of affine k-algebras to the category of affine k-schemes.*

DEFINITION 8.2.3. An *affine k-variety* is a *reduced* affine k-scheme Spec$_k A$, (i.e., A is reduced, i.e. there are no nonzero nilpotent elements in A).

REMARK 8.2.4. Let $\mathrm{Spec}_k A$ be an affine k-variety. Then writing A as a quotient $k[x_1, \ldots, x_n]/I$, I is a radical ideal in $k[x_1, \ldots, x_n]$ and $\mathrm{Spec}_k A$ may be identified with $V(I) \in \mathbb{A}^n$. Conversely, given a radical ideal I in $k[x_1, \ldots, x_n]$, $V(I)$ may be identified with $\mathrm{Spec}_k A$, where $A = k[x_1, \ldots, x_n]/I$. Thus affine k-varieties are simply the affine varieties as defined in §2.2.

DEFINITION 8.2.5. A *k-scheme* is a k-space $(X, \mathcal{O}_X)$ such that X has a finite cover by open sets U such that $(U, \mathcal{O}_X|_U)$ is an affine k-scheme.

REMARK 8.2.6. If $(X, \mathcal{O}_X)$ is a k-scheme, then X is a noetherian space.

DEFINITION 8.2.7. Let $(X, \mathcal{O}_X)$ be a k-scheme which is reduced, i.e. $\mathcal{O}_{X,x}$ has no nonzero nilpotent elements. Then $(X, \mathcal{O}_X)$ is called a *prevariety*; equivalently, a prevariety is a noetherian topological space X endowed with a sheaf $\mathcal{O}_X$ such that X is the union of finitely many open subsets U_i, such that

$$(U_i, \mathcal{O}_X|_{U_i}) \cong (V_i, \mathcal{O}_{V_i}), \quad V_i \text{ affine } k\text{-variety.}$$

Example 1. Let $X = \mathbb{P}^n$, and U_i as in §3.3. Let $x \in X$, and $x \in U_i$ for some i. Set

$$\mathcal{O}_{X,x} = \mathcal{O}_{U_i,x},$$

where the right hand side is as in Example 2, in §4.2. It can be checked easily that $\mathcal{O}_{X,x}$ is independent of i. For U open in X, define

$$\mathcal{O}_X(U) = \bigcap_{x \in U} \mathcal{O}_{X,x}.$$

Note that $(U_i, \mathcal{O}_X|_{U_i}) \cong (U_i, \mathcal{O}_{U_i})$.

Example 2. Let Y be an open or closed subset of a prevariety X. We can define a sheaf $\mathcal{O}_Y$ of functions on Y as follows:

(a) Let Y be open in X. Then $\mathcal{O}_X$ restricts to a sheaf of functions on the noetherian space Y, and Y is a union of affine open sets.

(b) Let Y be closed in X. Let $\mathcal{O}_Y$ be the sheaf on Y associated to the presheaf $U \cap Y \mapsto \mathcal{O}_X(U)/I_U(Y)$, where

$$I_U(Y) = \{f \in \mathcal{O}_X(U) \mid f = 0 \text{ on } U \cap Y\}.$$

Note that in case U is affine, then $U \cap Y$ is just $\mathrm{Spec}_k\big(\mathcal{O}_X(U)/I_U(Y)\big)$. Thus an affine covering of X induces an affine covering of Y.

REMARK 8.2.8. The projective varieties in $\mathbb{P}^n$ as defined in §3.2 are simply the closed subsets of $\mathbb{P}^n$ with the prevariety structure as defined in Example 2.

Example 3. Let Y be a locally closed subset of a prevariety, i.e. $Y = U \cap F$, with U open and F closed in X. Then by Example 2, Y inherits a prevariety structure from X. With this prevariety structure, Y is called a subprevariety of X. If X is affine (resp. projective), then Y is said to be *quasiaffine* (resp. *quasiprojective*).

8.3. The function field $k(X)$ of an irreducible prevariety

Let X be an irreducible prevariety, covered by affine open sets U_i. The irreducibility of X implies that $U_i \cap U_j \neq \emptyset$. From this it follows that U_i, U_j have the same function field (where for an irreducible affine variety $Y = \mathrm{Spec}A$, the *function field* of Y is defined to be the quotient field of the integral domain A), which is called the function field $k(X)$ of X.

8.4. Morphisms of prevarieties

Let X, Y be prevarieties. A morphism $\phi : X \to Y$ is a continuous mapping such that if $U \subset X$, $V \subset Y$ are open sets such that $\phi(U) \subset V$, then $f \circ \phi \in \mathcal{O}_X(U)$ for all $f \in \mathcal{O}_Y(V)$.

We denote the map $\mathcal{O}_Y(V) \to \mathcal{O}_X(U)$, $f \mapsto f \circ \phi$ by ϕ^{*V}_{U} (note that ϕ^{*V}_{U} is in fact an algebra homomorphism). The map ϕ^* is called the *comorphism*.

REMARK 8.4.1. Note that the above definition is equivalent to the earlier one when X and Y are affine varieties.

8.5. The algebra $k[X]$

A morphism $\phi : X \to \mathbb{A}^1 (= k)$ is called a regular function on X. The set of all regular functions on X is a k-algebra, called the algebra (or ring) of regular functions on X, and is denoted $k[X]$. .

EXAMPLE 8.5.1. If $X = \mathrm{Spec}\, A$ is affine, then $k[X] = A$.

DEFINITION 8.5.2. A morphism $\phi : X \to Y$ is *affine* if for every affine open subset V of Y, $\phi^{-1}(V)$ is affine.

DEFINITION 8.5.3. A morphism $\phi : X \to Y$ is *finite* if for every affine open subset V of Y, $\phi^{-1}(U)$ is affine and $k[\phi^{-1}(V)]$ is integral over $\phi^*(k[V])$ (an A-algebra B is said to be integral over A if every b in B satisfies a monic polynomial equation over A).

DEFINITION 8.5.4. A morphism $\phi : X \to Y$ is *proper* if for every prevariety Z, the morphism $\phi \times \mathrm{id}_Z : X \times Z \to Y \times Z$ is closed, i.e., it maps a closed set onto a closed set.

DEFINITION 8.5.5. A morphism $\phi : X \to Y$ is dominant if $\phi(X)$ is dense in Y.

8.6. Products of prevarieties

The product of two prevarieties, X and Y, is defined as in §2.9.
Existence. We saw in §2.9 the existence if X and Y are affine varieties.
(a) Let X, Y be both irreducible prevarieties. For all affine open sets $U \subset X$, $V \subset Y$, and all finite sets of polynomial functions f_i on U, g_i on V, consider $(U \times V)_{\sum f_i g_i}$. It is easy to check that this collection is a base for a topology on $X \times Y$; further, the induced topology on $U \times V$ coincides with the Zariski topology. Set $k(X \times Y)$ as the

quotient field of the integral domain $k(X) \otimes k(Y)$. For $(x, y) \in (X \times Y)$ set

$$\mathcal{O}_{X \times Y, (x,y)} = \mathfrak{m}_x \otimes \mathcal{O}_{Y, y} + \mathcal{O}_{X, x} \otimes \mathfrak{m}_y \subset k(X \times Y),$$

where $\mathfrak{m}_x$ (resp. $\mathfrak{m}_y$) is the maximal ideal $\mathcal{O}_{X, x}$ (resp. $\mathcal{O}_{Y, y}$).

For U open in $X \times Y$, define

$$\mathcal{O}_{X \times Y}(U) = \bigcap_{(x,y) \in U} \mathcal{O}_{X \times Y, (x,y)}.$$

Then $(X \times Y, \mathcal{O}_{X \times Y})$ is a prevariety (for details refer to [**124**]).
(b) Let X, Y be arbitrary prevarieties with irreducible components X_i, Y_j. Form the prevarieties $X_i \times Y_j$ as in (a). Define a topology on $X \times Y$ by declaring that a set in $X \times Y$ is open if and only if its intersection with all $X_i \times Y_j$ is open. The space $X \times Y$ is then endowed with a sheaf structure as in Example 2, §4.2.

8.7. Hausdorff axiom

An algebraic variety is a prevariety which satisfies the *Hausdorff axiom*, namely the diagonal $\Delta(X) = \{(x, x), x \in X\}$ is closed in $X \times X$. (In the category of topological spaces, with $X \times X$ given the usual product topology, this condition is equivalent to the usual Hausdorff separation axiom). The Hausdorff axiom is equivalent to the following condition: given $f, g : Y \to X$, Y a prevariety, $\{y \in Y \mid f(y) = g(y)\}$ is closed in Y.

EXAMPLE 8.7.1. Affine and projective varieties are varieties. A subprevariety of a variety is a variety (see [**124**] for details).

Note that if X, Y are varieties, then so is $X \times Y$.

8.8. Rational functions

Let X be an algebraic variety. The dense open subsets U in X form an inverse system (under inclusion). Hence their algebras of regular functions $k[U]$ form an inductive system. Their inductive limit is called the *algebra* (or *ring*) *of rational functions* on V and is denoted $k(X)$.

EXAMPLE 8.8.1. If $X = \operatorname{Spec}_k A$ is affine, then we have $k[V] = A$ and $k(V)$ is just the full ring of fractions of A.

Suppose now that X is irreducible. Then each nonempty open set U in X is dense and irreducible. If f is a nonzero element of $k[V]$, then $U_f = \{x \in U \mid f(x) \neq 0\}$ is nonempty and open (and dense). Now $\frac{1}{f} \in k[U_f]$. Hence we obtain that $k(X)$ is a field, called the *function field* of X.

EXAMPLE 8.8.2. If $X = \operatorname{Spec}_k A$ is irreducible, then A is an integral domain and $k(X)$ is simply the quotient field of A.

REMARK 8.8.3. If $\phi : X \to Y$ is a dominant morphism, i.e., $\phi(X)$ is dense in Y, then ϕ^* induces an injective map $\phi^* : k(Y) \to k(X)$.

DEFINITION 8.8.4. Let X, Y be irreducible algebraic varieties. A morphism $\phi : X \to Y$ is *birational* if ϕ is dominant and the induced map $\phi^* : k(Y) \to k(X)$ is an isomorphism of fields.

8.9. Coherent and quasi-coherent modules

Let A be an affine k-algebra. let M be an A-module, $\tilde{M}$ the associated sheaf of $\mathcal{O}_X$-modules on $\mathrm{Spec}_k(A)$. Let now X be a k-scheme. A sheaf $\mathcal{F}$ of $\mathcal{O}_X$-modules is said to be *quasi-coherent* if X has an affine covering $\{U_i = \mathrm{Spec}_k A_i\}$ such that $\mathcal{F}|_U \cong \tilde{M}_i$ for some A_i-module M_i. If the covering $\{U_i\}$ can be chosen such that M_i is finitely generated then $\mathcal{F}$ is said to be *coherent*.

THEOREM 8.9.1. *The map $M \longmapsto \tilde{M}$ is an equivalence from the category of A-modules to the category of quasi-coherent $\mathcal{O}_X$-modules, $X = \mathrm{Spec}_k A$. Further, $\tilde{M}$ is coherent if and only if M is finitely generated.*

See [**67**] for a proof.

8.10. Immersions

A morphism $f : X \to Y$ is called a *closed immersion* if f maps X homeomorphically onto a closed subspace of Y, and the local homomorphisms $\mathcal{O}_{Y,f(x)} \to \mathcal{O}_{X,x}$ are surjective for all $x \in X$. Given a closed subscheme Y of $X = \mathrm{Spec}_k A$, it determines a quasi-coherent sheaf $\mathcal{I}$ of ideals on X. Conversely, given a quasi-coherent sheaf $\mathcal{I}$ of ideals on X, $Y := \mathrm{Supp}\,(\mathcal{O}_X/\mathcal{I})$ $(= \{x \in X \mid (\mathcal{O}_X/\mathcal{I})_x \neq (0)\})$ is closed, and $\mathcal{O}_X/\mathcal{I}$ is the "extension by zero" of a sheaf $\mathcal{O}_Y$ on Y. Further, there is a natural closed immersion $(Y, \mathcal{O}_Y) \hookrightarrow (X, \mathcal{O}_X)$. The k-scheme Y is called the *closed subscheme of X defined by* $\mathcal{I}$. See [**67**] for details.

THEOREM 8.10.1. *The map $I \mapsto$ Spec A/I is a bijection from the ideals of A to the set of closed subschemes of* $\mathrm{Spec}_k A$. *In particular, every closed subscheme of* $\mathrm{Spec}_k A$ *is affine.*

DEFINITION 8.10.2. An *open immersion* is a morphism of the form

$$(U, \mathcal{O}_X(U)) \to (X, \mathcal{O}_X),$$

where X is a k-scheme and U is an open subset. The scheme $(U, \mathcal{O}_X(U))$ is called an *open subscheme* of $(X, \mathcal{O}_X)$. A closed subscheme of an open subscheme is called a *locally closed subscheme*.

8.11. Dimension of a topological space

For a topological space X, $\dim X$ is defined as the supremum of the lengths n of chains $F_0 \subset F_1 \subset \cdots \subset F_n$ of distinct irreducible closed sets in X.

If $X = X_1 \cup \cdots \cup X_r$, the X_i's being the irreducible components of X, then we have $\dim X = \max_i \dim X_i$.

Let $X = \mathrm{Spec}(A)$. Then clearly $\dim X = \dim A$ (where $\dim A$ is the Krull dimension of A (cf. §1.7)). If in addition A is an integral domain and a finitely generated k-algebra, then $\dim X = \text{tr.deg}_k K (= \text{tr.deg}_k A)$, where K is the quotient field of A. More generally, if X is an irreducible variety with function field $k(X)$, then $\dim X = \text{tr.deg}_k k(X)$.

EXAMPLE 8.11.1. We have $\dim \mathbb{A}^n = n$, $\dim \mathbb{P}^n = n$.

EXAMPLE 8.11.2. If X is irreducible, then for any affine open subset U, $\dim U = \dim X$ (since $k(U) = k(X)$).

PROPOSITION 8.11.3. *Let X be an irreducible variety, and Y a proper, closed subset. Then* $\dim Y < \dim X$. See **[124]** or **[67]** for a proof.

The *codimension* $\text{codim}_X Y$ *of Y in X* is defined to be $\dim X - \dim Y$.

8.12. Complete varieties

A variety V is *complete* if for any variety X, the projection $p_X : V \times X \to X$ is a closed map. (In the category of Hausdorff spaces, the analogous property characterizes compact spaces. Thus "complete" for varieties is the analogue of "compact" for topological spaces).

REMARK 8.12.1. As immediate consequences of the definition, we have
1. A closed subvariety of a complete variety is complete.
2. A product of complete varieties is complete.

THEOREM 8.12.2. *A projective variety is complete.*
Refer to **[124]** for proof.

8.13. Normal varieties

A point x on a variety X is said to be *normal on* X if the local ring $\mathcal{O}_{X,x}$ is normal, i.e. $\mathcal{O}_{X,x}$ is an integral domain integrally closed in its quotient field. A variety X is said to be *normal* if every point x of X is normal on X.

8.14. Cohen-Macaulay varieties

A variety X is said to be *Cohen-Macaulay* if $\mathcal{O}_{X,x}$ is Cohen-Macaulay for all $x \in X$ (i.e. $\text{depth}\mathcal{O}_{X,x} = \dim \mathcal{O}_{X,x}$).

8.15. Smooth varieties

A point x on a variety X is called a *simple* or *smooth* point if $\mathcal{O}_{X,x}$ is a regular local ring, and X is said to be *nonsingular* or *smooth* if every point x of X is a smooth point.

8.16. The Zariski tangent space

Let x be a point on a variety X. Let $k(x) = \mathcal{O}_{X,x}/\mathfrak{m}_x$ be the residue field of the local ring $\mathcal{O}_{X,x}$. Here $k(x)(= k)$ is viewed as an $\mathcal{O}_{X,x}$-module. The Zariski tangent space of X at x is the space $T(G)_x := \text{Der}_k(\mathcal{O}_{X,x}, k(x))$, i.e., the space

$$\{D : \mathcal{O}_{X,x} \to k(x),\ k\text{-linear such that } D(ab) = D(a)b + aD(b)\}.$$

This is canonically isomorphic to $(\mathfrak{m}_x/\mathfrak{m}_x^2)^*$, the linear dual of $\mathfrak{m}_x/\mathfrak{m}_x^2$.

8.17. The differential $(d\phi)_x$

Given a morphism $\phi : X \to Y$ of varieties, the comorphism $\phi^* : \mathcal{O}_{Y,\phi(x)} \to \mathcal{O}_{X,x}$ induces a natural map $(d\phi)_x : T(X)_x \to T(Y)_{\phi(x)}$, called the *differential* of ϕ at x.

CHAPTER 2

Algebraic groups

In this chapter, we briefly review the basic material on linear algebraic groups. For details, we recommend the readers to refer to [**22**], [**73**].

1. Generalities

1.1. Algebraic groups

DEFINITION 1.1.1. An *algebraic group* is an algebraic variety G together with a group structure on the set G such that the multiplication map $\mu : G \times G \to G$, $(x, y) \mapsto xy$ and the inverse map $i : G \to G$, $g \mapsto g^{-1}$ are morphisms of algebraic varieties.

EXAMPLE 1.1.2. The *general linear group:* Let GL_n, be the group of invertible $n \times n$ matrices with entries in k. Then GL_n is an algebraic group, the variety structure for GL_n being obtained from the identification $GL_n \cong X_f$, where X is the set of $n \times n$ matrices with entries in k (identified with $\mathbb{A}^{n^2}$), f is the determinant function, and $X_f = \{x \in X \mid f(x) \neq 0\}$.

1.1.3. Closed subgroups of G. Let G be an algebraic group. A closed subset H of G which is also a subgroup of G acquires an algebraic group structure in a natural way. We shall refer to the algebraic group H (with this induced structure) as a closed subgroup of G.

DEFINITION 1.1.4. A *morphism* of algebraic groups is a morphism of algebraic varieties which is also a homomorphism of groups.

DEFINITION 1.1.5. A closed subgroup of GL_n is called a *linear algebraic group*.

EXAMPLE 1.1.6. (1) The *special linear group* SL_n, the subgroup of GL_n consisting of matrices of determinant 1.

(2) The *symplectic group* $\mathrm{Sp}_{2n} = \{g \in GL_{2n} \mid {}^t gEg = E\}$, where $J =$ anti-diagonal$(1, \cdots, 1)$ of size n, and $E = \begin{pmatrix} 0 & J \\ -J & 0 \end{pmatrix}$.

(3) The *orthogonal group* $O_n = \{g \in GL_n \mid {}^t gJg = J\}$, J being as above.

REMARK 1.1.7. In the sequel, we shall also denote the groups GL_n, SL_n, O_n, Sp_{2n} by $GL(n)$, $SL(n)$, $O(n)$, $Sp(2n)$ respectively.

1.2. Hopf algebraic structure on $k[G]$

Let $G = \mathrm{Spec}_k A$ be an affine algebraic group. Let $A = k[G]$. The algebraic structure on G induces maps $\mu_0 : A \to A \otimes A$, $i_0 : A \to A$, $e : A \to k$ inducing a Hopf algebra structure on the associative, commutative k-algebra A (see **[22]** for details).

PROPOSITION 1.2.1. *Let G be an algebraic group. Then*

(1) *There exists a unique irreducible component of G containing e. Denote this component by G°.*

(2) *G° is a normal subgroup of finite index in G, whose cosets are the connected as well as the irreducible components of G.*

(3) *Each closed subgroup of finite index in G contains G°.*

For a proof, see **[22]**.

1.3. Lie algebra of an algebraic group

For our purpose, we shall define a *Lie algebra* over k as a subspace of an associative k-algebra which is closed under the bracket operation $[x, y] = xy - yx$.

Let G be an affine algebraic group. Let $A = k[G]$. Let $\mathrm{Der}_k A$ be the set of all k-derivations of A, namely

$$\mathrm{Der}_k A = \{D : A \to A,\ k\text{-linear such that } D(fg) = fD(g) + gD(f),\ f, g \in A\}.$$

Then $\mathrm{Der}_k A$ is a Lie algebra (for the bracket operation $[D_1, D_2] = D_1 D_2 - D_2 D_1$). Let

$$\mathrm{Lie}\, G = \{D \in \mathrm{Der}_k A \mid l_x \circ D = D \circ l_x \text{for all } x \in G\}$$

be the space of left invariant derivations of G (here, l_x is the comorphism of the automorphism $G \overset{\cong}{\to} G$ given by $g \mapsto x^{-1} g$); it is a Lie subalgebra of $\mathrm{Der}_k A$, and is defined as the Lie algebra of G.

1.4. Finite dimensionality of Lie G

Let $x \in G$, and $T(G)_x$ $(= \mathrm{Der}_k(A, k(x)))$ be the tangent space to G at x (here $k(x) = k$ (= the residue field of $\mathcal{O}_{G,x}$) is viewed as an $\mathcal{O}_{G,x}$-module). Let $\delta_x : A \to k(x)$ be the evaluation map at x, defined by $\delta_x(f) = f(x)$, $f \in A$, $x \in G$. Given $D \in \mathrm{Der}_k A$, let $D_x = \delta_x \circ D$. Then it is clear that $D_x \in T(G)_x$. Thus D can be viewed as a vector field assigning to each $x \in G$ an element of $T(G)_x$. We recall (see **[22]**, **[73]** for a proof) the following

THEOREM 1.4.1. *The map $\nu : \mathrm{Lie}\, G \to T(G)_e$, $\nu(D) = \delta_e \circ D$ is an isomorphism of k-vector spaces, where e is the identity element in G.*

In particular, we obtain the following

COROLLARY 1.4.2. $\dim \mathrm{Lie}\, G = \dim G$ *(note that since e is a simple point of G, $\dim T(G)_e = \dim G$).*

EXAMPLE 1.4.3. (1) For $G = GL(V)$, Lie $G = \mathrm{gl}(V)(= \mathrm{End}_k V)$, the Lie algebra of linear endomorphisms of V.

(2) For $G = SL(V)(= \{\text{linear automorphisms of } V \text{ of determinant } 1\})$, Lie G = sl(V), the Lie algebra of linear endomorphisms of V of trace zero.

1.5. The convolutions $*X$ and $X*$

Let $\eta : T(G)_e \to \operatorname{Lie} G$ be the inverse of the map ν. Then for $X \in T(G)_e$, $\eta(X)$ is a left invariant vector field on G, which is equal to X at e. The left invariant vector field $\eta(X)$ is often denoted by $*X$, and we write $f \mapsto f * X$ for its action on A. Given $f \in A$, let $\mu^*(f) = \sum_i u_i \otimes v_i$, where $u_i, v_i \in A$ (here μ^* is the comorphism of the morphism $\mu : G \times G \to G$ given by multiplication). We have

$$f * X(x) = \sum_i u_i(x) \cdot Xv_i, \quad x \in G.$$

Thus

$$f * X = \sum_i u_i \cdot Xv_i.$$

Similarly, one can define the right invariant vector field $X*$ which is equal to X at e. It is a derivation which commutes with right translations, and its action on A is given by

$$X * f = \sum_i Xu_i \cdot v_i$$

(u_i, v_i being as above, namely $\mu^* f = \sum_i u_i \otimes v_i$).

LEMMA 1.5.1. *Let $x \in G$, $f \in A$, $X \in T(G)_e$. Then*

$$(f * X)(x) = X(l_{x^{-1}} f),$$
$$(X * f)(x) = X(r_x f),$$

where r_x (resp. l_x) is the comorphism of the automorphism $G \overset{\cong}{\to} G$, $g \mapsto gx$ (resp. $g \mapsto x^{-1}g$).

PROOF. Let $\mu^* f = \sum_i u_i \otimes v_i$. Then from above, we have

(1) $$f * X = \sum_i u_i \cdot Xv_i,$$

(2) $$(f * X)(x) = \sum_i u_i(x) \cdot Xv_i.$$

On the other hand, we have

(3) $$l_{x^{-1}}(f)(y) = f(xy) = \sum_i u_i(x) v_i(y).$$

Hence we obtain

(5) $$l_{x^{-1}}(f) = \sum_i u_i(x) v_i, X(l_{x^{-1}}(f)) = \sum_i u_i(x) X v_i.$$

The first assertion follows from (2) and (5). The second assertion is proved in a similar way. □

PROPOSITION 1.5.2. *Let H be a closed subgroup of G, and I the ideal in A defining H. Then*

(1) $T(H)_e = \{X \in T(G)_e \mid XI = 0\}$.

(2) $\operatorname{Lie} H = \{X \in \operatorname{Lie} G \mid I * X \subset I\} = \{X \in \operatorname{Lie} G \mid X * I \subset I\}$.

PROOF. (1) Let $i : H \to G$ be the inclusion map. Then the map $L_i : T(H)_e \to T(G)_e$ is simply the map $\operatorname{Der}_k(A/I, k(e)) \to \operatorname{Der}_k(A, k(e))$ induced by the canonical map $A \to A/I$. Hence X belongs to the image of L_i if and only if $XI = 0$.

(2) Let $X \in \operatorname{Lie} H$. If $f \in I$, and $h \in H$, then clearly $l_{h^{-1}}(f) \in I$. Hence $X(l_{h^{-1}}(f))) = 0$ by (1). On the other hand, we have $(f * X)(h) = X(l_{h^{-1}}(f))$. Hence we obtain $f * X \in I$, for $f \in I$. This implies $I * X \subset I$. Conversely, let $X \in \operatorname{Lie} G$ be such that $I * X \subset I$. If $f \in I$, then $X(f) = X(l_e f) = (f * X)(e) = 0$ (since $f * X \in I$, and $e \in H$). Thus we obtain $I * X \subset I$, and hence $X \in \operatorname{Lie} H$ by (1). This proves $\operatorname{Lie} H = \{X \in \operatorname{Lie} G \mid I * X \subset I\}$.

The equality $\operatorname{Lie} H = \{X \in \operatorname{Lie} G \mid X * I \subset I\}$ is proved in a similar way. □

As a consequence, we have

THEOREM 1.5.3. *Let H be a closed subgroup of G, and I the ideal in A defining H. Then*

(1) $H = \{g \in G \mid r_g(I) = I\} = \{g \in G \mid l_g(I) = I\}$.

(2) $\operatorname{Lie} H = \{X \in \operatorname{Lie} G \mid I * X \subset I\} = \{X \in \operatorname{Lie} G \mid X * I \subset I\}$

(here, r_g,(resp. l_g), is the comorphism of the automorphism $G \overset{\cong}{\to} G$ given by $x \mapsto xg$, resp. $x \mapsto g^{-1}x$).

PROOF. (1) Let $h \in H$, and $f \in I$. Then for $x \in H$, $r_h(f)(x) = f(xh) = 0$ (note that $xh \in H$). Thus we obtain $r_h(f) \in I$, and hence $r_h(I) = I$. Conversely, let $g \in G$ be such that $r_g(I) = I$. Then for $f \in I$, we have $f(g) = f(eg) = r_g f(e) = 0$ (since $r_g f \in I$, and $e \in H$).

(2) follows from Proposition 1.5.2. □

DEFINITION 1.5.4. Let V be a finite dimensional k-vector space. A morphism $\phi : G \to GL(V)$ is called a *rational representation* of G and V is called a G-module.

REMARK 1.5.5. Given a rational representation $\phi : G \to GL(V)$, the differential $(d\phi)_e : \operatorname{Lie} G \to \mathrm{gl}(V)(= \operatorname{End}_k V)$ defines a $\operatorname{Lie} G$-module structure on V.

THEOREM 1.5.6. *(Chevalley) Let G be an affine algebraic group, and H a closed subgroup. There exists an immersive representation $L : G \to GL(V)$ (i.e. L is an injective morphism) and a line D in V such that $H = \{g \in G \mid g \cdot D = D\}$,* $\operatorname{Lie} H = \{x \in \operatorname{Lie} G \mid X \cdot D \subset D\}$.

PROOF. Let I be the ideal in $A = k[G]$ vanishing on H. Now A being a finitely generated k-algebra, A is noetherian, and hence I is finitely generated, say $I = (f_1, \ldots, f_r)$. Let $\mu^*(f_i) = \sum_j u_{ij} \otimes v_{ij}$, $u_{ij}, v_{ij} \in A$ (μ^* being the comorphism of the morphism $\mu : G \times G \to G$ given by multiplication). Then for

$x \in G$, we have, $r_x(f_i)(y) = f_i(yx) = \sum_j u_{ij}(y)v_{ij}(x)$, and hence we obtain $r_x(f_i) = \sum_j v_{ij}(x)u_{ij}$. Thus the span E_i of $\{u_{ij}\}$, i being fixed contains $r_x(f_i)$, $x \in G$. If we set $E = \sum_{i=1}^r E_i$, then E is a finite dimensional G-stable (for the action by right translations, i.e. $x \cdot f(g) = f(gx)$, $x, g \in G$, $f \in k[G]$) subspace of $k[G]$ containing f_i, $1 \leqslant i \leqslant r$. Set $M = E \cap I$. Then clearly M generates I. Now both E and I are H-stable (for the action by right translations). Hence M is H-stable.

Claim

(1) $H = \{g \in G \mid r_g \cdot M = M\}$.

(2) $\operatorname{Lie} H = \{X \in \operatorname{Lie} G \mid X * M \subset M\}$.

Proof of Claim. (1) We have (cf. Theorem 1.5.3)

$$H = \{g \in G \mid r_g \cdot I = I\}. \tag{*}$$

Let $h \in H$. Then $r_h \cdot M = M$ (since M is H-stable). Conversely, let $g \in G$ be such that $r_g \cdot M = M$. We have $r_g \cdot I = r_g(MA) = (r_g M)A = MA = I$. Thus we obtain $r_g \cdot I = I$, and hence $g \in H$ (cf. $(*)$ above).

(2) By Theorem 1.5.3, we have

$$\operatorname{Lie} H = \{X \in \operatorname{Lie} G \mid X * I \subset I\}. \tag{**}$$

Let $X \in \operatorname{Lie} H$. Then $X * M \subset M$ (since $X * E \subset E$, and $X * I \subset I$). Conversely, let $X \in \operatorname{Lie} G$ be such that $X * M \subset M$. This implies

$$X * I = X * MA \subset (X * M)A + M \cdot (X * A) \subset MA = I.$$

Thus we obtain $X * I \subset I$, and hence $X \in \operatorname{Lie} H$ (cf $(**)$ above). This completes the proof of the *Claim*.

Let $d = \dim M$. Take $V = \wedge^d E$, $D = \wedge^d M \subset V$. Now the representation $\rho : G \to GL(E)$ induces $\alpha : G \to GL(V)$. (In case α is not immersive, then replace V by $V \oplus U$ using any immersive representation.) The required results follow from the following Lemma. □

LEMMA 1.5.7. *Let M be a d-dimensional subspace of a finite dimensional vector space E. Let $G = GL(E)$, $D = \wedge^d M \subset V = \wedge^d E$. Then*

(1) *for $g \in GL(E)$, $(\wedge^d g)D = D \iff gM = M$,*

(2) *for $X \in gl(E)$, $(d\wedge^d)(X)D \subseteq D \iff XM \subseteq M$.*

(Here $d\wedge^d$ is the differential of $\wedge^d : G \to GL(V)$.)

PROOF. The implication $\Longleftarrow$ in both (1) and (2) is clear.

Let $\{e_1, \ldots, e_m\}$ be a basis for E such that $\{e_1, \ldots, e_d\}$ is a basis for M. To prove (1), we may further suppose that for some $n \geqslant 1$, $\{e_n, \ldots, e_{n+d-1}\}$ is a basis for $g \cdot M$. Now, $(\wedge^d g)(e_1 \wedge \ldots \wedge e_d)$ is a multiple of $e_n \wedge \ldots \wedge e_{n+d-1}$; hence $(\wedge^d g)D = D$ implies $n = 1$, i.e., $gM = M$.

To prove (2), choose a basis of M, and index this basis so that X maps the span of the first part of the basis back into M and sends the second part of the basis to vectors not belonging to M. Consider a $Y \in gl(E)$ agreeing with X on the first part of this basis, and acting as 0 on the remainder of the basis for M. Clearly Y leaves M stable, and hence $(d\wedge^d)Y$ leaves D stable. Now $X - Y$ satisfies the hypothesis that $(d\wedge^d)(X - Y)D \subset D$. Note that if we could show that M is stabilized by $X - Y$, then it would follow that M is stabilized by X (since M is stabilized by Y). Thus we may replace X by $X - Y$, and suppose that $M \cap X(M) = (0)$. Choose a basis $u_1, \ldots, u_d$ of M so that $X(u_1) = \cdots = X(u_r) = 0$, while $X(u_{r+1}), \ldots, X(u_d)$ form a basis for $X(M)$. We have

$$\begin{aligned}(d\wedge^d)(X)(u_1 \wedge \ldots \wedge u_d) &= \sum u_1 \wedge \ldots \wedge X(u_i) \wedge \ldots \wedge u_d \\ &= cu_1 \wedge \ldots \wedge u_d, \quad c \in k\end{aligned}$$

(by hypothesis). But now the vectors $u_1 \wedge \ldots \wedge X(u_i) \wedge \ldots \wedge u_d$ are either 0 or are linearly independent, and none of them is equal to $u_1 \wedge \ldots \wedge u_d$. Hence we conclude that $X(u_i) = 0$, $1 \leqslant i \leqslant d$, and hence $XM = (0)$, i.e. X stabilizes M. □

1.6. Adjoint action

Let $\mathfrak{g} = \operatorname{Lie} G$. Let $g \in G$, and $\operatorname{Int} g : G \to G$ defined by $\operatorname{Int} g(x) = gxg^{-1}$, $x \in G$. Then $\operatorname{Int} g$ is an algebraic automorphism of G, called an *inner automorphism*. Its differential at e, denoted by $\operatorname{Ad} g$ is an automorphism of $\mathfrak{g}$. The invertibility of $\operatorname{Ad} g$ follows from the fact that $(\operatorname{Ad} g)(\operatorname{Ad} g^{-1}) = d(\operatorname{Int} g) \circ d(\operatorname{Int} g^{-1}) = d(\operatorname{Int} e) = \mathrm{id}_{\mathfrak{g}}$. Thus $\mathfrak{g}$ has a G-module structure via the adjoint action $\operatorname{Ad} : G \to \operatorname{Aut}(\mathfrak{g})$. One refers to this as the *adjoint representation of* G.

1.7. Quasiprojective variety structure on G/H

Let H be a closed subgroup of G. As a consequence of Theorem 1.5.6, G/H acquires an algebraic variety structure; in fact with notations as in Theorem 1.5.6, G/H gets identified with the orbit $G \cdot x \subset \mathbb{P}(V)$, where x is the point of $\mathbb{P}(V)$ determined by D. Then using the fact that $G \cdot x$ is a smooth variety which is open in its closure in $\mathbb{P}(V)$ (cf. **[22]**, §1.8), we obtain that G/H is a smooth quasiprojective variety (cf. Chapter 1, §8.2, Example 3). Further, as a consequence of Theorem 1.5.6, $\pi : G \to G/H$ is a *quotient morphism*, i.e. π has the following two properties:

(1) π is surjective and open

(2) If $U \subset G$ is open, then π^* (the comorphism of π) induces an isomorphism of $k[\pi(U)]$ onto the set of $f \in k[U]$ which are constant on fibers of $\pi|_U$.

If in addition, H is normal in G, then G/H is an affine algebraic group, and π is a morphism of algebraic groups.

1.8. The flag variety

Let $V = k^n$. A sequence $(0) = V_0 \subset V_1 \subset \cdots \subset V_n = V$, such that $\dim V_i = i$, is called a *full flag* in V. Let $\mathcal{F}(V)$ denote the set of all flags in V. Let $\{e_i, 1 \leqslant i \leqslant n\}$ be the standard basis of k^n. The flag $F_0 := (V_0 \subset \cdots \subset V_i \subset \cdots \subset V_n)$, where V_i is the span of $\{e_1, \ldots, e_i\}$ for $1 \leqslant i \leqslant n$, is called the *standard flag*. We have a natural action of $G = GL(V)$ on $\mathcal{F}(V)$, which is transitive. The isotropy group at F_0 is precisely the subgroup B of upper triangular matrices. Thus through the identification of $\mathcal{F}(V)$ with G/B, $\mathcal{F}(V)$ acquires the structure of a quasiprojective variety (cf. 1.7); with this structure, $\mathcal{F}(V)$ is in fact projective. To see this, we consider the canonical embedding

$$\theta : \mathcal{F}(V) \hookrightarrow G_{1,n} \times G_{2,n} \times \cdots \times G_{n-1,n},$$

where $G_{d,n}$, $1 \leqslant d \leqslant n-1$, is the Grassmannian variety consisting of d-dimensional subspaces of V (see Chapter 6 for details). We have

$$\mathcal{F}(V) = \Big\{(U_1, \ldots, U_{n-1}) \in \prod_{i=1}^{n-1} G_{i,n} \mid U_1 \subset \cdots \subset U_{n-1}\Big\}.$$

It is not difficult to see that the incidence relations $U_i \subset U_{i+1}$, $1 \leqslant i \leqslant n-2$ identify $\mathcal{F}(V)$ as a closed subset of $\prod_{i=1}^{n-1} G_{i,n}$ (see **[73]** for details).

2. Borel subgroups and Parabolic subgroups

2.1. Solvable groups

Let G be an algebraic group. It is easily seen that the derived subgroups D^iG, defined inductively by $D^0G = G$, $D^{i+1}G = (D^iG, D^iG)$, are closed normal subgroups of G, and are connected if G is connected.

DEFINITION 2.1.1. G is *solvable* if its derived series

$$\cdots \supset D^iG \supset D^{i+1}G \supset \ldots$$

terminates in $\{e\}$.

DEFINITION 2.1.2. A maximal closed connected solvable subgroup of G is called a Borel subgroup.

THEOREM 2.1.3. *(Borel Fixed Theorem) If a connected solvable group H acts on a (nonempty) complete variety X, then the fixed point set X^H is nonempty.*
(see **[22]** or **[73]** for a proof). As a corollary, we have

THEOREM 2.1.4. *(Lie-Kolchin Theorem) Let H be a connected solvable subgroup of $GL(V)$, V being a finite dimensional k-vector space. Then there exists a common eigenvector in V for all the elements in H, i.e. there exists $v \in V \setminus \{0\}$ such that $h \cdot v = \chi(h) \cdot v$ for some $\chi \in X(H)$, where $X(H) =$ Hom(H, k^*), the character group of H.*

PROOF. For the action of H on the complete variety $\mathcal{F}(V)$, there is a flag $(0) = V_0 \subset V_1 \subset \cdots \subset V_n = V$ fixed by H. Let v_1 be a generator of V_1. Then v_1 is a common eigenvector for all elements in H. □

THEOREM 2.1.5. *(Conjugacy of Borel subgroups) Let B be a Borel subgroup of G. Then G/B is a projective variety, and all other Borel subgroups are conjugate to B.*

PROOF. Let S be a Borel subgroup of largest dimension. Represent G in some $GL(V)$ with an 1-dimensional subspace V_1, whose stabilizer in G is precisely S (cf. Theorem 1.5.6). By Lie-Kolchin Theorem, the induced action of S on V/V_1 is triagonalizable. Thus there is a full flag $F = (V_0 \subset V_1 \subset \cdots \subset V_n)$ stabilized by S. In fact, S is the isotropy of F in G, by our choice of V_1. Thus we obtain an isomorphism $G/S \cong G \cdot F$. Now, the stabilizer of any other flag is solvable and has dimension less than or equal to dim S. From this it follows that $G \cdot F$ has the smallest possible dimension, and hence it is closed in the complete variety $\mathcal{F}(V)$ (cf. [**22**], §1.8), and hence it is complete (cf. Chapter 1, Remark 8.12.1). Thus G/S is complete; also being quasiprojective, it follows that G/S is projective.

Now, for the action of the connected solvable group B on the complete variety G/S (given by left multiplication), there is a point xS fixed by B, i.e. $BxS = xS$, i.e. $x^{-1}Bx \subset S$. Hence $x^{-1}Bx = S$ (since both of these are Borel subgroups). Both the assertions now follow from this. □

DEFINITION 2.1.6. A closed subgroup of G containing a Borel subgroup is called a *parabolic subgroup*.

We saw earlier (cf. §1.7) that given a closed subgroup H of G the quotient G/H is quasi-projective. The Proposition below gives a characterization of parabolic subgroups in terms of the underlying variety structure of G/H. We first recall some basic facts on complete and projective varieties. Recall that a variety X is complete if for all varieties Y, the projection map $X \times Y \to Y$ is closed.

Some basic facts on Complete Varieties (see [**67**] for details)

(1) A closed subvariety of a complete (resp. projective) variety is complete (resp. projective).
(2) Let X be complete, and $f : X \to Y$ a morphism of varieties. Then the image is closed in Y and is complete.
(3) Let $Y \subset X$, and Y be complete. Then Y is closed in X.
(4) A complete affine variety has dimension 0.
(5) A projective variety is complete.
(6) A complete quasi-projective is projective.

PROPOSITION 2.1.7. *Let Q be a closed subgroup of G. Then Q is parabolic if and only if G/Q is projective.*

PROOF. Let $Q \supset B$, where B is a Borel subgroup. We have a surjective morphism $G/B \to G/Q$ with G/B complete. Hence G/Q is projective (being complete and quasiprojective). Let now G/Q be projective. Then B has a fixed point xQ (for the action given by left multiplication, by Theorem 2.1.3). Hence $BxQ = xQ$, i.e. $x^{-1}Bx \subset Q$. Thus Q contains a Borel subgroup. □

THEOREM 2.1.8. *Let Q be a parabolic subgroup of G. Then $N_G(Q) = Q$. Thus Q is its own normalizer.*

See [**22**], §11.16 for a proof.

2.2. The variety $\mathcal{B}$

Let $\mathcal{B}$ denote the set of all Borel subgroups of G. Fix an element $B \in \mathcal{B}$. For any $B' \in \mathcal{B}$, there is a point in G/B, say xB, fixed by B', i.e. $B'xB = xB$.

CLAIM . *xB is the unique B'-fixed point in G/B*

PROOF. Let yB be another B'-fixed point. Then $xBx^{-1} = B' = yBy^{-1}$, and hence $y^{-1}x \in N_G B = B$. Thus $xB = yB$, and the claim follows. □

The claim enables us to define a map $\mathcal{B} \to G/B$, $B' \mapsto xB$, the unique B'-fixed point in G/B. This map is clearly injective. It is also surjective, since for $x \in G$, xBx^{-1} maps to xB under the above map. Further, the above map is G-equivariant (the action of G on $\mathcal{B}$ being by conjugation). Thus $\mathcal{B}$ acquires the structure of a projective variety.

2.3. Semisimple and reductive groups

Let $H = \cap_{B \in \mathcal{B}} B$. Then H^o is a (closed) normal subgroup of G, and is in fact the (unique) largest connected normal solvable subgroup (if $H \subset H'$, where H' is a connected normal solvable subgroup, then $H' \subset B$, for some $B \in \mathcal{B}$. Normality of H' implies that for any $x \in G$, $H' = xH'x^{-1} \subset xBx^{-1}$. Hence $H' \subset \cap_{B \in \mathcal{B}} B$ (by the Conjugacy Theorem 2.1.5), which implies that $H' = H$).

H^o is called the *radical of* G, and is denoted $R(G)$.

DEFINITION 2.3.1. G is said to be *semisimple* if $R(G) = \{e\}$.

EXAMPLE 2.3.2. $SL_n(k)$ is semisimple.

DEFINITION 2.3.3. G is *reductive* if $R(G)$ is a torus (i.e., $R(G) = (k^*)^n$, for some n); equivalently, G has no nontrivial connected normal unipotent subgroups.

EXAMPLE 2.3.4. (1) $GL_n(k)$ is reductive.
(2) A torus $S = (k^*)^m$ is reductive.

3. Representation Theory

3.1. Right and left regular representations

Let G be an affine algebraic group. For $g \in G$, let

$$r_g : k[G] \longrightarrow k[G],$$
$$l_g : k[G] \longrightarrow k[G]$$

be defined by

$$(r_g(f))(x) = f(xg) \qquad \text{for all } x \in G,$$
$$(l_g(f))(x) = f(g^{-1}x) \qquad \text{for all } x \in G.$$

Then $g \mapsto r_g$ (resp. $g \mapsto l_g$) defines a representation $(r, k[G])$ (resp. $(l, k[G])$) of G on $k[G]$, called the *right* (resp. *left*) *regular representation of* G; it is infinite dimensional (unless G is finite). Let $f \in k[G]$ be a nonzero regular function on G.

Let $\varphi : G \times G \to G$ be the action of G on itself by right multiplication. Then for $f \in k[G]$, we have, $\varphi^*(f) \in k[G] \otimes k[G]$, say $\varphi^*(f) = \sum f_i \otimes g_i$, $f_i, g_i \in k[G]$. For $x, y \in G$, we have $(r_x f)(y) = f(yx) = \sum f_i(y) g_i(x)$, and hence $r_x f = \sum g_i(x) f_i$. Let V be the span of the f_i's. Then V is a finite dimensional subspace of $k[G]$; further, $r_x f \in V$ for all $x \in G$. The subspace of V spanned by $\{r_x f, x \in G\}$ is a G-submodule of $k[G]$ (for the action of G on $k[G]$ by right translations). Thus we obtain the following

THEOREM 3.1.1. *Every element f in $k[G]$ belongs to a finite dimensional G-stable subspace of the right regular representation $(r, k[G])$.*

As immediate consequences, we get the following

COROLLARY 3.1.2. *$k[G]$ is a union of finite dimensional G-stable subspaces.*

COROLLARY 3.1.3. *Every G-irreducible subspace of $(r, k[G])$ is finite dimensional. (A G-module M is irreducible if it does not have any nonzero proper G-submodules).*

REMARK 3.1.4. We have similar results for $(l, k[G])$.

As an application of these results, we obtain

THEOREM 3.1.5. *Every finite dimensional G-module is isomorphic to a submodule of the direct sum of a certain number of copies of $(r, k[G])$.*

PROOF. Let (π, V) be a rational representation of G. For $v \in V$, $f \in V^*$, let $c_{f,v} : G \to k$ be defined by $c_{f,v}(g) = f(\pi(g)v)$ for $g \in G$. Then $c_{f,v} \in k[G]$. We have for $v \in V$, $f \in V^*$, $x, g \in G$,

$$\begin{aligned} r_x \cdot c_{f,v}(g) &= c_{f,v}(gx) = f(\pi(gx)v) \\ &= f(\pi(g)\pi(x)v) = c_{f,\pi(x)v}(g). \end{aligned}$$

Thus fixing f, the map λ_f defined by $v \mapsto c_{f,v}$ is a G-equivariant homomorphism of V into $k[G]$. If π is irreducible, and $f \neq 0$, then $\mathrm{Im}\lambda_f \neq 0$, and λ_f is an isomorphism $V \cong \lambda_f(V)$ of G-modules. In the general case, let $\{f_1, \ldots, f_n\}$ be a basis of V^*, where $n = \dim V$. Then the map

$$\begin{aligned} \lambda : V &\longrightarrow k[G]^n \\ v &\longmapsto (c_{f_1,v}, \ldots, c_{f_n,v}) \end{aligned}$$

is a G-equivariant homomorphism of V into $k[G]^n$; further, λ is clearly injective. The required result now follows. □

CHAPTER 3

Generalities on G/B and G/Q

This chapter reviews the basic results on the *generalized flag varieties* $G/B, G/Q$, G being semisimple and B (resp. Q) a Borel (resp. parabolic) subgroup. We have included the review of the root system, parabolic subgroups, Schubert varieties, Bruhat decomposition, and line bundles.

1. Abstract Root Systems

For generalities on Abstract Root Systems, we recommend **[75]**. Fix a vector space V over $\mathbb{R}$ with a non degenerate inner product $(\,,\,)$. Define the *reflection* relative to a nonzero vector $\alpha \in V$ to be the linear transformation on V given by

(1)
$$s_\alpha(v) = v - \frac{2(v,\alpha)}{(\alpha,\alpha)}\alpha.$$

Note, s_α maps α to $-\alpha$ and fixes the hyperplane perpendicular to α. Following the literature, we will use the notation

(2)
$$\langle v,\alpha\rangle = \frac{2(v,\alpha)}{(\alpha,\alpha)}.$$

REMARK 1.0.1. In the sequel, we denote $\langle v,\alpha\rangle$ also by (v,α^*).

An abstract *root system* R in V is defined by the following axioms:

(1) R is finite, spans V and does not contain 0.
(2) If $\alpha \in R$, the only multiples of α in R are $\pm\alpha$.
(3) For each $\alpha \in R$, R is stable under s_α.
(4) (Integrality) If $\alpha, \beta \in R$, then $s_\alpha(\beta) - \beta = \langle\beta,\alpha\rangle\alpha$ is an integral multiple of α.

The *rank* of R is defined to be $l = \dim(V)$. A subset $S = \{\alpha_1, \dots, \alpha_l\}$ of R is called a *base of the root system* if S is a vector space basis of V and if for any $\alpha = \sum c_i\alpha_i \in R$, all the coefficients c_i are integers with the same sign. Using this property, R can be partitioned into *positive roots* R^+ and *negative roots* R^- with respect to S. An element $\alpha \in R^+$ (resp. R^-) is usually denoted by $\alpha >$ (resp. $<$) 0. The elements of S are called *simple roots*. The integers $\langle\alpha_i,\alpha_j\rangle$, $1 \le i,j \le l$, are called the *Cartan integers*, and these completely determine the root system up to isomorphism.

Example. Let V be the hyperplane in $\mathbb{R}^3$ consisting of $\{(x_1, x_2, x_3) \in \mathbb{R}^3 \mid \sum x_i = 0\}$ equipped with the usual inner product. Let $\{e_1, e_2, e_3\}$ denote the standard basis for $\mathbb{R}^3$. Let $R = \{\pm(e_1 - e_2), \pm(e_2 - e_3), \pm(e_1 - e_3)\}$. Then R is a root system (of rank 2) in V. The set $S := \{e_1 - e_2, e_2 - e_3\}$ is a base of the root system R. Denoting S by $\{\alpha_1, \alpha_2\}$, the Cartan integer $\langle \alpha_i, \alpha_j \rangle$ is 2 or -1 according as $i = j$ or $i \neq j$.

The *Weyl group* W of the root system R is the subgroup of Aut$V$ generated by the s_α for $\alpha \in R$. It follows from the axioms of root systems that W is also finite (in fact W is a subgroup of Aut$R$). W is minimally generated by the *simple reflections* $s_{\alpha_1}, \ldots, s_{\alpha_l}$, denoted $s_1, \ldots s_l$. For $w \in W$, *length*(w) is defined to be the minimum of the length of an expression for w as a product of simple reflections; if length$(w) = p$, then an expression $w = s_{a_1} s_{a_2} \cdots s_{a_p}$ is called a *reduced expression* for w and the sequence $a_1 a_2 \ldots a_p$ is called a *reduced word* for w. Note that for a given w, there can be more than one reduced expression. For instance, in the above example we have, $W =$ the symmetric group S_3, the simple reflections are $s_1 :=$ the transposition $(1\,2)$, $s_2 :=$ the transposition $(2\;3)$. For $w =$ the permutation $(3\;2\;1)$, $s_1 s_2 s_1$, $s_2 s_1 s_2$ are both reduced expressions.

Any vector $\lambda \in V$ such that $\langle \lambda, \alpha \rangle \in \mathbb{Z}$ for all $\alpha \in R$ is called an abstract *weight*. The set of all weights form a lattice Λ_{wt} which contains the lattice generated by R as a subgroup of finite index. If $S = \{\alpha_1, \ldots, \alpha_l\}$ is a basis of R, then Λ has a corresponding basis consisting of *fundamental weights* $\lambda_1, \ldots, \lambda_l$ defined by $\langle \lambda_i, \alpha_j \rangle = \delta_{ij}$ (Kronecker delta). A weight $\lambda = \sum c_i \lambda_i$ is *dominant* if all the c_i's are positive integers. A weight $\lambda = \sum c_i \lambda_i$ is *regular* if all the c_i's are non-zero. In the sequel, we shall denote the set of dominant weights by Λ^+_{wt}.

The *Weyl chambers* in V are the connected components in the complement of the union of hyperplanes orthogonal to the roots. These chambers are in one-to-one correspondence with the bases of the root system. The set of dominant weights with respect to S is also called the *dominant chamber*.

2. Root systems of semisimple algebraic groups

Let G be a semisimple, simply connected, algebraic group defined over an algebraically closed field k of arbitrary characteristic. Denote the Lie algebra of G by $\mathfrak{g}$ or Lie G. Let $T = (k^*)^l$ be a maximal torus in G (a torus in G is a closed subgroup isomorphic to $(k^*)^r$ for some r); l is defined as the *rank* of G. Let $N(T)$ be the normalizer of T in G, and $W := N(T)/T$ be the Weyl group. Let B be a Borel subgroup of G (a maximal, connected, solvable subgroup), where $B \supset T$.

Let $X(T)$ be the character group of T (written additively); note that $X(T) = \mathrm{Hom}(T, k^*)$.

Let V be any finite dimensional T-module. Then it is known that V is the direct sum of *weight spaces* , i.e. $V = \oplus V_\chi$ where $V_\chi = \{v \in V : tv = \chi(t)v\}$ and $\chi \in X(T)$. We say that $\chi \in X(T)$ is a *weight in* V if $V_\chi \neq 0$. The dimension of V_χ is called the *multiplicity* of the weight χ. A nonzero vector in V_χ is called a *weight vector* of weight χ.

Let $\Phi(T, V)$ be the set of non-zero weights in V. Then $\Phi(T, V)$ is called the *set of roots* of T in V. For any $g \in G$, consider $\text{Inn}(g) : G \to G$ given by $x \mapsto gxg^{-1}$. This induces the *adjoint action* on the Lie algebra $\text{Ad}(g)$: Lie $G \to$ Lie G. For $V =$ Lie G as a T-module via the adjoint action, the corresponding $\Phi(T, \text{Lie } G)$ is called the *root system of* G *relative to* T. Recall, $\Phi(T, \text{Lie } G)$ is a root system contained in the vector space $X(T) \otimes_{\mathbb{Z}} \mathbb{R}$ of rank l (= dim$T$) (cf. **[22, 73]**). We shall denote this root system by R. Further, for each root $\beta \in \Phi(T, \text{Lie } G)$, the multiplicity is 1 (see **[22]**, **[73]** for details).

The subset $\Phi(T, \text{Lie } B)$ of $\Phi(T, \text{Lie } G)$ is called the set of *positive roots of* R *relative to* B, denoted R^+. The indecomposable elements in R^+, i.e., the set of all α such that α cannot be expressed as a positive sum of other elements in R^+, is a set of simple roots for the root system R which we shall denote by S.

The elements in $X(T)$ which lie in the dominant chamber with respect to S are called *dominant characters*.

Let S be the set of simple roots relative to B and T. Let $\{\omega_i : 1 \leq i \leq l\}$ where $l = \text{rank}(G)$ be the associated fundamental weights. Given $\lambda \in X(T)$, say $\lambda = \sum a_i\omega_i$ for $a_i \in \mathbb{Z}$, λ is said to be *dominant* if $a_i \geq 0$ for all i and λ is said to be *regular* if all a_i are non-zero.

Example. Let $G = SL_n(k)$. Let T be the subgroup of diagonal matrices and B the subgroup of upper triangular matrices. Then T (resp. B) is a maximal torus (resp. a Borel subgroup). Further, $N(T)$ is the subgroup consisting of permutation matrices, and W is the symmetric group S_n. We may identify $X(T) \otimes_{\mathbb{Z}} \mathbb{R}$ with the hyperplane in $\mathbb{R}^n$ consisting of $\{(x_1, \cdots x_n) \in \mathbb{R}^n \mid \sum x_i = 0\}$. Let us denote by $\epsilon_i - \epsilon_j$ the character (in $X(T)$) sending a diagonal matrix $D = \text{diag}(t_1, \cdots t_n)$ to $t_i t_j^{-1}$. We have

$$\begin{aligned} \mathfrak{g} = \text{Lie } G = sl_n(k) \quad &(= \{n \times n \text{ traceless matrices with entries in } k\}) \\ = \mathfrak{h} \bigoplus_{\substack{1 \leqslant i,j \leqslant n, \\ i \neq j}} \mathfrak{g}_{i,j} \end{aligned}$$

where $\mathfrak{h}$ is the subspace of diagonal matrices in $sl_n(k)$ and for $1 \leqslant i, j \leqslant n$, $i \neq j$, $\mathfrak{g}_{i,j}$ is the one-dimensional span of the elementary matrix $E_{i,j}$ with 1 at the (i, j)-th place and zeros elsewhere. Thus we get an identification

$$\begin{aligned} R &= \{\epsilon_i - \epsilon_j,\ 1 \leqslant i, j \leqslant n,\ i \neq j\}, \\ R^+ &= \{\epsilon_i - \epsilon_j,\ 1 \leqslant i < j \leqslant n\} \end{aligned}$$

3. Root subgroups

Let $\mathfrak{g} = \text{Lie}\, G$. For $\beta \in R$, let $\mathfrak{g}_\beta$ be the (one-dimensional) T-weight space - called a *root space* - of weight β. There exists a unique connected T-stable subgroup U_β of G (i.e., U_β is normalized by T) with $\mathfrak{g}_\beta$ as its Lie algebra (see **[22, 73]** for details). The subgroup U_β is called a *root subgroup*. Let $\mathbb{G}_a$ denote the additive group k. There exists an isomorphism $\theta_\beta : \mathbb{G}_a \cong U_\beta$ such that for all $t \in T$, $x \in \mathbb{G}_a$, $t\theta_\beta(x)t^{-1} = \theta_\beta(\beta(t)x)$. Let U denote the unipotent part of B, i.e. the subgroup of B consisting of all the unipotent elements in B. Then we have an identification of varieties

$$U \cong \prod_{\beta \in R^+} U_\beta,$$

the product being taken in any fixed order. Further, U is normalized by T, and B is the semidirect product of U and T (see [**22, 73**] for details).

REMARK 3.0.1. In the above example (with $G = SL_n(k)$), $\mathfrak{g}_{i,j}$, $1 \leqslant i,j \leqslant n$, $i \neq j$ are all the root spaces. The root subgroup $U_{i,j}, 1 \leqslant i,j \leqslant n$, $i \neq j$ is the subgroup of G consisting of matrices with 1's along the diagonal and the only possible non-zero entry occurs at the (i,j)-th place, and U is the subgroup of G consisting of unipotent upper triangular matrices.

4. Parabolic subgroups

Let Q be a parabolic subgroup containing B. Let $R_u(Q)$ be the unipotent radical of Q (the subgroup of unipotent elements of $R(Q)$, the radical of Q), and R_Q^+ the subset of R^+ defined by $R^+ \setminus R_Q^+ = \{\beta \in R \mid U_\beta \subset R_u(Q)\}$.

Let $R_Q^- = \{-\beta, \beta \in R_Q^+\}$, $R_Q = R_Q^+ \cup R_Q^-$. Let $S_Q = \{\alpha \in S \mid \alpha \in R_Q\}$. Then R_Q is a sub root system of R called the *root system associated to* Q, with S_Q as a set of simple roots, and R_Q^+ (resp. R_Q^-) as the set of positive (resp. negative) roots of R_Q relative to S_Q.

Conversely, given a subset J of S, there exists a parabolic subgroup Q of G such that Q contains B. Namely, Q is generated by B and

$$\{U_{-\alpha} \mid \alpha \in R^+ \text{ and } \alpha = \sum_{\beta \in J} a_\beta \beta\}.$$

Thus the set of parabolic subgroups containing B is in bijection with the power set of S. Given any parabolic subgroup Q, we shall denote the associated set of simple roots by S_Q, $R_Q^+ = \{\alpha \in R^+ \text{ and } \alpha = \sum_{\beta \in S_Q} a_\beta \beta\}$ etc. Note that for $Q = B$ (resp. G), $S_Q =$ the empty set (resp. S).

The subgroup of Q generated by T and $\{U_\alpha \mid \alpha \in R_Q\}$ is called the *Levi subgroup* associated to S_Q, and is denoted L_Q. We have, Q is the semidirect product of $R_u(Q)$ and L_Q (see [**22, 73**] for details).

The set of *maximal parabolic subgroups* is in one-to-one correspondence with S, namely given $\alpha \in S$, the parabolic subgroup Q where $S_Q = S \setminus \{\alpha\}$ is a maximal parabolic subgroup, and conversely. In the sequel, we shall denote the maximal parabolic subgroup Q, where $S_Q = S \setminus \{\alpha_i\}$ by P_i or also by $P_{\widehat{\alpha_i}}$.

5. The Weyl group of a parabolic subgroup

Given a parabolic subgroup $Q \supset B$, let W_Q be the subgroup of W generated by $\{s_\alpha \mid \alpha \in S_Q\}$. W_Q is called the *Weyl group* of Q.

5.1. The set $W_Q^{\min}$ of minimal representatives of W/W_Q

In each coset $wW_Q \in W/W_Q$, there exists a unique element of minimal length (cf. [**74**, 1.10]). Let $W_Q^{\min}$ be this *set of minimal length representatives of* W/W_Q. $W_Q^{\min}$ is also denoted by W^Q in the literature. We have

$$W_Q^{\min} = \{w \in W \mid l(ww') = l(w) + l(w'), \text{ for all } w' \in W_Q\}. \tag{1}$$

In other words, each element $w \in W$ can be written uniquely as $w = u \cdot v$ where $u \in W^Q, v \in W_Q$ and $l(w) = l(u)+l(v)$. The set $W_Q^{\min}$ may be also be characterized as

$$W_Q^{\min} = \{w \in W \mid w(\alpha) > 0, \text{ for all } \alpha \in S_Q\}. \tag{2}$$

(Here by a root β being > 0 we mean $\beta \in R^+$).

5.2. The set $W_Q^{\max}$ of maximal representatives of W/W_Q

In each coset wW_Q there exists a unique element of maximal length. Let $W_Q^{\max}$ be the set of maximal length representatives of W/W_Q. We have

$$W_Q^{\max} = \{w \in W \mid w(\alpha) < 0 \text{ for all } \alpha \in S_Q\}.$$

Further, if we denote by w_Q the unique element of maximal length in W_Q, then we have

$$W_Q^{\max} = \{ww_Q \mid w \in W_Q^{\min}\}.$$

6. Schubert varieties

Let Q be a parabolic subgroup $Q \supset B$. Recall, $W = N(T)/T$ and $T \subset B \subset Q$ for any parabolic subgroup Q. Therefore, given any $w \in W$ there exists a well defined coset wQ in G/Q which we will denote by $e_{w,Q}$. Then the set of T-fixed points in G/Q for the action given by left multiplication is precisely $\{e_{w,Q} \mid w \in W^Q\}$. Let $w \in W^Q$, and let $X_Q(w)$ be the Zariski closure of $Be_{w,Q}$ in G/Q. Then $X_Q(w)$ with the canonical reduced structure is called the *Schubert variety* in G/Q associated to wW_Q. In particular, we have bijections between $W_Q^{\min}$ and the set of Schubert varieties in G/Q, and between $W_Q^{\max}$ and the set of Schubert varieties in G/Q. We have the extended version of the Bruhat decomposition

$$G/Q = \dot{\bigcup}_{w \in W^Q} Be_{w,Q},$$
$$X_Q(y) = \dot{\bigcup}_{\{w \in W^Q \mid e_{w,Q} \in X_Q(y)\}} Be_{w,Q},\ y \in W^Q,$$

If $Q = B$, we shall drop the suffix B and denote $e_{w,Q}, X_Q(w)$ etc., by just $e_w, X(w)$ etc.

6.1. dim$X(w)$

Consider the left multiplication by G on G/B. The isotropy subgroup of G at e_w is wBw^{-1}. Hence denoting the isotropy subgroup of B at e_w by B_w, we have $B_w = B \cap wBw^{-1}$, and hence B/B_w gets identified with $\prod_{\beta \in N(w)} U_\beta$, where $N(w) = \{\beta \in R^+ \mid w^{-1}(\beta) < 0\}$ (note that for $\beta \in R$, U_β fixes $e_w \iff U_\beta \subset wBw^{-1}$, i.e., $\iff \beta = w(\alpha)$ for some $\alpha > 0$, i.e., $\iff w^{-1}(\beta) > 0$). Hence denoting $n(w) = \#N(w)$ we obtain

$$\dim X(w) = n(w).$$

More generally, given a parabolic subgroup Q, and $w \in W^Q$, let

$$N_Q(w) = \{\beta \in R^+ \mid w^{-1}(\beta) \in R^- \setminus R_Q^-\}, \quad \text{and} \quad n_Q(w) = \#N_Q(w).$$

Then by similar considerations as above, we have,

$$\dim X_Q(w) = n_Q(w).$$

6.2. The big cell and the opposite big cell

From the above discussion, we have for $w = w_0$ (w_0 being the unique element of maximal length in W), $B_{w_0} = B \cap B^-$, where $B^-(= w_0 B w_0^{-1})$ is the Borel subgroup *opposite to* B; note that B^- is the subgroup of G generated by T and $\{U_\beta \mid \beta \in R^-\}$. We have, $B_{w_0} = T$, and hence B/B_{w_0} is simply U $(= \prod_{\beta \in R^+} U_\beta$, the unipotent part of B). Thus we obtain that under the projection

$$\pi_0 : G \longrightarrow G/B,$$
$$g \longmapsto g e_{w_0},$$

the restriction $\pi_0|_U : U \to \pi_0(U)$ is an isomorphism; further $\pi_0(U) = Be_{w_0}$. The B-orbit Be_{w_0} is called the *big cell* in G/B. It is a dense open subset of G/B. In a similar way, we have, under the projection

$$\pi : G \longrightarrow G/B,$$
$$g \longmapsto g e_{id},$$

the restriction $\pi|_{U^-} : U^- \to \pi(U^-)$ is an isomorphism (here, U^- is the unipotent part of B^-; note that $U^- = \prod_{\beta \in R^-} U_\beta$); further $\pi(U^-) = B^- e_{id}$. The B^--orbit $B^- e_{id}$ is called the *opposite big cell* in G/B. It is again a dense open subset of G/B. For example, for $G = SL(n)$, $T =$ the subgroup of diagonal matrices, $B =$ the subgroup of upper triangular matrices of G, we have an identification of the big cell with the subgroup of unipotent upper triangular matrices of G, and the opposite big cell with the subgroup of unipotent lower triangular matrices of G.

More generally, the B-orbit $Be_{w_0,Q}$ in G/Q is called the *big cell* in G/Q. It is a dense open subset of G/Q, and it gets identified with $R_u(Q)$, the unipotent radical of Q, namely the subgroup of B generated by $\{U_\alpha \mid \alpha \in R^+ \setminus R_Q^+\}$. The B^--orbit $B^- e_{\mathrm{id},Q}$ is called the *opposite big cell* in G/Q. This is again a dense open subset of G/Q, and it gets identified with the unipotent subgroup of B^- generated by

$\{U_\alpha \mid \alpha \in R^- \setminus R_Q^-\}$. Observe that both the big cell and the opposite big cell can be identified with $\mathbb{A}^{N_Q}$, where $N_Q = \#\{R^+ \setminus R_Q^+\}$.

While the big cell (being a B-orbit) has empty intersection with any Schubert variety, the opposite big cell has non-empty intersection with any Schubert variety! For a Schubert variety $X_Q(w) \subset G/Q$, $B^- e_{\mathrm{id},Q} \cap X_Q(w)$ is called the *opposite cell* in $X_Q(w)$ (by abuse of language). In general, it is not a cell. It is a nonempty affine open subvariety of $X_Q(w)$, and a closed subvariety of the affine space $B^- e_{\mathrm{id},Q} (= \mathbb{A}^{N_Q})$.

6.3. The Bruhat-Chevalley order

There is a partial order on W, known as the Bruhat-Chevalley order (or just Bruhat order) **[37]** determined by the structure of the Schubert varieties, namely, for $w_1, w_2 \in W$, $w_1 \geqslant w_2 \iff X(w_1) \supseteq X(w_2)$. This partial order plays an important role in the theory of Schubert varieties. Chevalley has shown (cf. **[37]**) that this order can be stated equivalently in terms of the Weyl group, namely $w_1 \geqslant w_2 \in W$ if every (equivalently, any) reduced expression for w_1 contains a subexpression which is a reduced expression for w_2. Another test for Bruhat-Chevalley order is given by the following: For $w \in W$, let $X(w^{(d)})$ denote the projection of $X(w)$ under $G/B \to G/P_d$, $1 \leqslant d \leqslant l$; then for $u, v \in W$, $u \geq v$ if and only if $X(u^{(d)}) \supseteq X(v^{(d)})$ for all $1 \leqslant d \leqslant l$.

7. P_α-stability

For a simple root α, let us denote the rank one parabolic with $\{\alpha\}$ as the associated set of simple roots by P_α. Then the canonical projection $\pi_\alpha : G/B \to G/P_\alpha$ is a $\mathbb{P}^1$ fibration (since the fibers are P_α/B and $P_\alpha/B \cong \mathbb{P}^1$).

PROPOSITION 7.0.1. *Let $w \in W$, $\alpha \in S$. The following are equivalent.*

(1) *The closed Bruhat cell $Z(w) := \overline{BwB}$ is stable for multiplication on the left (resp. right) by $U_{-\alpha}$ or equivalently by P_α.*
(2) *$l(s_\alpha w) < l(w)$ (resp. $l(ws_\alpha) < l(w)$).*
(3) *$n(s_\alpha w) < n(w)$ (resp. $n(ws_\alpha) < n(w)$).*
(4) *$w^{-1}(\alpha) < 0$ (resp. $w(\alpha) < 0$).*

(Here $n(w)$ is as in §6.1.)

PROOF. (3) $\iff$ (4). Recall that $n(w) = \#N(w)$ (cf. §6.1). Let $n(s_\alpha w) < n(w)$. If possible, let us assume $w^{-1}(\alpha) > 0$. Assumption implies that $\alpha \in N(s_\alpha w)$, $\alpha \notin N(w)$. It is easily checked that s_α induces a bijection of $N(w)$ onto $N(s_\alpha w) \setminus \{\alpha\}$. This implies that $n(w) = n(s_\alpha w) - 1$, which is a contradiction. Hence our assumption is wrong and we obtain the implication (3) $\Longrightarrow$ (4). Conversely, suppose that $w^{-1}(\alpha) < 0$. In this case, $\alpha \in N(w)$, $\alpha \notin N(s_\alpha w)$, and consequently s_α induces a bijection of $N(s_\alpha w)$ onto $N(w) \setminus \{\alpha\}$. Hence we obtain $n(s_\alpha w) = n(w) - 1$ from which we obtain the implication (4) $\Longrightarrow$ (3).

REMARK 7.0.2. Note that in the above proof we have also established that $w^{-1}(\alpha) >$ (resp. $<$) $0 \iff n(s_\alpha w) = n(w) + 1$ (resp. $n(w) - 1$). In particular, we have,

$$n(s_\alpha w) = n(w) \pm 1,\ \alpha \in S,\ w \in W.$$

(2) $\iff$ (3). We first prove one Lemma.

LEMMA 7.0.3. *Let $u \in W$. Then $n(u) = l(u)$ (here $l(u)$ denotes the length of u).*

Proof of Lemma. We prove the result by showing $n(u) \leqslant l(u) \leqslant n(u)$.

We shall first prove that $n(u) \leqslant l(u)$. We prove this by induction on $l(u)$. If $l(u) = 0$, then $u = id$ and $n(u)$ is also equal to 0. Let us now assume $l(u) > 0$ and the induction hypothesis that for all $v \in W$, $l(v) < l(u) \Longrightarrow n(v) \leqslant l(v)$. Let $u = s_1 \cdots s_r$, $r = l(u)$ be a reduced expression for u. Define $v = s_1 u$. We have, $l(v) = l(u) - 1$, and hence by induction hypothesis $n(v) \leqslant l(v) = l(u) - 1$. But now $u = s_1 v$, and hence by the above Remark, we have,
$n(u) = n(v) \pm 1 \leqslant n(v) + 1 \leqslant l(v) + 1 = l(u)$.

We next prove that $l(u) \leqslant n(u)$. We prove this by induction on $n(u)$. If $n(u) = 0$, then $u = id$ and $l(u)$ is also equal to 0. Assume $n(u) > 0$ and the induction hypothesis. Assumption implies that there exists a positive root β such that $w^{-1}(\beta) < 0$. We may suppose that β is simple. Now from the implication (4) $\Longrightarrow$ (3) and the Remark above, we get that $n(s_\beta u) = n(u) - 1$. Hence by induction hypothesis, we obtain, $l(s_\beta u) \leqslant n(s_\beta u) = n(u) - 1$. This means that we can write $s_\beta u = s_1 \cdots s_r$, $r \leqslant n(u) - 1$. Hence $u = s_\beta s_1 \cdots s_r$ which implies that $l(u) \leqslant r + 1 \leqslant n(u)$.

Combining the two parts we obtain $n(u) = l(u)$. This completes the proof of the Lemma.

As a direct consequence of Lemma 7.0.3, we obtain (2) $\iff$ (3).

(1) $\iff$ (4). Let $\pi_\alpha : G/B \to G/P_\alpha$ be the canonical projection. The fibers of π_α are P_α/B ($\cong \mathbb{P}^1$). We have $Z(w)$ is stable for multiplication on the right by P_α

$$\begin{aligned}
&\iff X(w) \text{ is saturated for the } \mathbb{P}^1 \text{ fibration } \pi_\alpha \\
&\iff X(w) = \pi_\alpha^{-1}(\pi_\alpha(X(w))) \\
&\iff \dim X(w) = \dim \pi_\alpha(X(w)) + 1 \\
&\iff n(w) = n_{P_\alpha}(w) + 1 \\
&\iff N(w) \neq N_{P_\alpha}(w) \ \text{(note that } N_{P_\alpha}(w) \subseteq N(w)) \\
&\iff w(\alpha) < 0.
\end{aligned}$$

□

7.1. The dimension of $X(w)$

Combining Lemma 7.0.3 with §6.1, we obtain

PROPOSITION 7.1.1. *We have,*

$$\dim X(w) = l(w),\ w \in W. \tag{1}$$

More generally, we have,

$$\dim X_Q(w) = l(w),\ w \in W^Q.$$

Another consequence of Lemma 7.0.3 is the following:

COROLLARY 7.1.2. *Let* $w \in W$. *Then* $n(w) = n(w^{-1})$.

PROOF. We have, $n(w) = l(w) = l(w^{-1}) = n(w^{-1})$. □

8. Line bundles on G/Q

8.1. Fiber bundles

Let G be an algebraic group acting (algebraically) on a variety F, i.e., we have an action of G on F and the map $G \times F \to F$, $(g, x) \mapsto gx$ is a morphism of varieties. Further, let the action be effective, i.e., if $g \cdot x = x$ for some $g \in G$ and all $x \in F$, then $g = e$ (the identity element of G).

DEFINITION 8.1.1. A *fiber bundle* over an algebraic variety X with *typical fiber* F and *structure group* G (G and F being as above) is a pair (V, π), where V is an algebraic variety and $\pi : V \to X$ is a morphism such that

(a) there exists an open covering $X = \bigcup_{i \in I} U_i$ and isomorphisms

$$h_i : \pi^{-1}(U_i) \longrightarrow U_i \times F$$

which map the fiber $\pi^{-1}(u)$ onto $u \times F$

(b) for $i, j \in I$, there exist morphisms $g_{ij} : U_i \cap U_j \to G$ such that $h_i h_j^{-1}$ (on $(U_i \cap U_j) \times F$) is given by $h_i h_j^{-1}(u, f) = (u, g_{ij}(u) f)$, $u \in U_i \cap U_j$, $f \in F$.

REMARK 8.1.2. Observe that since the action of G on F is effective, the morphism g_{ij} is uniquely determined by h_i and h_j, for all $i, j \in I$.

Examples. (1) The *trivial fiber bundle:* $V = X \times F$, and $\pi : V \to X$, the canonical projection. Any open covering satisfies (a), and $g_{ij} : U_i \cap U_j \to G$ is the map sending every u in $U_i \cap U_j$ to e.

(2) A vector bundle (E, π) of rank n over X is simply a fiber bundle with $G = GL_n(k)$ and $F = k^n$, the action of G on F being the natural action of $GL_n(k)$ on k^n.

8.2. Principal G-bundles

In the above definition, if $F = G$ and the action of G on itself is left translation, then (V, π) is called a *principal G-bundle* over X.

Example. Let G be an algebraic group and H a closed subgroup. Then (G, π) is a principal H-bundle over G/H, $\pi : G \to G/H$ being the canonical projection.

8.3. The fiber bundle associated to a principal G-bundle

Let (E, π) be a principal G-bundle over X. Consider now an action (not necessarily effective) of G on F. Define an equivalence relation $\sim$ on $E \times F$ by $(xg, gy) \sim (x, y)$, $g \in G, x \in E, y \in F$. Let $V = (E \times F)/\sim$ and $p : V \to X, p(x, y) = \pi(x)$. Then (V, p) is a fiber bundle over X with structure group G and fiber F.

We shall now see how the above considerations applied to G/B determine all line bundles on G/B. Let G be semisimple and simply connected.

Let $X = G/B$, and consider the canonical projection $\pi : G \to X$. Then $\pi : G \to X$ is a principal B-bundle with B as the structure group and fiber. Let $\lambda \in X(T)$. Then λ defines a character $\lambda_B : B \to \mathbb{G}_m$ obtained by composing the natural map $B \to T$ with $\lambda : T \to \mathbb{G}_m$ (recall that B is the semidirect product of T and U (= the unipotent part of B)). We shall denote λ_B also by just λ. Now $\lambda : B \to \mathbb{G}_m$ gives rise to an action of B on $k(= \mathbb{G}_a)$, namely $b \cdot c = \lambda(b)c$, $b \in B$, $c \in k$. Set $E = G \times k/\sim$, where $\sim$ is the equivalence relation defined by $(gb, b \cdot c) \sim (g, c)$, $g \in G, b \in B, c \in k$. Then E is the total space of a line bundle, say L_λ, over X. Thus we obtain a map

$$L : X(T) \longrightarrow \mathrm{Pic}(G/B),$$
$$\lambda \longmapsto L_\lambda,$$

where $\mathrm{Pic}(G/B)$ is the Picard group of G/B (namely the group of isomorphism classes of line bundles on G/B). We recall (cf. **[36]**) that the above map is in fact an isomorphism of groups (since G is simply connected).

On the other hand, consider the prime divisors $X(w_0 s_i)$, $1 \leqslant i \leqslant l = (\mathrm{rank} G)$ on G/B. Let $L_i = \mathcal{O}_{G/B}(X(w_0 s_i))$ be the line bundle defined by $X(w_0 s_i)$, $1 \leqslant i \leqslant l$ (see **[67]**, Chapter 2, §6 for the discussion on divisors and line bundles on an algebraic variety). Recall (cf. **[36]**) that $\mathrm{Pic}(G/B)$ is a free abelian group generated by the L_i's, and under the isomorphism $L : X(T) \cong \mathrm{Pic}(G/B)$, we have $L(\omega_i) = L_i$, $1 \leqslant i \leqslant l$. Thus for $\lambda = \sum_{i=1}^{l} (\lambda, \alpha_i^*)\omega_i$, we have $L_\lambda = \bigotimes_{i=1}^{l} L_i^{\otimes(\lambda, \alpha_i^*)}$. We shall write $L_\lambda \geqslant 0$ if $\lambda \geqslant 0$, i.e. if λ is dominant.

As above, let E denote the total space of the line bundle L_λ, over $X(= G/B)$. Let $\sigma : E \to X$ be the canonical map, namely $\sigma(g, c) = gB$. Let

$$M_\lambda = \{f \in k[G] \mid f(gb) = \lambda(b) f(g), g \in G, b \in B\}.$$

Then M_λ can be identified with *space of sections*

$$H^0(G/B, L_\lambda) := \{s : X \to E \mid \sigma \circ s = \mathrm{id}_X\}$$

as follows. Let $f \in M_\lambda$. To f, we associate a section $s : X \to E$ by setting $s(gB) = (g, f(g))$. To see that s is well defined, consider $g' = gb$, $b \in B$. Then $(g', f(g')) = (gb, f(gb)) = (gb, \lambda(b)f(g)) = (gb, b \cdot f(g)) \sim (g, f(g))$. From this, it follows that s is well defined. Conversely, given $s \in H^0(G/B, L_\lambda)$, consider $gB \in G/B$. Let $s(gB) = (g', f(g'))$, where $g' = gb$ for some $b \in B$ (note that $g'B = gB$, since $\sigma \circ s = \mathrm{id}_X$). Now the point $(g', f(g'))$ may also be represented by $(g, \lambda(b)^{-1} f(gb))$ (note that $(g', f(g')) = (gb, f(gb)) \sim (g, \lambda(b)^{-1} f(gb))$. Thus given $g \in G$, there exists a unique representative of the form $(g, f(g))$ for $s(gB)$. This defines a function $f : G \to k$. Further, this f has the property that for $b \in B$, $f(g) = \lambda(b)^{-1} f(gb)$, i.e. $f(gb) = \lambda(b) f(g)$, $b \in B$, $g \in G$. Thus we obtain an identification

$$M_\lambda = H^0(G/B, L_\lambda).$$

It can be easily checked that the above identification preserves the respective G-module structures.

Chevalley (cf. [**36**]) has shown that

(1) $H^0(G/B, L_\lambda) \neq 0$ if and only if λ is dominant.
(2) L_λ is ample if and only if L_λ is very ample if and only if λ is dominant and regular. (A line bundle L on a variety X is called *very ample* if there exists an immersion $i : X \hookrightarrow \mathbb{P}^N$ for some N, and $i^*\mathcal{O}_{\mathbb{P}^N}(1) = L$. A line bundle L is called *ample* if L^r is very ample for some positive integer r.)

9. Equations defining a Schubert variety

Given a projective variety X embedded in a projective space $\mathbb{P}^n$, let $I(X)$ denote the homogeneous ideal defining X for this embedding (cf. Chapter 1). By Hilbert's basis theorem, we can find a finite set of generators $f_1, \ldots, f_r$ for the ideal $I(X)$. We shall refer to f_i for $i = 1, \ldots, r$ as the *equations* defining X as a closed subvariety of $\mathbb{P}^n$.

Now take $X = G/Q$. Let L be an ample line bundle on G/Q. Then, in view of (2) above, L induces a projective embedding

$$G/Q \hookrightarrow \text{Proj}(H^0(G/Q, L)).$$

Here, $\text{Proj}(H^0(G/Q, L)) = \mathbb{P}(H^0(G/Q, L)^*)$ denotes $\text{Proj}(B)$ where B is the symmetric algebra of $H^0(G/Q, L)$ and $H^0(G/Q, L)^*$ denotes the dual of $H^0(G/Q, L)$. We recall (cf. [**137**]) the following:

Let

$$R = \bigoplus_{n \in \mathbb{Z}^+} H^0(G/Q, L^n)$$

$$R_Q(w) = \bigoplus_{n \in \mathbb{Z}^+} H^0(X_Q(w), L^n).$$

The natural map

$$S^n(H^0(G/Q, L)) \longrightarrow H^0(G/Q, L^n)$$

is surjective, and the kernel of the surjective map

$$\bigoplus_{n \in \mathbb{Z}^+} S^n(H^0(G/Q, L)) \longrightarrow \bigoplus_{n \in \mathbb{Z}^+} H^0(G/Q, L^n)$$

is generated (as an ideal) by elements of degree 2. Further, the restriction map $R \to R_Q(w)$ is surjective, and its kernel is generated as an ideal by elements of degree 1. Thus we obtain that the homogeneous ideal of G/Q for the embedding $G/Q \hookrightarrow$

Proj$(H^0(G/Q, L))$ is generated in degree 2, i.e. $I(G/Q)$ is generated as an ideal by the kernel of the surjective map $S^2(H^0(G/Q, L)) \to H^0(G/Q, L^2)$; further, the ideal defining the Schubert variety $X_Q(w)$ in G/Q is generated as an ideal by the kernel of the surjective map $H^0(G/Q, L) \to H^0(X_Q(w), L)$. Thus any Schubert variety $X_Q(w)$ in G/Q is scheme theoretically (even at the cone level) the intersection of G/Q with all the hyperplanes in Proj$(H^0(G/Q, L))$ containing X.

We have similar results for multi-cones over Schubert varieties (cf. [**85**]):

For a maximal parabolic subgroup P_i, let us denote the ample generator of Pic (G/P_i) ($\simeq \mathbb{Z}$) by L_i. Let us denote $S \setminus S_Q$ by $\{\alpha_1, \dots, \alpha_t\}$, for some t. Let

$$A = \bigoplus_{\underline{a}} H^0(G/Q, \bigotimes_i L_i^{a_i})$$

$$A_w = \bigoplus_{\underline{a}} H^0(X_Q(w), \bigotimes_i L_i^{a_i}),$$

where $\underline{a} = (a_1, \dots, a_t) \in \mathbb{Z}_+^t$. We recall (cf. [**85**]) that the natural map

$$\bigoplus \mathbb{S}^{a_1}(H^0(G/Q, L_1)) \otimes \cdots \otimes \mathbb{S}^{a_t}(H^0(G/Q, L_t)) \longrightarrow A$$

is surjective, and its kernel is generated as an ideal by elements of total degree 2. Further, the restriction map $A \to A_w$ is surjective, and its kernel is generated as an ideal by elements of total degree 1.

10. Geometric properties of Schubert varieties

Chevalley (cf. [**37**]) showed that Schubert varieties are nonsingular in codimension 1, i.e., the singular locus has codimension at least 2.

Demazure first proved that Schubert varieties are normal [**42**], characteristic of k being 0. Seshadri ([**150**]) proved the normality of Schubert varieties in arbitrary characteristic in the process of fixing a gap in Demazure's above cited paper. The normality of Schubert varieties in arbitrary characteristic was also proved by Anderson (cf. [**6**]) and Ramanan-Ramanathan (cf. [**135**]).

Recall that for a variety, the properties of being nonsingular in codimension 1 and Cohen-Macaulay imply normality (cf. Serre criterion, see [**67**], Theorem 8.22A for example). Arithmetic normality of a Schubert Variety X, that is the normality of the cone $\widehat{X}$, was obtained as a consequence of the standard monomial theory [**101**], [**106**] for G classical. The arithmetic normality, arithmetic Cohen-Macaulayness, and rational resolution of singularities for Schubert varieties (see [**84**] for the definition of "rational resolution of singularities" of a variety) for any semisimple G were obtained as consequences of Frobenius splitting of Schubert varieties [**135**], [**136**]. Recently Littelmann (cf. [**116**]) has given a "a Standard monomial theoretic" proof for the arithmetic normality and arithmetic Cohen-Macaulayness of Schubert varieties for any semisimple G. Further, Kumar and Littelmann ([**88**]) have given a Standard monomial theoretic proof of the Frobenius splitting of Schubert varieties. The details of this approach will appear in [**97**].

We conclude this chapter with a lemma that will frequently be used in the sequel.

DEFINITION 10.0.1. Let $w_1, w_2 \in W$, and let $X(w_2)$ be a Schubert divisor in $X(w_1)$. Then $X(w_2)$ is called a *moving divisor in* $X(w_1)$, and is said to be *moved by* α, if $w_2 = s_\alpha w_1$, α a *simple* root. (The terminology is à la Kempf (cf. **[83]**)).

LEMMA 10.0.2. *Let $\tau, \phi \in W$. Let $X(\phi)$ be a moving divisor in $X(\tau)$ moved by a simple root α. Then for any Schubert subvariety $X(w)$ of $X(\tau)$ we have, either $X(w) \subset X(\phi)$, or there exists a moving divisor $X(w')$ in $X(w)$ moved by α such that $X(w') \subset X(\phi)$.*

PROOF. The figure below is helpful to visualize what is happening.

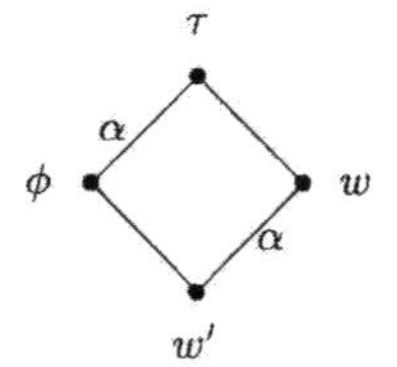

Let $\phi = s_{\beta_1} \cdots s_{\beta_r}$, β_i simple be a reduced expression for ϕ. Then $\tau = s_\alpha s_{\beta_1} \cdots s_{\beta_r}$ is a reduced expression for τ. Let us write $\tau = s_{\beta_0} s_{\beta_1} \cdots s_{\beta_r}$, where $\beta_0 = \alpha$. Let $w \leqslant \tau$; there exists a subsequence $0 \leqslant i_1 < \cdots < i_m \leqslant r$ such that $s_{\beta_{i_1}} \cdots s_{\beta_{i_m}}$ is a reduced expression for w. We have, $w \leqslant \phi \iff i_1 > 0$. Hence if $w \not\leqslant \phi$, then $w = s_\alpha w'$, where $w' \leqslant \phi$. The required result follows from this. □

CHAPTER 4

Representations of semisimple algebraic groups

In this chapter, we review the representation theory of semisimple algebraic groups. We bring out the relationship between the geometry of G/B and the representation theory of G where G is a semisimple algebraic group. We have also included the discussion of the various character formulas - Weyl character formula, Demazure character formula, Lakshmibai-Seshadri character formula.

1. Generalities on the Representation Theory of G

Let G be a semisimple algebraic group defined over an algebraically closed field of arbitrary characteristic. Let T be the maximal torus, B a Borel subgroup, $B \supset T$. Let l be the rank of G, namely $l = \dim T$. Let R be the root system of G relative to T, and S the set of simple roots of R relative to B.

For $\alpha \in R$, let U_α be the 1-dimensional unipotent subgroup of G. Let $T_\alpha = (\text{Ker } \alpha)^\circ$, and $Z_\alpha = Z_G(T_\alpha)$, the centralizer of T_α in G. Recall (cf. **[22]**, **[73]**) that Z_α is an algebraic group of semisimple rank 1, i.e. rk $Z_\alpha/R(Z_\alpha)$ is 1 (here $R(Z_\alpha)$ denotes the radical of Z_α).

1.1. Action of W on $X(T)$

Let $X(T) = \text{Hom}_{\text{alg. gp.}}(T, \mathbb{G}_m)$ be the character group of T (here $\mathbb{G}_m = k^*$). Let W be the Weyl group of T, namely $W = N_G(T)/T$, where $N_G(T)$ is the normalizer of T in G. We have an action of W on T, defined as follows. Let $w \in W$, say $w = nT$ for some $n \in N_G(T)$. We have $nTn^{-1} = T$; if $t \in T$, then we set $w{\cdot}t = ntn^{-1}$. Observe that ntn^{-1} depends only on the coset nT and not on the representative chosen. This action induces an action of W on $X(T)$, namely if $\chi \in X(T)$, then we define $w \circ \chi$, $w \in W$, as the character given by $w \circ \chi(t) = \chi(n^{-1}tn)$, where n is a representative in $N_G(T)$ of w (note that $w \circ \chi(t)$ has to be defined as $\chi(n^{-1}tn)$, rather than $\chi(ntn^{-1})$, in order to have the multiplicativity).

1.2. Weights in a module

Let V be a G-module, i.e. V is a finite dimensional k-vector space together with a morphism $\rho : G \to GL(V)$ of algebraic groups. For the action of T on V via ρ, V breaks up as a direct sum $V = \sum_\chi V_\chi$, where $\chi \in X(T)$, and

$$V_\chi = \{v \in V \mid t \cdot v = \chi(t) \cdot v\}.$$

Let $\Phi_\rho = \{\chi \mid V_\chi \neq (0)\}$. Then V being finite dimensional, Φ_ρ is a finite set.

DEFINITION 1.2.1. An element $\chi \in \Phi_\rho$ is called a *weight of* ρ. The subspace V_χ of V is called a *weight space*, and $\dim V_\chi$ is called the *multiplicity of* χ. A nonzero vector in V_χ is called a *weight vector* of weight χ.

Let $\chi \in \Phi_\rho$, and $v \in V_\chi, v \neq 0$. Let $w \in W$, say $w = nT$. Then

$$t \cdot nv = n(n^{-1}tn) \cdot v = w \cdot \chi(t)nv.$$

Thus we obtain $nv \in V_{w\circ\chi}$. In particular $w \circ \chi \in \Phi_\rho$. Thus, W permutes the weights of ρ.

1.3. Abstract weights

Let $E = X(T) \otimes \mathbb{R}$. Then R is a root system for E, with W as the Weyl group of the root system, and S as a base of the root system. W is generated by $\{s_\alpha, \alpha \in S\}$, where $s_\alpha : E \to E$ is the reflection given by $s_\alpha(\chi) = \chi - 2\frac{(\chi,\alpha)}{(\alpha,\alpha)}$, $(\ ,\ )$ being a W-invariant inner product on E. An element $\chi \in E$ is called an *abstract weight* if (χ, α^*) $(= 2\frac{(\chi,\alpha)}{(\alpha,\alpha)})$ is an integer.

LEMMA 1.3.1. *(cf.* [**73**], *§27.2) Let $\alpha \in R$. Then $\rho(U_\alpha)$ maps a weight space V_α into $\sum_{k\in\mathbb{Z}^+} V_{\lambda+k\alpha}$.*

As a consequence, we obtain that $\rho(Z_\alpha)$ stabilizes $\sum_{k\in\mathbb{Z}} V_{\lambda+k\alpha}$. In particular, we have $s_\alpha(\lambda) = \lambda + k\alpha$, for some $k \in \mathbb{Z}$. Thus $(\lambda, \alpha^*) \in \mathbb{Z}$ for all $\alpha \in R$. Hence we obtain

PROPOSITION 1.3.2. *A weight of ρ is in fact an abstract weight.*

1.4. Root lattice and weight lattice

The set of abstract weights in $E := X(T) \otimes \mathbb{R}$ form a lattice which we shall denote by Λ_{wt}. Note that Λ_{wt} is a free abelian group of rank l. Let Λ_{rt} be the root lattice, namely the free abelian group with the simple roots $\alpha_1, \ldots, \alpha_l$ as a basis. We define ω_i, $1 \leqslant i \leqslant l$, in Λ_{wt} by $(\omega_i, \alpha_j^*) = \delta_{ij}$. Then it is clear that $\{\omega_i, 1 \leqslant i \leqslant l\}$ is a $\mathbb{Z}$-basis of Λ_{wt}. The ω_i's are called the *fundamental weights*.

DEFINITION 1.4.1. An element $\lambda \in \Lambda_{\mathrm{wt}}$ is called a *dominant weight* if in the expression $\lambda = \sum_i c_i\omega_i$ for λ as a $\mathbb{Z}$-linear combination of the fundamental weights, $c_i \geqslant 0$ for all i. The set of dominant weights is denoted by Λ^+_{wt}.

DEFINITION 1.4.2. A G-module $V \neq (0)$ is said to be *irreducible* if the only G-stable submodules are V and (0).

Let now V be a G-module. Then Borel fixed point theorem (cf. Chapter 2, Theorem 2.1.3) applied for the action of a Borel subgroup B on $X = \mathbb{P}(V)$ implies that there exists a nonzero vector $v \in V$, and a character $\lambda : B \to \mathbb{G}_m$ such that $b \cdot v = \lambda(b)v$, $b \in B$. Let V_1 be the span of $G \cdot v$. Then V_1 is G-cyclic (generated as a G-module by v). Now $U^- \cdot B$ being dense in G, we obtain that V_1 is the k-span of

$U^- B \cdot v$, i.e. the k-span of $U^- v$ (since $B \cdot v \subseteq kv$). Hence, in view of Lemma 1.3.1, we obtain that the weights in V_1 are of the form $\lambda - \sum c_\alpha \alpha$, $\alpha \in R^+$, $c_\alpha \in \mathbb{Z}^+$. Thus for any weight χ of V_1, say $\chi = \lambda - \sum c_\alpha \alpha$, we have $\lambda \geqslant \chi$ (here the partial order $\geqslant$ on Λ_{wt} is given by $\theta \geqslant \mu$ if $\theta - \mu$ is a nonnegative integral linear combination of the positive roots). For this reason, λ is called a highest weight in V, and the vector v is said to be a *highest weight vector* with highest weight λ.

Further, for $w \in W$, $w \cdot \lambda$ is a weight in $\rho_1 : G \to GL(V_1)$, and hence $w \cdot \lambda \leqslant \lambda$. This implies that λ is dominant, i.e. $\lambda \in \Lambda^+_{\mathrm{wt}}$. In particular we have proved

PROPOSITION 1.4.3. *Let V be a G-module.*

(1) *There exist highest weight vectors.*
(2) *Let v be a highest weight vector in V with weight λ. Then $\lambda \in \Lambda^+_{\mathrm{wt}}$.*
(3) *If in addition V is G-cyclic, then any other weight is of the form*

$$\lambda - \sum_{\alpha \in R^+} c_\alpha \alpha, \quad c_\alpha \in \mathbb{Z}^+;$$

further, the multiplicity of the weight λ is 1.

PROPOSITION 1.4.4. *Let V, V' be two irreducible G-modules with highest weights λ, λ' respectively. Then $V \cong V'$ if and only if $\lambda = \lambda'$.*

PROOF. If $V \cong V'$, then clearly $\lambda = \lambda'$ (since the multiplicity of the highest weight is 1). Conversely, let $\lambda = \lambda'$. Let $U = V \oplus V'$. Fix a highest weight vector v in V (resp. v' in V' of weight $\lambda = \lambda'$). Then $u = v + v'$ is a highest weight vector of weight λ in U. Let V'' be the G-submodule of U generated by u. Then the multiplicity of the weight λ in U is 1. Now considering V, V' as submodules of U, we have $V \cap V''$, $V' \cap V''$ are proper submodules of V, V' (since v, v' are not in V'', neither of these being proportional to $u = v + v'$). Hence $V \cap V'' = (0) = V' \cap V''$ (by irreducibility of V, V'). This implies that the two projections $U \to V$, $U \to V'$ map V'' isomorphically onto (nonzero) submodules of V, V' that must be all of V, V' (by irreducibility). Thus V is isomorphic to V'. □

DEFINITION 1.4.5. Let V be a G-module (not necessarily finite dimensional). A nonzero vector $v \in V$ is said to be a *highest weight vector of weight* λ if B acts on v via λ, i.e. $b \cdot v = \lambda(b)v$, where $\lambda : B \to \mathbb{G}_m$ is a character of B.

2. Construction of irreducible G-modules

Given a finite dimensional irreducible G-module V, we have from above, the isomorphism class of V determines a unique element of Λ^+_{wt}. Conversely, given $\lambda \in \Lambda^+_{\mathrm{wt}}$, there exists a unique (up to isomorphism) finite dimensional irreducible G-module $V(\lambda)$ with highest weight λ, the construction of which is sketched below briefly.

Step 1: Reduction to the pair (V, v). It suffices to construct a G-module V containing a highest weight vector v of weight λ. For, suppose that such a pair (V, v) has been constructed. Let V_1 be the span of $G \cdot v$. Then V_1 is the span of $U^- Bv$, where U^- is the unipotent part of B^-, the Borel subgroup opposite to B (since $U^- B$ is dense in G), which is the same as the span of U^-v. Hence each proper G-submodule of V_1 does not contain v, and is a sum of T-weight spaces of weight $< \lambda$. Hence denoting by M the sum of all proper submodules of V_1, we obtain that M is the unique maximal proper G-submodule of V_1. From this it follows that V_1/M is an irreducible G-module with highest weight λ.

Step 2: Construction of the pair (V, v) corresponding to $d\lambda$ for a suitable d. First, we recall the following (cf. Chapter 2, Theorem 1.5.6):

Given a closed subgroup H of G, there exists an immersive representation $\rho : G \to GL(E)$, and a line $D \subset E$ such that

$$H = \{g \in G \mid \rho(g) \cdot D = D\}, \quad \operatorname{Lie} H = \{X \in \operatorname{Lie} G \mid d\rho(X) \cdot D \subset D\}.$$

For each maximal parabolic subgroup P_i, choose a rational representation $\rho_i : G \to GL(V_i)$ and a line $D_i = (v_i)$ in V_i such that

$$(*)\ P_i = \{g \in G \mid \rho_i(g) \cdot D_i = D_i\}, \quad \operatorname{Lie} P_i = \{X \in \operatorname{Lie} G \mid d\rho_i(X) \cdot D_i \subset D_i\}.$$

Let the weight of v_i be $\mu_i = \sum_j c_{ij}\omega_j$. Now s_α, $\alpha \neq \alpha_i$, fixes D_i. Let $\alpha = \alpha_k$, where $k \neq i$. Then μ_i equals the weight of $s_\alpha(v_i)$, and hence $\mu_i = s_\alpha(\mu_i) = \mu_i - (\mu_i, \alpha^*)\alpha$. Hence we obtain $(\mu_i, \alpha_k^*) = 0$, $k \neq i$, i.e. $c_{ik} = 0$, $k \neq i$. On the other hand G does not act trivially on v_i (cf. $(*)$). Hence we get $\mu_i = c_{ii}\omega_i = d_i\omega_i$ say, where $d_i \neq 0$. The pair (V_i, v_i) gives the construction of (V, v) corresponding to the weight μ_i. Letting $V = \bigotimes_{i=1}^l V_i$, and $v = v_1 \otimes \cdots \otimes v_l$, we get that v is a highest weight vector in v of weight $\sum \mu_i$.

Let $\lambda = \sum_{i=1}^l c_i\omega_i$. Let $d = \prod d_i$, $d_i' = d/d_i$, and $e_i = c_i d_i'$. Then letting $V = \bigotimes_{i=1}^l V_i^{e_i}$, and $v = \bigotimes_{i=1}^l v_i^{\otimes e_i}$, we obtain that v is a highest weight vector in V of weight $\sum e_i\mu_i = \sum c_i d_i' d_i \omega_i = d\lambda$.

Step 3: Realization of $V(\mu)$ inside $k[G]$. Let $\mu = d\lambda$, d being as above. By Steps 1 and 2, the irreducible G-module $V(\mu)$ with highest weight μ has been constructed. We now give a realization of $V(\mu)$ inside $k[G]$. Fix a highest weight vector $v_\mu = v$ say, in $V(\mu)$ of weight μ. Let us write $V(\mu) = (v) \oplus V'$, where V' is a subspace of $V(\mu)$ of codimension 1. Let f_μ (or just f) be the element of V^* which takes the value 1 at v, and 0 on V'. Define $c_\mu : G \to k$, $x \mapsto f(x \cdot v)$. Then $c_\mu \in k[G]$. Writing $x \cdot v = av + v'$, $a \in k$, $v' \in V'$, we have $f(x \cdot v) = a$. Thus we obtain $x \cdot v = c_\mu(x) \cdot v + v'$. Further, we have

(1) for $x \in B$, $x \cdot v = \mu(x) \cdot v$ and hence $c_\mu(x) = \mu(x)$.

For $b \in B$, $y \in G$, we have $b \cdot c_\mu(y) = c_\mu(yb) = \mu(b) \cdot c_\mu(y)$. Thus we obtain

(2) $b \cdot c_\mu = \mu(b) \cdot c_\mu$.

Thus c_μ is a highest weight vector of weight μ in the right regular representation $(r, k[G])$. As in *Step 1*, $V(\mu)$ may be realized as the unique irreducible quotient of G-span of c_μ.

Step 4: Further properties of c_μ. We have

(3) for $x \in B^-$, $xV' = V'$, and $x \cdot v \equiv \mu(x)v \pmod{V'}$.

Now for $x \in B^-$, $y \in G$, we have

$$\begin{aligned} xy \cdot v &= x(c_\mu(y)v + v_1), \quad v_1 \in V' \\ &= \mu(x)c_\mu(y)v + v_2, \quad v_2 \in V'. \end{aligned}$$

Thus we have

$$c_\mu(xy) = \mu(x)c_\mu(y), \quad x \in B^-, y \in G. \tag{4}$$

In particular, we have

$$c_\mu(xy) = \mu(y), \quad x \in U^-, y \in B. \tag{5}$$

Step 5: Construction of the pair (V, v) corresponding to an arbitrary $\lambda \in \Lambda^+_{\mathrm{wt}}$. Guided by the discussion in *Step 4*, we define a function $c_\lambda : U^- \cdot B \to k$ by $c_\lambda(xy) = \lambda(y)$, $x \in U^-$, $y \in B$. Denoting U^-B by Ω, we have $k(\Omega) = k(G)$ (since Ω is dense open in G). Thus $c_\lambda \in k(G)$. We have, $c_\mu = c_\lambda^d$ on Ω (μ and d being as in Step 3). This together with the fact that c_μ is regular on G implies that c_λ is regular on G. In fact, c_λ, as a regular function on G, can be explicitly described as follows. Let P_λ be the stabilizer of the line (c_λ), i.e. $P_\lambda = BW_\lambda B$, where W_λ $(= W_{P_\lambda})$ is the subgroup of W generated by $\{s_\alpha, \alpha \in S \mid s_\alpha(\lambda) = \lambda\}$. For each $w \in W$, let us fix a representative n_w in $N_T(G)$. Then using the Bruhat decomposition $G = \cup_w U^- wB$, any $g \in G$ can be written uniquely as $g = xn_wy$, with $x \in U^-$, $w \in W$, $y \in B$. Then

$$c_\lambda(g) = \begin{cases} \lambda(y), & \text{if } w \in W_\lambda \\ 0, & \text{otherwise} \end{cases}.$$

For $g \in G$, say $g = xn_wy$, $x \in U^-$, $w \in W$, $y \in B$, and $b \in B$, we have

$$b \cdot c_\lambda(g) = c_\lambda(gb) = c_\lambda(xn_wyb) = \lambda(yb) = \lambda(y)\lambda(b) = \lambda(b)c_\lambda(g).$$

Thus we obtain that c_λ is a highest weight vector of weight λ in the right regular representation $(r, k[G])$, and we may realize $V(\lambda)$ as the unique irreducible quotient of the G-span of c_λ.

This completes the construction of the irreducible G-module $V(\lambda)$, with highest weight λ, where $\lambda \in \Lambda^+_{\mathrm{wt}}$. Thus we obtain

THEOREM 2.0.1. *There is a bijection between Λ^+_{wt} and the set of isomorphism classes of finite dimensional irreducible G-modules.*

3. Geometric realization

3.1. The module F_λ

Given $\lambda \in \Lambda^+_{\mathrm{wt}}$, let

$$F_\lambda = \{f \in k[G] \mid f(xy) = \lambda(x)f(y) \text{ for } x \in B^-, y \in G\}.$$

First observe that F_λ is a G-stable subspace of $k[G]$ for the action of G on $k[G]$ by right translations; for, if $f \in F_\lambda$, $x \in B^-$, $y, g \in G$, then $g \circ f(xy) = f(xyg) = \lambda(x)f(yg) = \lambda(x)(g \circ f)(y)$. Thus $g \circ f(xy) = \lambda(x)(g \circ f)(y)$, $x \in B^-$, $y \in G$.

Claim. $c_\lambda \in F_\lambda$, where c_λ is as in Step 5 of §2.

PROOF. Let $b^- \in B^-, g \in G$, say $b^- = x_1 t$, $x_1 \in U^-$, $t \in T$, and $g = xn_w y$, $x \in U^-$, $w \in W$, $y \in B$. Then

$$c_\lambda(b^- g) = c_\lambda(x_1 t x n_w y) = c_\lambda(x' t n_w y) = c_\lambda(x' n_w t_1 y),$$

where $x' \in U^-$, and $tn_w = n_w t_1$; we have $\lambda(t_1) = \lambda(n_w^{-1} t n_w) = w(\lambda)(t)$ (cf. §1.1) $= \lambda(t)$, if $w \in W(P_\lambda)$, P_λ being as in Step 5 of §2. Hence we obtain that if $w \in W(P_\lambda)$, then $c_\lambda(b^- g) = \lambda(t_1 y) = \lambda(t_1)\lambda(y) = \lambda(t)\lambda(y) = \lambda(b^-)c_\lambda(g)$. If $w \notin W(P_\lambda)$, then $c_\lambda(b^- g) = 0 = \lambda(b^-)c_\lambda(g)$ (note that $c_\lambda(b^- g)$ and $c_\lambda(g)$ are 0 if $w \notin W(P_\lambda)$). The Claim now follows. □

Let $X = B^- \backslash G$, and $\pi : G \to X$, the canonical projection. Then $\pi : G \to X$ is a principal B^- bundle (with B^- as the structure group and fiber). Now given $\lambda : B^- \to \mathbb{G}_m$, λ gives rise to an action (on the right) of B^- on $k(\cong \mathbb{G}_a)$, namely $c \cdot x = \lambda(x)c$, $x \in B^-, c \in k$. Set $E = k \times G/\sim$, where $\sim$ is the equivalence relation defined by $(c \cdot x, xg) \sim (c, g)$, $c \in k, x \in B^-, g \in G$. Then E is the total space of a line bundle, say $L(\lambda)$, over X. Let $\sigma : E \to X$ be the canonical map, namely $\sigma(c, g) = B^- g$. We shall now see how F_λ can be identified with $H^0(B^- \backslash G, L(\lambda))$ $(= \{s : X \to E \mid \sigma \circ s = \mathrm{id}_X\})$. Let $f \in F_\lambda$. To f, we associate a section $s : X \to E$, by setting $s(B^- g) = (f(g), g)$. To see that s is well defined, consider $g' = xg$, $x \in B^-$. Then

$$(f(g'), g') = (f(xg), xg) = (\lambda(x) f(g), xg) = (f(g) \cdot x, xg) \sim (f(g), g).$$

From this it follows that s is well defined.

Conversely, given $s \in H^0(B^- \backslash G, L(\lambda))$, consider $B^- g \in B^- \backslash G$. Let $s(B^- g) = (f(g'), g')$, where $g' = xg$ for some $x \in B^-$ (note that $B^- g' = B^- g$, since $\sigma \circ s = \mathrm{id}_X$). Now the point $(f(g'), g')$ may also be represented by $(f(xg)\lambda(x)^{-1}, g)$ (note that $(f(g'), g') = (f(xg), xg) \sim (f(xg)\lambda(x)^{-1}, g)$). Thus given $g \in G$, there exists a unique representative of the form $(f(g), g)$ for $s(B^- g)$. This defines a function $f : G \to k$. Further, this f has the property that for $x \in B^-$, $f(g) = f(xg)\lambda(x)^{-1}$, i.e. $f(xg) = \lambda(x) f(g)$, $x \in B^-$, $g \in G$. Thus we obtain an identification

THEOREM 3.1.1. $F_\lambda = H^0(B^- \backslash G, L(\lambda))$.

It can be checked easily that the above identification preserves the respective G-module structure. Thus we obtain (in view of the identification in Theorem 3.1.1)

THEOREM 3.1.2. *Let $\lambda \in \Lambda^+_{\mathrm{wt}}$. Then F_λ is finite dimensional, and is isomorphic to $H^0(B^- \backslash G, L(\lambda))$. Further, $c_\lambda \in F_\lambda$, and is a highest weight vector of weight λ.*

We now recall the following result, due to Chevalley (cf. [**36**]).

THEOREM 3.1.3. *The line kc_λ is the unique B-fixed line in F_λ.*

PROOF. Let (f) a B-fixed line in F_λ, i.e. f is a highest weight vector in F_λ of weight μ say. We shall now show that $\mu = \lambda$, and $f = ac_\lambda$, for some $a \in k^*$.

For $x \in G, y \in B$, we have

$$y \cdot f(x) = f(xy) = \mu(y) f(x).$$

Further,

$$(*) \qquad f(e) = f(tet^{-1}) = \lambda(t)\mu(t^{-1})f(e), \text{ for all } t \in T.$$

Claim. $f(e) \neq 0$.

Proof of Claim. Let $\Omega = U^- B$. Let $x \in U^-, y \in B$. Then $f(xy) = \lambda(x) \cdot f(y) = f(y) = f(ey) = \mu(y) f(e)$. Hence $f(e) = 0$ would imply $f \equiv 0$ on Ω, and hence we would obtain $f \equiv 0$, which is not true. This proves the Claim.

Now Claim and $(*)$ imply that $\lambda(t) = \mu(t)$ for all $t \in T$, i.e. $\lambda = \mu$. Now for $x \in U^-, y \in B$, we have as above

$$f(xy) = f(y) = f(ey) = \lambda(y) f(e),$$

and

$$c_\lambda(xy) = \lambda(y)$$

(by the definition of c_λ, cf. §2). Hence we obtain

$$f = ac_\lambda \text{ on } \Omega,$$

where $a = f(e) \in k^*$. Thus we get $f = ac_\lambda$ on G, and the required result follows. □

As an immediate consequence, we obtain

THEOREM 3.1.4. *The G-module F_λ is indecomposable (i.e., F_λ can not be written as a direct sum of two nonzero proper submodules).*

PROOF. If M is a G-submodule of F_λ, then there exists a B-fixed line in M (by Borel fixed point Theorem (cf. Chapter 2, Theorem 2.1.3)). By Theorem 3.1.3, this B-fixed line is none other than the line kc_λ. The required result now follows from this. □

THEOREM 3.1.5. *Let* char$k = 0$. *Then the set of isomorphism classes of all finite dimensional irreducible G-modules is given by $\{[F_\lambda], \lambda \in \Lambda^+_{\text{wt}}\}$ (here for a G-module M, $[M]$ denotes its isomorphism class).*

PROOF. In characteristic 0, a reductive group is completely reducible (see for example [**73**]), i.e. any finite dimensional G-module M breaks up as the direct sum of irreducible G-modules. In particular, we get that in characteristic 0, F_λ, $\lambda \in \Lambda^+_{\text{wt}}$ is both completely reducible and indecomposable. Hence we obtain that F_λ, $\lambda \in \Lambda^+_{\text{wt}}$, is irreducible in characteristic 0 with associated highest weight λ. The result now follows from Theorem 2.0.1. □

REMARK 3.1.6. F_λ need not be irreducible in positive characteristic, as shown by the following counterexample. Let $G = SL(2)$. We have, $F_r = \mathcal{S}^r(k^2)$. Let char$k = p$, and $r = np$. Let $E = \{f^p, f \in \mathcal{S}^n(k^2)\}$. Then E is a G-stable subspace of F_r, and thus F_r is not irreducible.

REMARK 3.1.7. The representation theory of G in positive characteristic (unlike the characteristic 0 case) is quite complicated; for example, the problem of finding the character of $V(\lambda)$, $\lambda \in \Lambda^+_{\text{wt}}$, is still an open problem (though much progress has been achieved, thanks to the work in [**118**], [**82**], [**79**], [**7**]). Nevertheless the G-modules F_λ, $\lambda \in \Lambda^+_{\text{wt}}$, are well understood in all characteristics, as given by the following

THEOREM 3.1.8. *The dimension and the character of F_λ are independent of the characteristic.*

PROOF. By Theorem 3.1.2, we have the identification $F_\lambda \cong H^0(B^-\backslash G, L(\lambda))$. On the other hand, we have (cf. [**83**])

$$H^i(B^-\backslash G, L(\lambda)) = 0, \quad i \geqslant 1, \quad \lambda \in \Lambda^+_{\text{wt}}.$$

Hence the result follows by semicontinuity Theorem (cf. [**125**]) and "reduction mod p" (since $B^-\backslash G$ is defined over $\mathbb{Z}$ (cf. [**36**]), etc.). □

Note. In the above discussion, we have considered right module structure on $H^0(B^-\backslash G, L(\lambda))$. It is sometimes convenient to have the corresponding statements for the left module structure on $H^0(G/B, L(\lambda))$ handy. The maps $B^-\backslash G \to B\backslash G$, $B^-g \mapsto Bw_0gw_0$, and $B\backslash G \to G/B$, $Bg \mapsto g^{-1}B$ are well defined , and in fact are isomorphisms. Using these two isomorphisms, the above results on $B^-\backslash G$ yield corresponding results on G/B as described below:

Let $\lambda \in \Lambda^+_{\text{wt}}$, and $L(\lambda)$ be the associated line bundle on G/B. Let us denote G/B by Y. Let F be the total space of $L(\lambda)$, and $\sigma : F \to Y$ the canonical map, namely $\sigma(g, c) = gB$, $g \in G$, $c \in k$ (note that $F = G \times k/\sim$, where $\sim$ is the equivalence relation $(gb, bc) \sim (g, c)$, $g \in G$, $b \in B$, $c \in k$). Let

$$M_\lambda = \{f \in k[G] \mid f(gb) = \lambda(b)f(g), g \in G, b \in B\}.$$

Then we have an identification $M_\lambda = H^0(G/B, L(\lambda))(= \{s : Y \to F \mid \sigma \circ s = \text{id}_Y\})$. Further, this identification preserves the respective (left) G-module structures. For $w \in W$, fix a representative n_w in $N_T(G)$. Let $g \in G$, say $g = un_wb$, for unique u, w, b, $u \in U$, $w \in W$, $b \in B$ (cf. Bruhat decomposition $G = \cup_w UwB$) (here, U is the unipotent part of B). Define $d_\lambda : G \to k$ by

$$d_\lambda(un_wb) = \begin{cases} \lambda(b), & \text{if } w_0w \in W_\lambda \\ 0, & \text{otherwise,} \end{cases}$$

where W_λ is the subgroup of W generated by $\{s_\alpha, \alpha \in S \mid s_\alpha(\lambda) = \lambda\}$.

PROPOSITION 3.1.9. *Let $\lambda \in \Lambda^+_{\text{wt}}$. Then $d_\lambda \in H^0(G/B, L(\lambda))$, and is a highest weight vector of weight $i(\lambda)$, where i is the element (the Weyl involution) $-w_0$ of* Aut$X(T)$, *i.e.* $i(\chi) = -w_0(\chi)$, $\chi \in X(T)$.

THEOREM 3.1.10. *Let $\lambda \in \Lambda^+_{\text{wt}}$. The line kd_λ is the unique B-fixed line in* $H^0(G/B, L(\lambda))$.

THEOREM 3.1.11. *Let* $\operatorname{char} k = 0$, *and* $\lambda \in \Lambda_{\mathrm{wt}}^+$. *Then* $H^0(G/B, L(\lambda))$ *is isomorphic to* $V(i(\lambda))$, *the irreducible G-module with highest weight* $i(\lambda)$.

THEOREM 3.1.12. *Let* $\lambda \in \Lambda_{\mathrm{wt}}$. *Then* $H^0(G/B, L(\lambda)) \neq 0$ *if and only if* $\lambda \in \Lambda_{\mathrm{wt}}^+$.

3.2. Summary

Summarizing, let $\lambda \in \Lambda_{\mathrm{wt}}$, and $L(\lambda)$ be the associated line bundle on G/B. We have,

1. $H^0(G/B, L(\lambda)) = \{f \in k[G] \mid f(gb) = \lambda(b)f(g), g \in G, b \in B\}$.
2. $H^0(G/B, L(\lambda)) \neq 0$ if and only if $\lambda \in \Lambda_{\mathrm{wt}}^+$.
3. There exists a unique B-stable line in $H^0(G/B, L(\lambda))$ if λ is dominant. As a consequence we have, $H^0(G/B, L(\lambda))$ is an indecomposable G-module.
4. In characteristic 0, $H^0(G/B, L(\lambda))$ is G-irreducible (with highest weight $i(\lambda)$, $i(= -w_0) \in \operatorname{Aut}X(T)$ being the Weyl involution) and conversely. Thus in characteristic 0, $\{H^0(G/B, L(\lambda)),\ \lambda \in \Lambda_{\mathrm{wt}}^+\}$ gives (up to isomorphism) *all* finite dimensional irreducible G-modules.
5. $L(\lambda)$ is ample if and only if $L(\lambda)$ is very ample if and only if λ is dominant and regular.

4. Extremal weight vectors, Weyl and Demazure modules

4.1. Extremal weight vectors

For a dominant weight λ, let $V(\lambda)$ be the irreducible $G_{\mathbb{C}}$-module (over **C**) with highest weight λ (here, $G_{\mathbb{C}}$ is the group over $\mathbb{C}$ with the same root data as G). Writing $V(\lambda)$ as a direct sum of T weight spaces, we have (see **[75]** for example) that the multiplicity of the weight $w(\lambda)$, $w \in W$ is 1. One usually refers to $w(\lambda)$ as the *extremal weight* in $V(\lambda)$ corresponding to w; a generator of the one dimensional weight space (of weight $w(\lambda)$) is referred to as the *extremal weight vector* (which is unique up to scalars) in $V(\lambda)$ of weight $w(\lambda)$.

4.2. Weyl and Demazure modules

Let $\{H_i, 1 \leqslant i \leqslant l, X_\alpha, \alpha \in R\}$ be the *Chevalley Basis* of $\mathfrak{g}_{\mathbb{C}} = \operatorname{Lie} G_{\mathbb{C}}$, l being the rank of G (see **[75]** for example); for example, for $G = SL(n)$, the Chevalley basis for $\mathfrak{g}_{\mathbb{C}}$ is given by

$$\{E_{ii} - E_{i+1,i+1}, 1 \leqslant i \leqslant n-1, E_{ij}, 1 \leqslant i, j \leqslant n, i \neq j\},$$

where for $1 \leqslant l, m \leqslant n$, E_{lm} denotes the elementary matrix with 1 at the (l, m)-th place and zeros elsewhere. Let $U(\mathfrak{g}_{\mathbb{C}})$ be the universal enveloping algebra of $\mathfrak{g}_{\mathbb{C}}$, $U^+(\mathfrak{g}_{\mathbb{C}})$ the subalgebra of $U(\mathfrak{g}_{\mathbb{C}})$ generated by $\{X_\alpha, \alpha \in S\}$. Let $U_{\mathbb{Z}}^+(\mathfrak{g}_{\mathbb{C}})$ be the Kostant $\mathbb{Z}$-form of $U^+(\mathfrak{g}_{\mathbb{C}})$, namely the $\mathbb{Z}$-subalgebra of $U^+(\mathfrak{g}_{\mathbb{C}})$ generated by

$$\{\frac{X_\alpha^n}{n!}, \alpha \in R^+, n \in \mathbb{Z}^+\}$$

(cf. [75]). Let λ be a dominant weight. Let us fix a highest weight vector u in $V(\lambda)$. For $w \in W$, fix a representative n_w for w in $N_T(G)$ (the normalizer of T in G), and set $u_w = n_w \cdot u$, and $V_{w,\mathbb{Z}} = U_{\mathbb{Z}}^{+}(\mathfrak{g}_{\mathbb{C}})u_w$. For any field k, let $V_{w,\lambda} = V_{w,\mathbb{Z}} \otimes k,\ w \in W$. Then $V_k(\lambda) = V_{w_0,\lambda}\ (= V_{w_0,\mathbb{Z}} \otimes k)$, is the *Weyl module* corresponding to λ, and for $w \in W,\ V_{w,\lambda}$ is the *Demazure module* corresponding to w (and λ). Note that for $\tau, w \in W,\ u_\tau \in V_{w,\lambda} \Longleftrightarrow w \geqslant \tau$ (assuming λ to be regular. If λ is not regular, then one should replace W by a set of representatives of $W/W_\lambda,\ W_\lambda$ being as in §2, Step 5). We have (see [42])

$$(*) \qquad H^0(G/B, L_\lambda) \simeq V_k(\lambda)^*, \qquad H^0(X(w), L_\lambda) \simeq V_{w,\lambda}^*,$$

where L_λ is the line bundle on G/B associated to λ.

4.3. The vectors $p_w,\ w \in W$

In view of the equation (*), we have, $\{-w(\lambda),\ w \in W\}$ are weights in the T-module $H^0(G/B, L_\lambda)$ of multiplicity 1. Let us fix a generator e for the one-dimensional weight space in $H^0(G/B, L_\lambda)$ of weight $-\lambda$ (note that e is a lowest weight vector). For $w \in W$, as above fix a representative n_w for w in $N_T(G)$, and set $p_w = n_w \cdot e$. Note that $H^0(G/B, L_\lambda) = V_k(i(\lambda))$, where $i(= -w_0)$ is the Weyl involution and $p_w,\ w \in W$ are simply the extremal weight vectors in $V_k(i(\lambda))$. Note also that the weight of p_w is $-w(\lambda)$.

Let λ be a dominant weight, say $\lambda = \sum a_i\omega_i$. Let $J = \{i, 1 \leqslant i \leqslant l \mid a_i \neq 0\}$, and let Q be the parabolic subgroup with $S \setminus J$ as the associated set of simple roots. (Note that Q is simply P_λ (cf. §2, Step 5)). Consider the projective embedding

$$G/Q \hookrightarrow \mathbb{P}(V_k(\lambda))\ (= \operatorname{Proj}(H^0(G/Q, L_\lambda)))$$

induced by the very ample line bundle L_λ on G/Q. The surjectivity of $H^0(G/Q, L_\lambda) \to H^0(X_Q(w), L_\lambda),\ w \in W^Q$, implies that $\mathbb{P}(V_{w,\lambda})$ is the smallest linear subspace of $\mathbb{P}(V_k(\lambda))$ that contains $X_Q(w)$.

LEMMA 4.3.1. *Let notations be as above. For $\tau, w \in W^Q$, we have*

$$p_\tau\,|\,_{X_Q(w)} \neq 0 \Longleftrightarrow w \geqslant \tau.$$

PROOF. Let $\mathbf{B}(,)$ be the $\mathfrak{g}$-invariant, non-degenerate bilinear on $V_k(\lambda) \times V_k(\lambda)^*$ (cf. [27]). Let x, x' be two weight vectors of weight χ, χ' respectively. Since $\mathbf{B}(,)$ is $\mathfrak{g}$-invariant, we have

$$\mathbf{B}(Hx, x') + \mathbf{B}(x, Hx') = 0,\ H \in \mathfrak{g}.$$

Taking for H an element of the Cartan sub-algebra $\mathfrak{h}$ of $\mathfrak{g}$, we have,

$$\chi(H)\mathbf{B}(x, x') + \chi'(H)\mathbf{B}(x, x') = 0,$$

$$\text{i.e., } (\chi(H) + \chi'(H))\mathbf{B}(x, x') = 0,$$

$$\text{i.e., } (\chi + \chi')(H)\mathbf{B}(x, x') = 0.$$

Suppose now that $\chi + \chi' \neq 0$. Then choosing H such that $(\chi + \chi')(H) \neq 0$, we obtain $\mathbf{B}(x, x') = 0$. Now choosing $x = u_\phi,\ x' = p_\tau$, we have, $\mathbf{B}(u_\phi, p_\tau) = 0$, if $\phi \neq \tau$; further, if $\phi = \tau$, then $\mathbf{B}(u_\tau, p_\tau) \neq 0$ (since $\mathbf{B}(,)$ is non-degenerate, and $p_\tau \neq 0$, there exists a $y \in V_k(\lambda)$ such that $\mathbf{B}(y, p_\tau) \neq 0$. Such a y is a weight vector of weight $= -\text{weight}p_\tau = \text{weight}u_\tau$, i.e., y is a scalar multiple of u_τ). Hence for $\tau, w \in W^Q$, we have, $p_\tau \mid X_{Q(w)} \neq 0 \Longleftrightarrow u_\tau \in V_{w,\lambda}$, i.e., if and only if $w \geqslant \tau$. □

5. Character formulas

5.1. Character of a T-module

Let us denote $X(T)$ by just X. Consider the group ring $\mathbb{Z}X$. We shall denote by e^λ or $\exp\lambda$ the element of $\mathbb{Z}X$ corresponding to λ in X. Now given a G-module V, as seen in §1.2, we can write

$$V = \bigoplus_{\chi \in X} V_\chi \quad \text{(as } T\text{-modules)},$$

where

$$V_\chi = \{v \in V \mid tv = \chi(t)v, t \in T\}.$$

Let $\Phi_V = \{\chi \in X \mid V_\chi \neq 0\}$, i.e. the set of weights of V.

DEFINITION 5.1.1. Let $m_\chi = \dim V_\chi$. The *character* of V is defined to be the element $\sum m_\chi e^\chi$ of $\mathbb{Z}X$, and is denoted ch V (or char V).

THEOREM 5.1.2. *Let* char $k = 0$. *Let V_1, V_2 be two finite dimensional G-modules. Then $V_1 \cong V_2$ (as G-modules) if and only if* ch $V_1 =$ ch V_2.

See **[75]** for a proof.

Consider the left G-module structure on $H^0(G/B, L(\lambda))$ given by $g \cdot f(x) = f(g^{-1}x)$. By §3.2, we have, in *characteristic* 0, $H^0(G/B, L(\lambda))$ is isomorphic to $V(i(\lambda))$. Hence ch $H^0(G/B, L(\lambda))$ is given by the Weyl character formula:

5.2. Weyl character formula

Let char $k = 0$. Let $\lambda \in \Lambda^+_{\text{wt}}$. Then we have the celebrated Weyl character formula (see for example **[75]**)

$$\operatorname{ch} V(\lambda) = \frac{\sum_{w \in W} \epsilon_w e^{w(\lambda+\rho)}}{\sum_{w \in W} \epsilon_w e^{w(\rho)}},$$

where $\epsilon_w = (-1)^{l(w)}$, and $\rho = \frac{1}{2}\sum_{\alpha \in R^+} \alpha$ (note that $\rho = \sum_{i=1}^{l} \omega_i$, and thus $\rho \in \mathbb{Z}X$).

By computing the sum of the coefficients of ch $V(\lambda)$, we get (see for example **[75]**)

$$\dim V(\lambda) = \prod_{\alpha > 0} \frac{(\lambda + \rho, \alpha^*)}{(\rho, \alpha^*)}.$$

Hence we obtain

$$\operatorname{ch} H^0(G/B, L(\lambda)) = \frac{\sum_{w \in W} \epsilon_w e^{w(i(\lambda)+\rho)}}{\sum_{w \in W} \epsilon_w e^{w(\rho)}}$$

DEFINITION 5.3.1. Denote by L_{s_α}, $\alpha \in S$, the linear operator $L_{s_\alpha} : \mathbb{Z}X \to \mathbb{Z}X$ defined by

$$L_{s_\alpha}(\exp\lambda) = \frac{\exp\lambda - \exp\big(s_\alpha(\lambda)\big)}{1-\exp\alpha}$$

Let $M_{s_\alpha} : \mathbb{Z}X \to \mathbb{Z}X$ be the operator defined by

$$M_{s_\alpha}(\exp\lambda) = \exp\rho \cdot L_{s_\alpha}(\lambda - \rho).$$

5.3. Demazure character formula

We have

$$M_{s_\alpha}(e^\mu) = \begin{cases} \sum_{j=0}^{m} e^{\mu - j\alpha} & \text{if } (\mu,\alpha^*) = m \geqslant 0, \\ 0 & \text{if } (\mu,\alpha^*) = -1, \\ -\sum_{j=1}^{m} e^{\mu + j\alpha} & \text{if } (\mu,\alpha^*) = -(m+1) \leqslant -2. \end{cases}$$

Let $w \in W$, and let $s_{i_1} \dots s_{i_r}$ be a reduced expression for w. Define L_w as $L_{i_1} \cdots L_{i_r}$, M_w as $M_{i_1} \cdots M_{i_r}$, where for $\alpha = \alpha_i$ (a simple root), $L_i = L_{s_{\alpha_i}}$, $M_i = M_{s_{\alpha_i}}$.

REMARK 5.3.2. (cf. [**42**]) L_w, M_w depend only on w and not on the reduced expression chosen.

THEOREM 5.3.3. *(cf.* [**42**]*) If λ is a dominant weight and $\tau, w \in W$ such that $w < \tau$ and $w = s_\alpha \tau$, $\alpha \in S$, then*

$$\operatorname{char} H^0\big(X(\tau), L(\lambda)\big) = M_{s_\alpha}\big(\operatorname{char} H^0\big(X(w), L(\lambda)\big)\big).$$

COROLLARY 5.3.4. *Let $\tau \in W$, and let λ be a dominant weight.*
If $\tau = s_{\alpha_1} \cdots s_{\alpha_r}$ is a reduced decomposition, then

$$\operatorname{char} H^0\big(X(\tau), L(\lambda)\big) = M_{s_{\alpha_1}} \cdots M_{s_{\alpha_r}}\big(\exp(-\lambda)\big).$$

(Note that the right hand side is independent of the reduced decomposition.)
The Corollary follows from the fact that char $H^0\big(X(\mathrm{id}), L(\lambda)\big) = \big(\exp(-\lambda)\big)$.

5.4. Lakshmibai-Seshadri Character Formula

Both Weyl and Demazure character formulas involve massive cancellations. Now we describe a character formula which is positive-termed.

Let $\lambda \in \Lambda^+_{\mathrm{wt}}$, and let W_λ be the subgroup of W generated by

$$\{s_\alpha, \alpha \in S \mid s_\alpha(\lambda) = \lambda\}.$$

Let W^λ denote the set of minimal representatives of W/W_λ. Let $\mathbf{c} = \{\mu_0, \mu_1, \dots, \mu_r\}$ be a chain in W^λ, i.e. $\mu_i \geqslant \mu_{i-1}$, $l(\mu_i) = l(\mu_{i-1}) + 1$. Then one knows that $\mu_{i+1} = \mu_i s_{\beta_i}$ for some $\beta_i \in R^+$. We define $m_\lambda(\mu_{i-1}, \mu_i)$, the *$\lambda$-multiplicity of $X(\mu_{i-1})$ in $X(\mu_i)$* as $m_\lambda(\mu_{i-1}, \mu_i) = (\lambda, \beta_i^*)$. Let us denote $m_\lambda(\mu_{i-1}, \mu_i)$ by just m_i.

The sets $I(\lambda), I_w(\lambda)$.

DEFINITION 5.4.1. A *weighted λ-chain* is a pair $(\mathbf{c}, \mathbf{n})$, where $\mathbf{c} = \{\mu_0, \dots, \mu_r\}$ is a chain, and $\mathbf{n} = \{n_1, \dots, n_r\}$, $n_i \in \mathbb{Z}^+$.

DEFINITION 5.4.2. A weighted λ-chain $(\mathbf{c}, \mathbf{n})$ is said to be *admissible* if

$$1 \geqslant \frac{n_1}{m_1} \geqslant \frac{n_2}{m_2} \geqslant \dots \geqslant \frac{n_r}{m_r} \geqslant 0.$$

EXAMPLE 5.4.3. $G = SL(3)$, $\lambda = 6\epsilon_1 + 2\epsilon_2 = 4\omega_1 + 2\omega_2$

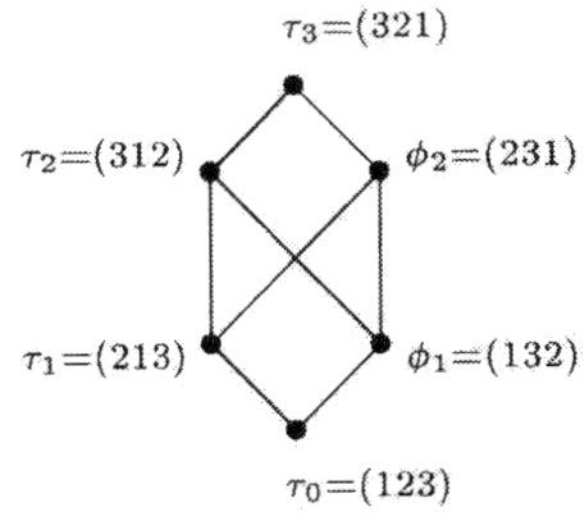

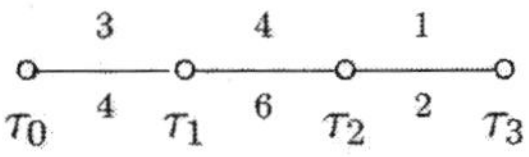

Let $(\mathbf{c}, \mathbf{n})$ be admissible. Let us denote the distinct values in $\frac{n_1}{m_1}, \frac{n_2}{m_2}, \dots, \frac{n_r}{m_r}$ by $a_1, a_2, \dots, a_s$, so that $1 \geqslant a_1 > a_2 > \dots > a_s \geqslant 0$. Let $i_0, \dots, i_s$ be defined by $i_0 = 0, i_s = r, \frac{n_j}{m_j} = a_t, i_{t-1} + 1 \leqslant j \leqslant i_t$.

We set $D_{\mathbf{c},\mathbf{n}} = \{(a_1, \dots, a_s); (\mu_{i_0}, \dots, \mu_{i_s})\}$.

EXAMPLE 5.4.4.

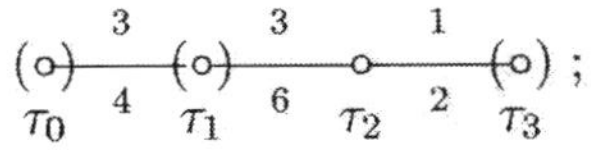

$$D_{\underline{c},\underline{n}} = \left\{\left(\frac{3}{4}, \frac{1}{2}\right); (\tau_0, \tau_1, \tau_3)\right\}$$

DEFINITION 5.4.5. Let $(\mathbf{c}, \mathbf{n})$, $(\mathbf{c}', \mathbf{n}')$ be two admissible weighted λ-chains. We say $(\mathbf{c}, \mathbf{n}) \sim (\mathbf{c}', \mathbf{n}')$ if $D_{\mathbf{c},\mathbf{n}} = D_{\mathbf{c}',\mathbf{n}'}$.

EXAMPLE 5.4.6.

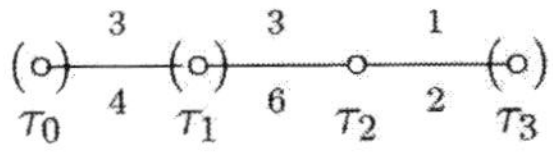

is $\sim$ to

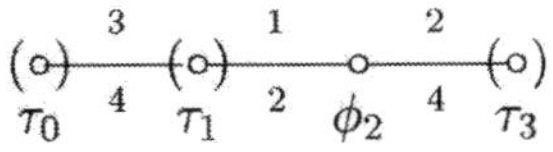

Let $C(\lambda)$ be the set of all admissible weighted λ-chains, and $I(\lambda) = C(\lambda)/\sim$.

Let $\pi \in I(\lambda)$, and let $(\mathbf{c}, \mathbf{n})$ be a representative of π. We set $i(\pi) = \mu_{i_s}$ (notations being as above; note that $i(\pi)$ depends only on π and not on the representative chosen). For $w \in W^\lambda$, let $I_w(\lambda) = \{\pi \in I(\lambda) \mid i(\pi) \leqslant w\}$.

Littelmann's Root Operators

Each element $\pi = \{(a_1, \dots, a_s); (\mu_{i_0}, \dots, \mu_{i_s})\}$ in I_λ can be thought of as a certain piecewise linear path

$$\pi : [0, 1] \longrightarrow X \otimes \mathbb{R},$$

as follows: if $a_r \leqslant t \leqslant a_{r-1}, 0 \leqslant i \leqslant s$ (here $a_0 = 1, a_{s+1} = 0$), then

$$\pi(t) = \sum_{j=r}^{s} (a_j - a_{j+1})\mu_{i_j}(\lambda) + (t - a_r)\mu_{i_{r-1}}(\lambda).$$

Let $\pi \in I(\lambda)$, and let $(\mathbf{c}, \mathbf{n})$ be a representative of π. With notations as above, we set $\mathrm{wt}(\pi) = \sum_{j=0}^{s}(a_j - a_{j+1})\mu_{i_j}(\lambda)$, where $a_0 = 1$ and $a_{s+1} = 0$. Note that $\mathrm{wt}(\pi)$ depends only on π and not on the representative chosen; also note that $\mathrm{wt}(\pi) = \pi(1)$, π being considered as a path as above. Following **[114]**, we shall refer to the elements of $I(\lambda)$ as *Lakshmibai-Seshadri paths of shape* λ.

Let $X = X(T)$, and $X_{\mathbb{R}} = X(T) \otimes_{\mathbb{Z}} \mathbb{R}$. Let Π be the set of all piecewise linear paths $\pi : [0,1] \to X_{\mathbb{R}}$ such that $\pi(0) = 0$, modulo the equivalence relation $\pi \sim \pi'$ if $\pi = \pi'$ up to a reparametrization. For $\pi_1, \pi_2 \in \Pi$, we denote by $\pi_1 * \pi_2$ the concatenation of π_1 and π_2, i.e., $\pi_1 * \pi_2$ is the path defined by

$$\pi_1 * \pi_2(t) = \begin{cases} \pi_1(2t), & 0 \leqslant t \leqslant 1/2 \\ \pi_1(1) + \pi_2(2t-1), & 1/2 \leqslant t \leqslant 1. \end{cases}$$

For a simple root α, let $s_\alpha(\pi)$ be the path given by $s_\alpha(\pi)(t) = s_\alpha(\pi(t))$.

For a simple root α, the operator f_α on Π is defined as follows. Let $\pi \in \Pi$. Let Q be the minimum value of the function $h_\alpha : [0,1] \to \mathbb{R}$, $h_\alpha(t) = (\pi(t), \alpha^*)$. Let $t_1 \in [0,1]$ be the largest t for which this minimum is attained, and let $t_2 \in [0,1]$ be the smallest $t > t_1$, where the function attains $Q+1$ (if there exists any such t). Now write π as the concatenation of three segments $\pi = \pi_1 * \pi_2 * \pi_3$ corresponding to $t \in [0, t_1]$, $t \in [t_1, t_2]$, and $t \in [t_2, 1]$. If t_2 exists, then define

$$f_\alpha(\pi) = \pi_1 * s_\alpha \pi_2 * \pi_3,$$

where $s_\alpha \pi_2$ is the path $t \mapsto s_\alpha(\pi_2(t))$. If t_2 does not exist, then $f_\alpha \pi$ is undefined.

The operator e_α is similarly defined, namely, for $\pi \in I(\lambda)$, let t'_2 be the smallest t where the function h_α attains its minimum Q, and let t'_1 be the largest $t < t'_2$ where the function h_α attains $Q+1$ (if there exists any such t). Write $\pi = \pi'_2 * \pi'_2 * \pi'_3$ as the concatenation of the three segments corresponding to $t \in [0, t'_1]$, $t \in [t'_1, t'_2]$, and $t \in [t'_2, 1]$. If t'_1 exists, then define

$$e_\alpha = \pi'_1 * s_\alpha \pi'_2 * \pi'_3.$$

If t'_1 does not exist, then $e_\alpha \pi$ is undefined. The following facts are easily checked:

(1) $e_\alpha(f_\alpha \pi) = \pi$, whenever $f_\alpha(\pi)$ is defined.
(2) For $\pi \in \Pi$, define $\mathrm{wt}\,\pi = \pi(1)$. We have
 (a) $\mathrm{wt}\, f_\alpha \pi = \mathrm{wt}\,\pi - \alpha$.
 (b) $\mathrm{wt}\, e_\alpha \pi = \mathrm{wt}\,\pi + \alpha$.

THEOREM 5.4.7. *(cf.* **[114]***) Let $\lambda \in X^+$ and $I(\lambda)$ (as above) the set of Lakshmibai-Seshadri paths of shape λ. Let e_λ the path $t \mapsto t \cdot \lambda$, i.e. the line segment joining* 0 *and λ. Then*

(1) *$I(\lambda)$ is stable under the operators e_α, f_α.*
(2) $I(\lambda) = \{\pi \in \Pi \mid \pi = f_{\alpha_{i_1}} \circ \cdots \circ f_{\alpha_{i_r}}(e_\lambda)$ for some $i_1, \ldots, i_r\}$.

L-S character formula (Lakshmibai-Seshadri character formula):

(1) $\mathrm{ch}\, V_\lambda = \sum_{\pi \in I(\lambda)} e^{\mathrm{wt}(\pi)}$.

Note that the above formula also gives a formula for the multiplicity of a weight ψ.

(2) For $w \in W^\lambda$, $\operatorname{ch} V_w(\lambda) = \sum_{\pi \in I_w(\lambda)} e^{\mathrm{wt}(\pi)}$.

(Here, V_λ is the Weyl module associated to λ, and $V_w(\lambda)$ is the Demazure module associated to w (cf. §4.2))

REMARK 5.4.8. The above character formula was conjectured by Lakshmibai and Seshadri (cf. [**107**]) and was proved by Littelmann (cf. [**114**]).

The details of the path model theory will appear in [**97**].

CHAPTER 5

The groups $SL(n)$ and $GL(n)$

In this chapter, we discuss the general linear group $GL(n)$ and the special linear group $SL(n)$ and their representation theory over $\mathbb{C}$.

1. The symmetric group $\mathfrak{S}_n$

Let $\mathfrak{S}_n$ be the group of permutations of the set $\{1, \ldots, n\}$, i.e. the set of bijections $w : \{1, \ldots, n\} \to \{1, \ldots, n\}$. An element $w \in \mathfrak{S}_n$ is defined by the numbers $w(i)$, $1 \leqslant i \leqslant n$, and if $w(i) = a_i$, $1 \leqslant i \leqslant n$, then w will be written as the sequence of n numbers

$$w = (a_1 \ldots a_n).$$

For $1 \leqslant i \leqslant n-1$, the permutation $(1, \ldots, i-1, i+1, i, i+2, \ldots, n)$, switching the elements i and $i+1$, is denoted by s_i. The s_i's are called *simple transpositions*. Let $S = \{s_i \mid 1 \leqslant i \leqslant n-1\}$ be the set of simple transpositions. Then the symmetric group $\mathfrak{S}_n$ is generated by S, and the following relations are satisfied

(a) $$s_i^2 = \text{id}, \qquad 1 \leqslant i \leqslant n-1$$

(b) $$s_i s_j = s_j s_i, \qquad \text{if } |i-j| > 1$$

(c) $$s_i s_j s_i = s_j s_i s_j, \qquad \text{if } |i-j| = 1.$$

In fact this gives a presentation of the symmetric group, i.e. $\mathfrak{S}_n$ is the group generated by $n-1$ elements $\{s_i \mid 1 \leqslant i \leqslant n-1\}$, subject to the relations (a), (b) and (c).

An expression $s_{j_1} \ldots s_{j_r}$ is called a subexpression of $s_{i_1} \ldots s_{i_s}$ if $\{j_1, \ldots j_r\}$ is a subsequence of $\{i_1, \ldots i_s\}$. For example $s_1 s_2 s_1$ is a subexpression of $s_2 s_1 s_2 s_1 s_2$.

1.1. The length of w

DEFINITION 1.1.1. Given $w \in \mathfrak{S}_n$, the smallest integer r such that w can be written as $w = s_{i_1} \ldots s_{i_r}$ is called the *length* of w, and denoted by $l(w)$. An expression $w = s_{i_1} \ldots s_{i_r}$, with $r = l(w)$, is called a *reduced expression* of w.

EXAMPLE 1.1.2. Let $w = (231)$ in $\mathfrak{S}_3$. Then w can be expressed as $w = s_1 s_2$, and also as $w = s_2 s_1 s_2 s_1$, in view of the relation (c). The first expression is reduced, while the second one is not.

REMARK 1.1.3. An element w can have more than one reduced expression. For example $s_3s_1s_2s_1$ and $s_3s_2s_1s_2$ are both reduced expressions for (4213) in $\mathcal{S}_4$.

DEFINITION 1.1.4. Given $w \in \mathcal{S}_n$, an expression $s_{i_1} \dots s_{i_r}$ is called a *minimal expression* of w if $w = s_{i_1} \dots s_{i_r}$, and w is not equal to any subexpression of $s_{i_1} \dots s_{i_r}$.

REMARK 1.1.5. An expression of $w \in \mathcal{S}_n$ is reduced if and only if it is minimal.

LEMMA 1.1.6. *Let $w = (a_1 \dots a_n)$ be an element of $\mathcal{S}_n$. For $1 \leqslant i \leqslant n-1$, let $r_i = \#\{j > i \mid a_j < a_i\}$, and $\sigma_i = s_{r_i+i-1} \dots s_i$ if $r_i \geqslant 1$, $\sigma_i = \mathrm{id}$ if $r_i = 0$ (note that $r_i + i - 1 \geqslant i$ if $r_i \geqslant 1$). Then $w = \sigma_1 \dots \sigma_{n-1}$, and this is a reduced expression for w*

PROOF. We have

$$\begin{aligned} a_1 &= r_1 + 1 \\ a_2 &= \text{the } (r_2+1)\text{-th entry in } \{1, \dots, n\} \setminus \{a_1\} \\ a_3 &= \text{the } (r_3+1)\text{-th entry in } \{1, \dots, n\} \setminus \{a_1, a_2\} \\ &\dots \\ a_i &= \text{the } (r_i+1)\text{-th entry in } \{1, \dots, n\} \setminus \{a_1, \dots, a_{i-1}\} \\ &\dots \end{aligned}$$

Now

$$\begin{aligned} \sigma_1 &= ((r_1+1)12 \dots \widehat{(r_1+1)} \dots n) = (a_1 12 \dots \widehat{a_1} \dots n) \\ \sigma_2 &= (1(r_2+1)23 \dots \widehat{(r_2+2)} \dots n) \\ &\dots \\ \sigma_i &= (1 \dots (i-1)(r_i+i)i(i+1) \dots \widehat{(r_i+i)} \dots n) \\ &\dots \end{aligned}$$

We have,

$$\sigma_1\sigma_2 = (a_1a_2 12 \dots \widehat{a_1} \dots \widehat{a_2} \dots n),$$

and inductively

$$\sigma_1 \dots \sigma_i = (a_1 \dots a_i 1 \dots \widehat{a_1} \dots \widehat{a_i} \dots n).$$

For $i = n-1$, this means $\sigma_1 \dots \sigma_{n-1} = (a_1 \dots a_n) = w$.

It remains to show that $\sigma_1 \dots \sigma_{n-1}$ is a reduced expression for w. We may assume that $a_1 > 1$ (if $a_1 = 1$, then $w \in \mathcal{S}_{n-1}$ and the result follows by induction on n). Now $a_1 > 1$ implies $\sigma_1 = (a_1 123 \dots \widehat{a_1} \dots n)$. Write $w = \sigma_1 w_1$, where $w_1 = \sigma_2 \dots \sigma_{n-1}$. Then w_1 is of the form $(1b_2 \dots b_n)$. By induction on n, we have $\sigma_2 \dots \sigma_{n-1}$ is a reduced expression for w_1. If $\sigma_1 w_1$ is not reduced, let $\sigma_1' w_1'$ be a reduced expression for w, where σ_1' (resp. w_1') is a subexpression of σ_1 (resp. w_1). The facts that $a_1 > 1$ and w_1' does not involve s_1 imply that σ_1' involves s_1. Let r be the largest number $\leqslant r_1$ such that s_t is present in σ_1' for all

$1 \leqslant t \leqslant r$. Then σ_1' has the form $\sigma_1' = \sigma_1'' s_r \ldots s_1$. Now σ_1'' commutes with $s_r \ldots s_1$ and hence $w = s_r \ldots s_1 \sigma_1'' w_1'$. Also $\sigma_1'' w_1'$ is of the form $(1c_2 \ldots c_n)$ and hence $w (= s_r \ldots s_1 \sigma_1'' w_1') = ((r+1)12 \ldots r\widehat{(r+1)} \ldots n)(1c_2 \ldots c_n)$. This implies $r + 1 = a_1 = r_1 + 1$, and hence $r = r_1, \sigma_1' = \sigma_1, w_1' = w_1$. □

COROLLARY 1.1.7. (**Length Formula**)

$$l(w) = \sum_{r=1}^{n} r_i = \#\{(i,j) \mid 1 \leqslant i < j \leqslant n,\ a_j < a_i\}.$$

REMARK 1.1.8. A pair (i, j) as in the set on the right hand side of the Length Formula is called an *inversion* of w.

LEMMA 1.1.9. *Let $w \in \mathcal{S}_n$. Then $l(w) = l(w^{-1})$.*

PROOF. Let $l(w) = r$, and let $w = s_{i_1} \ldots s_{i_r}$ be a reduced expression for w. Then $w^{-1} = s_{i_r} \ldots s_{i_1}$, which shows that $l(w^{-1}) \leqslant r = l(w)$. The result follows from this by switching the roles of w and w^{-1}. □

LEMMA 1.1.10. *Let $w_1, w_2 \in \mathcal{S}_n$. Then*

$$l(w_1) - l(w_2) \leqslant l(w_1 w_2) \leqslant l(w_1) + l(w_2).$$

PROOF. Let $w_1 = s_{i_1} \ldots s_{i_r}, w_2 = s_{j_1} \ldots s_{j_s}$ be reduced expressions for w_1, w_2 respectively. Then $w_1 w_2 = s_{i_1} \ldots s_{i_r} s_{j_1} \ldots s_{j_s}$. Hence

$$l(w_1 w_2) \leqslant r + s = l(w_1) + l(w_2).$$

Writing this inequality for $w_1 w_2$ and w_2^{-1} instead of w_1, w_2 respectively, we obtain $l(w_1 w_2 w_2^{-1}) \leqslant l(w_1 w_2) + l(w_2^{-1})$, i.e. $l(w_1) \leqslant l(w_1 w_2) + l(w_2)$. Hence we obtain $l(w_1) - l(w_2) \leqslant l(w_1 w_2)$. □

1.2. The signature map

Define $\theta : \mathcal{S}_n \to \{\pm 1\}$ as follows. Let $w \in \mathcal{S}_n$, and let $w = s_{i_1} \ldots s_{i_r}$ be an expression for w as a product of transpositions. Set $\theta(w) = (-1)^r$. The fact that θ is well defined follows from the fact that if $w = s_{i_1} \ldots s_{i_r}, w = s_{j_1} \ldots s_{j_t}$ are two expressions for w, as a product of transpositions, then r and t are of the same parity. It is easily seen that θ is a group homomorphism, called the *signature map*. For $w \in \mathcal{S}_n$, $\theta(w)$ is also denoted by sign(w).

REMARK 1.2.1. Note that sign(w) is simply $(-1)^{l(w)}$.

LEMMA 1.2.2. *Let $w \in \mathcal{S}_n$ and $s \in S$. Then $l(ws) = l(sw) = l(w) \pm 1$.*

PROOF. Let $l(w) = r$. Then the multiplicativity of the signature map implies that $l(ws)$ is of the same parity as $r + 1$. This, together with Lemma 1.1.10 implies the required result. □

1.3. Permutation matrices

To an element $w \in \mathcal{S}_n$, we associate a *permutation matrix* A, i.e. a matrix having precisely one nonzero entry in each row and each column, as follows:

$$a_{ij} = \begin{cases} 1, & \text{if } i = w(j) \\ 0, & \text{otherwise.} \end{cases}$$

Then $\mathrm{sign}(w)$ is simply $\det(A)$.

Let N be the set of permutation matrices in G, where G is either $GL(n)$ or $SL(n)$. Let T be the subgroup consisting of diagonal matrices. Then it can be easily seen that N is the normalizer of T in G, and $N/T \cong \mathcal{S}_n$.

1.4. Partial order on $\mathcal{S}_n$

Given $w_1 = (a_1 \dots a_n)$, $w_2 = (b_1, \dots, b_n)$ in $\mathcal{S}_n$, we define

$$w_1 \geqslant w_2 \quad \text{if} \quad \{a_1, \dots, a_i\} \uparrow \geqslant \{b_1, \dots, b_i\} \uparrow \text{ for all } 1 \leqslant i \leqslant n-1$$

(here $\{a_1, \dots, a_i\} \uparrow$ denotes the ordered set $\{a_1, \dots, a_i\}$ with the entries arranged in ascending order). This is in fact the Bruhat-Chevalley order on $W = \mathcal{S}_n$ (W being considered as the Weyl group of $SL(n)$).

2. Representation theory of finite groups

In characteristic zero, the representation theory of $SL(n)$ (and $GL(n)$) is closely linked to that of the symmetric group $\mathcal{S}_n$. We first recall some basic results on the representation theory of finite groups (see **[55]** for example).

2.1. The group algebra kG

Let k be a field. Let G be a finite group of order m. Let kG be the group algebra of G over k. The group G is canonically identified with a subgroup of the group of all units in kG. Given a representation ρ of G, i.e. a group homomorphism $\rho : G \to \mathrm{Aut}\, V$, where V is a finite dimensional k-vector space and $\mathrm{Aut}\, V$ is the group of linear automorphisms of V, ρ induces a representation of kG on V, namely $\rho : kG \to \mathrm{End}\, V$ given by

$$\rho(\sum \lambda_i g_i) = \sum \lambda_i \rho(g_i),$$

and vice versa, $\mathrm{End}\, V$ being the k-algebra of linear endomorphisms of V.

EXAMPLE 2.1.1. The left module structure of kG over itself given by left multiplication (called the *left regular representation* of G over k).

Let A be a ring, and M a left A-module.

DEFINITION 2.1.2. M is *simple* or *irreducible* if M has no nontrivial proper submodules. M is *semisimple* or *completely reducible* if M breaks up as a direct sum of simple A-modules. The ring A is *semisimple* if A considered as a left A-module is semisimple.

We have the following

PROPOSITION 2.1.3. (1) *A is semisimple $\iff$ every A-module is semisimple.* (2) *kG is semisimple $\iff$ $m \neq 0$ in k (i.e. if char $k = p$, then either $p = 0$, or p does not divide m).* See **[55]** for a proof.

2.2. Classification of irreducible G-modules

Let us suppose that k is algebraically closed, and $m \neq 0$ in k. Given a G-module, $\rho : G \to \operatorname{Aut} V$, one would like to describe ρ in some explicit way. Now, for each $g \in G$, $\rho(g)$ is a linear automorphism of V. The *character of the representation* ρ is $\psi_V : G \to k$ defined by

$$\psi_V(g) = \operatorname{trace} \rho(g).$$

We have

$$\psi_V(hgh^{-1}) = \psi_V(g), \quad g, h \in G.$$

Thus ψ_V is a *class function*, i.e., ψ_V is constant on each conjugacy class. We have

PROPOSITION 2.2.1. (1) $\psi_{V_1 \oplus V_2} = \psi_{V_1} + \psi_{V_2}$.
(2) $V_1 \cong V_2 \iff \psi_{V_1} = \psi_{V_2}$. See **[55]** for a proof.

We can now completely describe the representation theory of G as follows. In view of Propositions 2.1.3 and 2.2.1, the representation theory of kG will be completely known by knowing the irreducible characters (i.e. characters of irreducible G-modules). Towards this, we have the following

THEOREM 2.2.2. (1) *The number of nonisomorphic simple (i.e. irreducible) G-modules is given by the number of conjugacy classes in G.*

(2) *Let the number in* (1) *be r, and let $\{V_i, 1 \leqslant i \leqslant r\}$ be all the nonisomorphic simple G-modules. Let* $\dim V_i = n_i$. *Then*

$$kG = \bigoplus_{i=1}^{r} V_i^{\oplus n_i} \quad \text{and} \quad \psi_{kG} = \sum_{i=1}^{r} n_i \psi_i,$$

where $\psi_i = \psi_{V_i}$ (here kG is the left regular representation, i.e. kG considered as a kG-module by left multiplication).

For the rest of this section, we assume $k = \mathbb{C}$. Let $\mathbb{C}_{\text{class}}(G)$ be the set of complex valued class functions on G. We define a Hermitian inner product $\langle , \rangle$ on $\mathbb{C}_{\text{class}}(G)$ by

$$\langle \alpha, \beta \rangle = \frac{1}{|G|} \sum \overline{\alpha(g)} \beta(g).$$

(Here $|G|$ denotes the cardinality of G and for a complex number z, $\bar{z}$ denotes its complex conjugate.)

THEOREM 2.2.3. *The irreducible characters form an orthonormal basis for* $\mathbb{C}_{\text{class}}(G)$.

See **[55]** for proofs of Theorems 2.2.2 and 2.2.3

2.3. The representation ring $R(G)$

The *representation ring* $R(G)$ is the free abelian group generated by all isomorphism classes of representations of G modulo the subgroup generated by elements of the form $[V] + [W] - [(V \oplus W)]$, multiplication being given by tensor product.

Note that in view of complete reducibility, we have

$$R(G) = \{\sum a_i[V_i] \mid a_i \in \mathbb{Z},\ V_i \text{ simple}\}.$$

Now ψ induces a natural map

$$\psi : R(G) \otimes \mathbb{C} \to \mathbb{C}_{\text{class}}(G).$$

Using Theorem 2.2.3, we obtain

THEOREM 2.3.1. *$\psi : R(G) \otimes \mathbb{C} \to \mathbb{C}_{\text{class}}(G)$ is an isomorphism.*

3. Representation Theory of the symmetric group $\mathbb{S}_d$

Let us now describe all of the above results for the symmetric group $\mathbb{S}_d$. Given a permutation $\sigma \in \mathbb{S}_d$, let $\sigma = \sigma_1 \dots \sigma_r$ be the expression of σ as a product of disjoint cycles of lengths $k_1, \dots, k_r$, with $k_1 \geqslant k_2 \geqslant \dots \geqslant k_r$. Thus σ determines a partition of d where $d = \sum k_i$. We represent this diagrammatically as an array of d boxes arranged in r rows of lengths $k_1, \dots, k_r$, as shown below.

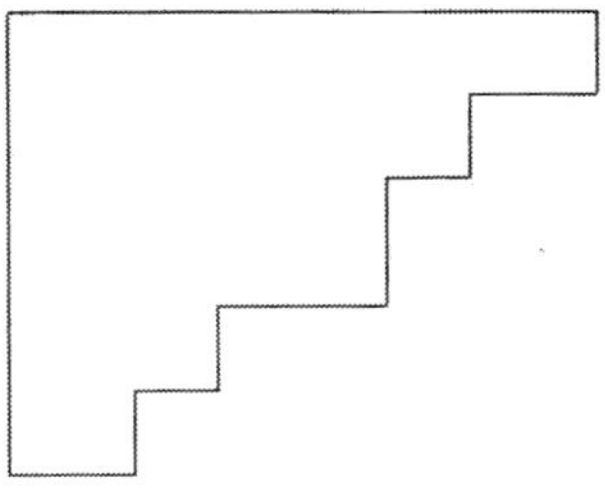

, e.g. $\lambda = (3, 3, 2) =$

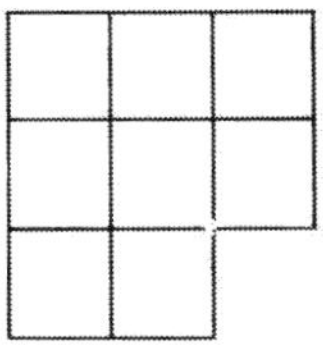

Such a diagram is called a *Young diagram*.

Given a Young diagram $\lambda = (\lambda_1, \dots, \lambda_r)$ ($\lambda_1 \geqslant \lambda_2 \geqslant \dots \geqslant \lambda_r$), the numbers $\lambda_1, \dots, \lambda_r$ are called the *parts* of λ. The *length* of λ, denoted $l(\lambda)$, is the number of parts of λ. The *weight* of λ, denoted $|\lambda|$, is defined as $\sum \lambda_i$. If $|\lambda| = n$, then we say that λ is a *partition of n*, and this is denoted by $\lambda \vdash n$. The *conjugate of a partition* λ is the partition λ' whose diagram is the transpose of λ.

EXAMPLE 3.0.1. If $\lambda = (3, 2, 2)$, then $\lambda' = (3, 3, 1)$.

$\lambda = (3, 2, 2) =$ 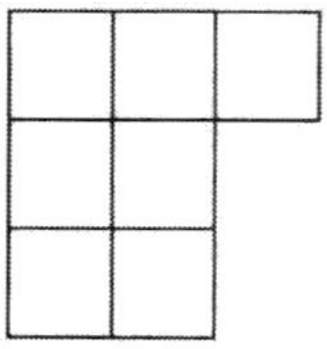 , $\lambda' = (3, 3, 1) =$ 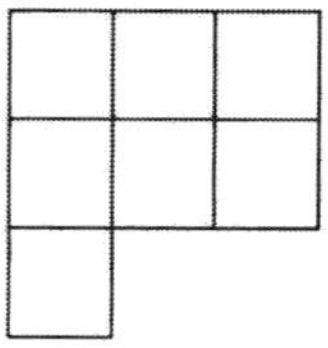 .

Given $\lambda \vdash d$, let us label the boxes consecutively, for example,

1	2	3	4
5	6	7	8
9	10	11	
12			

Let

$$P_\lambda = \{\sigma \in \mathbb{S}_d \mid \sigma \text{ preserves the rows of } \lambda\},$$

and

$$Q_\lambda = \{\sigma \in \mathbb{S}_d \mid \sigma \text{ preserves the columns of } \lambda\}.$$

Let us define $a_\lambda, b_\lambda \in k\mathbb{S}_d$ as
$a_\lambda = \sum_{\tau \in P_\lambda} \tau$, $b_\lambda = \sum_{\tau \in Q_\lambda} (\operatorname{sign} \tau)\tau$. Let $c_\lambda = a_\lambda b_\lambda$ (called a *Young symmetrizer*).

Recall the following well-known result

PROPOSITION 3.0.2. *Let $\sigma, \tau \in \mathbb{S}_d$. Then σ and τ are conjugates if and only if they determine the same Young diagram.*

Let Λ_d be the set of partitions $\lambda \vdash d$ (so that the number of conjugacy classes in $\mathbb{S}_d$ is the cardinality of Λ_d). Let us denote $\mathbb{C}\mathbb{S}_d$ by R.

THEOREM 3.0.3. (1) *$\{Rc_\lambda, \lambda \in \Lambda_d\}$ gives all the irreducible representations of $\mathbb{S}_d$.*

(2) *For $\lambda = (\lambda_1, \ldots, \lambda_r)$, let V_λ denote the associated simple $\mathbb{S}_d$-module. Then*

$$\dim V_\lambda = \frac{d!}{l_1! \ldots l_r!} \prod_{i<j} (l_i - l_j),$$

where $l = (l_1, \ldots, l_r) = \lambda + \rho$, ρ being the r-tuple $(r-1, r-2, \ldots, 0)$.

(3) *Consider the variables $x_1, \ldots, x_r$. Let $P_j(x) = x_1^j + \cdots + x_r^j$, and $\Delta(x) = \prod_{i<j}(x_i - x_j)$ the Vandermonde determinant. Given $\mu = (\mu_1, \ldots, \mu_s)$, $\mu \vdash d$, let $\mu(j) = \#\{\mu_t \mid \mu_t = j\}$. Let C_μ be the conjugacy class of μ. Let $\sigma \in C_\mu$. Let $\psi_\lambda = \operatorname{char} V_\lambda$. Then we have the Frobenius formula*

$$\psi_\lambda(\sigma) = \left[\Delta(x) \prod_j P_j(x)^{\mu(j)}\right]_{(l_1, \ldots, l_r)},$$

where $(l_1, \ldots, l_r)$ *is as above, and for a polynomial* $f(x)$ *in* $x_1, \ldots, x_r$, *and an* r*-tuple* $(l_1, \ldots, l_r)$, $l_i \in \mathbb{Z}^+$, $[f(x)]_{(l_1,\ldots,l_r)}$ *denotes the coefficient of* $x_1^{l_1} \ldots x_r^{l_r}$ *in* $f(x)$.

See [**55**] for a proof.

REMARK 3.0.4. Given a Young diagram λ, let $|\lambda| = d$. A *Young tableau of shape* λ is a filling of λ with the numbers $1, \ldots, d$. A Young tableau is *standard* (resp. *semistandard*) if the rows and columns are strictly increasing (resp. the columns are strictly increasing and the rows are non-decreasing). We have $\dim V_\lambda = \#\{\text{standard Young tableaux of shape } \lambda\}$.

See [**55**] for a proof.

EXAMPLE 3.0.5. $\lambda = (d)$; $V_\lambda = \mathbb{C}c_\lambda$ is 1-dimensional.

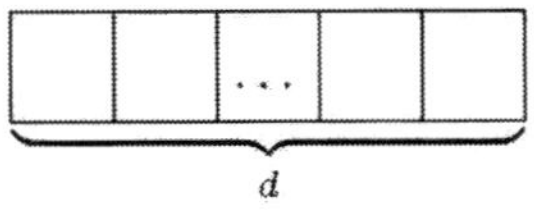

EXAMPLE 3.0.6. $\lambda = (d-1, 1)$; $V_\lambda = \{(x_1, \ldots, x_d) \in \mathbb{C}^d \mid x_1 + \ldots x_d = 0\}$, the action of $\mathbb{S}_d$ being the natural action.

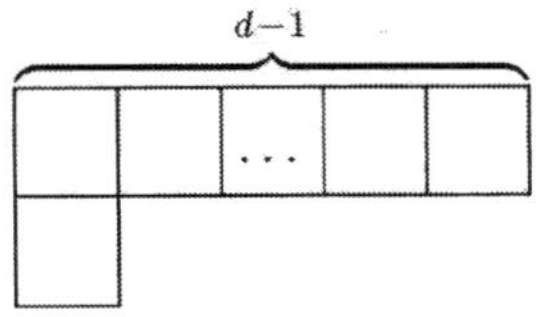

4. Representation theory of $GL_n(\mathbb{C})$

The representation theory of $GL_n(\mathbb{C})$ is closely linked to that of $\mathbb{C}\mathbb{S}_d$. For example, let $V = \mathbb{C}^n$ and $G = GL(V)$. Then for the canonical actions of G and $\mathbb{S}_d$ on $V^{\otimes d}$, we have

$$\mathrm{End}_G(V^{\otimes d}) = \mathbb{C}\mathbb{S}_d, \quad \mathrm{End}_{\mathbb{S}_d}(V^{\otimes d}) = \mathbb{C}G$$

(i.e. the algebras of operators spanned by $\mathbb{S}_d$ and $GL_n(\mathbb{C})$ on $V^{\otimes d}$ are each the full centralizer of the other). In particular, for $\lambda \vdash d$, setting $S_\lambda(V) = c_\lambda V^{\otimes d}$, we have, $S_\lambda(V)$ is a $GL_n(\mathbb{C})$-module.

THEOREM 4.0.1. *Let* $\lambda = (\lambda_1, \ldots, \lambda_r)$, *with* $\lambda_i > 0$ *for all* i. *Then*

(1) $S_\lambda(V)$ *is an irreducible* $GL_n(\mathbb{C})$*-module for* $r \leqslant n$, *and* $S_\lambda(V) = 0$ *for* $r > n$.

(2) $\dim S_\lambda V = \displaystyle\prod_{1 \leqslant i < j \leqslant r} \frac{\lambda_i - \lambda_j + j - i}{j - i}$.

(3) $V^{\otimes d} = \sum_\lambda (S_\lambda V)^{\oplus m_\lambda}$, m_λ *being the dimension of* V_λ, *the irreducible* $\mathbb{S}_d$*-module corresponding to* λ.

(4) *More generally, for any* $c \in R(= \mathbb{C}\mathfrak{S}_d)$, *let* $cR = \sum_\lambda V_\lambda^{\oplus r_\lambda}$ *(as* $\mathfrak{S}_d$*-modules). Then*

$$cV^{\otimes d} = \sum_\lambda (S_\lambda V)^{\oplus r_\lambda}.$$

(5) $\psi_{S_\lambda V}$ *(the character of* $S_\lambda V$*) is given by*

$$\psi_{S_\lambda V}(g) = S_\lambda(x_1, \dots, x_n),$$

where $x_1, \dots, x_n$ *are the eigenvalues of* g *on* V, *and* S_λ *is the Schur polynomial, defined as follows:*

Given $\alpha = (\alpha_1, \dots, \alpha_n)$, *let* $x^\alpha = x_1^{\alpha_1} \dots x_n^{\alpha_n}$. *Let* $a_\alpha = \sum_{\sigma \in \mathfrak{S}_n} (\operatorname{sign} \sigma)\sigma(x^\alpha)$. *Then* a_α *is skew symmetric, i.e.* $\sigma(a_\alpha) = (\operatorname{sign} \sigma) a_{\sigma(\alpha)}$. *Hence* $a_\alpha \neq 0$ *if and only if* $\alpha_1, \dots, \alpha_n$ *are all distinct. Now given* $\lambda \vdash n$, *let* $\alpha = \lambda + \delta$, *where* $\delta = (n-1, \dots, 1, 0)$. *Then*

$$a_\alpha = \det(x_i^{\lambda_j + n - j}).$$

This is divisible by $x_i - x_j$, $i < j$, *and hence by the Vandermonde determinant* $a_\delta = \det(x_i^{n-j})$. *The* Schur polynomial S_λ *is defined as*

$$S_\lambda = \frac{a_{\lambda+\delta}}{a_\delta}.$$

Note that S_λ is symmetric. See [**55**] for a proof of Theorem 4.0.1.

EXAMPLE 4.0.2. $\lambda = (d)$.

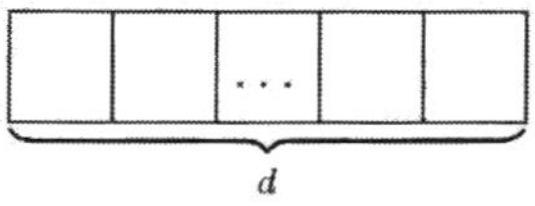

$S_\lambda V = \operatorname{Sym}^d V$, S_λ is the complete symmetric polynomial of degree d in $x_1, \dots, x_n$, i.e. the sum of all distinct monomials of degree d (also denoted by $H_d(x_1, \dots, x_n)$).

EXAMPLE 4.0.3. $\lambda = (1, 1, \cdots, 1)$ (d factors).

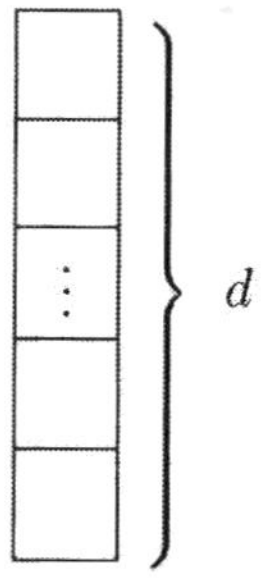

$S_\lambda V = \wedge^d V$, S_λ is the elementary polynomial of degree d in $x_1, \dots, x_n$ (also denoted by $E_d(x_1, \dots, x_n)$).

REMARK 4.0.4. $\dim S_\lambda V = \#\{$ semistandard Young tableaux of shape $\lambda\}$. See [**55**] for a proof.

4.1. The irreducible representations of $GL_n(\mathbb{C})$

Let D_r denote the 1-dimensional representation of $GL_n(\mathbb{C})$ given by the r-th power of the determinant, $r \in \mathbb{Z}$. We have the following

PROPOSITION 4.1.1. (1) *If* $r \geqslant 0$, *then* $D_r = (\wedge^n V)^{\otimes r}$, *where* $V = \mathbb{C}^n$.
(2) D_{-r} *is the dual of* D_r.

Given $\lambda = (\lambda_1, \ldots, \lambda_n)$, $\lambda_i \in \mathbb{Z}$, $\lambda_1 \geqslant \ldots \geqslant \lambda_n$, where we allow λ_i to be negative also, choose r so that $\lambda_n + r > 0$. Set

$$\chi_\lambda = S_{\lambda+r} V \otimes D_{-r}$$

(here $\lambda + r = (\lambda_1 + r, \ldots, \lambda_n + r)$).

THEOREM 4.1.2. $\{\chi_\lambda \mid \lambda = (\lambda_1, \ldots, \lambda_n), \lambda_i \in \mathbb{Z}, \lambda_1 \geqslant \ldots \geqslant \lambda_n\}$ *give all the finite dimensional irreducible representations of* $GL_n(\mathbb{C})$.

5. Representation Theory of $SL_n(\mathbb{C})$

Let k be any base field. Let $GL(n)$ be the group of invertible $n \times n$ matrices with entries in k, and $SL(n)$ the subgroup of $GL(n)$ consisting of $n \times n$ matrices of determinant 1.

5.1. The Lie algebras $\mathfrak{gl}(n)$ and $\mathfrak{sl}(n)$

We have Lie$GL(n) = \mathfrak{gl}(n)$, Lie$SL(n) = \mathfrak{sl}(n)$, where $\mathfrak{gl}(n)$, $\mathfrak{sl}(n)$ are given by

$$\mathfrak{gl}(n) = \{n \times n \text{ matrices with entries in } k\},$$
$$\mathfrak{sl}(n) = \{n \times n \text{ matrices with trace } 0\}.$$

Let $\mathfrak{g}$ be $\mathfrak{gl}(n)$ or $\mathfrak{sl}(n)$, and let

$$\mathfrak{h} = \{\text{diagonal matrices in } \mathfrak{g}\},$$

$$\mathfrak{n}_+ = \bigoplus_{1 \leqslant i < j \leqslant n} (E_{ij}), \qquad \mathfrak{n}_- = \bigoplus_{1 \leqslant j < i \leqslant n} (E_{ij}),$$

where E_{ij} is the elementary $n \times n$ matrix with 1 at the (i, j)-th place and zeros elsewhere. We have

$$\mathfrak{g} = \mathfrak{h} \oplus \mathfrak{n}_+ \oplus \mathfrak{n}_-.$$

5.2. The character group $X(T)$

Let D_n (resp. T) be the maximal torus in $GL(n)$ (resp. $SL(n)$) consisting of diagonal matrices. Let $t \in D_n$, say $t = \operatorname{diag}(t_1, \ldots, t_n)$, and let $\epsilon_i : D_n \to \mathbb{G}_m$ be the character defined by $\epsilon_i(t) = t_i$, $1 \leqslant i \leqslant n$. We have, the character group $X(D_n) = \operatorname{Hom}(D_n, \mathbb{G}_m)$ (written additively) can be identified with the free abelian group generated by ϵ_i. Let $X(T)$ be the character group $\operatorname{Hom}(T, \mathbb{G}_m)$ of T (written additively). For ϵ_i as above, we shall denote $\epsilon_i|_T$ also by just ϵ_i. We have, on T, $\sum \epsilon_i = 0$ (since $\prod t_i = \det t = 1$, for $t \in T$). Further $\{\epsilon_i - \epsilon_j, 1 \leqslant i, j \leqslant n, i \neq j\}$ is a basis for the $\mathbb{R}$-vector space $X(T) \otimes \mathbb{R}$.

5.3. Root system of $SL(n)$

The action of T on the 1-dimensional root space kE_{ij} is given by $t \cdot E_{ij} = t_i t_j^{-1} E_{ij}$, $t \in T$ (here t is a diagonal matrix with diagonal entries $t_1, \ldots, t_n$). Thus the root system R of $SL(n)$ relative to T may be identified with $\{\epsilon_i - \epsilon_j, 1 \leqslant i, j \leqslant n, i \neq j\}$.

Denoting the standard basis for $\mathbb{R}^n$ by $\{e_i,\ 1 \leqslant i \leqslant n\}$, let H be the hyperplane in $\mathbb{R}^n$ consisting of $\{\sum a_i e_i \mid \sum a_i = 0\}$. Then the map $\epsilon_i - \epsilon_j \mapsto e_i - e_j$ defines an isomorphism $X(T) \otimes \mathbb{R} \cong H$. We have, R is a root system for H, and is of type A_l, where $l = n - 1$.

The set of positive roots of R relative to B (the Borel subgroup of $SL(n)$ consisting of upper triangular matrices) consists of $\{\epsilon_i - \epsilon_j, 1 \leqslant i < j \leqslant n\}$, and the corresponding system of simple roots consists of $\{\epsilon_i - \epsilon_{i+1}, 1 \leqslant i \leqslant n-1\}$. The fundamental weights are given by $\omega_i = \epsilon_1 + \cdots + \epsilon_i, 1 \leqslant i \leqslant n-1$.

The Weyl group may be identified with the symmetric group $\mathbb{S}_n$. The reflection s_i (corresponding to the simple root $\alpha_i = \epsilon_i - \epsilon_{i+1}$) as a permutation is simply the transposition $(i, i+1)$. For $w \in \mathbb{S}_n$, say $w = (a_1, \ldots, a_n)$, we have $w(\epsilon_i - \epsilon_j) = \epsilon_{a_i} - \epsilon_{a_j}$. Let $(\,,\,)$ denote the inner product on H induced by the standard inner product on $\mathbb{R}^n$. Then $(\,,\,)$ is clearly W-invariant, and for $\lambda \in \Lambda_{\mathrm{wt}}$, we have $(\lambda, \alpha^*) = 2\dfrac{(\lambda, \alpha)}{(\alpha, \alpha)}$.

5.4. Indexing of irreducible representations of $SL(n)$ by Young diagrams

Let $G = SL(n)$. Given a dominant weight $\lambda \in \Lambda_{\mathrm{wt}}$, say $\lambda = \sum_{i=1}^{n-1} a_i \omega_i$, $a_i \in \mathbb{Z}_+$, we write $\lambda = \sum_{i=1}^{n-1} \lambda_i \epsilon_i$, where $\lambda_i = a_i + \cdots + a_{n-1}$, $1 \leqslant i \leqslant n-1$. Then $\lambda = (\lambda_1, \ldots, \lambda_{n-1})$ determines a Young diagram (note that $\lambda_1 \geqslant \lambda_2 \geqslant \ldots \geqslant \lambda_{n-1} \geqslant 0$).

Conversely, a Young diagram $\lambda = (\lambda_1 \geqslant \lambda_2 \geqslant \ldots \geqslant \lambda_{n-1} \geqslant 0)$, determines a dominant weight, namely $\sum_{i=1}^{n-1} a_i \omega_i$, where $a_i = \lambda_i - \lambda_{i+1}$, $1 \leqslant i \leqslant n-1$ (here, $\lambda_n = 0$). (Note that there is no restriction on the length of a row in λ). Further, two Young diagrams $\lambda = \lambda_1 \geqslant \lambda_2 \geqslant \ldots \geqslant \lambda_{n-1} \geqslant 0$, $\mu = \mu_1 \geqslant \mu_2 \geqslant \ldots \geqslant \mu_{n-1} \geqslant 0$ determine the same dominant weight if and only if $\lambda_i - \mu_i$ is the same for all i. Thus we obtain a bijection between Λ_{wt}^+(= the set of dominant weights) and the set of Young diagrams with length of any column $\leqslant n-1$, where two Young diagrams $\lambda = \lambda_1 \geqslant \lambda_2 \geqslant \ldots \geqslant \lambda_{n-1} \geqslant 0$, $\mu = \mu_1 \geqslant \mu_2 \geqslant \ldots \geqslant \mu_{n-1} \geqslant 0$ are identified if $\lambda_i - \mu_i$ is the same for all i.

Thus we obtain (in view of Chapter 4, Theorem 2.0.1) an indexing of the set of finite dimensional irreducible $SL(n)$-modules by the set of Young diagrams with length of any column $\leqslant n-1$, two Young diagrams $\lambda = \lambda_1 \geqslant \lambda_2 \geqslant \ldots \geqslant \lambda_{n-1} \geqslant 0$, $\mu = \mu_1 \geqslant \mu_2 \geqslant \ldots \geqslant \mu_{n-1} \geqslant 0$ being identified if $\lambda_i - \mu_i$ is the same for all i.

5.5. The irreducible representations of $SL_n(\mathbb{C})$

THEOREM 5.5.1. (1) *$S_\lambda V$, where $\lambda = (\lambda_1, \ldots, \lambda_n)$, $\lambda_i \in \mathbb{Z}, \lambda_1 \geqslant \ldots \geqslant \lambda_n \geqslant 0$ is an irreducible $SL_n(\mathbb{C})$-module.*

(2) *$S_\lambda V \cong S_\mu V$ (as $SL_n(\mathbb{C})$-modules) if and only if $\lambda_i - \mu_i$ is the same for all i.*

(3) *Any finite dimensional irreducible* $SL_n(\mathbb{C})$*-module is of the form* $S_\lambda V$, λ *being as in* (1).

See [**55**] for a proof.

CHAPTER 6

The Grassmannian variety $G_{d,n}$

In this chapter, we study the Schubert varieties in the Grassmannian. After introducing the Grassmannian and its Schubert varieties in §1 and §2, we present the Standard monomial theory for Schubert varieties in the Grassmannian in §3 and §4. We then present a proof of the "vanishing theorems" for Schubert varieties in §5, and deduce normality and Cohen-Macaulayness for Schubert varieties in §6.

1. The Plücker embedding

Let us fix the integers $1 \leqslant d < n$ and let $V = k^n$. The *Grassmannian* $G_{d,n}$ is the set of d-dimensional subspaces $U \subset V$. Let U be an element of $G_{d,n}$, and $a_1, \ldots, a_d$ a basis of U, where

$$a_j = \begin{pmatrix} a_{1j} \\ a_{2j} \\ \cdots \\ a_{nj} \end{pmatrix}, \text{ with } a_{ij} \in k, \text{for } 1 \leqslant i \leqslant n,\ 1 \leqslant j \leqslant d.$$

(here, each vector a_j is written as a column vector with respect to the standard basis of k^n). Thus the basis $\{a_1, \ldots, a_d\}$ gives rise to an $n \times d$ matrix $A = (a_{ij})$ of rank d, whose columns are the vectors $a_1, \ldots, a_d$. Similarly, another basis $a'_1, \ldots, a'_d$ gives rise to an $n \times d$ matrix A'. Then there exists a change of basis matrix $C \in GL_d(k)$, such that $A' = AC$. Conversely, given two $n \times d$ matrices A and A' of rank d, such that $A' = AC$, for some matrix $C \in GL_d(k)$, the columns of both A and A' are d linearly independent vectors in V, generating one and the same d-dimensional subspace of V. Identifying an $n \times d$ matrix with a point in the affine space $\mathbb{A}^{nd}$, we see that $G_{d,n}$ can be viewed as $(\mathbb{A}^{nd} \setminus Z)/\sim$, where Z is the set of $n \times d$ matrices of rank less than d, and the equivalence relation $\sim$ is defined by

$$A \sim A' \quad \text{if there exists a } C \in GL_d(k) \text{ such that } A' = AC.$$

(Note the similarity with the definition of the projective space $\mathbb{P}^{n-1}$; in fact it is easily seen that $\mathbb{P}^{n-1} = G_{1,n}$.)

Thus denoting $M^0_{n\times d} = \mathbb{A}^{nd} \setminus Z$, $G_{d,n}$ is simply the orbit space $M^0_{n\times d}/GL_d(k)$, for the action of $GL_d(k)$ on $M^0_{n\times d}$ by multiplication on the right.

1.1. The partially ordered set $I_{d,n}$

Define the set

$$I_{d,n} := \{\underline{i} = (i_1, \ldots, i_d) | 1 \leqslant i_1 < \cdots < i_d \leqslant n\},$$

Define a partial order $\geqslant$ on $I_{d,n}$ by defining $\underline{i} \geqslant \underline{j} \Leftrightarrow i_t \geqslant j_t, \forall t$.

1.2. Plücker embedding and Plücker coordinates

Let $N = \# I_{d,n}$; then $N = \binom{n}{d}$, and the coordinates of the affine space $\wedge^d V = k^N$ will be indexed by the set $I_{d,n}$. Let

$$X = \underbrace{V \oplus \cdots \oplus V}_{d \text{ times}} = k^{nd}.$$

The exterior product map

$$\wedge^d : X \longrightarrow \wedge^d V$$

sends the element $(a_1, \ldots, a_d)$ to the point $a_1 \wedge \cdots \wedge a_d$ in $\wedge^d V$. Identifying $\wedge^d V$ with k^N, the $\underline{i}$-th coordinate of $a_1 \wedge \cdots \wedge a_d$ is given by the determinant of the d-minor of the $n \times d$ matrix $A = (a_{ij})$ (where the columns of the matrix $A = (a_{ij})$ are the vectors $a_1, \ldots, a_d$), with row indices $i_1, \ldots, i_d$. Note that $a_1 \wedge \cdots \wedge a_d = 0$ if and only if the matrix A belongs to the set Z of matrices of rank less than d. Moreover, it is easily seen that given $(a_1, \ldots, a_d)$, $(a'_1, \ldots, a'_d)$ in X, such that the associated $n \times d$ matrices A, A', are equivalent (i.e., there exists a matrix $C \in GL_d(k)$ such that $A' = AC$), then

$$a'_1 \wedge \cdots \wedge a'_d = \det(C)\, a_1 \wedge \cdots \wedge a_d.$$

This shows that the map $\wedge^d$ induces a well defined map

$$p : G_{d,n} \longrightarrow \mathbb{P}(\wedge^d V) = \mathbb{P}^{N-1}$$

called the *Plücker map*. For $\underline{i} \in I_{d,n}$, the $\underline{i}$-th component of p is denoted by $p_{\underline{i}}$, or by $p_{i_1,\ldots,i_d}$, where $\underline{i} = (i_1, \ldots, i_d)$; the $p_{\underline{i}}$'s, with $\underline{i} \in I_{d,n}$, are called the *Plücker coordinates*. If a point U in $G_{d,n}$ is represented by the $n \times d$ matrix A, then $p_{i_1,\ldots,i_d}(U) = \det(A_{i_1,\ldots,i_d})$, where $A_{i_1,\ldots,i_d}$ denotes the $d \times d$ sub matrix of A consisting of the rows with indices $i_1, \ldots, i_d$. The projective coordinates of points in $\mathbb{P}^{N-1}$ will also be denoted by $p_{\underline{i}}$, with $\underline{i} \in I_{d,n}$.

Given a point U in $G_{d,n}$ represented by the $n \times d$ matrix A, and numbers $1 \leqslant i_1, \ldots, i_d \leqslant n$ (not necessarily distinct, nor in increasing order), $A_{i_1,\ldots,i_d}$ will denote the matrix whose rows are the rows of A with indices $i_1, \ldots, i_d$, taken in the order in which they appear in A. Also, we define $p_{i_1,\ldots,i_d}(U) = \det(A_{i_1,\ldots,i_d})$. It is easy to see that $p_{i_1,\ldots,i_d} = 0$ if $i_1, \ldots, i_d$ are not distinct, and $p_{i_1,\ldots,i_d} = (-1)^{l(\sigma)} p_{i_{\sigma(1)},\ldots,i_{\sigma(d)}}$ if $i_1, \ldots, i_d$ are distinct and σ is the permutation of the set $\{1, \ldots, d\}$, such that $\sigma(1) < \cdots < \sigma(d)$.

For each $\underline{i} \in I_{d,n}$ consider the point $e_{\underline{i}}$ of $G_{d,n}$ represented by the $n \times d$ matrix whose entries are all 0, except the ones in the i_j-th row and j-th column, for each

$1 \leqslant j \leqslant d$, which are equal to 1. Clearly, for $\underline{i}, \underline{j} \in I_{d,n}$,

$$p_{\underline{i}}(e_{\underline{j}}) = \begin{cases} 1, & \text{if } \underline{i} = \underline{j}; \\ 0, & \text{otherwise.} \end{cases}$$

THEOREM 1.2.1. *The Plücker map is injective.*

PROOF. Let U, respectively U' be two points in $G_{d,n}$ such that $p(U) = p(U')$. Then there exists $\underline{l} = (l_1, \ldots, l_d) \in I_{d,n}$ such that $p_{\underline{l}}(U) = p_{\underline{l}}(U') \neq 0$, and we may assume that $p_{\underline{l}}(U) = p_{\underline{l}}(U') = 1$. If U and U' are represented by the $n \times d$ matrices A, respectively A', we have $\det(A_{l_1,\ldots,l_d}) = \det(A'_{l_1,\ldots,l_d}) = 1$. Replacing A and A' with AC, respectively $A'C'$, where $C = A^{-1}_{l_1,\ldots,l_d}$, $C' = A'^{-1}_{l_1,\ldots,l_d}$, we can assume that $A_{l_1,\ldots,l_d} = A'_{l_1,\ldots,l_d} = \mathrm{Id}_d$. For $1 \leqslant i \leqslant n$ and $1 \leqslant j \leqslant d$, we have

$$\begin{aligned} a_{ij} &= \det A_{l_1,\ldots,l_{j-1},i,l_{j+1},\ldots,l_d} = p_{l_1,\ldots,l_{j-1},i,l_{j+1},\ldots,l_d}(U) \\ &= p_{l_1,\ldots,l_{j-1},i,l_{j+1},\ldots,l_d}(U') = \det A'_{l_1,\ldots,l_{j-1},i,l_{j+1},\ldots,l_d} \\ &= a'_{ij}. \end{aligned}$$

Thus $A = A'$, hence $U = U'$, which shows that p is injective. □

In view of Theorem 1.2.1, we identify $G_{d,n}$ with $\operatorname{Im} p$. Thus, we consider $G_{d,n} \subset \mathbb{P}^{N-1}$. Next we will show that $G_{d,n}$ is closed in $\mathbb{P}^{N-1}$.

1.3. Plücker quadratic relations

THEOREM 1.3.1. *The Grassmannian $G_{d,n} \subset \mathbb{P}^{N-1}$ consists of the zeroes in $\mathbb{P}^{N-1}$ of the following quadratic polynomials:*

$$(*) \qquad \sum_{\lambda=1}^{d+1} (-1)^{\lambda} p_{i_1,\ldots,i_{d-1},j_\lambda} p_{j_1,\ldots,\widehat{j_\lambda},\ldots,j_{d+1}},$$

where $i_1, \ldots, i_{d-1}$ and $j_1, \ldots, j_{d+1}$ are any numbers between 1 and n.
(A relation obtained by equating an expression given by (*) to 0 is usually referred to as a Plücker relation*).*

PROOF. Fix $1 \leqslant i_1, \ldots, i_{d-1}, j_1, \ldots, j_{d+1} \leqslant n$. Let $U \in G_{d,n}$ be represented by the $n \times d$ matrix A. Then (*) evaluated at U equals

$$\sum_{\lambda=1}^{d+1} (-1)^{\lambda} \det(A_{i_1,\ldots,i_{d-1},j_\lambda}) \det(A_{j_1,\ldots,\widehat{j_\lambda},\ldots,j_{d+1}}).$$

Expanding the first determinants along their last rows, the above expression equals

$$\sum_{\lambda=1}^{d+1} (-1)^{\lambda} \sum_{\mu=1}^{d} (-1)^{d+\mu} a_{j_\lambda,\mu} \det(A^{\widehat{\mu}}_{i_1,\ldots,i_{d-1}}) \det(A_{j_1,\ldots,\widehat{j_\lambda},\ldots,j_{d+1}}),$$

where $A^{\widehat{\mu}}_{i_1,\ldots,i_{d-1}}$ is obtained from $A_{i_1,\ldots,i_{d-1}}$ by deleting the μ-th column. The last expression can be written in the form

$$\sum_{\mu=1}^{d}(-1)^{d+\mu}\det(A^{\widehat{\mu}}_{i_1,\ldots,i_{d-1}})\sum_{\lambda=1}^{d+1}(-1)^{\lambda}a_{j_\lambda,\mu}\det(A_{j_1,\ldots,\widehat{j_\lambda},\ldots,j_{d+1}}),$$

or

$$\sum_{\mu=1}^{d}(-1)^{d+\mu}\det(A^{\widehat{\mu}}_{i_1,\ldots,i_{d-1}})(-1)\det({}^{\mu}A_{j_1,\ldots,\widehat{j_\lambda},\ldots,j_{d+1}})$$

where ${}^{\mu}A_{j_1,\ldots,\widehat{j_\lambda},\ldots,j_{d+1}}$ is obtained from the matrix $A_{j_1,\ldots,\widehat{j_\lambda},\ldots,j_{d+1}}$ by adding the μ-th column of A as its first column. Note that this matrix has two identical columns, and therefore its determinant is 0. This implies that the last expression is 0 and hence we obtain that the point U in $G_{d,n}$ is a zero of the polynomials given by (*).

Conversely, let $q = (q_{\underline{i}})_{\underline{i}\in I_{d,n}}$ be a zero of the polynomials given by (*). Let $q_{l_1,\ldots,l_d} = 1$, for some $(l_1,\ldots,l_d) \in I_{d,n}$. For $1 \leqslant i \leqslant n$, $1 \leqslant j \leqslant d$, $A = (a_{ij})$ where

$$a_{ij} = q_{l_1,\ldots,l_{j-1},i,l_{j+1},\ldots,l_d}.$$

Note that $A_{l_1,\ldots,l_d} = \mathrm{Id}_d$. Clearly, rank $A = d$ (since $\det A_{l_1,\ldots,l_d} = 1$). For $\underline{j} \in I_{d,n}$, let us denote $p_{\underline{j}}(A) = \det A_{j_1,\ldots,j_d}$. We shall now show that for $\underline{j} \in I_{d,n}$, we have $p_{\underline{j}}(A) = q_{\underline{j}}$, from which it will follow that the point U in $G_{d,n}$ represented by A satisfies $p_{\underline{j}}(U) = q_{\underline{j}}$, $\underline{j} \in I_{d,n}$, i.e., $q \in G_{d,n}$.

We prove this by decreasing induction on $\#\{\underline{l}\cap\underline{j}\}$, the starting point of induction being $\#\{\underline{l}\cap\underline{j}\} = d-1$ (i.e. $\underline{l}$ and $\underline{j}$ differ in just one place), and the result in this case follows from our definition of A. In the Plücker relation $q_{\underline{l}}q_{\underline{j}} + \sum \pm q_{\underline{l}'}q_{\underline{j}'} = 0$, $\underline{l}'$ differs from $\underline{l}$ in just one place (namely the d-th place), while $\#\{\underline{j}'\cap\underline{l}\} > \#\{\underline{j}\cap\underline{l}\}$ (since some entry in $\{\underline{j}\setminus\underline{l}\}$ has been replaced by l_d). Thus we obtain $q_{\underline{l}'} = p_{\underline{l}'}(A)$, $q_{\underline{j}'} = p_{\underline{j}'}(A)$. Also, from the first part, the d-minors of A satisfy the Plücker relations. Hence we obtain $p_{\underline{l}}(A)q_{\underline{j}}(A) = q_{\underline{l}}q_{\underline{j}}$, which implies $p_{\underline{j}}(A) = q_{\underline{j}}$ (since $p_{\underline{l}}(A) = q_{\underline{l}} = 1$). □

1.4. More general quadratic relations

The Plücker coordinates satisfy a more general type of quadratic relations described as follows: Let $\underline{i},\underline{j} \in I_{d,n}$. Suppose that $\underline{i},\underline{j}$ are not comparable, say, $\underline{i} \not\geqslant \underline{j}$. This implies that there exists a t, $t \leqslant d$ such that $i_r \geqslant j_r$, $1 \leqslant r \leqslant t-1$, and $i_t < j_t$. Let $[\underline{i},\underline{j}]$ denote the set of permutations σ of the set $\{i_1,\cdots,i_t,j_t,\cdots,j_d\}$ such that $i^\sigma_1 < \cdots < i^\sigma_t$; $j^\sigma_t < \cdots < j^\sigma_d$. Then the Plücker coordinates $p_{\underline{i}}$, $p_{\underline{j}}$ satisfy the following relation:

$$(**)\qquad Q_{\underline{i},\underline{j}} : \sum_{\sigma}\mathrm{sign}(\sigma)\,p_{\underline{i}^\sigma}p_{\underline{j}^\sigma} = 0,$$

where $\underline{i}^\sigma = (i^\sigma_1,\cdots,i^\sigma_t,i_{t+1},\cdots,i_d)\uparrow$, $\underline{j}^\sigma = (j_1,\cdots,j_{t-1},j^\sigma_t,\cdots,j^\sigma_d)\uparrow$, and σ runs over $[\underline{i},\underline{j}]$ (here, for a d-tuple $(l_1,\cdots,l_d)$, $l_i \in \mathbb{Z}$, $(l_1,\cdots,l_d)\uparrow$ denotes the

d-tuple obtained from $(l_1, \cdots, l_d)$ by arranging the entries in the ascending order). The proof of this is very similar to the above proof.

For $\sigma \neq id$, $\underline{i}^\sigma$ is either $= 0$ (namely, if there is a repetition in $\{i_1^\sigma, \cdots, i_t^\sigma, i_{t+1}, \cdots, i_d\}$) or $\underline{i}^\sigma >$ both $\underline{i}$ and $\underline{j}$ (since in $\{i_1, \cdots, i_t, j_t, \cdots, j_d\}$ any $j >$ any i, and $i_r \geqslant j_r$, $1 \leqslant r \leqslant t-1$); similarly, for $\sigma \neq id$, $\underline{j}^\sigma$ is either $= 0$ (namely, if there is a repetition in $\{j_1, \cdots, j_{t-1}, j_t^\sigma, \cdots, j_d^\sigma\}$) or $\underline{j}^\sigma <$ both $\underline{i}$ and $\underline{j}$. Hence (**) may be reformulated as

$$(***) \qquad p_{\underline{i}} p_{\underline{j}} = \sum \pm p_\alpha p_\beta,$$

where α, β run over a certain subset of $I_{d,n}$ such that $\alpha >$ both $\underline{i}$ and $\underline{j}$, and $\beta <$ both $\underline{i}$ and $\underline{j}$.

REMARK 1.4.1. The image of $G_{d,n}$ under the Plücker embedding may be identified with

$$\{[v] \in \mathbb{P}(\wedge^d V) \mid v \text{ is a decomposable vector}\} = \{[v] \in \mathbb{P}(\wedge^d V) \mid v = v_1 \wedge \ldots \wedge v_d,\ v_i \in V\}$$

(note that $v = \sum a_{\underline{i}} e_{\underline{i}}$ in $\wedge^d V$ is decomposable if and only if $a_{\underline{i}}$, $\underline{i} \in I_{d,n}$, are all the d-minors of a matrix A).

1.5. The cone $\widehat{G_{d,n}}$ over $G_{d,n}$

Let $A = k[G_{d,n}]$, the homogeneous coordinate ring of $G_{d,n}$ for the Plücker embedding. Let $\widehat{G_{d,n}} = \operatorname{Spec} A$, the cone over $G_{d,n}$. Let $X = V \oplus \cdots \oplus V$ (d copies). With respect to the standard basis $\{e_1, \ldots, e_n\}$ of $V(= k^n)$, X may be identified with $M_{n\times d}$, the space of $n \times d$ matrices with entries in k. With this identification we have $k[X] = k[x_{ij},\ 1 \leqslant i \leqslant n, 1 \leqslant j \leqslant d]$. Consider the canonical morphism $\pi : X \to \wedge^d V$, $\pi(u_1, \ldots, u_d) = u_1 \wedge \ldots \wedge u_d$. The morphism π corresponds to the algebra homomorphism $\pi^* : k[x_{\underline{i}},\ \underline{i} \in I_{d,n}] \to k[x_{ij}]$, $\pi^*(x_{\underline{i}}) = p_{\underline{i}}$. The image of π is precisely $\widehat{G_{d,n}}$, the cone over $G_{d,n}$ (note that $\pi(X)$ is precisely the set of decomposable vectors in $\wedge^d V$). Thus $k[G_{d,n}]$ gets identified with the subring of $k[x_{ij}]$ generated by $\{p_{\underline{i}} \mid \underline{i} \in I_{d,n}\}$.

1.6. Identification of G/P_d with $G_{d,n}$

Let $G = SL(n)$. Let P_d be the maximal parabolic subgroup with $S \setminus \{\alpha_d\}$ as the associated set of simple roots. Then

$$P_d = \left\{A \in G \,\middle|\, A = \begin{pmatrix} * & * \\ 0_{(n-d)\times d} & * \end{pmatrix}\right\},$$

$$W_{P_d} = S_d \times S_{n-d}.$$

For the natural action of G on $\mathbb{P}(\wedge^d V)$, we have, the isotropy at $[e_1 \wedge \cdots e_d]$ is P_d while the orbit through $[e_1 \wedge \cdots e_d]$ is $G_{d,n}$. Thus we obtain a surjective morphism $\pi : G \to G_{d,n}$, $g \mapsto g \cdot e$, where $e = [e_1 \wedge \cdots e_d]$. Further, the differential $d\pi_e$: Lie$G \to T(G_{d,n})_e$ (= the tangent space to $G_{d,n}$ at e) is easily seen to be surjective.

Hence we obtain an identification $f_d : G/P_d \cong G_{d,n}$ (cf. [22], Proposition 6.7). With this identification, we have $H^0(G/P_d, L_{\omega_d}) = (\wedge^d V)^*$ (here ω_d is the fundamental weight associated to P_d and L_{ω_d} is the associated line bundle on G/P_d; note that L_{ω_d} is the ample generator of Pic G/P_d ($\cong \mathbb{Z}$)).

2. Schubert varieties of $G_{d,n}$

For $1 \leqslant t \leqslant n$, let V_t be the subspace of V spanned by $\{e_1, \ldots, e_t\}$. For each $\underline{i} \in I_{d,n}$, the *Schubert variety associated to* $\underline{i}$ is defined to be

$$X_{\underline{i}} = \{U \in G_{d,n} \mid \dim(U \cup V_{i_t}) \geqslant t\,,\ 1 \leqslant t \leqslant d\}.$$

For the action of G on $\mathbb{P}(\wedge^d V)$ the T-fixed points are precisely the points corresponding to the T-eigenvectors in $\wedge^d V$. Now

$$\wedge^d V = \bigoplus_{\underline{i} \in I_{d,n}} k e_{\underline{i}}, \quad \text{as } T\text{-modules},$$

where for $\underline{i} = (i_1, \cdots, i_d)$, $e_{\underline{i}} = e_{i_1} \wedge \cdots \wedge e_{i_d}$. Thus the T-fixed points in $\mathbb{P}(\wedge^d V)$ are precisely $[e_{\underline{i}}]$, $\underline{i} \in I_{d,n}$, and these points, obviously, belong to $G_{d,n}$. Further, the Schubert variety $X_{\underline{i}}$ associated to $\underline{i}$ is simply the Zariski closure of the B-orbit $B\,[e_{\underline{i}}]$ through the T-fixed point $[e_{\underline{i}}]$ (with the canonical reduced structure), B being the subgroup of upper triangular matrices in G.

2.1. Bruhat decomposition

Let $\underline{i} \in I_{d,n}$. Let $C_{\underline{i}} = B[e_{\underline{i}}]$ be the Schubert cell associated to $\underline{i}$. The $C_{\underline{i}}$'s provide a cell decomposition of $G_{d,n}$. Let $X = V \oplus \cdots \oplus V$ be as above. Let

$$\begin{aligned} \pi :\quad & X \longrightarrow \wedge^d V, \\ (u_1, \ldots, u_d) & \longmapsto u_1 \wedge \ldots \wedge u_d, \end{aligned} \qquad \text{and} \qquad \begin{aligned} p :\ & \wedge^d V \setminus \{0\} \longrightarrow \mathbb{P}(\wedge^d V), \\ & u_1 \wedge \ldots \wedge u_d \longmapsto [u_1 \wedge \ldots \wedge u_d]. \end{aligned}$$

Let $v_{\underline{i}}$ denote the point $(e_{i_1}, \ldots, e_{i_d}) \in X$. As above, identifying X with $M_{n \times d}$, $v_{\underline{i}}$ gets identified with the $n \times d$ matrix whose entries are all zero except the ones in the i_j-th row and j-th column, $1 \leqslant j \leqslant d$, which are equal to 1. We have

$$B \cdot v_{\underline{i}} = \{A \in M_{n \times d} \mid x_{ij} = 0, i > i_j, \text{ and } \prod_t x_{i_t t} \neq 0\}.$$

Denoting $\overline{B \cdot v_{\underline{i}}}$ by $D_{\underline{i}}$, we have $D_{\underline{i}} = \{A \in M_{n \times d} \mid x_{ij} = 0, i > i_j\}$. Further, $\pi(B \cdot v_{\underline{i}}) = p^{-1}(C_{\underline{i}})$, $\pi(D_{\underline{i}}) = \widehat{X_{\underline{i}}}$. From this, we obtain

THEOREM 2.1.1. (1) $X_{\underline{i}} \subseteq X_{\underline{j}}$ *if and only if* $\underline{i} \leqslant \underline{j}$. (2) *Bruhat Decomposition:* $X_{\underline{j}} = \dot{\bigcup}_{\underline{i} \leqslant \underline{j}} B e_{\underline{i}}$.

In view of Theorem 2.1.1, we have $X_{\underline{i}} \subseteq X_{\underline{j}}$ if and only if $\underline{i} \leqslant \underline{j}$. Thus, under the set-theoretic bijection between the set of Schubert varieties and the set $I_{d,n}$, the partial order on the set of Schubert varieties given by inclusion induces the partial order $\geqslant$ on $I_{d,n}$. Note that $I_{d,n}$ can be identified with $\mathcal{S}_n/\mathcal{S}_d \times \mathcal{S}_{n-d}$.

In the sequel, we shall identify $I_{d,n}$ with the set of "minimal representatives" of $\mathcal{S}_n/\mathcal{S}_d \times \mathcal{S}_{n-d}$ in $\mathcal{S}_n$; to be very precise, a d-tuple $\underline{i} \in I_{d,n}$ will be identified with the element $(i_1, \ldots, i_d, j_1, \ldots, j_{n-d}) \in \mathcal{S}_n$, where $\{j_1, \ldots, j_{n-d}\}$ is the complement of $\{i_1, \ldots, i_d\}$ in $\{1, \ldots, n\}$ arranged in increasing order.

2.2. Dimension of $X_{\underline{i}}$

Let $E_{\underline{i}} = B^u \cdot v_{\underline{i}}$, where B^u is the subgroup of unipotent upper triangular matrices in G. Let U_d be the subgroup of unipotent upper triangular matrices in $GL_d(k)$. Then $E_{\underline{i}}$ is U_d-stable and the canonical surjective map $\pi_0 : E_{\underline{i}} \to C_{\underline{i}}$, $Av_{\underline{i}} \mapsto A \cdot [e_{\underline{i}}]$, $A \in B^u$, is in fact the orbit map (note that $C_{\underline{i}} = B \cdot [e_{\underline{i}}] = B^u \cdot [e_{\underline{i}}]$, as $[e_{\underline{i}}]$ is a T-fixed point). For $M \in E_{\underline{i}}$, it is easily checked that the isotropy subgroup in U_d at M is $\{\mathrm{id}\}$. Thus all the U_d-orbits in $E_{\underline{i}}$ are isomorphic to U_d. Thus $\pi_0 : E_{\underline{i}} \to C_{\underline{i}}$ is a principal fibration with fiber U_d. Hence we obtain

$$\begin{aligned}\dim C_{\underline{i}} &= \dim E_{\underline{i}} - \frac{1}{2}d(d-1) = \sum_{t=1}^{d} i_t - d - \frac{1}{2}d(d-1) \\ &= \sum_{t=1}^{d} i_t - \sum_{t=1}^{d} t = \sum_{t=1}^{d}(i_t - t).\end{aligned}$$

Thus we obtain

PROPOSITION 2.2.1. $\dim X_{\underline{i}} = \sum_{1 \leqslant t \leqslant d} (i_t - t)$.

REMARK 2.2.2. We have $G_{d,n} = X_{\underline{i}}$, where $\underline{i} = (n+1-d, n+2-d, \ldots, n)$. In particular, $\dim G_{d,n} = d(n-d)$.

REMARK 2.2.3. Each $\underline{i} \in I_{d,n}$ determines a Young diagram $\lambda = (i_d - d, \ldots, i_2 - 2, i_1 - 1)$ and $\dim X_{\underline{i}}$ is simply $|\lambda| (= \sum \lambda_i,\ \lambda_i = i_{d+1-i} - (d+1-i))$.

2.3. More results on Schubert varieties

LEMMA 2.3.1. *Let $\underline{i}, \underline{j} \in I_{d,n}$. Then*

$$p_{\underline{j}}\big|_{X_{\underline{i}}} \neq 0 \iff \underline{i} \geqslant \underline{j}.$$

PROOF. From §1.2, we have $p_{\underline{j}}|_{X_{\underline{i}}} \neq 0 \iff e_{\underline{j}} \in X_{\underline{i}}$. In view of Bruhat decomposition, we have $e_{\underline{j}} \in X_{\underline{i}} \iff \underline{j} \leqslant \underline{i}$. The result follows from this. □

LEMMA 2.3.2. *Let X_1, X_2 be two Schubert varieties in $G_{d,n}$. Then $X_1 \cap X_2$ is irreducible, i.e. $X_1 \cap X_2$ is a Schubert variety (set-theoretically).*

PROOF. Let $X_1 = X(\tau_1)$, $X_2 = X(\tau_2)$, where $\tau_1 = (a_1, \ldots, a_d)$, $\tau_2 = (b_1, \ldots, b_d)$. Let $X_1 \cap X_2 = \bigcup X(w_i)$, where $w_i < \tau_1$, $w_i < \tau_2$. Let $c_j = \min\{a_j, b_j\}$, $1 \leqslant j \leqslant d$, and $\tau = (c_1, \ldots, c_d)$. Then, clearly $\tau \in I_{d,n}$, and $\tau < \tau_i$, $i = 1, 2$. We have $w_i \leqslant \tau$, and hence $X_1 \cap X_2 = X(\tau)$. □

LEMMA 2.3.3. *Let X_1, X_2 be two Schubert divisors in a Schubert variety $X(\tau)$. Then $X_1 \cap X_2$ is irreducible of codimension 2 in X.*

PROOF. Let $X_1 = X(\tau_1)$, $X_2 = X(\tau_2)$. Let $\tau = (c_1, \ldots, c_d)$, $\tau_1 = (a_1, \ldots, a_d)$, $\tau_2 = (b_1, \ldots, b_d)$. Since $X(\tau_i)$ has codimension 1 in $X(\tau)$, the entries in τ_i are the same as the entries in τ, but for one place, where the entry is one less than the corresponding entry in τ, $i = 1, 2$. Let then j such that $a_t = c_t$, $t \neq j$, $a_j = c_j - 1$, and k such that $b_t = c_t$, $t \neq k$, $b_k = c_k - 1$. Let $w = (f_1, \ldots, f_d)$, where $f_t = c_t$, $t \neq j, k$, $f_j = c_j - 1$, $f_k = c_k - 1$. Then $X_1 \cap X_2 = X(w)$, and $\dim X(w) = \dim X(\tau) - 2$ (clearly, in view of Proposition 2.2.1). □

3. Standard monomial theory for Schubert varieties in $G_{d,n}$

3.1. Standard monomials

Let R_0 be the homogeneous coordinate ring of $G_{d,n}$ for the Plücker embedding, and for $\tau \in I_{d,n}$, let $R(\tau)$ be the homogeneous coordinate ring of the Schubert variety $X(\tau)$ (we shall denote $X_{P_d}(\tau)$ by just $X(\tau)$). In this section, we present a Standard monomial theory for $X(\tau)$ (cf. [**128**]), and as a consequence obtain a basis for $R(\tau)_m$, $m \in \mathbb{Z}^+$, prove the vanishing theorems, arithmetic normality and arithmetic Cohen-Macaulayness for $X(\tau)$. Standard monomial theory consists in constructing explicit bases for $R(\tau)$, $\tau \in I_{d,n}$.

DEFINITION 3.1.1. A monomial $f = p_{\tau_1} \cdots p_{\tau_m}$ is said to be *standard* if

$$(*) \qquad \tau_1 \geqslant \cdots \geqslant \tau_m.$$

Such a monomial is said to be *standard on* $X(\tau)$, if in addition to condition (*), we have $\tau \geqslant \tau_1$.

REMARK 3.1.2. Note that in the presence of condition (*), the standardness of f on $X(\tau)$ is equivalent to the condition that $f\big|_{X(\tau)} \neq 0$. Thus given a standard monomial $f = p_{\tau_1} \cdots p_{\tau_m}$, $f\big|_{X(\tau)}$ is either 0 or remains standard on $X(\tau)$.

3.2. Linear independence of standard monomials

THEOREM 3.2.1. *The standard monomials on $X(\tau)$ of degree m are linearly independent in $R(\tau)$.*

PROOF. We proceed by induction on $\dim X(\tau)$.

If $\dim X(\tau) = 0$, p_{id} is the only standard monomial on $X(\tau)$, and the result is obvious. Let $\dim X(\tau) > 0$. Let

$$(*) \qquad \sum_{i=1}^{r} c_i F_i,\ c_i \in k\,,$$

be a linear relation of standard monomials F_i of degree m. Let $F_i = p_{w_{i1}} \cdots p_{w_{im}}$. Suppose that $w_{i1} < \tau$ for some i. For simplicity, assume that $w_{11} < \tau$, and w_{11} is a minimal element of $\{w_{j1} \mid w_{j1} < \tau\}$. Let us denote w_{11} by ϕ. Then for $i \geqslant 2$,

$F_i|_{X(\phi)}$ is either 0, or is standard on $X(\phi)$. Hence restricting $(*)$ to $X(\phi)$, we obtain a nontrivial standard sum on $X(\phi)$ being zero, which is not possible (by induction hypothesis). Hence we conclude that $w_{i1} = \tau$ for all i, $1 \leqslant i \leqslant m$. Canceling p_τ, we obtain a linear relation among standard monomials on $X(\tau)$ of degree $m-1$. Using induction on m, the required result follows. □

3.3. Generation by standard monomials

THEOREM 3.3.1. *Let $F = p_{w_1} \dots p_{w_m}$ be any monomial in the Plücker coordinates of degree m. Then F is a linear combination of standard monomials of degree m.*

PROOF. For $F = p_{w_1} \dots p_{w_m}$, define

$$N_F = l(w_1)N^{m-1} + l(w_2)N^{m-2} + \dots l(w_m),$$

where $N \gg 0$, say $N > d(n-d)(= \dim G_{d,n})$ and $l(w) = \dim X(w)$. If F is standard, there is nothing to prove. Let t be the first violation of standardness, i.e., $p_{w_1} \cdots p_{w_{t-1}}$ is standard, but $p_{w_1} \cdots p_{w_t}$ is not. Hence $w_{t-1} \not\geq w_t$, and using the quadratic relation (cf. the relation (***) in §1.4)

$$(*) \qquad p_{w_{t-1}} p_{w_t} = \sum_{\alpha,\beta} \pm p_\alpha p_\beta,$$

F can be expressed as $F = \sum F_i$, with $N_{F_i} > N_F$ (since $\alpha > w_{t-1}$ for all α on the right hand side of $(*)$). Now the required result is obtained by decreasing induction on N_F (the starting point of induction, i.e. the case when N_F is the largest, corresponds to standard monomial $F = p_\theta^m$, where $\theta = (n+1-d, n+2-d, \cdots, n)$, in which case F is clearly standard; note that $X(\theta) = G_{d,n}$). □

Combining Theorems 3.2.1 and 3.3.1, we obtain

THEOREM 3.3.2. *Standard monomials on $X(\tau)$ of degree m give a basis for $R(\tau)_m$.*

As a consequence of Theorem 3.3.2, we have a qualitative description of a typical quadratic relation on a Schubert variety $X(w)$ as given by the following

PROPOSITION 3.3.3. *Let $w, \tau, \phi \in I_{d,n}$, $w > \tau, \phi$. Further let τ, ϕ be non-comparable (so that $p_\tau p_\phi$ is a non-standard degree 2 monomial on $X(w)$). Let*

$$(*) \qquad p_\tau p_\phi = \sum_{\alpha,\beta} c_{\alpha,\beta} p_\alpha p_\beta, \; c_{\alpha,\beta} \in k^*$$

be the expression for $p_\tau p_\phi$ as a sum of standard monomials on $X(w)$. Then for every α, β on the right hand side we have, $\alpha >$ both τ and ϕ, and $\beta <$ both τ and ϕ. Such a relation as in (*) is called a *straightening relation.*

PROOF. Among all the α's choose a minimal one, call it α_0. Restricting (*) to $X(\alpha_0)$, we have, the restriction of the right hand side is a non-zero sum of standard monomials on $X(\alpha_0)$ (note that the restriction of $p_\alpha p_\beta$ to $X(\alpha_0)$ is non-zero if and only if $\alpha = \alpha_0$ - by the minimality of α_0, and there is at least one term, namely $p_{\alpha_0}p_\beta$ whose restriction to $X(\alpha_0)$ is non-zero). Hence in view of linear independence of standard monomials on $X(\alpha_0)$, we obtain that the restriction of the left hand side to $X(\alpha_0)$ is non-zero. From this we conclude $\alpha_0 \geqslant$ both τ and ϕ; in fact, $\alpha_0 >$ both τ and ϕ (for, if α_0 equals one of them, say, $\alpha_0 = \tau$, then this would imply (by weight considerations) that $\beta = \phi$ which in turn would imply τ and ϕ are comparable (since α and β are), a contradiction. Any α is $\geqslant$ some minimal α_0. Hence $\alpha >$ both τ and ϕ.

Now replacing τ and ϕ by $w_0\tau$ and $w_0\phi$, we conclude from above $w_0\beta >$ both $w_0\tau$ and $w_0\phi$, i.e., $\beta <$ both τ and ϕ. □

3.4. Equations defining Schubert varieties in the Grassmannian

Let $\tau \in I_{d,n}$. Let π_τ be the map $R_0 \to R(\tau)$ (the restriction map). Let Ker $\pi_\tau = J_\tau$. Let $Z_\tau = \{$all standard monomials $F \mid F$ starts with p_ϕ for some $\phi \nleq \tau\}$. We shall now give a set of generators for J_τ in terms of Plücker coordinates.

LEMMA 3.4.1. *Let $I_\tau = (p_w, \tau \ngeqslant w)$ (ideal in R). Then Z_τ is a basis for I_τ.*

PROOF. Let $F \in I_\tau$. Then writing F as a linear combination of standard monomials

$$F = \sum a_i F_i + \sum b_j G_j,$$

where in the first sum each F_i starts with $p_{\tau_{i1}}$, with $\tau \ngeqslant \tau_{i1}$, and in the second sum each G_j starts with ϕ_{j1}, with $\tau \geqslant \phi_{j1}$. This implies (cf. Lemma 2.3.1) that $\sum a_i F_i \in I_\tau$, and hence we obtain

$$\sum b_j G_j \in I_\tau.$$

This now implies that considered as an element of $R(\tau)$, $\sum b_j G_j$ is equal to 0 (note that $I_\tau \subset J_\tau$). Now the linear independence of standard monomials on $X(\tau)$ implies that $b_j = 0$ for all j. The required result now follows. □

PROPOSITION 3.4.2. *Let $\tau \in I_{d,n}$. Then $R(\tau) = R_0/I_\tau$.*

PROOF. We have, $R(\tau) = R_0/J_\tau$ (where J_τ is as above). We shall now show that the inclusion $I_\tau \subset J_\tau$ is in fact an equality. Let $F \in R_0$. Writing F as a linear combination of standard monomials

$$F = \sum a_i F_i + \sum b_j G_j,$$

where in the first sum each F_i starts with $p_{\tau_{i1}}$, with $\tau \ngeqslant \tau_{i1}$, and in the second sum each G_j starts with ϕ_{j1}, with $\tau \geqslant \phi_{j1}$, we have, $\sum a_i F_i \in I_\tau$, and hence we obtain

$$\begin{aligned}
F \in J_\tau &\Longleftrightarrow \sum b_j G_j \in J_\tau \text{ since } \sum a_i F_i \in I_\tau, \text{ and } I_\tau \subset J_\tau) \\
&\Longleftrightarrow \pi_\tau(F)(= \sum b_j G_j) \quad \text{is zero} \\
&\Longleftrightarrow \sum b_j G_j \quad (= \text{a sum of standard monomials on } X(\tau)) \text{ is zero on} X(\tau)
\end{aligned}$$

$$\iff b_j = 0 \quad \text{for all } j \text{ (in view of the linear independence of standard monomials on } X(\tau))$$

$$\iff F = \sum a_i F_i$$

$$\iff F \in I_\tau .$$

Hence we obtain $J_\tau = I_\tau$. □

Equations defining Schubert varieties:

Let $\tau \in I_{d,n}$. We have (cf. Lemma 2.3.1), the kernel of $(R_0)_1 \to (R(\tau))_1$ has a basis given by $\{p_w, \tau \not\geq w\}$. Combining this fact with Proposition 3.4.2 and Lemma 3.4.1 we obtain that the ideal J_τ (= the kernel of the restriction map $R_0 \to R(\tau)$) is generated by $\{p_w \mid \tau \not\geq w, w \in I_{d,n}\}$. Hence J_τ is generated by the kernel of $(R_0)_1 \to (R(\tau))_1$. Thus we obtain that $X(\tau)$ is scheme-theoretically (even at the cone level) the intersection of $G_{d,n}$ with all hyperplanes in $\mathbb{P}(\wedge^d k^n)$ containing $X(\tau)$. Further, as a closed subvariety of $G_{d,n}$, $X(\tau)$ is defined (scheme-theoretically) by the vanishing of $p_w, \tau \not\geq w,\ w \in I_{d,n}$.

4. Standard monomial theory for a union of Schubert varieties

In this section, we prove results similar to Theorems 3.2.1, 3.3.1, 3.3.2 for a union of Schubert varieties.

Let X_i be Schubert varieties in $G_{d,n}$. Let $X = \cup X_i$.

DEFINITION 4.0.1. A monomial F in the Plücker coordinates is *standard* on the union $X = \bigcup X_i$ if it is standard on some X_i.

4.1. Linear independence of standard monomials on $X = \bigcup X_i$

THEOREM 4.1.1. *Monomials standard on $X = \bigcup X_i$ are linearly independent.*

PROOF. If possible, let

$$(*) \qquad \sum_{i=1}^{r} a_i F_i = 0$$

be a nontrivial relation among standard monomials on X. Let $F_i = p_{\tau_{i1}} \cdots p_{\tau_{ir_i}}$. For simplicity, denote τ_{11} by just τ. Then restricting $(*)$ to $X(\tau)$, we obtain a nontrivial relation among standard monomials on $X(\tau)$, which is a contradiction (note that for any i, $F_i|_{X(\tau)}$ is either 0 or remains standard on $X(\tau)$; further, $F_1|_{X(\tau)}$ is non-zero). □

4.2. Standard monomial basis

THEOREM 4.2.1. *Let $X = \bigcup_{i=1}^{r} X(\tau_i)$, and S the homogeneous coordinate ring of X. Then the standard monomials on X give a basis for S.*

PROOF. For $\tau \in I_{d,n}$, let I_τ be as in Lemma 3.4.1. Let us denote $I_t = I_{\tau_t}$, $1 \leqslant t \leqslant r$. We have $R(\tau_t) = R_0/I_t$ (cf. Proposition 3.4.2). Let $S = R_0/I$. Then

$I = \bigcap I_t$ (note that being the intersection of reduced ideals, I is also reduced, and hence the set theoretic equality $X = \bigcup X_i$ is also scheme theoretic). Let $F \in I$. Further, let us write F as a linear combination of standard monomials

$$F = \sum c_k F_k.$$

Then in view of Lemma 3.4.1, we obtain that each F_k starts with $p_{\tau_{k1}}$, where $\tau_{k1} \not\leqslant \tau_i$, $1 \leqslant i \leqslant r$. A typical element in R_0/I may be written as $\pi(f)$, for some $f \in R_0$, where π is the canonical projection $R_0 \to R_0/I$. Let us write f as a sum of standard monomials

$$f = \sum a_j G_j + \sum b_l H_l,$$

where each G_j (resp. H_l) starts with p_τ such that $\tau \not\leqslant \tau_i$, $1 \leqslant i \leqslant r$ (resp. $\tau \leqslant \tau_i$ for some i). We have $\pi(f) = \sum b_l H_l$ (since $\sum a_j G_j \in I$). Thus we obtain that S (as a vector space) is generated by monomials standard on X. This together with the linear independence of standard monomials on X implies the required result. □

4.3. Consequences

THEOREM 4.3.1. *Let X_1, X_2 be two Schubert varieties in $G_{d,n}$. Then*

(1) *$X_1 \cup X_2$ is reduced.*

(2) *$X_1 \cap X_2$ is reduced.*

PROOF. (1). For a closed subscheme Y in $G_{d,n}$, let $I(Y)$ denote the ideal defining Y in $G_{d,n}$. We have $I(X_1 \cup X_2) = I(X_1) \cap I(X_2)$, and hence $I(X_1 \cup X_2)$ is reduced (since $I(X_1), I(X_2)$ are reduced).

(2) Let $X_1 = X(w_1)$, $X_2 = X(w_2)$. Let A be the homogeneous coordinate ring of $X_1 \cap X_2$. Let $A = R_0/I$. Then $I = I_1 + I_2$. Let $X_1 \cap X_2 = \bigcup_{i=1}^r X(\tau_i)$ (set theoretically). Let $F \in I$. Then in the expression for F as a linear combination of standard monomials

$$F = \sum a_j F_j,$$

each F_j starts with $p_{\tau_{j1}}$, where $\tau_{j1} \not\leqslant$ either w_1 or w_2. If $B = R_0/\sqrt{I}$, then by Theorem 4.2.1, under $\pi : R_0 \to B$, Ker π consists of all f such that $f = \sum c_k f_k$, f_k being standard monomials such that each f_k starts with $p_{\phi_{k1}}$, where $\phi_{k1} \not\leqslant \tau_i$, for all i, $1 \leqslant i \leqslant r$. Hence $\phi_{k1} \not\leqslant$ either w_1 or w_2. Hence $\sqrt{I} = I$, and the required result follows from this. □

THEOREM 4.3.2. *(Pieri's formula)*

$$X(\tau) \cap \{p_\tau = 0\} = \bigcup_{i=1}^{r} X(\tau_i), \quad \textit{scheme theoretically},$$

where the varieties $X(\tau_i)$, $1 \leqslant i \leqslant r$ are all the Schubert divisors in $X(\tau)$.

PROOF. Let $X = \bigcup_{i=1}^r X(\tau_i)$, and let B be the homogeneous coordinate ring of X. Let $B = R(\tau)/I$. Clearly, $(p_\tau) \subseteq I$, (p_τ) being the principal ideal in $R(\tau)$

generated by p_τ. Let $f \in I$. Writing f as

$$f = \sum b_i G_i + \sum c_j H_j,$$

where each G_i is a standard monomial in $R(\tau)$ starting with p_τ and each H_j is a standard monomial in $R(\tau)$ starting with $p_{\theta_{j1}}$, where $\theta_{j1} < \tau$, we have, $\sum b_i G_i \in I$. This now implies $\sum c_j H_j$ is zero on $\bigcup_{i=1}^r X(\tau_i)$. But now $\sum c_j H_j$ being a sum of standard monomials on $\bigcup_{i=1}^r X(\tau_i)$, we have by Theorem 4.1.1, $c_j = 0$, for all j. Thus we obtain $f = \sum b_i G_i$, and hence $f \in (p_\tau)$. This implies $I = (p_\tau)$. Hence we obtain $B = R(\tau)/(p_\tau)$, and the result follows from this. □

5. Vanishing Theorems

Cohomology of sheaves

Let X be an algebraic scheme, $\mathcal{A}b(X)$ the category of sheaves of abelian groups on X. Let $\mathcal{A}b$ denote the category of abelian groups. Let $\Gamma(X, \cdot)$ be the global section functor. It is left exact, i.e., if $0 \to \mathcal{F} \to \mathcal{G} \to \mathcal{H} \to 0$ is a short exact sequence of sheaves - namely $0 \to \mathcal{F}_x \to \mathcal{G}_x \to \mathcal{H}_x \to 0$ is exact for every $x \in X$ - then $0 \to \Gamma(X, \mathcal{F}) \to \Gamma(X, \mathcal{G}) \to \Gamma(X, \mathcal{H})$ is exact. The cohomology functors $H^i(X, \cdot)$ are defined to be the right derived functors of $\Gamma(X, \cdot)$. For any sheaf $\mathcal{F}$, the groups $H^i(X, \mathcal{F})$ are defined as the cohomology groups of $\mathcal{F}$. Thus $H^i(X, \mathcal{F})$ are just the cohomology groups of the complex obtained by applying $\Gamma(X, \cdot)$ to an injective resolution of $\mathcal{F}$ (see **[67]** for details).

In the discussion below, we will be using the following well-known fact from cohomology theory (see **[153]** for example):

PROPOSITION 5.0.1. *Let $0 \to \mathcal{F} \to \mathcal{G} \to \mathcal{H} \to 0$ be a short exact sequence of sheaves on an algebraic scheme X. This gives rise to a cohomology long exact sequence*

$$\cdots \to H^{i-1}(X, \mathcal{H}) \to H^i(X, \mathcal{F}) \to H^i(X, \mathcal{G}) \to H^i(X, \mathcal{H}) \to \cdots$$

Let now X be a union of Schubert varieties. Let $S(X, m)$ be the set of standard monomials on X of degree m, and $s(X, m)$ the cardinality of $S(X, m)$. If X=$X(\tau)$ for some τ, $S(X, m)$ and $s(X, m)$ will also be denoted by just $S(\tau, m)$, respectively $s(\tau, m)$.

LEMMA 5.0.2. (1) *Let $Y = Y_1 \cup Y_2$. Then*

$$s(Y, m) = s(Y_1, m) + s(Y_2, m) - s(Y_1 \cap Y_2, m).$$

(2) $s(\tau, m) = s(\tau, m-1) + s(H_\tau, m)$, *where H_τ is the union of all the Schubert divisors in X_τ.*

(1) and (2) are easy consequences of the results of the previous section.

PROPOSITION 5.0.3. *Let r be an integer $\leqslant d(n-d)$. Suppose that all Schubert varieties X in $G_{d,n}$ of dimension at most r satisfy the following two properties:*

(1) $H^i(X, L^m) = 0$, *for* $i \geqslant 1$, $m \geqslant 0$, *where* $L = L_{\omega_d}$.

(2) *The set* $S(X, m)$ *is a basis for* $H^0(X, L^m)$, $m \geqslant 0$.

Then all unions and intersections of Schubert varieties of dimension at most r *satisfy* (1) *and* (2).

PROOF. We will prove the result by induction on r. Let S_r denote the set of Schubert varieties X in $G_{d,n}$ of dimension at most r. Let $Y = \bigcup_{j=1}^{t} X_j$, $X_j \in S_r$. Let $Y_1 = \bigcup_{j=1}^{t-1} X_j$, and $Y_2 = X_t$. Consider the exact sequence

$$0 \to \mathcal{O}_Y \to \mathcal{O}_{Y_1} \oplus \mathcal{O}_{Y_2} \to \mathcal{O}_{Y_1 \cap Y_2} \to 0,$$

where $\mathcal{O}_Y \to \mathcal{O}_{Y_1} \oplus \mathcal{O}_{Y_2}$ is the map $f \mapsto (f|_{Y_1}, f|_{Y_2})$ and $\mathcal{O}_{Y_1} \oplus \mathcal{O}_{Y_2} \to \mathcal{O}_{Y_1 \cap Y_2}$ is the map $(f, g) \mapsto (f - g)|_{Y_1 \cap Y_2}$. Tensoring with L^m, we obtain the long exact sequence

$$\cdots \to H^{i-1}(Y_1 \cap Y_2, L^m) \to H^i(Y, L^m) \to$$
$$H^i(Y_1, L^m) \oplus H^i(Y_2, L^m) \to H^i(Y_1 \cap Y_2, L^m) \to \cdots$$

Now $Y_1 \cap Y_2$ is reduced (cf. Theorem 4.3.1) and $Y_1 \cap Y_2 \in S_{r-1}$. Hence, by the induction hypothesis (1) and (2) hold for $Y_1 \cap Y_2$. In particular, if $m \geqslant 0$, then (2) implies that the map $H^0(Y_1, L^m) \oplus H^0(Y_2, L^m) \to H^0(Y_1 \cap Y_2, L^m)$ is surjective. Hence we obtain that the sequence

$$0 \to H^0(Y, L^m) \to H^0(Y_1, L^m) \oplus H^0(Y_2, L^m) \to H^0(Y_1 \cap Y_2, L^m) \to 0$$

is exact. This implies $H^0(Y_1 \cap Y_2, L^m) \to H^1(Y, L^m)$ is the zero map; we have, $H^1(Y, L^m) \to H^1(Y_1, L^m) \oplus H^1(Y_2, L^m)$ is also the zero map (since by induction $H^1(Y_1, L^m) = 0 = H^1(Y_2, L^m)$). Hence we obtain $H^1(Y, L^m) = 0$, $m \geqslant 0$, and for $i \geqslant 2$, the assertion that $H^i(Y, L^m) = 0$, $m \geqslant 0$ follows from the long exact cohomology sequence above (and induction hypothesis). This proves the assertion (1) for Y.

To prove assertion (2) for Y, we observe

$$h^0(Y, L^m) = h^0(Y_1, L^m) + h^0(Y_2, L^m) - h^0(Y_1 \cap Y_2, L^m)$$
$$= s(Y_1, m) + s(Y_2, m) - s(Y_1 \cap Y_2, m).$$

Hence Lemma 5.0.2 and induction hypothesis imply that

$$h^0(Y, L^m) = s(Y, L^m).$$

This together with linear independence of standard monomials on Y proves assertion (2) for Y. □

PROPOSITION 5.0.4. *Let* X *be a Schubert variety in* $G_{d,n}$. *Let* $Y = \bigcup_{j=1}^{t} X_j$ *be a union of Schubert divisors in* X. *Suppose* $H^i(X_j, L^m) = 0$, *for* $m < 0$, $0 \leqslant i \leqslant$ *dim*$Y - 1$(= *dim*$X_j - 1$). *Then* $H^i(Y, L^m) = 0$, *for* $m < 0$, $0 \leqslant i \leqslant$ *dim*$Y - 1$.

PROOF. We will prove the result by induction on t and dim$Y$ (=dim$X - 1$). As in the proof of Proposition 5.0.3, we set $Y_1 = \bigcup_{j=1}^{t-1} X_j$, $Y_2 = X_t$, and consider the long exact cohomology sequence

$$\cdots \to H^{i-1}(Y_1 \cap Y_2, L^m) \to H^i(Y, L^m) \to$$
$$H^i(Y_1, L^m) \oplus H^i(Y_2, L^m) \to H^i(Y_1 \cap Y_2, L^m) \to \cdots .$$

We have, $Y_1 \cap Y_2 = \bigcup_{j=1}^{t-1} X_j \cap X_t$. By Lemma 2.3.3, $X_j \cap X_t$, $1 \leqslant j \leqslant t-1$ is irreducible of codimension 1 in X_j (and X_t). Hence by induction on dim$Y$, we have, $H^i(Y_1 \cap Y_2, L^m) = 0$, for $m < 0$, $0 \leqslant i \leqslant$ dim$Y_1 - 2(=$ dim$Y - 2)$; further, by induction on t, $H^i(Y_1, L^m) = 0$, $m < 0$, $0 \leqslant i <$ dimY_1, and by hypothesis, $H^i(Y_2, L^m) = 0$, $m < 0$, $0 \leqslant i <$ dimY_1. The required result now follows. □

THEOREM 5.0.5. **Vanishing Theorems** *Let X be a Schubert variety in $G_{d,n}$. Then*

(1) $H^i(X, L^m) = 0$ *for* $i \geqslant 1$, $m \geqslant 0$.

(2) $H^i(X, L^m) = 0$ *for* $0 \leqslant i < \dim X$, $m < 0$.

(3) $S(X, m)$ *is a basis for* $H^0(X, L^m)$.

PROOF. We prove the result by induction on m, and $\dim X$.

If $\dim X = 0$, X is just a point, and the result is obvious. Assume now that $\dim X \geqslant 1$. Let $X = X(\tau)$. Let $X_1, \ldots, X_s$ be all the Schubert divisors in X, and $Y = \bigcup_{i=1}^{s} X_i$. Then by Pieri's formula (cf. §4.3.2), we have,

$$Y = X(\tau) \cap \{p_\tau = 0\} \quad \text{(scheme theoretically)}.$$

Hence the sequence

$$0 \to \mathcal{O}_X(-1) \to \mathcal{O}_X \to \mathcal{O}_Y \to 0$$

is exact. Tensoring it with L^m, and writing the cohomology exact sequence, we obtain the long exact cohomology sequence

$$\cdots \to H^{i-1}(Y, L^m) \to H^i(X, L^{m-1}) \to H^i(X, L^m) \to H^i(Y, L^m) \to \cdots .$$

Let $m \geqslant 0$, $i \geqslant 2$. Then the induction hypothesis on $\dim X$ implies (in view of Proposition 5.0.3) that $H^i(Y, L^m) = 0$, $i \geqslant 1$. Hence we obtain that the sequence $0 \to H^i(X, L^{m-1}) \to H^i(X, L^m)$, $i \geqslant 2$, is exact. If $i = 1$, again the induction hypothesis implies the surjectivity of $H^0(X, L^m) \to H^0(Y, L^m)$. This in turn implies that the map $H^0(Y, L^m) \to H^1(X, L^{m-1})$ is the zero map, and hence we obtain that the sequence $0 \to H^1(X, L^{m-1}) \to H^1(X, L^m)$ is exact. Thus we obtain that $0 \to H^i(X, L^{m-1}) \to H^i(X, L^m)$, $m \geqslant 0$, $i \geqslant 1$ is exact. But $H^i(X, L^m) = 0$, $m \gg 0$, $i \geqslant 1$ (cf. [**147**]). Hence we obtain

(1) $$H^i(X, L^m) = 0 \text{ for } i \geqslant 1,\ m \geqslant 0,$$

and

(2) $$h^0(X, L^m) = h^0(X, L^{m-1}) + h^0(Y, L^m).$$

In particular, assertion (a) follows from (1). The induction hypothesis on m implies that $h^0(X, L^{m-1}) = s(X, m-1)$. On the other hand, the induction hypothesis on

$\dim X$ implies (in view of Proposition 5.0.3) that $h^0(Y, L^m) = s(Y, m)$. Hence we obtain

$$h^0(X, L^m) = s(X, m-1) + s(Y, m). \tag{3}$$

Now (3) together with Lemma 5.0.2,(2) implies $h^0(X, L^m) = s(X, m)$. Hence (c) follows in view of the linear independence of standard monomials on $X(\tau)$ (cf. Theorem 3.2.1).

To prove (b), consider the long exact cohomology sequence

$$\cdots \to H^{i-1}(Y, L^m) \to H^i(X, L^{m-1}) \to H^i(X, L^m) \to \cdots$$

We have, $H^i(Y, L^m) = 0$, $m \in \mathbb{Z}$, $1 \leqslant i \leqslant \dim Y - 1$ (by induction hypothesis and Propositions 5.0.3 and 5.0.4). Hence we obtain the exact sequence

$$0 \to H^i(X, L^{m-1}) \to H^i(X, L^m), \qquad 2 \leqslant i < \dim X.$$

But $H^i(X, L^m) = 0$, $m \gg 0$ (cf. **[147]**). Hence we obtain, $H^i(X, L^m) = 0$, for $2 \leqslant i < \dim X$, $m \in \mathbb{Z}$ (in particular for $m < 0$).

It remains to prove (b) for $i = 0, 1$. Let then $m < 0$. Induction hypothesis together with Proposition 5.0.4 implies that $H^0(Y, L^m) = 0$. Hence we obtain the exact sequence

$$0 \to H^1(X, L^{m-1}) \to H^1(X, L^m), \qquad m < 0$$

Thus we obtain inclusions

$$H^1(X, L^{m-1}) \subset H^1(X, L^m) \subset \cdots \subset H^1(X, L^{-1}).$$

But now the isomorphism $H^0(X, \mathcal{O}_X) \cong H^0(Y, \mathcal{O}_Y)$ together with the fact that $H^1(X, \mathcal{O}_X) = 0$ implies that $H^1(X, L^{-1}) = 0$. This now implies (in view of the above inclusions) that $H^1(X, L^m) = 0$, $m < 0$. Now $H^0(Y, L^m) = 0$ implies

$$H^0(X, L^{m-1}) \cong H^0(X, L^m), \qquad m < 0. \tag{4}$$

Again the isomorphism $H^0(X, \mathcal{O}_X) \cong H^0(Y, \mathcal{O}_Y)$ implies that $H^0(X, L^{-1}) = 0$, and it follows that $H^0(X, L^m) = 0$, $m < 0$ (note that in view of (4), we have, $H^0(X, L^m) \cong H^0(X, L^{-1}) = 0$). □

COROLLARY 5.0.6. *We have*

(1) $R_0 = \bigoplus_{m \in \mathbb{Z}^+} H^0(X, L^m)$, $X = G_{d,n}$.

(2) $R(\tau) = \bigoplus_{m \in \mathbb{Z}^+} H^0(X(\tau), L^m)$, $\tau \in I_{d,n}$.

The assertions (1) and (2) follow immediately from Theorems 3.3.2, 5.0.5, (c).

6. Arithmetic normality and Cohen-Macaulayness for Schubert varieties in $G_{d,n}$

6.1. Arithmetic normality

PROPOSITION 6.1.1. *$X(\tau)$ is arithmetically normal, i.e., the cone $\widehat{X(\tau)}$(= Spec $R(\tau)$) is normal.*

PROOF. Let $\mathbb{P} = \mathbb{P}(\wedge^d V)$. The canonical homomorphism

$$\phi_r : H^0(\mathbb{P}, L^m) \longrightarrow H^0(X(\tau), L^m)$$

is surjective for $m \gg 0$. Let $S_m = H^0(X(\tau), L^m)$. Let R be the homogeneous coordinate ring of $X(\tau)$ for the Plücker embedding. We have (see **[67]**, Chapter II, Ex. 5.14 for example) $\operatorname{Im}\phi_m = R_m$, and $S = \oplus S_m$ is the integral closure of R in its quotient field. Now by Corollary 5.0.6, we have $R_m = S_m$ for all m, and hence we obtain that R is normal. Therefore $X(\tau)$ is arithmetically normal. □

6.2. Arithmetic Cohen-Macaulayness

LEMMA 6.2.1. *Let $X \subset \mathbb{P}$ be a closed subscheme. Then $\widehat{X}$ is Cohen-Macaulay if and only if $\widehat{X}$ is Cohen-Macaulay at its vertex.*

PROOF. Suppose that the vertex (0) is a Cohen-Macaulay point of $\widehat{X}$. Then $\widehat{X}$ is Cohen-Macaulay in a neighborhood U of (0). It is clear that each fiber of the natural projection $\pi : \widehat{X} \setminus \{(0)\} \to X$ meets U. But these fibers are simply the $\mathbb{G}_m$-orbits in $\widehat{X}\setminus\{(0)\}$, and hence $\widehat{X}$ is Cohen-Macaulay at each point of $\pi^{-1}(x)$ for all $x \in X$. □

PROPOSITION 6.2.2. *$X(\tau)$ is arithmetically Cohen-Macaulay, i.e., the cone $\widehat{X(\tau)}(= \operatorname{Spec} R(\tau))$ is Cohen-Macaulay.*

PROOF. We first recall the following result.

Let $X \hookrightarrow \mathbb{P}^N$ be a closed subscheme of $\mathbb{P}^N$. Let I be the homogeneous ideal defining X. Let A be the homogeneous coordinate ring of X, i.e. $A = k[x_0, \ldots, x_N]/I$. Let L denote the tautological line bundle on $\mathbb{P}^N$ (cf. Chapter 1 §7.10), as well as its restriction to X. Let (0) denote the vertex of the cone $\widehat{X}$, and $A_{(0)}$ denote the local ring $\mathcal{O}_{\widehat{X},(0)}$ at the vertex of $\widehat{X}$. Then we have

THEOREM 6.2.3. (**Serre-Grothendieck Criterion**) *The statements (1) and (2) below are equivalent:*

(1) (a) *The canonical homomorphisms $\phi_r : H^0(\mathbb{P}^N, L^r) \to H^0(X, L^r)$ are surjective for all $r \geqslant 0$.*
(b) *$H^i(X, L^r) = 0$ for $1 \leqslant i \leqslant d-2$ and $r \in \mathbb{Z}$.*

(2) depth $A_{(0)} \geqslant d$.

Proof of Serre-Grothendieck Criterion. This result is an immediate consequence of the local cohomology theory (cf. **[143]**). Let $U = \widehat{X} \setminus \{(0)\}$, and $p : U \to X$ the canonical projection. For a coherent sheaf F on $\widehat{X}$, we have an exact sequence of local cohomology:

$$0 \to H^0_{(0)}(\widehat{X}, F) \to H^0(\widehat{X}, F) \to H^0(U, F) \to H^1_{(0)}(\widehat{X}, F) \to 0,$$

$$H^{i-1}(U, F) \cong H^i_{(0)}(\widehat{X}, F),\ i \geqslant 2.$$

Further, we have

(1) $\operatorname{depth}_{(0)} F \geqslant d \iff H^i_{(0)}(\widehat{X}, F) = 0,\ i < d.$

(2) For $d \geqslant 2$, and $F|_U = p^*L$, $\text{depth}_{(0)}F \geqslant d \iff H^0(\widehat{X}, F) = \sum_{r\in\mathbb{Z}} H^0(X, L^r)$, and $H^{i-1}(X, L^r) = 0, 2 \leqslant i < d,\ r \in \mathbb{Z}$.
Thus the equivalence in the Theorem is simply the equivalence (2) above.
Continuation of proof of Proposition. Let $X = X(\tau)$ be a Schubert variety in $G_{d,n}$. Let $\dim X = t$. Then the condition (a) in (1) (in Serre-Grothendieck Criterion) holds (in view of Theorem 5.0.5, (c)) , and (b) in (1) holds (in view of Theorem 5.0.5, (a) and (b)), with $d - 2 = t - 1$, i.e., $d = t + 1$. Hence we obtain $\text{depth}_{(0)}\widehat{X(\tau)} \geqslant t + 1$. But $\dim \widehat{X(\tau)} = t + 1$. The result follows from this. □

We include below two more proofs for the arithmetic Cohen-Macaulayness of $X(\tau)$ (cf. [**128**]).

Second Proof. The result is proved in the same spirit as the proof of Theorem 5.0.5. We first prove the following

PROPOSITION 6.2.4. *Let* $\dim X(\tau) = t$, *and let* $X = \bigcup_{j=1}^{s} X_j$, *where* $X_1, \ldots, X_s$ *are some of the Schubert divisors in* $X(\tau)$. *Suppose that* $\widehat{X(w)}$ *is Cohen-Macaulay at its vertex for all Schubert varieties* $X(w)$ *of dimension* $< t$. *Then* $\widehat{X}$ *is Cohen-Macaulay at its vertex.*

PROOF. We prove the result by induction on s and t. Let us write $X = Y \cup Z$, where $Y = \bigcup_{j=1}^{s-1} X_j$, $Z = X_s$. By induction on s, we have, $\widehat{Y}$ is Cohen-Macaulay at its vertex; and by induction on t, $\widehat{Z}$ is Cohen-Macaulay at its vertex. The Cohen-Macaulayness of $\widehat{X}$ at its vertex follows from the following Lemma. □

LEMMA 6.2.5. *Let* $A = \bigoplus_{n\geqslant 0} A_n$ *be a Noetherian graded ring with* $A_0 = k$. *Let* I, J *be two homogeneous ideals of* A *such that*

$$\dim A/I = \dim A/J = d = \dim(A/I + J) + 1.$$

If A/I, A/J *and* $A/I + J$ *are Cohen-Macaulay, then* $A/I \cap J$ *is Cohen-Macaulay of dimension* d.

PROOF. Let $\mathfrak{m} = A^+$ be the irrelevant ideal. It suffices to show (cf. [**120**]) that $(A/I \cap J)_\mathfrak{m}$ is Cohen-Macaulay. Consider the exact sequence

$$0 \to (A/I \cap J)_\mathfrak{m} \to (A/I)_\mathfrak{m} \oplus (A/J)_\mathfrak{m} \to (A/I + J)_\mathfrak{m} \to 0\,.$$

Taking local cohomology, we get,

$$\cdots \to H^{i-1}_\mathfrak{m}((A/I + J)_\mathfrak{m}) \to H^i_\mathfrak{m}((A/I \cap J)_\mathfrak{m})$$
$$\to H^i_\mathfrak{m}((A/I)_\mathfrak{m}) \oplus H^i_\mathfrak{m}((A/J)_\mathfrak{m}) \to \cdots.$$

The depth of a finitely generated $A_\mathfrak{m}$-module N is characterized by

$$\text{depth}\, N = \min\{i \mid H^i_\mathfrak{m}(N) \neq 0\}.$$

By hypothesis, $H^i_{\mathfrak{m}}((A/I)_{\mathfrak{m}}) = H^i_{\mathfrak{m}}((A/J)_{\mathfrak{m}}) = 0,\ i < d$, and $H^i_{\mathfrak{m}}((A/I+J)_{\mathfrak{m}}) = 0$, for $i < d-1$. Hence we obtain $H^i_{\mathfrak{m}}((A/I \cap J)_{\mathfrak{m}}) = 0,\ i < d$. This implies

$$d \leqslant \operatorname{depth}(A/I \cap J)_{\mathfrak{m}} \leqslant \dim(A/I \cap J)_{\mathfrak{m}} \leqslant d$$

It follows that $(A/I \cap J)_{\mathfrak{m}}$ is Cohen-Macaulay. □

Continuation of second Proof. Let H be the union of all the Schubert divisors in $X(\tau)$. Let $R(H)$ be the homogeneous coordinate ring of H. Then $R(H)_{(0)} = R(\tau)_{(0)}/(p_\tau)$ (cf. Theorem 4.3.2), where $R(H)_{(0)}$ (resp. $R(\tau)_{(0)}$) is the local ring $\mathcal{O}_{\widehat{H},0}$ (resp. $\mathcal{O}_{\widehat{X(\tau)},(0)}$) at the vertex of H (resp. $X(\tau)$). Now by Proposition 6.2.4, we have, $R(H)_{(0)}$ is Cohen-Macaulay. This together with the fact that p_τ is a nonzero divisor in $R(\tau)_{(0)}$ implies the Cohen-Macaulayness of $R(\tau)_{(0)}$.

Third Proof. Let $\dim X(\tau) = t$. For $0 \leqslant j \leqslant t$, set

$$f_j(\tau) = \sum p_{\phi_{kj}},$$

where the summation runs over all Schubert subvarieties $X(\phi_{kj})$ in $X(\tau)$ of codimension j. We shall now show that $f_0(\tau), f_1(\tau), \cdots, f_t(\tau)$ form a system of parameters at the vertex. We shall show this by induction on t. It $t = 0$ or 1, the result is clear (note that if $t = 1$, then $X(\tau) \cong \mathbb{P}^1$). Let $X_1, \ldots, X_r$ be all the Schubert divisors in $X(\tau)$. Let $Y_s = \bigcup_{j=1}^{s} X_j,\ 1 \leqslant s \leqslant r$. We define $f_{js},\ 0 \leqslant j \leqslant t-1$ in the same manner as $f_j(\tau)$ (note that $\dim Y_s = t-1$), namely, $f_{js} = \sum p_{\phi_{kj}}$ where the summation runs over all Schubert subvarieties $X(\phi_{kj})$ in Y_s of codimension j in Y_s. We shall first show that $f_{js},\ 1 \leqslant j \leqslant t-1$ form a system of parameters at the vertex for $\widehat{Y_s}$. We shall prove this by induction on s. If $s = 1$, then the result holds (by induction on t). Let then $r \geqslant s > 1$. Let us denote $X = X_s, Y = Y_{s-1}$ so that $Y_s = X \cup Y$. Observe that $X \cap Y$ is a union of Schubert subvarieties of $X(\tau)$ of codimension 2 (cf. Lemma 2.3.3). The Cohen-Macaulayness of $\widehat{Y}$ at its vertex follows by induction on s in view of the following Lemma:

LEMMA 6.2.6. *Let A be a noetherian local domain with maximal ideal* $\mathfrak{m}$. *Let I, J be two ideals in A such that* depth $A/I+J \geqslant d-1$, depth $A/I \geqslant d$, depth $A/J \geqslant d$. *Suppose that $f_1, \ldots, f_d \in \mathfrak{m}$ are such that $f_1, \ldots, f_d \pmod I$ (resp. $\pmod J$) form a A/I (resp. A/J) regular sequence, and $f_1, \ldots, f_{d-1} \pmod{I+J}$ is an $A/I+J$-regular sequence. Then $f_1, \cdots, f_d \pmod{I \cap J}$ is an $A/I \cap J$-regular sequence.*

PROOF. Since the conclusion of the Lemma is only about $A/I \cap J$, we may assume that $I \cap J = (0)$. We have an inclusion

$$A \hookrightarrow A/I \times A/J,$$

under which A gets identified with

$$\{(a_1, a_2) \in A/I \times A/J \mid a_1 \equiv a_2 \pmod{I+J}\}.$$

Let $((f_1, f_1))$ denote the ideal in $A/I \times A/J$ generated by (f_1, f_1). We have, $f_1 A \subset A \cap ((f_1, f_1))$ (clearly). Now we show the reverse inclusion. Let $(a_1, a_2) \in A \cap ((f_1, f_1))$. Then $(a_1, a_2) = (\alpha f_1, \beta f_1)$ for some $(\alpha, \beta) \in A/I \times A/J$. Now $a_1 \equiv a_2 \pmod{I+J}$ implies $\alpha f_1 - \beta f_1 \equiv 0 \pmod{I+J}$. But by hypothesis, $f_1 \pmod{I+J}$ is not a zerodivisor in $A/I + J$. Hence we obtain $\alpha \equiv \beta \pmod{I+J}$. Hence $f_1(\alpha, \beta)\,(= (a_1, a_2)) \in f_1 A$. Thus we obtain

$$f_1 A = A \cap ((f_1, f_1)).$$

Also, we have,

$$A \cap ((f_1, f_1)) = (f_1, I) \cap (f_1, J).$$

Hence we obtain

$$(*) \qquad f_1 A = (f_1, I) \cap (f_1, J).$$

By hypothesis, we have, $f_2, \ldots, f_d \pmod{(f_1, I)}$ (resp. $\pmod{(f_1, J)}$) form a $A/(f_1, I)$ (resp. $A/(f_1, J)$) regular sequence, and $f_2, \ldots, f_{d-1} \pmod{(f_1, I+J)}$ is an $A/(f_1, I+J)$-regular sequence. Hence by induction on d, we obtain that $f_2, \cdots, f_d \pmod{(f_1, I) \cap (f_1, J)}$ is a $A/(f_1, I) \cap (f_1, J)$-regular sequence, i.e., $f_2, \cdots, f_d \pmod{f_1 A}$ is a $A/f_1 A$-regular sequence (in view of (*)); note that for $d = 1$, the result holds, since f_1 is not a zero-divisor in A. The assertion in Lemma now follows. □

Finally, to prove the result for $X(\tau)$, we have, by Pieri's formula:

$$Z := X(\tau) \cap \{p_\tau = 0\} = \bigcup_{i=1}^{r} X_i,$$

where X_i, $1 \leqslant i \leqslant r$ are all the Schubert divisors in $X(\tau)$. From our discussion above we have, $\{f_1(\tau), \cdots, f_t(\tau)\}$ is a system of parameters at the vertex for $\widehat{Z}$. Now the stalk $\mathcal{O}_{Z,o}$ is simply $R(\tau)_{(o)}/(p_\tau)$. Hence we obtain that $f_0(\tau), f_1(\tau), \cdots, f_t(\tau)$ is a system of parameters at the vertex for $\widehat{X(\tau)}$.

CHAPTER 7

The flag variety

In this chapter, we study the flag variety $SL(n)/B$ and its Schubert subvarieties. After introducing the flag variety in §1, we present the Standard Monomial Theory for Schubert varieties in $SL(n)/B$ in §2-4. We then present a proof of the " vanishing theorems" for Schubert varieties in §4, and deduce ideal theory, arithmetic normality and arithmetic Cohen-Macaulayness for Schubert varieties in §5 and §6.

1. The flag variety $SL(n)/B$

Let $G = SL(n)$, the special linear group of rank $n-1$. Let T be the maximal torus consisting of all the diagonal matrices in G, and B the Borel subgroup consisting of all the upper triangular matrices in G. We have $N_G(T) = \{\text{all permutation matrices in } G\}$, i.e., all matrices in G with a unique non-zero entry in each row and column. Then W gets identified with S_n, the symmetric group on n letters. Any $w \in S_n$ is usually written as $(w(1), w(2), \cdots w(n))$.

1.1. The flag variety

Let $V = k^n$. A sequence $(0) = V_0 \subset V_1 \subset \cdots \subset V_n = V$, such that $\dim V_i = i$, is called a *flag* in V. Let $\mathcal{F}(V)$ denote the set of all flags in V. Let $\{e_i, 1 \leqslant i \leqslant n\}$ be the standard basis of k^n. The flag $F_0 = (V_0 \subset \cdots \subset V_i \subset \cdots \subset V_n)$, where V_i is the span of $\{e_1, \dots, e_i\}$ for $1 \leqslant i \leqslant n$, is called the *standard flag*. We have a natural action of G on $\mathcal{F}(V)$, which is transitive. The isotropy group at F_0 is precisely B. Thus through the identification of $\mathcal{F}(V)$ with G/B, $\mathcal{F}(V)$ acquires the structure of a projective variety (cf. Chapter 2, Proposition 2.1.7). We have a canonical closed embedding

$$\theta : \mathcal{F}(V) \hookrightarrow G_{1,n} \times G_{2,n} \times \cdots \times G_{n-1,n}.$$

Under this embedding we have

$$\mathcal{F}(V) = \{(U_1, \dots, U_{n-1}) \in \prod_{i=1}^{n-1} G_{i,n} \mid U_1 \subset \cdots \subset U_{n-1}\}.$$

The incidence relations $U_i \subset U_{i+1}$, $1 \leqslant i \leqslant n-2$ identify $\mathcal{F}(V)$ as a closed subset of $\prod_{i=1}^{n-1} G_{i,n}$.

Following [27], we denote the simple roots by $\epsilon_i - \epsilon_{i+1}$, $1 \leqslant i \leqslant n-1$ (note that $\epsilon_i - \epsilon_{i+1}$ is the character sending $\operatorname{diag}(t_1, \dots, t_n)$ to $t_i t_{i+1}^{-1}$). Then $R = \{\epsilon_i - \epsilon_j \mid 1 \leqslant$

$i, j \leqslant n,\ i \neq j\}$, and the reflection $s_{\epsilon_i - \epsilon_j}$ may be identified with the transposition (i, j) in S_n.

1.2. The Bruhat-Chevalley order on S_n

Let $w_1, w_2 \in S_n$, $w_1 = (a_1, \dots, a_n)$, $w_2 = (b_1, \dots, b_n)$. We have

$$X(w_1) \subset X(w_2) \iff \pi_d(X(w_1)) \subset \pi_d(X(w_2)),\ \text{for all } 1 \leqslant d \leqslant n-1,$$

where π_d is the canonical projection $G/B \to G/P_d$. Let $X(w_i^{(d)}) = \pi_d(X(w_i))$, $i = 1, 2$; then the corresponding d-tuples in $I_{d,n}$ are $(a_1 \dots a_d) \uparrow$, $(b_1 \dots b_d) \uparrow$ (here, for a d-tuple $(t_1 \dots t_d)$ of distinct integers, $(t_1 \dots t_d) \uparrow$ denotes the ordered d-tuple obtained from $\{t_1, \dots, t_d\}$ by arranging its elements in ascending order). From Chapter 6, we have $X(w_1^{(d)}) \subset X(w_2^{(d)})$ if and only if $(a_1 \dots a_d) \uparrow \leqslant (b_1 \dots b_d) \uparrow$. Hence we obtain that for $(a_1 \dots a_n), (b_1 \dots b_n) \in S_n$,

$$(a_1 \dots a_n) \leqslant (b_1 \dots b_n) \iff (a_1 \dots a_d) \uparrow \leqslant (b_1 \dots b_d) \uparrow,\ \text{for all } 1 \leqslant d \leqslant n-1.$$

1.3. The partially ordered set $I_{a_1,\dots,a_k}$

Let Q be a parabolic subgroup in $SL(n)$. First, let $Q = P_d$, the maximal parabolic subgroup with $S \setminus \{\alpha_d\}$ as the associated set of simple roots. Then

$$P_d = \left\{ A \in G \;\middle|\; A = \begin{pmatrix} * & * \\ 0_{(n-d)\times d} & * \end{pmatrix} \right\},$$
$$W_{P_d} = S_d \times S_{n-d}.$$

Hence

$$W_{P_d}^{\min} = \{(a_1 \dots a_n) \in W \mid a_1 < \cdots < a_d,\quad a_{d+1} < \cdots < a_n\}.$$

Thus $W_{P_d}^{\min}$ may be identified with

$$I_{d,n} := \{\underline{i} = (i_1, \dots, i_d) \mid 1 \leqslant i_1 < \cdots < i_d \leqslant n\}.$$

In the sequel, we shall denote $W_{P_d}^{\min}$ by just W^d.

Let now Q be any parabolic subgroup. Let $1 \leqslant a_1 < \cdots < a_k \leqslant n$ be such that $S_Q = S \setminus \{\alpha_{a_1}, \dots, \alpha_{a_k}\}$. Then $Q = P_{a_1} \cap \cdots \cap P_{a_k}$, and consists of all the elements of G of the form

$$\begin{pmatrix} A_1 & * & * & \cdots & * & * \\ 0 & A_2 & * & \cdots & * & * \\ \vdots & \vdots & \vdots & & \vdots & \vdots \\ 0 & 0 & 0 & \cdots & A_k & * \\ 0 & 0 & 0 & \cdots & 0 & A \end{pmatrix},$$

where A_t is a matrix of size $c_t \times c_t$, $c_t = a_t - a_{t-1}$, $1 \leqslant t \leqslant k$ (here $a_0 = 0$), A is a matrix of size $a \times a$ ($a = n - a_k$), and $x_{ml} = 0$, $m > a_t$, $l \leqslant a_t$, $1 \leqslant t \leqslant k$. We have, $W_Q = \mathbb{S}_{a_1} \times \mathbb{S}_{a_2 - a_1} \times \cdots \times \mathbb{S}_{n-a_k}$. Let

$$I_{a_1,\dots,a_k} = \{(\underline{i}_1, \dots, \underline{i}_k) \in I_{a_1,n} \times \cdots \times I_{a_k,n} \mid \underline{i}_t \subset \underline{i}_{t+1} \text{ for all } 1 \leqslant t \leqslant k-1\}.$$

Then it is easily seen that $W_Q^{\min}$ may be identified with $I_{a_1,\ldots,a_k}$.

The partial order on the set of Schubert varieties in G/Q (given by inclusion) induces a partial order $\geqslant$ on $I_{a_1,\ldots,a_k}$, namely, for $\mathbf{i} = (\underline{i}_1, \ldots, \underline{i}_k), \mathbf{j} = (\underline{j}_1, \ldots, \underline{j}_k) \in I_{a_1,\ldots,a_k}$, $\mathbf{i} \geqslant \mathbf{j} \iff \underline{i}_t \geqslant \underline{j}_t$ for all $1 \leqslant t \leqslant k$.

1.4. The minimal and maximal representatives as permutations

Let $w \in W$, and let $w_Q^{\min}$ be the element in $W_Q^{\min}$ which represents the coset wW_Q. Under the identification of $W_Q^{\min}$ with $I_{a_1,\ldots,a_k}$, let $\mathbf{i} = (\underline{i}_1, \ldots, \underline{i}_k)$ be the element in $I_{a_1,\ldots,a_k}$ which corresponds to $w_Q^{\min}$. As a permutation, the element $w_Q^{\min}$ is given by $\underline{i}_1$, followed by $\underline{i}_2 \setminus \underline{i}_1$ arranged in ascending order, and so on, ending with $\{1, \ldots, n\} \setminus \underline{i}_k$ arranged in ascending order.

Similarly, if $w_Q^{\max}$ is the element in $W_Q^{\max}$ which represents the coset wW_Q, then as a permutation, the element $w_Q^{\max}$ is given by $\underline{i}_1$ arranged in descending order, followed by $\underline{i}_2 \setminus \underline{i}_1$ arranged in descending order, etc..

2. Standard Young tableaux and standard monomials

Let $X(T)$ be the character group of T. For $\lambda \in X(T)$, write $\lambda = \sum_{i=1}^{n-1} a_i\omega_i$, $a_i \in \mathbb{Z}_+$, ω_i being the fundamental weights (cf. Chapter 3, §1). Let $L(\lambda)$ denote the line bundle on the flag variety G/B associated to λ. Recall (cf. Chapter 4, §3.2) that

$$H^0(G/B, L(\lambda)) = \{f : G \to k \mid f(gb) = \lambda(b)f(b), \text{ for all } g \in G \text{ and } b \in B\}.$$

Recall that λ is *dominant* (written $\lambda \geqslant 0$) if and only if $a_i \geqslant 0$ for all i, which also is equivalent to $H^0(G/B, L(\lambda)) \neq (0)$. Further, $L(\lambda)$ is *ample* if and only if $a_i > 0$ for all i.

Let $P \supset B$ be a maximal parabolic subgroup of G with ω as the associated fundamental weight. Recall (cf. **[27]**) that P (or ω) is *minuscule* if

$$(\omega, \beta^*)(= 2(\omega, \beta)/(\beta, \beta)) \leqslant 1$$

for all positive roots β, where (,) denotes a W-invariant inner product on $X(T)$; equivalently, all the weights of the Weyl module $V_k(\omega)$ are *extremal*, i.e. they are of the form $\tau(\omega)$, $\tau \in W$, and are parametrized by $\tau \in W/W_P$.

DEFINITION 2.0.1. A *Young tableau* Λ *of shape* λ is defined to be a filling of λ with integers. A Young tableau is said to be *standard* if the entries along any row (resp. column) are non-decreasing (resp. strictly increasing).

Let λ be a Young diagram with length of any column $\leqslant n-1$. Let Λ_n be the set of all Young tableaux of shape λ, where the entries of any tableau belong to $\{1, \ldots, n\}$. Given a Young diagram of shape λ, let Λ_n^{std} be the set of all standard Young tableaux of shape λ, where the entries of any tableau belong to $\{1, \ldots, n\}$.

2.1. The element p_Λ of $H^0(G/B, L(\lambda))$

Given a Young tableau $\Lambda \in \Lambda_n$, let $\tau = \{i_1, \ldots, i_d\}$ be a typical column in Λ, where $1 \leqslant i_1 < \cdots < i_d \leqslant n$ for some d, $1 \leqslant d \leqslant n-1$. To the column τ, we associate the Plücker coordinate p_τ (cf. Chapter 6, §1.2). We set $p_\Lambda = \prod_\tau p_\tau$, where the product is taken over all the columns of Λ. We say that p_Λ is a *standard monomial* on G/B if Λ is standard.

Given a Young diagram λ (with length of any column $\leqslant n-1$), we shall denote the associated dominant weight also by just λ.

In this Chapter we give a proof of the following well-known

THEOREM 2.1.1. *Let λ be a Young diagram with length of any column $\leqslant n-1$. Then $\{p_\Lambda, \Lambda \in \Lambda_n^{\text{std}}$ is a basis for $H^0(G/B, L(\lambda))$.*

In fact, we prove a similar result (cf. Theorem 4.3.1) for a union of Schubert varieties in G/B, and Theorem 2.1.1 is obtained as a special case of Theorem 4.3.1. In order to state Theorem 4.3.1, we first need to define the notion of a monomial p_Λ being standard on a Schubert variety.

Let S be a collection of distinct standard monomials on $G_{d,n}(= G/P_d)$. Let $X(w)$ be a Schubert variety in $G_{d,n}$. Given a standard monomial F on $G_{d,n}$, $F|_{X(w)}$ is either 0 or remains standard on $X(w)$ (cf. Chapter 6, Remark 3.1.2). Let $S_w = \{F|_{X(w)} \mid F \in S,\ F|_{X(w)} \neq 0\}$. Then in view of Chapter 6, Theorem 3.2.1, we get that S_w is linearly independent. This phenomenon does not hold in G/B, as explained by the following example.

EXAMPLE 2.1.2. Let $G = SL(3)$, $w = (312)$,

$$T_1 = \begin{array}{|c|c|}\hline 1 & 3 \\ \hline 2 \\ \cline{1-1}\end{array}\,, \qquad T_2 = \begin{array}{|c|c|}\hline 1 & 2 \\ \hline 3 \\ \cline{1-1}\end{array}\,.$$

Then it can be easily checked that $p_{T_1}\big|_{X(w)} = p_{T_2}\big|_{X(w)}$. Now taking $S = \{p_{T_1}, p_{T_2}\}$, S is linearly independent on G/B, but $S_w\ (= p_{T_1}\big|_{X(w)}, p_{T_2}\big|_{X(w)})$ is linearly dependent on $X(w)$.

In particular, the above example shows that we cannot define the standardness of a monomial p_T on a Schubert variety $X(w) \subset G/B$ as we did in Chapter 6. Thus we *cannot* define the standardness of p_T on $X(w)$ by saying that T is standard and $p_T\big|_{X(w)} \neq 0$. Hence, in order to define the standardness of p_T on $X(w)$, we need to refine the notion of standardness of p_T as defined above.

2.2. The parabolic subgroup Q_λ

Given a Young diagram λ, consider a typical column $\underline{c}$; let d be the length of $\underline{c}$. Thus $\underline{c}$ determines a maximal parabolic subgroup of G, namely P_d. Let $P_{i_1}, \ldots, P_{i_r}$ be the distinct maximal parabolic subgroups determined by the columns of λ. We denote by

Q_λ the parabolic subgroup with $S_{Q_\lambda} = S \setminus \{\alpha_{i_1}, \dots, \alpha_{i_r}\}$. In other words, we have $Q_\lambda = \beta cap_{t=1}^r P_{i_t}$.

EXAMPLE 2.2.1. Let

$\lambda =$ 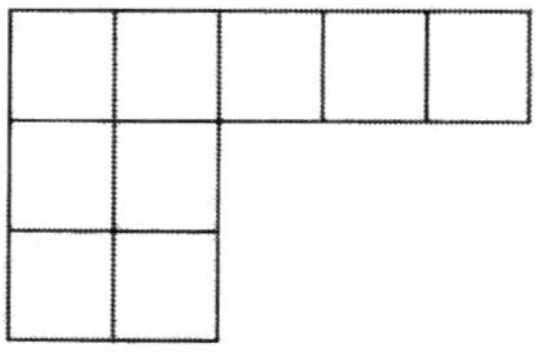

Then P_1 and P_3 are the maximal parabolic subgroups determined by the columns of λ, and $Q_\lambda = P_1 \cap P_3$.

Let T be a Young tableau of shape λ. For $1 \leqslant d \leqslant n-1$, let m_d be the number of columns in λ of length d, where $m_d \geqslant 0$. Let the m_d columns be denoted by $\tau_{d1}, \dots, \tau_{dm_d}$, $1 \leqslant d \leqslant n-1$.

DEFINITION 2.2.2. Given $w \in W^{Q_\lambda}$, T (or p_T) is said to be a *Young tableau on* $X(w)$ if $w^{(d)} \geqslant \tau_{dj}$, $1 \leqslant j \leqslant m_d$, $1 \leqslant d \leqslant n-1$.
(here, $w^{(d)}$ is given by $X(w^{(d)}) = \pi_d(X(w))$, π_d being the projection $G/B \to G/P_d$).

DEFINITION 2.2.3. Let T be a Young tableau of shape λ. Given $w \in W^{Q_\lambda}$, T (or p_T) is said to be a *standard tableau on* $X(w)$ if there exists a sequence in G/Q_λ

$$X(w) \supseteq X(\phi_{11}) \supseteq \cdots \supseteq X(\phi_{1m_1}) \supseteq X(\phi_{21}) \supseteq \cdots \supseteq X(\phi_{n-1m_{n-1}})$$

such that

$$(*) \qquad \pi_d(X(\phi_{dj})) = X(\tau_{dj}),\ 1 \leqslant j \leqslant m_d, 1 \leqslant d \leqslant n-1,$$

π_d being as in §1.2 (here, τ_{dj}, $1 \leqslant j \leqslant m_d$, $1 \leqslant d \leqslant n-1$ is the dj-th column of T; also, if $m_d = 0$ for some d, $1 \leqslant d \leqslant n-1$, then the family $\{\phi_{dj}, 1 \leqslant j \leqslant m_d\}$ is understood to be empty).

REMARK 2.2.4. For $X(w) = G/B$ the above definition coincides with the classical notion (cf. Definition 2.0.1).

REMARK 2.2.5. Going back to the example in 2.1.2, corresponding to

$T =$

1	2
3	

,

$X(231) \supset X(132)$ is the unique sequence satisfying $(*)$. Hence $X(321)$, $X(231)$ are the only two subvarieties on which p_T is standard. Thus p_T is standard on G/B; further, $p_T|_{X(312)} \neq 0$ (since the Plücker coordinates p_2, p_{13}, coming from G/P_1 and G/P_2 respectively, are nonzero on $X(312)$), but p_T is not standard on $X(312)$.

2.3. Notion of standard Young tableaux with respect to any indexing of the maximal parabolic subgroups

In Definition 2.2.2, we took the indexing of the maximal parabolic subgroups as in **[27]**. We shall now see how one can define the notion of standard Young tableaux with respect to any indexing of the maximal parabolic subgroups. To this end, we first observe that given $\tau_1 = (i_1, \ldots, i_{d_1})$, $\tau_2 = (j_1, \ldots, j_{d_2})$, if T is the Young tableau with two columns, where the first column is given by τ_2 and the second one by τ_1 (here we suppose that $d_2 \geqslant d_1$), then $\pi_{d_2}(\pi_{d_1}^{-1}(X(\tau_1))) \supseteq X(\tau_2)$ if and only if T is standard in the classical sense (here, π_d is the canonical projection $G/B \to G/P_d$). This leads to a partial order $\geqslant$ on $\bigcup_{d=1}^{n-1} W^d$ (here $W^d = W_{P_d}^{\min}$), namely given $\tau_1 = (i_1, \ldots, i_{d_1})$, $\tau_2 = (j_1, \ldots, j_{d_2})$,

$$\tau_1 \geqslant \tau_2 \quad \text{if} \quad d_1 \leqslant d_2, \quad \text{and} \quad i_t \geqslant j_t \quad \text{for all } t, \quad 1 \leqslant t \leqslant d_1.$$

2.3.1. Partial order on $\cup_P W^P$. Consider $\cup_P W^P$, where P runs over all the maximal parabolic subgroups,. Let $\tau \in W^P$, $\phi \in W^Q$, P, Q being two maximal parabolic subgroups.

We define $\tau \geqslant \phi$ if $\pi_Q(\pi_P^{-1}(X(\tau))) \supset X(\phi)$, where $\pi_P : G/B \to G/P$, $\pi_Q : G/B \to G/Q$ are the canonical projections. Let us take an arbitrary indexing of the maximal parabolic subgroups of G, say $P_1, \ldots, P_{n-1}$. Let Q be any parabolic subgroup. By changing the indexing if necessary, we may suppose that $Q = P_1 \cap P_2 \cap \cdots \cap P_r$.

DEFINITION 2.3.2. A *Young tableau in W^Q of type $\underline{m} = (m_1, \ldots, m_r)$*, where $m_i \geqslant 0$, $1 \leqslant i \leqslant r$, is a sequence $\lambda = (\lambda_{ij})$, with $\lambda_{ij} \in W^i$, $1 \leqslant j \leqslant m_i$, $1 \leqslant i \leqslant r$ (here $W^i = W_{P_i}^{\min}$).

DEFINITION 2.3.3. A Young tableau $\lambda = (\lambda_{ij})$ is said to be a *Young tableau on a Schubert variety $X(\phi) \subset G/Q$* if $\phi_i \geqslant \lambda_{ij}$ for all $1 \leqslant i \leqslant r$, $1 \leqslant j \leqslant m_i$, where $X(\phi_i)$ is the image of $X(\phi)$ under $G/Q \to G/P_i$.

DEFINITION 2.3.4. A Young tableau $\lambda = (\lambda_{ij})$ on $X(\phi)$, $\phi \in W^Q$ is said to be *standard* on $X(\phi)$ if there exists a sequence $\theta = (\theta_{ij})$ (which we call a *defining sequence for* λ) such that

(1) θ_{ij} is in W^Q, $1 \leqslant j \leqslant m_i$, $1 \leqslant i \leqslant r$

(2) each θ_{ij} is a lifting of λ_{ij} under $W/W_Q \to W/W_i$

(3) $X(\phi) \supseteq X(\theta_{11}) \supseteq X(\theta_{12}) \supseteq \cdots \supseteq X(\theta_{1m_1}) \supseteq X(\theta_{21}) \supseteq \cdots \supseteq X(\theta_{rm_r})$ (in G/Q).

DEFINITION 2.3.5. More generally, a Young tableau $\lambda = (\lambda_{ij})$ is said to be *standard on a union of Schubert varieties $Z = \cup_{i=1}^{t} X(\phi_i)$* in G/Q if λ is standard on $X(\phi_i)$ for some i, $1 \leqslant i \leqslant t$. If $m_t = 0$ for any t, $1 \leqslant t \leqslant r$, the family $\{\theta_{tj}, 1 \leqslant j \leqslant m_t\}$ is understood to be empty.

3. Linear independence of standard monomials

3.1. Maximal and minimal defining chains

We shall make use of the following result, due to V. V. Deodhar.

LEMMA 3.1.1. *Let Q be a parabolic subgroup of G, W_Q the Weyl subgroup of Q and W^Q the set of minimal representatives of W/W_Q in W. Then given $\sigma, \sigma' \in W^Q$ and $w \in W_Q$ such that $\sigma' \leqslant \sigma$, there exists a unique $w' = \eta(\sigma', \sigma, w)$ in W_Q such that $\sigma' w' \leqslant \sigma w$ and w' is maximal for this property, i.e. for any $w'' \in W_Q$, if $\sigma' w'' \leqslant \sigma w$, then $w'' \leqslant w'$ in W_Q.*

PROOF. We proceed by induction on $l(\sigma)$. If $l(\sigma) = 0$, then $\sigma' = \sigma = \text{id}$ and it is clear that $w' = w$ is the required element. Now let $l(\sigma) > 0$. Choose an $\alpha \in S$ such that $s\sigma < \sigma$, where $s = s_\alpha$. Note that $s\sigma \in W^Q$ (since $\sigma(\beta) \neq \alpha$, for any $\beta \in R_Q^+$, as $\sigma^{-1}(\alpha) < 0$). Now we distinguish the following cases.
Case 1: $s\sigma' < \sigma'$
In this case $s\sigma' < s\sigma$, and $s\sigma' \in W^Q$. Hence, by induction hypothesis $\eta(s\sigma', s\sigma, w)$ is defined. Define $\eta(\sigma', \sigma, w) = \eta(s\sigma', s\sigma, w)$. We shall now show that this element has the required properties. Call this element τ'. First note that for any τ in W_Q, $\sigma'\tau \leqslant \sigma w \iff s\sigma'\tau \leqslant s\sigma w$. Hence if $\sigma' w'' \leqslant \sigma w$, then $s\sigma' w'' < s\sigma w$, and we obtain $w'' \leqslant \tau'$. Also note that $\sigma'\tau' \leqslant \sigma w$ (since $s\sigma'\tau' \leqslant s\sigma w$).
Case 2: $s\sigma' > \sigma'$
By Chapter 3, Lemma 10.0.2, we get that $\sigma' \leqslant s\sigma$. Hence, by induction hypothesis $\eta(\sigma', s\sigma, w)$ is defined. Now define $\eta(\sigma', \sigma, w) = \eta(\sigma', s\sigma, w)$. We shall now show that this has the required properties. Call this element τ'. First note that for any τ in W_Q, $\sigma w \geqslant \sigma'\tau \iff s\sigma w \geqslant \sigma'\tau$. Hence if $\sigma' w'' \leqslant \sigma w$, then $\sigma' w'' < s\sigma w$, and we obtain $w'' < \tau'$. Also note that $\sigma'\tau' < \sigma w$ (since $\sigma'\tau' \leqslant s\sigma w$). □

COROLLARY 3.1.2. *Let $X(\phi)$ be a Schubert variety in G/Q, $\phi \in W/W_Q$. Let λ be a standard tableau on $X(\phi)$. Then we can find a unique maximal defining sequence θ^+ relative to $X(\phi)$ for λ, i.e. θ is a defining sequence for λ relative to $X(\phi)$ (i.e. $\phi \geqslant \theta_{11}$) implies that $\theta^+ \geqslant \theta$ (here, if $\theta^+ = (\theta^+_{ij})$, $\theta = (\theta_{ij})$, by $\theta^+ \geqslant \theta$ we mean $\theta^+_{ij} \geqslant \theta_{ij}$, $1 \leqslant j \leqslant m_i$, $1 \leqslant i \leqslant r$).*

PROOF. This is an immediate consequence of the Lemma above. Let $\lambda = (\lambda_{ij})$ and $\theta = (\theta_{ij})$. Define $\theta^+ = (\theta^+_{ij})$ inductively as follows. We are given that $\phi \geqslant \theta_{11}$ and the image of θ_{11} is λ_{11}. By the lemma, we can find a unique maximal element, say θ^+_{11}, such that $\phi \geqslant \theta^+_{11}$ and the image of $\theta^+_{11}$is λ_{11}. Clearly, we have $\theta^+_{11} \geqslant \theta_{11}$. Now choose θ^+_{12} to be maximal within θ^+_{11} such that the image of θ^+_{12} is λ_{12}, etc.. This completes the proof. □

We note that the above Lemma and its Corollary admit dual versions as follows

LEMMA 3.1.3. *Let Q be any parabolic subgroup of G. Let W_Q and W^Q be as in the above Lemma. Then given $\sigma, \sigma' \in W^Q$ and $w \in W_Q$ with $\sigma' \geqslant \sigma$, there exists* *a unique $w' = \mu(\sigma', \sigma, w)$ in W_Q such that $\sigma' w' \geqslant \sigma w$ and w' is minimal for this property, i.e. for any $w'' \in W_Q$, if $\sigma' w'' \geqslant \sigma w$, then $w'' \geqslant w'$ in W_Q.*

COROLLARY 3.1.4. *Let $X(\phi)$ be a Schubert variety in G/Q, $\phi \in W/W_Q$. Let λ be a standard Young tableau on $X(\phi)$. Then we can find a unique minimal defining sequence θ^- for λ, i.e. if θ is any defining sequence for λ, then $\theta^- \leqslant \theta$. In fact, this minimal defining sequence for λ depends only on λ, but not on $X(\phi)$ (In other words, θ^- is an* absolute *minimal defining sequence for λ, unlike the relative maximal defining sequence θ^+ for λ which depends on $X(\phi)$. In particular, $\theta^- \leqslant \theta^+$).*

PROOF. The proof is just the same as that of Corollary 3.1.2. Let $\theta = (\theta_{ij})$ be any defining sequence for λ. Now define $\theta^- = (\theta^-_{ij})$ inductively as follows. We are given that the canonical image of θ_{ij} is λ_{ij}. Take $\theta^-_{rm_r}$ to be the minimal representative of λ_{rm_r} and obviously $\theta_{rm_r} \geqslant \theta^-_{rm_r}$. Now by Lemma 3.1.3, we choose the unique minimal element, say $\theta^-_{rm_r-1}$, such that $\theta^-_{r,m_r-1} \geqslant \theta_{r,m_r}$ and the image of θ^-_{r,m_r-1} is λ_{r,m_r-1}. Clearly $\theta_{r,m_r-1} \geqslant \theta^-_{r,m_r-1}$, and the proof is completed by a repeated use of the Lemma. □

Almost always, the relative maximal defining sequence θ^+ is different from the absolute minimal defining sequence for λ.

EXAMPLE 3.1.5. Let $G = SL(5)$, $\phi = (54312)$. Let $\lambda = (\lambda_1, \lambda_2, \lambda_3)$, where $\lambda_1 = 5$, $\lambda_2 = (14)$, $\lambda_3 = (124)$. Then $\theta^- = (\theta_1^-, \theta_2^-, \theta_3^-)$ is given by $\theta_1^- = (54123)$, $\theta_2^- = (14235)$, $\theta_3^- = (12435)$, and $\theta^+ = (\theta_1^+, \theta_2^+, \theta_3^+)$, the unique maximal defining sequence relative to $X(\phi)$, is given by $\theta_1^+ = (54312)$, $\theta_2^+ = (41532)$, $\theta_3^+ = (41253)$.

REMARK 3.1.6. A purely combinatorial description of θ^- may be found in **[112]**.

3.2. The element $p(\lambda)$

Let L_i be the ample generator of Pic (G/P_i). As seen in Chapter 6, the extremal weight vectors in $H^0(G/P_i, L_i)$ give a k-basis for $H^0(G/P_i, L_i)$, which we shall denote by $\{p_\tau, \tau \in W^i\}$ (these are simply the Plücker coordinates). Given a Young diagram $\lambda = (\lambda_{ij})$, we shall set

$$p(\lambda) = \prod_{i=1}^{r} \prod_{j=1}^{m_i} p_{\lambda_{ij}}$$

(note that $p(\lambda) \in H^0(G/Q, L_{\underline{m}})$, where $L_{\underline{m}} = L_1^{m_1} \otimes \cdots \otimes L_r^{m_r}$). Such a monomial will be called *standard on Z* if λ is standard on Z (cf. Definition 2.3.5).

THEOREM 3.2.1. *Let G, Q, Z as above. Then the standard monomials on Z of length $\underline{m} = (m_1, \ldots, m_r)$ form a basis of $H^0(Z, L_{\underline{m}})$, where $L_{\underline{m}} = L_1^{m_1} \otimes \cdots \otimes L_r^{m_r}$.*

The rest of this chapter is devoted to proving the above Theorem and deducing the geometric consequences – normality, Cohen-macaulayness and ideal theory. We carry out the proof in several steps.

DEFINITION 3.2.2. Let $\phi_1, \ldots, \phi_s \in W$ and $Y = X(\phi_1) \cup \cdots \cup X(\phi_s)$ be the schematic union of Schubert varieties (note that Y is reduced). For $\mathbf{a} = (a_1, \ldots, a_l) \in \mathbb{Z}_+^l$ (l being the rank of G), let $S(Y, \mathbf{a})$ denote the set of standard monomials on Y of multidegree $\mathbf{a}$, i.e. the set of all monomials of multidegree $\mathbf{a}$ standard on $X(\phi_t)$ for some t, $1 \leqslant t \leqslant s$. Let $s(Y, \mathbf{a})$ be the cardinality of $S(Y, \mathbf{a})$, and $h^0(Y, \mathbf{a}) = \dim H^0(Y, L(\pi))$, $\pi = \sum_i a_i \omega_i$.

PROPOSITION 3.2.3. *The set of standard monomials on Y is linearly independent; in particular, for all $\mathbf{a} = (a_1, \ldots, a_n) \in \mathbb{Z}_+^n$, we have*

$$h^0(Y, L(\pi)) \geqslant s(Y, \mathbf{a}), \quad \pi = \sum_i a_i \omega_i.$$

PROOF. We have $Y = X(\phi_1) \cup \cdots \cup X(\phi_s)$, $\phi_i \in W$. The proof is by induction on a_1. Suppose that $a_1 = 0$. Let $Q = P_2 \cap \cdots \cap P_l$, and let $X(\phi_i')$ be the image of $X(\phi_i)$ under $\eta : G/B \to G/Q$. Let $Y' = X(\phi_1') \cup \cdots \cup X(\phi_s') = \eta(Y)$. We find that any standard monomial of multidegree $(0, a_2, \ldots, a_n)$ on $X(\phi_i)$ can be canonically identified with a standard monomial of multidegree $(a_2, \ldots, a_n)$ on $X(\phi_i')$. This is true for every $i = 1, \ldots, s$. Thus, in order to prove the result for Y, it suffices to prove it for Y'. But Y' is a union of Schubert varieties in G/Q, and so by decreasing induction on the rank of Q, we may assume that the result is true for Y'. (If $a_2 = a_3 = \cdots = a_{l-1} = 0$, then $Q = P_l$ is a maximal parabolic subgroup and the result follows from the results in Chapter 6; in fact, our proof below includes this case also). Thus we may assume $a_1 \geqslant 1$. We have, by induction hypothesis, that

$$h^0(Y, L(\pi')) \geqslant s(Y, \mathbf{a}'), \text{where } \mathbf{a}' = (a_1 - 1, a_2, \ldots, a_l).$$

Let $p(\lambda_{ij}^k)$, $1 \leqslant k \leqslant t$ be a minimal set of linearly dependent standard monomials on Y, where for $1 \leqslant k \leqslant t$, $\lambda_{ij}^k \in W^i$, $1 \leqslant j \leqslant a_i$, $1 \leqslant i \leqslant l$, and

$$p(\lambda_{ij}^k) = \prod_{i=1}^{n} \prod_{j=1}^{a_i} p_{\lambda_{ij}^k}.$$

Let (θ_{ij}^k) be a defining sequence for $p(\lambda_{ij}^k)$. We note that for each k, $\theta_{11}^k \leqslant \phi_m$ for some m, $1 \leqslant m \leqslant s$. We distinguish two cases.
Case 1:
Let $\lambda_{11}^1 = \lambda_{11}^2 = \cdots = \lambda_{11}^t = \lambda$, say. By definition, each $p(\lambda_{ij}^k)$, $1 \leqslant k \leqslant t$, is standard on a Schubert variety Y_k which is an irreducible component of Y. Let $Z = Y_1 \cup \cdots \cup Y_t$. Observe that $p(\lambda_{ij}^k)$, $1 \leqslant k \leqslant t$, is standard on Z.

Write $p(\lambda_{ij}^k) = p_\lambda p'(\lambda_{ij}^k)$ in an obvious way and observe that the $p'(\lambda_{ij}^k)$ are standard monomials of multidegree $\mathbf{a}' = (a_1 - 1, a_2, \ldots, a_l)$ on Z, and hence are linearly independent (by induction). Since Z is reduced and $p_\lambda \neq 0$ on Y_k ($1 \leqslant k \leqslant t$), we find that a non-trivial dependency relation for the $p(\lambda_{ij}^k)$'s gives one for the $p'(\lambda_{ij}^k)$'s, which is a contradiction.
Case 2:

Let λ be a minimal element in the set $\{\lambda_{11}^1, \dots, \lambda_{11}^t\}$. Let $\lambda = \lambda_{11}^1 = \cdots = \lambda_{11}^{t_0}$ and $\lambda \not\geqslant \lambda_{11}^{t_0+1}, \dots, \lambda_{11}^t$, where $1 \leqslant t_0 < t$. Let $Y_0 = X(\theta_{11}^1) \cup \cdots \cup X(\theta_{11}^{t_0})$. By definition, $\eta_1(Y_0) = X(\lambda)$ in G/P_1, where $\eta_1 : G/B \to G/P_1$ and $p_{\lambda_{11}^k} = 0$ on $X(\lambda)$ for $t_0 + 1 \leqslant k \leqslant t$. Hence we get

$$p(\lambda_{ij}^k) = 0, \text{ on } Y_0, \text{ for } t_0 + 1 \leqslant k \leqslant t.$$

Moreover, $p(\lambda_{ij}^k)$, $1 \leqslant k \leqslant t_0$, are standard on Y_0 and are linearly independent on Y_0, by *case 1*. Now start with a dependency relation on Y, say

$$c_1 p(\lambda_{ij}^1) + \cdots + c_t p(\lambda_{ij}^t) = 0, \; c_i \in k^*.$$

Restricting this relation to Y_0, we get the nontrivial relation

$$c_1 p(\lambda_{ij}^k) + \cdots + c_{t_0} p(\lambda_{ij}^{t_0}) = 0 \text{ on } Y_0,$$

which implies $c_i = 0$, $1 \leqslant i \leqslant t_0$. But this contradicts the minimality assumption on $\{p(\lambda_{ij}^k)\}$. This completes the proof. □

4. A basis for $H^0(X, L)$ and vanishing of higher cohomology groups

4.1. Pieri's Formula

Let $\mathrm{Ch}(G/B)$ denote the *Chow ring* of G/B. For $\tau \in W$, let $[X(\tau)]$ denote the element of $\mathrm{Ch}(G/B)$ determined by the Schubert variety $X(\tau)$. We have (cf. **[37]**), the elements $[X(\tau)]$, $\tau \in W$ form a basis of $\mathrm{Ch}(G/B)$ considered as a $\mathbb{Z}$-module. Let $P \supset B$ be a maximal parabolic subgroup with ω (resp. α) as the associated fundamental weight (resp. simple root). Let H_P be the unique codimension one Schubert subvariety of G/P. Then, $X(s_\alpha w_0)$ (w_0 being the unique element of maximal length in W) is the inverse image of H_P under the canonical projection $\pi : G/B \to G/P$. Consider now the multiplication $[X(\tau)] \cdot [X(s_\alpha w_0)]$ in the ring $\mathrm{Ch}(G/B)$. Let $[X(\tau)] \cdot [X(s_\alpha w_0)] = \sum d_i [X(\tau_i)]$, where the sum runs over all the Schubert divisors $X(\tau_i)$ in $X(\tau)$. Let $\tau_i = \tau s_{\beta_i}$, for some positive root β_i. Chevalley has shown (cf. **[37]**) that $d_i = (\omega, \beta_i^*)$. Since all of the fundamental weights of $SL(n)$ are minuscule (cf. §2), it follows that $0 \leqslant d_i \leqslant 1$. Thus we obtain

$$(*) \qquad [X(\tau)] \cdot [X(s_\alpha w_0)] = \sum d_i [X(\tau_i)], \; d_i \neq 0,$$

where the summation runs over a certain set of Schubert divisors in $X(\tau)$.

Let $\pi(X(\tau)) = X_P(\theta)$, where $\theta \in W^P$. Suppose $\dim X_P(\theta) > 0$. Then one knows (cf. Chapter 6, §4.3.2) that $X_P(\theta) \cap \{p_\theta = 0\}$ is precisely the union of all Schubert divisors in $X_P(\theta)$. Hence we obtain
$X(\tau) \cap \{p_\theta = 0\} = X(\tau) \cap (\pi^{-1}(X_P(\theta) \cap \{p_\theta = 0\}))$. Thus we get

$$(**) \qquad X(\tau) \cap \{p_\theta = 0\} = \bigcup_i X(\tau_i), \; \tau_i \text{ as in } (*).$$

Further, from (*) it follows that the intersection multiplicity of the left hand side of (**) along any $X(\tau_i)$ is 1 (refer to **[67]** for the definition of intersection multiplicity).

This together with the normality of $X(\tau)$ implies that the scheme-theoretic intersection $X(\tau) \cap \{p_\theta = 0\}$ is *reduced* (cf. [**148**]). Thus we obtain that the equality in (**) is scheme-theoretic.

4.2. Preparatory lemmas

If Y is a union of Schubert varieties, let $S(Y, \mathbf{a})$ denote the set of standard monomials on Y of mutidegree $\mathbf{a}$, and let $s(Y, \mathbf{a}) = \#S(Y, \mathbf{a})$.

LEMMA 4.2.1. *Let* $\phi_i, \psi_j \in W$, $1 \leqslant i \leqslant s$, $1 \leqslant j \leqslant t$, $Y_1 = \bigcup_i X(\phi_i)$, $Y_2 = \bigcup_j X(\psi_j)$, *and* $\mathbf{a} = (a_1, \ldots, a_n) \in \mathbb{Z}_+^n$. *Then we have*

$$s(Y_1 \cup Y_2, \mathbf{a}) = s(Y_1, \mathbf{a}) + s(Y_2, \mathbf{a}) - s\big((Y_1 \cap Y_2)_{\text{red}}, \mathbf{a}\big).$$

PROOF. Recall that $(Y_1 \cap Y_2)_{\text{red}}$ is a union of Schubert varieties, say

$$(Y_1 \cap Y_2)_{\text{red}} = \bigcup_{i,j} X(\sigma_{ij}),$$

for suitable $\sigma_{ij} \in W$ such that $\phi_i, \psi_j \geqslant \sigma_{ij}$. We see that we only have to check that if $p(\theta_{ij})$ is a standard monomial of multidegree a on $X(\phi_\alpha)$ and also on $X(\psi_\beta)$ for some α, β with $1 \leqslant \alpha \leqslant s$ and $1 \leqslant \beta \leqslant t$, then $p(\theta_{ij})$ is standard on $X(\sigma_{\alpha\beta})$ for some $\sigma_{\alpha\beta}$. Look at the minimal defining sequence (λ_{ij}^-) for the standard Young tableau (θ_{ij}), which depends only on (θ_{ij}), but not on ϕ_α or ϕ_β (cf. Corollary 3.1.4). We have

$$\phi_\alpha \geqslant \lambda_{11}^- \text{and } \psi_\beta \geqslant \lambda_{11}^- \quad (\text{in } W).$$

But then we know that $\phi_\alpha, \psi_\beta \geqslant \sigma_{\alpha\beta} \geqslant \lambda_{11}^-$ for some $\sigma_{\alpha\beta}$, and hence $p(\theta_{ij})$ is standard on $X(\sigma_{\alpha\beta})$, as required. □

LEMMA 4.2.2. *Let* $\mathbf{a} = (a_1, \ldots, a_l) \in \mathbb{Z}_+^n$ *be such that* $a_1 \geqslant 1$. *Let* $\mathbf{a}' = (a_1 - 1, a_2, \ldots, a_l)$. *Let* $P = P_1$, *and* τ, τ_i *be as in §4.1. Then we have*

$$s(X(\tau), \mathbf{a}) - s(X(\tau), \mathbf{a}') = s\Big(\bigcup_{i=1}^{r} X(\tau_i), \mathbf{a}\Big).$$

PROOF. It is clear that by the mapping $q \mapsto p_\tau q$, we can identify $S(X(\tau), \mathbf{a}')$ with the subset of elements of $S(X(\tau), \mathbf{a})$ which begin with p_τ. It is also clear, since $X(\tau_i)$ are the maximal lifts in $X(\tau)$ of the set of all Schubert divisors of $X_P(\theta)$, θ being as in §4.1 that the subset of elements of $S(X(\tau), \mathbf{a})$ which do not begin with p_τ can be identified with $S\big(\bigcup_{i=1}^r X(\tau_i), \mathbf{a}\big)$. The required result now follows. □

If Z is a closed subscheme of G/B, let us denote by $\mathcal{O}_Z(a)$ and $\mathcal{O}_Z(m)$ the invertible sheaves on Z, associated respectively to the line bundle $L(\lambda)|_Z$ and $L(m\rho)|_Z$, where $\lambda = \sum a_i\omega_i$, $\rho = \sum \omega_i$ and $m \in \mathbb{Z}$. Note that $\mathcal{O}_Z(1)$ is a very ample sheaf on Z. If Y is a union of Schubert varieties in G/B (endowed with its canonical reduced structure), we write $S(Y, m\rho)$, $s(Y, m\rho)$, etc. for $S\big(Y, (m, \ldots, m)\big)$, etc..

LEMMA 4.2.3. *Let Y_1, Y_2 be unions of Schubert varieties of G/B. Suppose that for $m \gg 0$ we have*

$$h^0(Y_i, m\rho) = s(Y_i, m\rho), \quad i = 1, 2.$$

Then we have

(a) the scheme-theoretic intersection $Y_1 \cap Y_2$ is reduced.

(b) $h^0(Y_1 \cup Y_2, m\rho) = s(Y_1 \cup Y_2, m\rho)$, and $h^0(Y_1 \cap Y_2, m\rho) = s(Y_1 \cap Y_2, m\rho)$, for $m \gg 0$.

PROOF. We have the following exact sequence of sheaves of $\mathcal{O}_{G/B}$-modules

$$0 \to \mathcal{O}_{Y_1 \cup Y_2} \to \mathcal{O}_{Y_1} \oplus \mathcal{O}_{Y_2} \to \mathcal{O}_{Y_1 \cap Y_2} \to 0,$$

where $Y_1 \cap Y_2$ (resp. $Y_1 \cup Y_2$) denotes the scheme-theoretic intersection (resp. union). This gives the exact sequence

$$0 \to \mathcal{O}_{Y_1 \cup Y_2}(m) \to \mathcal{O}_{Y_1}(m) \oplus \mathcal{O}_{Y_2}(m) \to \mathcal{O}_{Y_1 \cap Y_2}(m) \to 0.$$

Consider the cohomology exact sequence, and using Serre's vanishing theorem (cf. [**147**]), we obtain for $m \gg 0$

$$(*) \qquad h^0(Y_1 \cap Y_2, m\rho) = h^0(Y_1, m\rho) + h^0(Y_2, m\rho) - h^0(Y_1 \cup Y_2, m\rho).$$

We have, by Lemma 3.2.3

$$h^0(Y_1 \cup Y_2, m\rho) \geqslant s(Y_1 \cup Y_2, m\rho).$$

Thus, in view of hypothesis, we get for $m \gg 0$,

$$h^0(Y_1, m\rho) + h^0(Y_2, m\rho) - h^0(Y_1 \cup Y_2, m\rho) \leqslant s(Y_1, m\rho) + s(Y_2, m\rho) - s(Y_1 \cup Y_2, m\rho).$$

Now by Lemma 4.2.1, we have,

$$s(Y_1, m\rho) + s(Y_2, m\rho) - s(Y_1 \cup Y_2, m\rho) = s((Y_1 \cap Y_2)_{\text{red}}, m\rho).$$

Thus we conclude that

$$(**) \qquad h^0(Y_1 \cap Y_2, m\rho) \leqslant s\big((Y_1 \cap Y_2)_{\text{red}}, m\rho\big).$$

On the other hand, we see that if $Y_1 \cap Y_2$ is not reduced, then

$$h^0(Y_1 \cap Y_2, m\rho) > h^0\big((Y_1 \cap Y_2)_{\text{red}}, m\rho\big), \qquad \text{for } m \gg 0.$$

Also, by Lemma 3.2.3, we have

$$s\big((Y_1 \cap Y_2)_{\text{red}}, m\rho\big) \leqslant h^0\big((Y_1 \cap Y_2)_{\text{red}}, m\rho\big),$$

and hence we obtain

$$h^0(Y_1 \cap Y_2, m\rho) > s((Y_1 \cap Y_2)_{\text{red}}, m\rho), \qquad \text{for } m \gg 0.$$

This contradicts (**), and we conclude that $Y_1 \cap Y_2$ is reduced.

The fact that $Y_1 \cap Y_2$ is reduced together with (**) and linear independence of standard monomials on $Y_1 \cap Y_2$ (cf. Proposition 3.2.3) implies that $h^0(Y_1 \cap Y_2, m\rho) = s(Y_1 \cap Y_2, m\rho)$, for $m \gg 0$. Also, (*) is satisfied with h^0 replaced by s (cf. Lemma

4.2.1). From this we obtain (in view of hypothesis) that $h^0(Y_1 \cup Y_2, m\rho) = s(Y_1 \cup Y_2, m\rho)$, $m \gg 0$. $\square$

LEMMA 4.2.4. *Let S_d denote the set of all reduced subschemes Y of G/B such that every irreducible component of Y is a Schubert variety of dimension $\leqslant d$. Suppose that for every Schubert variety $X \in S_d$ we have*

$$(*) \qquad s(X, \mathbf{a}) = h^0(X, L(\pi)),\ \pi = \sum a_i\omega_i,\ \mathbf{a} = (a_1, \ldots, a_l), a_i \geqslant 0.$$

Then

(i) for any $Y_1, Y_2 \in S_d$, $Y_1 \cap Y_2$ is reduced

(ii) the assertion $(*)$ holds for any $Y \in S_d$.

PROOF. Suppose that $Y \in S_d$. Then by induction on the number of components of Y, it follows, (as an immediate extension of the argument of Lemma 4.2.3), that $h^0(Y, m\rho) = s(Y, m\rho)$ for $m \gg 0$, and by Lemma 4.2.3, the assertion (i) follows. Thus, it only remains to prove the assertion (ii).

We observe that (ii) obviously holds for S_0. We now prove (ii) by induction on t. Let $t \leqslant d$ and suppose that (ii) holds for every $X \in S_m$, $m < t$; then we shall show that (ii) holds also for every $X \in S_t$. We shall make another inductive argument, namely, if $C(X)$ denotes the number of irreducible components of $X \in S_t$, we suppose that (ii) is true for $X \in S_t$ such that $C(X) \leqslant r-1$. We now take $X \in S_t$ such that $C(X) = r$ and will show that (ii) holds for X. Let then

$$X = X_1 \cup \cdots \cup X_r,$$

where X_i are the distinct irreducible components of X. We set

$$Y_1 = X_1 \cup \cdots \cup X_{r-1} \quad \text{and} \quad Y_2 = X_r.$$

By (i), $Y_1 \cap Y_2$ is reduced, and we get the following exact sequence of $\mathcal{O}_{G/B}$-modules

$$(1) \qquad 0 \to \mathcal{O}_X(\mathbf{a}) \to \mathcal{O}_{Y_1}(\mathbf{a}) \oplus \mathcal{O}_{Y_2}(\mathbf{a}) \to \mathcal{O}_{Y_1 \cap Y_2}(\mathbf{a}) \to 0.$$

We see that $Y_1 \cap Y_2 \in S_{t-1}$ since X_i are the distinct irreducible components of X. By our inductive hypothesis, (ii) holds for Y_1, Y_2 and $Y_1 \cap Y_2$. Hence $H^0(Y_1 \cap Y_2, L(\pi))$ has a basis of standard monomials, where $\pi = \sum a_i\omega_i$. This implies that the canonical mapping

$$H^0(\mathcal{O}_{Y_1}(\mathbf{a}) \oplus \mathcal{O}_{Y_2}(\mathbf{a})) \longrightarrow H^0(\mathcal{O}_{Y_1 \cap Y_2}(\mathbf{a}))$$

is surjective. Writing the cohomology exact sequence of (1), we obtain

$$(2) \qquad h^0(Y_1, L(\pi)) + h^0(Y_2, L(\pi)) = h^0(X, L(\pi)) + h^0(Y_1 \cap Y_2, L(\pi)).$$

On the other hand, we have

$$h^0(Y_i, L(\pi)) = s(Y_i, \mathbf{a}), \qquad i = 1, 2,$$
$$h^0(Y_1 \cap Y_2, L(\pi)) = s(Y_1 \cap Y_2, \mathbf{a}).$$

Hence, from (2) we get

$$(3) \qquad h^0(X, L(\pi)) = s(Y_1, \mathbf{a}) + s(Y_2, \mathbf{a}) - s(Y_1 \cap Y_2, \mathbf{a}).$$

By Lemma 4.2.1, we have

$$s(Y_1, \mathbf{a}) + s(Y_2, \mathbf{a}) - s(Y_1 \cap Y_2, \mathbf{a}) = s(Y_1 \cup Y_2, \mathbf{a}) \ (X = Y_1 \cup Y_2).$$

Hence we get that

$$h^0(X, L(\pi)) = s(X, \mathbf{a}).$$

This completes the proof of the Lemma. □

REMARK 4.2.5. In Lemma 4.2.4, if in addition to $(*)$ we suppose also that for every Schubert variety $X \in S_d$

$$H^i(X, L(\pi)) = 0, \ i > 0, \pi \geqslant 0,$$

then the same proof gives also, in addition,
(iii) $H^i(Y, L(\pi)) = 0, \ i > 0, \pi \geqslant 0, \ Y \in S_d$.

4.3. Basis for $H^0(X, L)$

We shall now state and prove the main basis theorem:

THEOREM 4.3.1. *Let X be a union of Schubert subvarieties of G/B, where $G = SL(n)$. Then we have*

$$(*) \qquad h^0(X, L(\pi)) = s(X, \mathbf{a}), \ \pi \geqslant 0, \ \pi = \sum a_i \omega_i, \ \mathbf{a} = (a_1, \ldots, a_l),$$

i.e. $H^0(X, L(\pi))$ has a basis consisting of its standard monomials.

PROOF. We prove the Theorem by induction on $\dim X$. If $\dim X = 0$, then clearly the Theorem is true. We suppose that the Theorem is true for every $Y \in S_t$, where $t < \dim X$. By Lemma 4.2.4, it suffices to prove $(*)$ for the case when X is a Schubert variety. Suppose then that $X = X(\tau)$, $\tau \in W$.

Let now $\pi = \sum a_i \omega_i$ $(\pi \geqslant 0)$ be such that $a_k = 0$, $k < i$. Let Q_i be the parabolic subgroup $Q_i = P_i \cap P_{i+1} \cap \cdots \cap P_l$ and $g : G/B \to G/Q_i$ the canonical morphism of G/B onto G/Q_i. We see that $L(\pi)$ descends to a line bundle on G/Q_i, which again we denote by $L(\pi)$. Let Y be the Schubert variety in G/Q_i which is the image of X (under g). We see that standard monomials on X of type $\mathbf{a}$ are precisely standard monomials on Y of type $\mathbf{a}$. One knows that Y is normal (cf. **[150]**). From this it follows that if $f : X \to Y$ is the canonical morphism, then $f_* \mathcal{O}_X = \mathcal{O}_Y$ (there is an open subset of X, namely the big cell which is a "product" over the big cell in Y, which shows that the function field of Y is algebraically closed in that of X, etc.; see also **[150]**). This fact implies that $H^0(X, L(\pi))$ is isomorphic to $H^0(Y, L(\pi))$. Thus it suffices to prove the theorem for the Schubert variety Y in G/Q_i. Suppose that $i = l$ so that $Q_l = P_l$. Then we are reduced to the case of one maximal parabolic subgroup, where $(*)$ is known (cf. Chapter 6, Theorem 5.0.5). We do one more inductive argument, namely assume $(*)$ for $\pi = \sum a_i \omega_i$, $\pi \geqslant 0$ of the form

$$a_k = 0, \ k < (i+1)$$

or what can be termed as assuming $(*)$ for Schubert varieties in G/Q_k, $k \geqslant (i+1)$. Thus we may suppose that $(*)$ holds when $a_1 = 0$.

Let $X_{P_1}(\theta)$ be the image of $X(\tau)$ in G/P_1 and $X(\tau_i)$ the Schubert divisors in $X(\tau) \cap \{p_\theta = 0\}$ (cf. §4.1). Let $H(\tau)$ be the union of the Schubert varieties $X(\tau_i)$

$$H(\tau) = \bigcup_i X(\tau_i).$$

Then by the scheme-theoretic Pieri's formula (cf. 4.1), we get the exact sequence

$$(1) \qquad 0 \to \mathcal{O}_{X(\tau)}(-1, 0, 0, \ldots, 0) \to \mathcal{O}_{X(\tau)} \to \mathcal{O}_{H(\tau)} \to 0.$$

Tensoring with $L(\pi)$, we obtain the exact sequence

$$(2) \qquad 0 \to \mathcal{O}_{X(\tau)}(\mathbf{a}') \to \mathcal{O}_{X(\tau)}(\mathbf{a}) \to \mathcal{O}_{H(\tau)}(\mathbf{a}) \to 0,$$

where $\mathbf{a}' = (a_1 - 1, a_2, \ldots, a_l)$. By our induction hypothesis, we have $s(H(\tau), \mathbf{a}) = h^0(H(\tau), \mathbf{a})$, which implies, in particular, that the canonical mapping

$$H^0\big(X(\tau), L(\pi)\big) \to H^0\big(H(\tau), L(\pi)\big)$$

is surjective. Writing the cohomology exact sequence of (2), we get the exact sequence

$$0 \to H^0\big(X(\tau), L(\pi')\big) \to H^0\big(X(\tau), L(\pi)\big) \to H^0\big(H(\tau), L(\pi)\big) \to 0,$$

where $\pi' = (a_1 - 1)\omega_1 + a_2\omega_2 + \cdots + a_l\omega_l$. This gives

$$(3) \qquad h^0\big(X(\tau), L(\pi)\big) = h^0\big(X(\tau), L(\pi')\big) + h^0\big(H(\tau), L(\pi)\big).$$

Now we prove $(*)$ by induction on a_1. If $a_1 = 0$, as we observed above, $(*)$ holds. Let then $a_1 > 0$. Then, by induction, we may suppose that

$$h^0\big(X(\tau), L(\pi')\big) = s(X(\tau), \mathbf{a}'),$$

π' being as above. Besides, by our induction hypothesis, we also have (as observed above)

$$h^0\big(H(\tau), L(\pi')\big) = s(H(\tau), \mathbf{a}).$$

Hence (3) gives

$$(4) \qquad h^0\big(X(\tau), L(\pi)\big) = s(X(\tau), \mathbf{a}') + s(H(\tau), \mathbf{a}).$$

On the other hand, by Lemma 4.2.2 we have

$$s(X(\tau), \mathbf{a}) = s(X(\tau), \mathbf{a}') + s(H(\tau), \mathbf{a}),$$

which implies that

$$s(X(\tau), \mathbf{a}) = h^0\big(X(\tau), L(\pi)\big).$$

This proves $(*)$ for $X(\tau)$ and concludes the proof of the theorem. □

COROLLARY 4.3.2. *Let X_1, X_2 be unions of Schubert varieties in G/B. Then the scheme theoretic intersection $X_1 \cap X_2$ is reduced.*

This is an immediate consequence of Lemma 4.2.4 and Theorem 4.3.1.

4.4. Vanishing Theorems

THEOREM 4.4.1. *Let X be a union of Schubert varieties in G/B, endowed with its canonical reduced structure. Then we have*

$$(*) \qquad H^i(X, L(\pi)) = 0, \quad i > 0, \ \pi \geqslant 0.$$

PROOF. In view of Remark 4.2.5, we see that it suffices to prove $(*)$ when X is a Schubert variety $X(\tau)$, $\tau \in W$. The proof is carried out in the same spirit as that of Theorem 5.0.5, Chapter 6. Consider the following exact sequence (the exact sequence (2) in the proof of Theorem 4.3.1)

$$(1) \qquad 0 \to \mathcal{O}_{X(\tau)}(\mathbf{a}') \to \mathcal{O}_{X(\tau)}(\mathbf{a}) \to \mathcal{O}_{H(\tau)}(\mathbf{a}) \to 0,$$

where $\pi = \sum a_i \pi_i$ and $\mathbf{a}' = (a_1 - 1, a_2, \dots, a_l)$. Setting $\pi' = \sum a_i{}' \omega_i$, by a suitable induction hypothesis, we can suppose that

(a) $H^i(X(\tau), L(\pi')) = 0, \quad i > 0,$

(b) $H^i(H(\tau), L(\pi)) = 0, \quad i > 0$. Then writing the cohomology sequence for (1), $(*)$ would follow: but there is one crucial point, namely that we may *not* have $\pi' \geqslant 0$. But in this case $a_1 = 0$, and this case could be treated by taking the image of $X(\tau)$ in G/Q, Q being a suitable parabolic subgroup, on the same lines as in the proof of Theorem 4.3.1. □

5. Ideal theory of Schubert varieties

In view of Corollary 4.3.2, we obtain that the intersection of a family of Schubert varieties is reduced. (Also note that a union of Schubert varieties is obviously reduced.) In particular this enables us to compute the ideal sheaves of Schubert varieties as given by Theorem 5.1.3 below.

Let P_i, $1 \leqslant i \leqslant l$ be the maximal parabolic subgroups of G with ω_i as the associated fundamental weight. We have the Plücker embedding of the Grassmannian, namely

$$G/P_i \hookrightarrow \mathbb{P}(H^0(G/P_i, L(\omega_i))^*),$$

$H^0(G/P_i, L(\omega_i))^*$ being the dual of $H^0(G/P_i, L(\omega_i))$. The homogeneous coordinate ring R_i of G/P_i for the above embedding is given by

$$R_i = \bigoplus_{r \geqslant 0} H^0(G/P_i, L(r\omega_i))$$

(cf. Chapter 6, Corollary 5.0.6). By the "diagonal embedding" of the flag variety

$$G/B \longrightarrow \prod_{1 \leqslant i \leqslant n} G/P_i,$$

we consider the closed immersion

$$G/B \hookrightarrow \prod_{1 \leqslant i \leqslant l} \mathbb{P}^{m_i}, \qquad \mathbb{P}^{m_i} = \mathbb{P}(H^0(G/P_i, L(\omega_i))^*.$$

We denote by L_i the ample generator of Pic $(\mathbb{P}^{m_i})$. The restriction of L_i to G/P_i is $L(\omega_i)$. We denote by S the multi-homogeneous coordinate ring of $Z := \prod_i \mathbb{P}^{m_i}$, i.e.

$$S = \oplus_{\mathbf{a}} S_{\mathbf{a}},\ S_{\mathbf{a}} = H^0(Z, L_1^{a_1} \otimes \cdots \otimes L_l^{a_l}),\ \mathbf{a} = (a_1, \dots, a_l),\ a_i \geqslant 0.$$

Let $\mathbb{A} = \prod_i \mathbb{A}^{m_i+1}$. We have

$$\begin{aligned} S &= k[\dots, x_0^i, \dots, x_{m_i}^i, \dots] \\ \mathbb{A} &= \operatorname{Spec} S \\ \mathbb{A}^{m_i+1} &= \operatorname{Spec} k[x_0^i, \dots, x_{m_i}^i]. \end{aligned}$$

We denote by T the torus group $T = (t_1, \dots, t_l),\ t_i \in \mathbb{G}_m$. We have a canonical action of T on $\mathbb{A}$, namely multiplication by t_i on the component $\mathbb{A}^{m_i+1}$. We denote by $\mathbb{A}^\circ$ the open subscheme formed of points $x = (x_i),\ x_i \in \mathbb{A}^{m_i+1}$ such that $x_i \neq 0$ for all i. Then T operates freely on $\mathbb{A}^\circ$ and Z identifies with the orbit space $\mathbb{A}^\circ/T$.

Let X be a closed subscheme of the multi-projective space Z. We denote by $I(X)$ the ideal of S generated by all $f \in S_{\mathbf{a}}$ (for varying $\mathbf{a}$) such that f vanishes on X (f considered, canonically as above, as a section of a line bundle on Z). We call $I(X)$ the *ideal of* X (in S). Obviously, $I(X)$ is a multigraded ideal in S. If $\widehat{X} = \operatorname{Spec} S/I(X)$, we call $\widehat{X}$ the *multicone over* X. On the other hand, let J be a multigraded ideal in S. We can then canonically associate a closed subscheme $V(J)$ of Z, called the *variety* associated to J, in the following way. The ideal J determines a T-stable sheaf $\tilde{J}$ of ideals in $\mathbb{A}$. The restriction of $\tilde{J}$ to $\mathbb{A}^\circ$ goes down to a sheaf of ideals in Z and hence defines a closed subscheme $V(J)$ of Z. More concretely, Z can be covered by open subsets of the form $U_1 \times \cdots \times U_l$, where U_i is the open subset of $\mathbb{P}^{m_i}$ defined by $x_{k_i}^i \neq 0$, i.e. $U_i = \operatorname{Spec} k[x_0^i/x_{k_i}^i, x_1^i/x_{k_i}^i, \dots]$ and the restriction of $V(J)$ to $U_1 \times \cdots \times U_l$ is given by the ideal (in the coordinate ring of $U_1 \times \cdots \times U_l$), generated by elements of the form

$$\frac{F}{\prod_{1 \leqslant i \leqslant l} (x_{k_i}^i)^{a_i}}, \quad F \in S_{\mathbf{a}}.$$

The mapping $J \mapsto V(J)$ has the property that a union of ideals is taken to the corresponding scheme theoretic intersection (in Z) and an intersection of ideals to the corresponding scheme theoretic union. We observe that $J \subseteq I(V(J))$. Note however that if J_1, J_2 are two multigraded ideals of S such that $V(J_1) = V(J_2)$, it does *not* necessarily follow that $(J_1)_{\mathbf{a}} = (J_2)_{\mathbf{a}}$ for all but a finite number of $\mathbf{a}$'s.

Let $R = \oplus R_{\mathbf{a}}$ be the multigraded ring where

$$R_{\mathbf{a}} = H^0(G/B, L(\pi)),\ \pi = \sum a_i \omega_i,\ a_i \geqslant 0.$$

One knows that the canonical mapping

$$H^0(Z, L_1^{a_1} \otimes \cdots \otimes L_l^{a_l}) \longrightarrow H^0(G/B, L(\pi)),$$

is surjective (for example, by Theorem 4.3.1). Hence we see that the ideal $I(G/B)$ of G/B is precisely the kernel of the canonical homomorphism $S \to R$. Let now I be a multigraded ideal of R. Then $V(I)$ is a closed subscheme of G/B; conversely, if X

is a closed subscheme of G/B, the ideal in R is generated by all multihomogeneous $f \in R$ vanishing on X is called the ideal of X (in R) and is the image of $I(X)$ under the canonical map $S \to R$. We denote this ideal in R by just $I(X)$.

Let now J_i be a graded ideal of the homogeneous coordinate ring R_i of G/P_i and $\tilde{J}_i$ the multigraded ideal of R generated by J_i. We denote (as usual) by $V(J_i)$ the closed subscheme G/P_i defined by J_i. Then we observe that if $\eta : G/B \to G/P_i$ is the canonical projection, we have

$$V(\tilde{J}_i) = \eta^{-1}(V(J_i))$$

in the scheme theoretic sense. This is simple to check; for example, use the concrete description, given above, for the restriction of $V(\tilde{J}_i)$ to affine open subsets of a suitable covering of Z.

A subset T_i of W^i is called a *right half space* if for $\alpha, \beta \in W^i$, $\alpha \in T_i$ and $\beta \geqslant \alpha$, then $\beta \in T_i$.

THEOREM 5.0.1. *Let J be an ideal of R generated by Plücker coordinates p_ϕ, $\phi \in T$, where $T = \bigcup_{1 \leqslant i \leqslant l} T^i$ such that T^i, $1 \leqslant i \leqslant l$ is a right half space in W^i (some T^i could be empty). Then the closed subscheme $V(J)$ of G/B is reduced and is in fact a union of Schubert varieties.*

PROOF. Let J_i (resp. $\tilde{J}_i$) be the ideal in the homogeneous coordinate ring R_i (resp. R), generated by p_ϕ, $\phi \in T^i$. One knows (by the results of Ch 6) that the closed subscheme $V(J_i)$ of G/P_i is reduced and is a union of Schubert varieties (in G/P_i). Then obviously $\eta^{-1}(V(J_i))$ is reduced, η being the canonical mapping $G/B \to G/P_i$. As we observed above, we have $V(\tilde{J}_i) = \eta^{-1}(V(J_i))$, so that $V(\tilde{J}_i)$ is reduced and is a union of Schubert varieties in G/B. Now one has

$$V(J) = V(\tilde{J}_1) \cap \cdots \cap V(\tilde{J}_l).$$

Now by Corollary 4.3.2 it follows that $V(J)$ is reduced and is a union of Schubert varieties in G/B. □

REMARK 5.0.2. If Q is any parabolic subgroup of G, then, as we did above for the case of B, we can define the multigraded ring of G/Q. We see also that Theorem 5.0.1 has an obvious extension to this case.

REMARK 5.0.3. Note that we do *not* claim that $R/V(J)$ or $R/V(\tilde{J}_i)$ is reduced. This is probably false, unlike the case of G/P, P being a maximal parabolic subgroup. Examples of other difficulties, which crop up in the case when Q is not maximal, are as follows:

(i) Let X_1, X_2 be two Schubert varieties in G/B. Then

$$I(X_1 \cap X_2) = I(X_1) + I(X_2)\ ?$$

(ii) Let $X = X(\tau)$ be a Schubert variety in G/B and let $X_P(\theta)$ be the image of X in G/P, P being a maximal parabolic subgroup. Let $R(\tau) = R/I(X(\tau))$. Let $H(\tau) = X(\tau) \cap \{p_\theta = 0\}$. Let $I_\tau(H(\tau))$ be the ideal of $H(\tau)$ in $R(\tau)$, i.e. the image of $I(H(\tau))$ in $R(\tau)$. Then

$$I_\tau(H(\tau)) = p_\theta R(\tau)\ ?$$

If $f \in R(\tau)$ ($\mathbf{a} = (a_1, \ldots, a_l)$) such that $a_1 > 0$, then by applying Theorem 4.3.1, it follows that $f = p_\theta g$, $g \in R(\tau)$. However, it seems very likely that

$$I_\tau(H(\tau)) \cap R(\tau)_{\mathbf{a}} \neq 0$$

with $a_1 = 0$, in which case the above equality does not hold.

Thus, while the ideal theory of Schubert varieties in G/B is as good as in the case of G/P (P being a maximal parabolic subgroup), it doesn't seem to extend in the same manner to the multicone over G/B.

REMARK 5.0.4. A closed subscheme X of G/B is said to be of *product type* if it is of the form $V(J)$ as in Theorem 5.0.1. We shall now show that a Schubert variety in G/B is always of product type. This is a consequence of the following more general

REMARK 5.0.5. One notes that an intersection of subschemes of G/B of product type is of product type and that the subscheme $H(\tau)$ of $X(\tau)$ (see (ii) of Remark 5.0.3 above) is of product type. An arbitrary union of Schubert varieties in G/B is perhaps not of product type.

5.1. Equations defining Schubert varieties

LEMMA 5.1.1. *Let $w \in W$. For $1 \leq i \leq l$, let $X(\varphi_i)$ be the pull-back under $\pi_i : G/B \to G/P_i$ of $\pi_i(X(w))$. Let $\theta \in W$ be such that $\theta \leq \varphi_i$, $1 \leq i \leq l$ (where $l = \text{rank } G$). Then $\theta \leq w$.*

PROOF. (By induction on $l(w)$). If $l(w) = 0$, then $w = id$, and φ_i is simply the element of maximal length in W_{P_i}. Now if $\theta \in W$ is such that $\theta \leq \varphi_i$, $1 \leq i \leq l$, then θ is in fact the identity element (since $\bigcap_{i=1}^{l} P_i = B$). Thus $\theta = w$.

Now let $l(w) \geq 1$. Fix a simple root α such that $s_\alpha w < w$. Set $s_\alpha = s$. For $1 \leq i \leq l$, let $\sigma_i \in W_{P_i}$ be such that $\varphi_i = w\sigma_i$ with $l(w\sigma_i) = l(w) + l(\sigma_i)$.

Let $\theta \leqslant \varphi_i$, $1 \leqslant i \leqslant l$. This implies that $\theta \leqslant w$ in view of the following:

Lemma. Let $\sigma, \sigma_1, \sigma_2 \in W$ be such that $l(\sigma\sigma_i) = l(\sigma) + l(\sigma_i)$, $i = 1, 2$. Then for all $\theta \in W$, we have $\theta \leqslant \sigma\sigma_i$, $i = 1, 2$ if and only if $\theta \leqslant \sigma\sigma_0$ for some σ_0 such that $\sigma_0 \leqslant \sigma_i$, $i = 1, 2$, and $l(\sigma\sigma_0) = l(\sigma) + l(\sigma_0)$.

For a proof, see [**100**], Lemma 3.11.

Note that in view of the Lemma, we have $\theta \leqslant \varphi_i$, $1 \leqslant i \leqslant l$ implies that there exists a σ such that $\theta \leqslant w\sigma$, and $\sigma \leqslant \sigma_i$, $1 \leqslant i \leqslant l$. Hence $\sigma \in W_{P_i}$, $1 \leqslant i \leqslant l$. But $\bigcap W_{P_i} = \text{id}$, and hence $\sigma = \text{id}$. This implies that $\theta \leqslant w$. □

As an immediate consequence of the above Lemma, we have

COROLLARY 5.1.2. *Notations being as in the Lemma above, we have set-theoretically $X(w) = \bigcap_{i=1}^{l} X(\varphi_i)$.*

THEOREM 5.1.3. *For a Schubert variety $X = X(w)$ in G/B, G being $SL(n)$, the ideal sheaf of X in G/B is generated by $\{p_\tau, \tau \in W^{P_i}, 1 \leq i \leq l \mid w_i \not\geq \tau\}$, where, for $1 \leq i \leq l, w_i$ is the minimal representative in W of the coset wW_{P_i}.*

PROOF. By Corollary 5.1.2, we have

$$(*)\qquad X(w) = \bigcap_{i=1}^{l} \pi_i^{-1}(X(w_i)) \text{ (set-theoretically).}$$

If J_i denotes the ideal in R_i generated by $\{p_\tau, \tau \in W^{P_i}, 1 \leq i \leq l \mid w_i \not\geq \tau\}$, then we have (cf. Chapter 6, Theorem 3.4.2)

$$V(J_i) = X(w_i),\ V(\tilde{J}_i) = \pi_i^{-1}(X(w_i))$$

where $\tilde{J}_i$ denotes the multigraded ideal of R generated by J_i (notations being as above). This implies that the relation (*) above is in fact scheme-theoretic from which the required result follows. □

6. Cohen-Macaulayness

6.1. Lexicographic shellability

This discussion is à la Björner-Wachs (cf. [**19**]).

DEFINITION 6.1.1. A *graded poset* is a finite partially ordered set P with a unique maximal and a unique minimal element (denoted by $\widehat{1}$, $\widehat{0}$ respectively) in which all maximal chains $\widehat{1} = x_0 > x_1 > \cdots > x_r = \widehat{0}$ have the same length r. This common length r is called the *rank of* P.

REMARK 6.1.2. Given a finite partially ordered set A, there is a canonically defined total order on A^n, n being some positive integer, namely the lexicographic order: $(\alpha_1, \ldots, \alpha_n)$ is lexicographically greater than $(\beta_1, \ldots, \beta_n)$, if there exists a $t \leqslant n$ such that $\alpha_i = \beta_i$, $i < t$, and $\alpha_t > \beta_t$.

DEFINITION 6.1.3. A graded poset P is said to be *lexicographically shellable* (or L-shellable) if every maximal chain $\mathbf{m}$: $\widehat{1} = x_0 > x_1 > \cdots > x_r = \widehat{0}$ can be labeled, say, $\lambda(\mathbf{m}) = (\lambda_1(\mathbf{m}), \lambda_2(\mathbf{m}), \ldots, \lambda_r(\mathbf{m}))$, where $\lambda_i(\mathbf{m})$ are elements of another partially ordered set in such a way that the following two conditions hold:
(L1) If two maximal chains $\mathbf{m}, \mathbf{m}'$ have the same first d-edges, then the corresponding labels of the first d-edges are the same (here, we think of $\lambda_i(\mathbf{m})$ as being associated with the edge $x_{i-1} \to x_i$).
(L2) Given an interval $[x, y]$ together with a path $\mathbf{c}$ from $\widehat{1}$ to y (we refer to this as a *rooted interval*), among all the maximal chains in $[x, y]$ there exists a unique chain whose label is increasing; further, this unique increasing label is lexicographically smaller than the label of any other chain (here, a label for a maximal chain in $[x, y]$ is that induced by the maximal chain obtained by following $\mathbf{c}$ by the maximal chain from y to x under consideration, followed by an arbitrary chain from x to $\widehat{0}$).

EXAMPLE 6.1.4. The Bruhat order in $\mathfrak{S}_3$. Let $s_1 = (1, 2)$, $s_2 = (2, 3)$. Then

$$(321) = s_1 s_2 s_1.$$

We label the maximal chains as shown in the diagram below

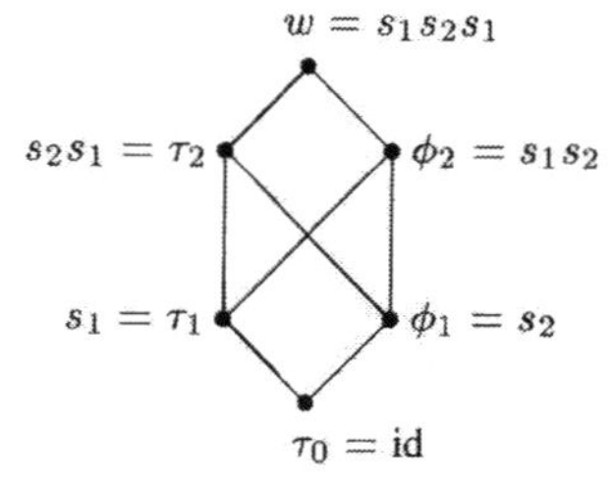

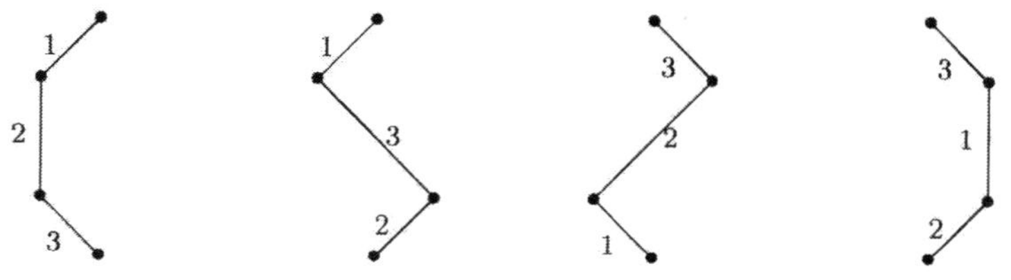

REMARK 6.1.5. The same edge occurring in two different maximal chains may have different indexing. For instance, in the above example the edge $\tau_1 \to \tau_0$ has the index 3 when considered as an edge in $\mathbf{m}$: $w > \tau_2 > \tau_1 > \tau_0$ and has index 1 when considered as an edge in $\mathbf{m}'$: $w > \phi_2 > \tau_1 > \tau_0$.

6.2. Lexicographic shellability for Bruhat orders

Let (W, S) be a Coxeter group. For $J \subset S$, let W^J be the set of minimal representatives of W/W_J. Recall (cf. Chapter 3) that $W^J = \{w \in W \mid w(\alpha) > 0, \text{ for } \alpha \in J\}$. For $w, w' \in W^J$, $w' \geqslant w$, we shall describe a labeling of the maximal chains in the interval $[w, w'] = \{u \in W^J \mid w \leqslant u \leqslant w'\}$. Let us fix a reduced expression $w' = s_1 s_2 \ldots s_q$. Let $\mathbf{m}$: $w' = w_0 > w_1 > \cdots > w_r = w$ be a maximal chain in $[w, w']$, where $r = l(w') - l(w)$. Now since $w_1 < w'$ and $l(w_1) = l(w') - 1$, a reduced expression for w_1 is obtained by omitting a reflection s_i in the above reduced expression for w', and the deleted reflection is uniquely determined. We set $\lambda_1(\mathbf{m}) = i$. Proceeding thus, at each step we label an edge by the position of the reflection that is deleted in the chosen reduced expression for w'.

THEOREM 6.2.1. *(cf.* **[19]***) With notations as above, the interval* $[w, w']$ *in* W^J *is lexicographic shellable for the labeling of the maximal chains in* $[w, w']$ *as described above.*

DEFINITION 6.2.2. Given a finite poset P, the simplicial complex obtained by taking a q-simplex to be a chain of length q in P is called *the order complex of* P, and is denoted by $\Delta(P)$.

Let Δ be a simplicial complex. A maximal face of Δ is called a *facet*. Δ is said to be of *pure dimension* d if all facets are of dimension d. For a facet F, $\overline{F}$ denotes $\{G \mid G \subseteq F\}$, and ∂F denotes $\{G \mid G \subset F\}$

DEFINITION 6.2.3. Let Δ be a pure d-dimensional simplicial complex. An ordering $F_1, F_2, \ldots, F_t$ of the facets of Δ is said to be a *shelling* if for $2 \leqslant j \leqslant t$, $\overline{F_j} \cap \left(\bigcup_{i=1}^{j-1} \overline{F_i}\right)$, is a $(d-1)$-complex having a shelling which extends to a shelling of ∂F_j, i.e. ∂F_j has a shelling in which the facets of $\overline{F_j} \cap \left(\bigcup_{i=1}^{j-1} \overline{F_i}\right)$ come first. We say that Δ is *shellable* if it admits a shelling.

THEOREM 6.2.4. *If P is lexicographic shellable, then the order complex $\Delta(P)$ is shellable.*

PROOF. P being L-shellable, we consider the total ordering $\prec$ of the maximal chains in $\Delta(P)$ given by $\mathbf{m}_1 \prec \mathbf{m}_2 \prec \cdots \prec \mathbf{m}_s$, where $\lambda(\mathbf{m}_1) < \lambda(\mathbf{m}_2) < \cdots < \lambda(\mathbf{m}_r)$. It can be shown that this total ordering gives a shelling for $\Delta(P)$ (see **[18]**). □

6.3. Consequences for Schubert varieties

Let G be a connected semisimple algebraic group, T a maximal torus in G, B a Borel subgroup of G, $B \supset T$, and Q a parabolic subgroup containing B. Let W be the Weyl group of G relative to T. For $\tau \in W^Q$, let $X(\tau)$ be the Schubert variety in G/Q associated to τ.

DEFINITION 6.3.1. Let Z be a union of Schubert varieties in G/Q. Further, let Z be pure of dimension d, i.e., all irreducible components of Z have dimension d. We say that Z *admits a nice indexing* if there exists an indexing, say $X_1, \ldots, X_r$, of the components of Z in such a way that

(1) $X_j \cap \left(\bigcup_{i=1}^{j-1} X_i\right)$ is of pure dimension $d-1$ for $2 \leqslant j \leqslant r$.

(2) $X_j \cap \left(\bigcup_{i=1}^{j-1} X_i\right)$ admits a nice indexing which extends to a nice indexing of the union Y of Schubert divisors in X_j for $2 \leqslant j \leqslant r$ (i.e. Y admits a nice indexing in which the components of $X_j \cap \left(\bigcup_{i=1}^{j-1} X_i\right)$ come first).

THEOREM 6.3.2. *Let $\tau \in W^Q$ and let Z be the union of the Schubert divisors in $X(\tau)$. Then Z admits a nice indexing.*

The result follows immediately from Theorems 6.2.1 and 6.2.4.

THEOREM 6.3.3. *Let $\tau \in W^Q$ and let P be a parabolic subgroup of G containing Q. Let $\pi : G/Q \to G/P$ be the canonical projection. Let Y be the union of all Schubert divisors $X(w)$, $w \in W^Q$, in $X(\tau)$ such that $\pi(X(w)) \subset \pi(X(\tau))$. Then Y admits a nice indexing.*

PROOF. Let

$$\mathcal{B}_\tau = \{w \in W^Q \mid X(w) \text{ is a divisor in } X(\tau) \text{ and } \pi(X(w)) \subset \pi(X(\tau))\}.$$

so that $Y = \bigcup_{w\in\mathcal{B}_\tau} X(w)$. Let us write $\tau = \tau_0\theta_0$, where $\tau_0 \in W^P$, $\theta_0 \in W_P$ and $l(\tau) = l(\tau_0) + l(\theta_0)$. Let $\tau_0 = s_1 \dots s_t$ and $\theta_0 = s_{t+1} \dots s_{t+k}$ be reduced expressions for τ_0 and θ_0 so that $\tau = s_1 \dots s_t s_{t+1} \dots s_{t+k}$ is a reduced expression for τ. Working with this reduced expression for τ, we index the maximal chains in $[\text{id}, \tau]$. Now if $X(w)$ is a divisor in $X(\tau)$, then $w \in \mathcal{B}_\tau$ if and only if $w = s_1 \dots \widehat{s_m} \dots s_{t+k}$ for some $1 \leqslant m \leqslant t$. Hence the maximal chains in $[\text{id}, \tau]$ which start with an edge $\tau \to w$, $w \in \mathcal{B}_\tau$, come earlier (in the lexicographic order) than those which start with an edge $\tau \to w$, $w \notin \mathcal{B}_\tau$. Now the result follows from Theorems 6.2.4 and 6.3.2. □

6.4. Cohen-Macaulayness of cones and multi-cones over Schubert varieties in $SL(n)/Q$

THEOREM 6.4.1. *Let $\tau \in W$ and let $L(\pi)$ be any ample line bundle on G/B, say $\pi = \sum_{i=1}^{l} a_i\omega_i$, where $\pi > 0$ (i.e., $a_i > 0$ for all i). Let $\widehat{X(\tau)}$ be the cone over $X(\tau)$ defined by the projective embedding of $X(\tau)$ associated to the very ample line bundle $L(\pi)$. Then $\widehat{X(\tau)}$ is Cohen-Macaulay.*

PROOF. First observe that $\widehat{X(\tau)} = \text{Spec}\Big(\bigoplus_{m\in\mathbb{Z}^+} H^0(X, L(m\pi))\Big)$. Hence by Serre-Grothendieck Criterion (cf. Chapter 6, Theorem 6.2.3) for Cohen-Macaulayness of $\widehat{X}$ at the vertex in terms of the vanishing of $H^i(X, L)$, it is enough to show that

$$(*) \qquad H^i(X(\tau), L(-\pi)) = 0, \text{ for } 0 \leqslant i < \dim X(\tau),\ \pi > 0.$$

Let us fix a simple root α so that $\tau > \tau s_\alpha$ (note that $X(\tau)$ is stable for the $\mathbb{P}^1$ fibration $G/B \to G/P_\alpha$, where P_α is the rank one parabolic associated to α). Let $X(\overline{\tau})$ be the projection of $X(\tau)$ under $G/B \to G/P$, where $P = P_{\widehat{\alpha}}$ is the maximal parabolic subgroup associated to α, and let $H(\tau) = X(\tau) \cap \{p_{\overline{\tau}} = 0\}$. Then $H(\tau) = \bigcup_{w\in\mathcal{B}_\tau} X(w)$, where $\mathcal{B}_\tau$ is as in the proof of Theorem 6.3.3. Let us take an indexing of the simple roots so that $\alpha_1 = \alpha$. Let $L = L(\omega_1)$. Consider the exact sequence

$$(\dagger) \qquad 0 \to L^{-1} \to \mathcal{O}_X \to \mathcal{O}_H \to 0,$$

where $X = X(\tau)$ and $H = H(\tau)$. We will now show that

$$(**) \qquad H^i(H, L(-\pi)) = 0, \text{ for } 0 \leqslant i < \dim X - 1.$$

In view of Theorem 6.3.3, H admits a nice indexing $X_1, \dots, X_r$ of its components. Hence using induction on r and induction on $\dim X(\tau)$, we may assume that $(*)$ holds for $Y = \bigcup_{i=1}^{r-1} X_i$, $Z = X_i$ and $Y \cap Z$ (note that $Y \cap Z$ admits a nice indexing). Now $(**)$ follows by considering the exact sequence

$$0 \to \mathcal{O}_H \to \mathcal{O}_Y \oplus \mathcal{O}_Z \to \mathcal{O}_{Y\cap Z} \to 0$$

and the cohomology exact sequence

$$\cdots \to H^{i-1}(Y \cap Z, L(-\pi)) \to H^i(H, L(-\pi)) \to$$
$$H^i(Y, L(-\pi)) \oplus H^i(Z, L(-\pi)) \to \cdots$$

Now tensoring (†) by $L(-\pi)$ and writing down the cohomology exact sequence, we obtain an exact sequence

$$\cdots \to H^{i-1}(H, L(-\pi)) \to H^i(X, L(-\pi')) \to$$
$$H^i(X, L(-\pi)) \to H^i(H, L(-\pi)) \to \cdots$$

where $\pi' = (a_1 + 1, a_2, \ldots, a_l)$. In view of $(**)$, for $1 \leqslant i < \dim X$, we obtain the exact sequence

$$0 \to H^i(X, L(-\pi')) \to H^i(X, L(-\pi)).$$

When $a_1 = 1$, $H^i(X, L(-\pi)) = 0$ for all i (since X is stable for the $\mathbb{P}^1$-fibration $G/B \to G/P_\alpha$, P_α being the rank one parabolic subgroup associated to the simple root α). Hence we obtain

(1) $$H^i(X, L(-\pi)) = 0 \qquad \text{for } 1 \leqslant i \leqslant \dim X, \quad \pi > 0.$$

When $i = 0$, since $H^0(H, L(-\pi)) = 0$ (cf. (**)), we obtain the exact sequence

$$0 \to H^0(X, L(-\pi')) \to H^0(X, L(-\pi)).$$

Hence by iteration, we obtain $0 \to H^0(X, L(-\pi)) \to H^0(X, L(-\eta))$ is exact, where η denotes $(1, a_2, \ldots, a_l)$. The stability of $X(\tau)$ for the $\mathbb{P}^1$-fibration $G/B \to G/P_{a_1}$ implies $H^i(X, L(-\eta)) = 0$, for all i. From this it follows

(2) $$H^0(X, L(-\pi)) = 0, \ \pi > 0.$$

From (1) and (2) we obtain

$$H^i(X, L(-\pi)) = 0 \text{ for } 0 \leqslant i < \dim X, \ \pi > 0.$$

This completes the proof. □

THEOREM 6.4.2. *Let $\tau \in W$. Then the ring $R(\tau) = \bigoplus_{\pi \geqslant 0} H^0(X, L(\pi))$ is Cohen-Macaulay. Thus the multi-cone for the multi-projective embedding*

$$X(\tau) \hookrightarrow \prod_{1 \leqslant i \leqslant l} \mathbb{P}(H^0(G/P_i, L(\pi_i))^*)$$

is Cohen-Macaulay.

PROOF. We prove the result by induction on $\dim X(\tau)$. Let α be a simple root such that $\tau > \tau s_\alpha$. Let $P = P_{\widehat{\alpha}}$ be the maximal parabolic subgroup corresponding to α. Let $X(\overline{\tau})$ be the image of $X(\tau)$ under $G/B \to G/P$, and let $H(\tau) = X(\tau) \cap \{p_{\overline{\tau}} = 0\}$. Let us take an indexing of the simple roots so that $\alpha_1 = \alpha$. We first show that

(*) $$I(H(\tau)) = p_{\overline{\tau}} R(\tau).$$

We clearly have $I(H(\tau)) \supset p_{\overline{\tau}} R(\tau)$. Let now $f \in R(\tau)_{\mathbf{a}}$, where $\mathbf{a} = (a_1, \ldots, a_l)$, be such that $f = 0$ on $H(\tau)$. The facts that $f \neq 0$ on $X(\tau)$ and $f = 0$ on $H(\tau)$

imply that $a_1 \neq 0$ (note that $X(\tau s_\alpha) \subseteq H(\tau)$, and that for all $\mathbf{b} = (0, b_2, \ldots, b_l)$, we have, $H^0(X(\tau), L_{\mathbf{b}}) \cong H^0(X(\tau s_\alpha), L_{\mathbf{b}})$). Let us write $f = F_1 + F_2$, where each of F_1 and F_2 is a sum of standard monomials on $X(\tau)$ of type $\mathbf{a}$ and each monomial in F_1 involves $p_{\overline{\tau}}$, while each monomial in F_2 does not involve $p_{\overline{\tau}}$. Hence we obtain $F_1 \in I(H(\tau))$, and therefore $F_2 \in I(H(\tau))$, i.e., $F_2 = 0$ on $H(\tau)$. On the other hand, each monomial in F_2 remains standard on $H(\tau)$, since none of them involves $p_{\overline{\tau}}$. Hence the linear independence of standard monomials on $H(\tau)$ implies that $F_2 = 0$. This implies $f = F_1$, and hence $f \in p_{\overline{\tau}}R(\tau)$ (since $F_1 \in p_{\overline{\tau}}R(\tau)$, clearly). This establishes the equality in $(*)$.

Next we show that the multigraded ring $R(\tau)/p_{\overline{\tau}}R(\tau)$ is Cohen-Macaulay. In fact we prove the following more general statement:

"Let Z be a union of Schubert varieties such that Z is pure of dimension less than $\dim X(\tau)$, admitting a nice indexing. Then $R(\tau)/I(Z)$ is Cohen-Macaulay".
We prove this by induction on the number of components of Z and on $\dim Z$. Let $X_1, \ldots, X_r$ be a "nice" indexing of the components of Z. Let $Z = X \cup Y$, where $X = \bigcup_{i=1}^{r-1} X_i$ and $Y = X_r$. Denoting the ideals of X, Y, Z in $X(\tau)$ by $I(X)$, $I(Y)$, $I(Z)$ respectively, the assertion that $R(\tau)/I(Z)$ is Cohen-Macaulay follows from Chapter 6, Lemma 6.2.5, by taking $A = R(\tau)$, $I = I(X)$, $J = I(Y)$.

Now, taking $Z = H(\tau)$, we have $R(\tau)/I(Z) = R(\tau)/p_{\overline{\tau}}R(\tau)$, and $p_{\overline{\tau}}$ being a nonzero divisor in $R(\tau)$, we conclude that $R(\tau)$ is Cohen-Macaulay. □

CHAPTER 8

Singularities of Schubert varieties

In this Chapter, we introduce the tangent space to a Schubert variety at a T-fixed point and discuss the smoothness at a T-fixed point using the Jacobian criterion, in §1-6. We recall the generalities on the multiplicity at a given point on an algebraic variety in §7. In §8, we give some explicit formulas for the degree of a Schubert variety. In §9, we describe a basis for the tangent cone at a singular point on a Schubert variety in the Grassmannian, and as a consequence obtain a recursive formula for the multiplicity at a singular point. In §10, §11, we present similar results for Schubert varieties in a minuscule G/P, the symplectic Grassmannian respectively. In §12, §13, we determine the irreducible components of the singular locus of a Schubert variety in the Grassmannian, the symplectic and orthogonal Grassmanians respectively. In §14, we present a closed formula (due to Rosenthal, Rosenthal-Zelevinsky) for the multiplicity at a singular point on a Schubert variety in the Grassmannian. In §15, we present a criterion for smoothness for Schubert varieties in the flag variety. In §16, we present a conjecture on the singular locus of a Schubert variety in the flag variety.

1. Tangent space and Smoothness

1.1. The Zariski tangent space

Let x be a point on a variety X. Let $\mathfrak{m}_x$ be the maximal ideal of the local ring $\mathcal{O}_{X,x}$ with residue field $k(x)(= \mathcal{O}_{X,x}/\mathfrak{m}_x)$. Note that $k(x) = k$ (since k is algebraically closed). The Zariski tangent space to X at x is defined as

$$\begin{aligned} T_x(X) &= \mathrm{Der}_k(\mathcal{O}_{X,x}, k(x)) \\ &= \{D : \mathcal{O}_{X,x} \to k(x),\ k\text{-linear such that } D(ab) = D(a)b + aD(b)\} \end{aligned}$$

(here $k(x)$ is regarded as an $\mathcal{O}_{X,x}$-module). It can be seen easily that $T_x(X)$ is canonically isomorphic to $\mathrm{Hom}_{k\text{-mod}}(\mathfrak{m}_x/\mathfrak{m}_x^2, k)$.

1.2. Smooth and non-smooth points

A point x on a variety X is said to be a *simple* or *smooth* or *nonsingular point of* X if $\mathcal{O}_{X,x}$ is a regular local ring. A point x which is not simple is called a *multiple* or *non-smooth* or *singular point* of X. The set $\mathrm{Sing}\,X = \{x \in X \mid x \text{ is a singular point}\}$ is called the *singular locus of* X. A variety X is said to be *smooth* if $\mathrm{Sing}X = \emptyset$.

THEOREM 1.2.1. *Let $x \in X$. Then* $\dim_k T_x(X) \geqslant \dim \mathcal{O}_{X,x}$ *(*$\dim \mathcal{O}_{X,x}$ *is also denoted* $\dim_x X$*) with equality if and only if x is a simple point of X.*

PROOF. We have (cf. Chapter 1, §1.9)

$$\dim_k T_x(X) = \dim_k(\mathfrak{m}_x/\mathfrak{m}_x^2) \geqslant \dim \mathcal{O}_{X,x},$$

with equality if and only if $\mathcal{O}_{X,x}$ is regular. The result now follows. □

2. The space $T(w, \tau)$

Let G be a semisimple algebraic group. Let T, B, W, R, S etc., be as in Chapter 3. For $\alpha \in R$, let X_α be the element of the Chevalley basis for $\mathfrak{g}(= \text{Lie}G)$, corresponding to α. We follow **[27]** for denoting elements of R, R^+, S etc.

For $\tau \leq w$, let $T(w,\tau)$ be the Zariski tangent space to $X(w)$ at e_τ. Let w_0 be the element of largest length in W. Now the tangent space to G at e_{id} is $\mathfrak{g}$, and hence the tangent space to G/B at e_{id} is $\bigoplus_{\beta \in R^+} \mathfrak{g}_{-\beta}$. For $\tau \in W$, identifying G/B with $G/{}^\tau B$ (where ${}^\tau B = \tau B \tau^{-1}$) via the map $gB \mapsto (n_\tau g n_\tau^{-1})^\tau B$, n_τ being a fixed lift of τ in $N_G(T)$, we have, the tangent space to G/B at e_τ is $\bigoplus_{\beta \in \tau(R^+)} \mathfrak{g}_{-\beta}$, i.e.,

$$T(w_0, \tau) = \bigoplus_{\beta \in \tau(R^+)} \mathfrak{g}_{-\beta}.$$

Set

$$N(w,\tau) = \{\beta \in \tau(R^+) \mid X_{-\beta} \in T(w,\tau)\}.$$

Since $T(w,\tau)$ is a T-stable subspace of $T(w_0,\tau)$, we have

$$T(w,\tau) = \text{ the span of } \{X_{-\beta},\ \beta \in N(w,\tau)\}.$$

3. A canonical affine neighborhood of a T-fixed point

Let $\tau \in W$. Let U_τ^- be the unipotent subgroup of G generated by the root subgroups $U_{-\beta}$, $\beta \in \tau(R^+)$ (note that U_τ^- is the unipotent part of the Borel sub group ${}^\tau B^-$, opposite to ${}^\tau B$ $(= \tau B \tau^{-1})$). We have

$$U_{-\beta} \cong \mathbb{G}_a,\ U_\tau^- \cong \prod_{\beta \in \tau(R^+)} U_{-\beta}.$$

Now, $U_\tau^- e_\tau$ gets identified with $\mathbb{A}^N$, where $N = \#R^+$ via the above identification and we shall denote the induced coordinate system on $U_\tau^- e_\tau$ by $\{x_{-\beta},\ \beta \in \tau(R^+)\}$. In the sequel, we shall denote $U_\tau^- e_\tau$ by $\mathcal{O}_\tau^-$. Thus we obtain that $\mathcal{O}_\tau^-$ is an affine neighborhood of e_τ in G/B.

3.1. The Affine Variety $Y(w,\tau)$

For $w \in W$, $w \geqslant \tau$, let us denote $Y(w,\tau) := \mathcal{O}_\tau^- \cap X(w)$. It is a nonempty affine open subvariety of $X(w)$, and a closed subvariety of the affine space $\mathcal{O}_\tau^-$.

3.2. Equations defining $Y(w,\tau)$ in $\mathcal{O}_\tau^-$

Let $w \in W$. For $1 \leqslant i \leqslant l(= \text{rank}G)$, fix a basis D_i for the kernel of the surjective map $H^0(G/B, L_i) \to H^0(X(w), L_i)$ (given by restriction), L_i being the ample generator of Pic(G/P_i). We have (cf. Chapter 3, §9), the ideal sheaf of $X(w)$ in G/B is generated by $\{f \in D_i,\ 1 \leqslant i \leqslant l\}$. Let $\tau \in W$ be such that $\tau \leq w$. Let $Y(w,\tau)$ be as in §3.1, $I(w,\tau)$ the ideal defining $Y(w,\tau)$ as a closed subvariety of $\mathcal{O}_\tau^-$. We have

$$I(w,\tau) \text{ is generated by } \{f|_{\mathcal{O}_\tau^-}, f \in D_i,\ 1 \leqslant i \leqslant l\}.$$

For example, take $G = SL(4)$. The ideal of $Y((3412), id)$ in $\mathcal{O}_{\text{id}}^-$ is generated by $\{p_4, p_{234}\}$, and the ideal of $Y((2413), id)$ is generated by $\{p_3, p_4, p_{34}, p_{134}, p_{234}\}$.

4. Jacobian Criterion for smoothness

Let Y be an affine variety in $\mathbb{A}^n$, and let $I(Y)$ be the ideal defining Y in $\mathbb{A}^n$. Let $I(Y)$ be generated by $\{f_1, f_2, \dots, f_r\}$. Let J be the Jacobian matrix $(\frac{\partial f_i}{\partial x_j})$. Then we have (cf. Theorem 1.2.1) the dimension of the tangent space to Y at a point P is greater than or equal to the dimension of Y, with equality if and only if P is a smooth point. Equivalently, rank $J_P \leq \text{codim}_{\mathbb{A}^n} Y$ with equality if and only if P is a smooth point of Y (here J_P denotes J evaluated at P).

For example, take $X(2413)$ and $X(3412)$. As stated in §3.2, the equations defining $Y((3412), id)$ in $\mathcal{O}_{id}^-$ are $\{p_4, p_{234}\}$, and the equations defining $Y((2413), \text{id})$ are $\{p_3, p_4, p_{34}, p_{134}, p_{234}\}$. Identifying $\mathcal{O}_{id}^-$ with the group of unipotent lower triangular matrices in SL_4, we have

$$\begin{aligned}
p_3 &= x_{31}\\
p_4 &= x_{41}\\
p_{34} &= x_{42}x_{31} - x_{41}x_{32}\\
p_{134} &= x_{43}x_{32} - x_{42}\\
p_{234} &= x_{21}(x_{43}x_{32} - x_{42}) - (x_{43}x_{31} - x_{41})
\end{aligned}$$

Hence, the equations defining $Y((3412), \text{id})$ are $\{x_{41},\ x_{21}(x_{43}x_{32} - x_{42}) - (x_{43}x_{31} - x_{41})\}$. The Jacobian matrix is

$$\begin{bmatrix} 0 & 0 & 0 & 1 & 0 & 0 \\ (x_{43}x_{32} - x_{42}) & -x_{43} & x_{21}x_{43} & 1 & -x_{21} & -x_{31} \end{bmatrix}.$$

Here, the columns are indexed by $x_{21}, x_{31}, x_{32}, x_{41}, x_{42}, x_{43}$. The rank of this matrix at $P = $id (*i.e.* all $x_{ij} = 0$) is $1 < 2 = 6 - l(3412) = \text{codim}Y((3412), \text{id})$. Therefore, $X(3412)$ is singular at e_{id}.

In contrast, the equations defining $Y((2413),\mathrm{id})$ are $\{x_{31},\ x_{41},\ x_{31}x_{42}-x_{41}x_{32},\ x_{42}-x_{43}x_{32},\ x_{21}(x_{43}x_{32}-x_{42})-(x_{43}x_{31}-x_{41})\}$. The Jacobian matrix is

$$\begin{bmatrix} 0 & 1 & 0 & 0 & 0 & 0 \\ 0 & 0 & 0 & 1 & 0 & 0 \\ 0 & x_{42} & -x_{41} & -x_{32} & x_{31} & 0 \\ 0 & 0 & -x_{43} & 0 & 1 & -x_{32} \\ (x_{43}x_{32}-x_{42}) & -x_{43} & x_{21}x_{43} & 1 & -x_{21} & (x_{32}x_{21}-x_{31}) \end{bmatrix}.$$

This matrix is easily seen to have rank 3 when all variables are set to zero; $3 = 6 - l(2413) = \mathrm{codim}Y((2413),\mathrm{id})$ which implies $X(2413)$ smooth at e_{id}.

5. Discussion of smoothness at a T-fixed point

Let $w, \tau \in W$, $w \geqslant \tau$. The problem of determining whether or not e_τ is a smooth point of $X(w)$ is equivalent to determining whether or not e_τ is a smooth point of $Y(w,\tau)$ (since $Y(w,\tau)$ is an open neighborhood of e_τ in $X(w)$). Let D_i, $1 \leqslant i \leqslant l$ be as in §3.2. In view of Jacobian criterion, the problem is reduced to computing $\partial f/\partial x_\beta, f \in D_i, 1 \leqslant i \leqslant l$. To carry out this computation, we first observe the following:

Let λ be a dominant character of T (or B), and L_λ be the associated line bundle on G/B. Let V be the G-module $H^0(G/B, L_\lambda)$. Then as seen in Chapter 3, §3.2, V can be identified as

$$V = \{f : G \to k \mid f(gb) = \lambda(b)f(g),\ b \in B,\ g \in G\}.$$

Now V is also a $\mathfrak{g}$-module. Given X in $\mathfrak{g}$, we identify X with the corresponding right invariant vector field D_X on G. Thus, if $v \in V$ corresponds to a function f on G as above, then we have $D_X f = Xf$. If $f \in V$, then denoting the restriction of f to $\mathcal{O}_\tau^-$ also by just f, we note that the evaluations of $\partial f/\partial x_\beta$ and $X_\beta f$, $\beta \in \tau(R^-)$, at e_τ coincide.

Summarizing, given $w, \tau \in W$, $w \geqslant \tau$, the problem of determining whether or not e_τ is a smooth point of $X(\tau)$ is reduced to constructing bases D_i for the kernel of $H^0(G/B, L_i) \to H^0(X(w), L_i)$ and computing $X_\beta f$, for $\beta \in \tau(R^-)$, and $f \in D_i$, $1 \leqslant i \leqslant l$.

5.1. T-stable curves

For a root $\alpha > 0$, let Z_α denote the $SL(2)$-copy in G corresponding to α; note that Z_α is simply the subgroup of G generated by U_α and $U_{-\alpha}$. Given $x \in W$, precisely one of $\{U_\alpha, U_{-\alpha}\}$ fixes the point e_x. Thus $Z_\alpha \cdot e_x$ is a T-stable curve in G/B (note that $Z_\alpha \cdot e_x \cong \mathbf{P}^1$), and conversely any T-stable curve in G/B is of this form (cf. **[34]**). Corresponding to each $\alpha > 0$, the number of T-stable curves in G/B is $\frac{1}{2}(\#W)$, and thus the number of T-stable curves in G/B is $\frac{1}{2}(\#W)l(w_0)$, w_0 being the element of largest length in W. Now a T-stable curve $Z_\alpha \cdot e_x$ is contained in a Schubert variety $X(w)$ if and only if $e_x, e_{s_\alpha x}$ are both in $X(w)$. Given $y \leqslant w$, let

$\mathcal{R}(y,w) = \{\alpha > 0 \mid s_\alpha y \leqslant w\}$. Then there are precisely $r(y,w)$ T-stable curves in $X(w)$ passing through e_y, where $r(y,w) = \#\mathcal{R}(y,w)$.
Deodhar's Inequality: (see [**34**] for example) For $x \leqslant y \leqslant w$,

$$\#\{\alpha \in R^+ \mid x \leqslant s_\alpha y \leqslant w\} \geqslant l(w) - l(x).$$

In view of Deodhar's Inequality, we have $r(y,w) \geqslant l(w)$ (taking $x = id$ in Deodhar's Inequality). Thus there are at least $l(w)$ T-stable curves in $X(w)$ passing through e_y.

LEMMA 5.1.1. *Let $w, \tau \in W$, $w \geqslant \tau$. Let $\beta \in \tau(R^+)$. If $w \geqslant s_\beta \tau$, then $X_{-\beta} \in T(w,\tau)$.*

PROOF. The hypothesis that $w \geqslant s_\beta\tau$ implies that the curve $Z_\beta \cdot e_\tau$ is contained in $X(w)$. Now the tangent space to $Z_\beta \cdot e_\tau$ at e_τ is the one-dimensional span of $X_{-\beta}$. The required result now follows. □

We shall show in the following section that w, τ being as above, $T(w,\tau)$ is precisely the span of $\{X_{-\beta},\ \beta \in \tau(R^+) \mid w \geqslant s_\beta\tau\}$, G being $SL(n)$.

6. Singularities of Schubert varieties

Let $G = SL(n)$, T the maximal torus in G consisting of diagonal matrices, and B the Borel subgroup consisting of upper triangular matrices.

6.1. Ideal of $Y(w,\tau)$

Let $w, \tau \in W$, $w \geqslant \tau$. Let $Y(w,\tau)$ be as in §3.1, and $I(w,\tau)$ the ideal defining $Y(w,\tau)$ as a closed subvariety of $\mathcal{O}_\tau^-$. For $1 \leqslant d \leqslant l$, we have

$$H^0(G/B, L(\omega_d)) = (\wedge^d k^n)^*,$$

the linear dual of $\wedge^d k^n$. Further, the Plücker coordinates p_θ, $\theta \in I_{d,n}$ form a basis of $H^0(G/B, L(\omega_d))$. The kernel of $H^0(G/B, L(\omega_d)) \to H^0(X(w), L(\omega_d))$ has a basis consisting of $\{p_\theta,\ \theta \in I_{d,n} \mid w^{(d)} \not\geqslant \theta\}$, where $w^{(d)}$ is the element of $I_{d,n}$ given by $\pi_d(X(w)) = X_{P_d}(w^{(d)})$, π_d being the canonical projection $G/B \to G/P_d$. Hence we obtain (in view of Theorem 5.1.3, Chapter 7)

THEOREM 6.1.1. *The ideal $I(w,\tau)$ is generated by*

$$\{p_\theta \mid_{\mathcal{O}_\tau^-}, \theta \in I_{d,n},\ 1 \leqslant d \leqslant n-1 \mid w^{(d)} \not\geqslant \theta\}.$$

Let w, τ be as above. We shall now give a criterion for $X(w)$ to be smooth at e_τ. First we note that

(1) For $\theta, \mu \in W^d$, $1 \leqslant d \leqslant n-1$, $p_\theta(e_\mu) \neq 0 \iff \theta = \mu$, where, recall that for $\theta = (i_1 \cdots i_d) \in W^d\ (= I_{d,n})$, e_θ denotes the vector $e_{i_1} \wedge \cdots \wedge e_{i_d}$ in $\wedge^d k^n$, and p_θ denotes the Plücker coordinate associated to θ.

(2) Let X_α be the element of the Chevalley basis of $\mathfrak{g}$, corresponding to $\alpha \in R$. If $X_\alpha p_\mu \neq 0$, $\mu \in W^d$, then $X_\alpha p_\mu = \pm p_{s_\alpha \mu}$, where s_α is the reflection corresponding to the root α.

(3) For $\alpha \neq \beta$, if $X_\alpha p_\mu$, $X_\beta p_\mu$ are non-zero, then $X_\alpha p_\mu \neq X_\beta p_\mu$.

The first remark is obvious, since $\{p_\theta \mid \theta \in W^d\}$ is the basis of $(\wedge^d k^n)^*(= H^0(G/P_d, L(\omega_d)))$, dual to the basis $\{e_\phi,\ \phi \in I_{d,n}\}$ of $\wedge^d k^n$.

The second remark is a consequence of SL_2 theory, using the following facts

(a) $|\langle\chi, \alpha^*\rangle| = \left|\dfrac{2(\chi,\alpha)}{(\alpha,\alpha)}\right| = 0$ or 1, χ being the weight of p_μ.

(b) p_μ is the lowest weight vector for the Borel subgroup ${}^\mu B = \mu B \mu^{-1}$.

The third remark follows from weight considerations (note that if $X_\alpha p_\mu \neq 0$, then $X_\alpha p_\mu$ is a weight vector (for the T-action) of weight $\chi + \alpha$, χ being the weight of p_μ).

THEOREM 6.1.2. *(cf.* **[105]***) Let $\tau, w \in W$, $\tau \leqslant w$. Then*

$$\dim T(w,\tau) = \#\{\gamma \in \tau(R^+) \mid w \geqslant s_\gamma \tau\}.$$

PROOF. By Theorem 6.1.1, we have, $I(w,\tau)$ is generated by $\{p_\theta \mid_{\mathcal{O}_\tau^-}, \theta \in I_{d,n},\ 1 \leqslant d \leqslant n-1 \mid w^{(d)} \not\geqslant \theta\}$. Denoting the affine coordinates on $\mathcal{O}_\tau^-$ by x_β, $\beta \in \tau(R^-)$, we have (cf. §5) the evaluations of $\dfrac{\partial p_\theta}{\partial x_\beta}$ and $X_\beta p_\theta$ at e_τ coincide. Let J_w denote the Jacobian matrix of $Y(w,\tau)$ (considered as a subvariety of the affine space $\mathcal{O}_\tau^-$). We shall index the rows of J_w by

$$\{p_\theta \mid_{\mathcal{O}_\tau^-}, \theta \in I_{d,n},\ 1 \leqslant d \leqslant n-1 \mid w^{(d)} \not\geqslant \theta\}$$

and the columns by x_β, $\beta \in \tau(R^-)$. Let $J_w(\tau)$ denote J_w evaluated at τ. Now in view of (1) & (2) above, the (p_θ, x_β)-th entry in $J_w(\tau)$ is non-zero if and only if $X_\beta p_\theta = \pm p_\tau$. Hence in view of (3) above, we obtain that in each row of $J_w(\tau)$, there is at most one non-zero entry. Hence $\operatorname{rank} J_w(\tau) =$ the number of non-zero columns of $J_w(\tau)$. Now, $X_\beta p_\theta = \pm p_\tau$ if and only if $\theta \equiv s_\beta \tau (\bmod W_{P_d})$. Thus the column of $J_w(\tau)$ indexed by x_β is non-zero if and only if $w \not\geqslant s_\beta \tau$. Hence $\operatorname{rank} J_w(\tau) = \#\{\gamma \in \tau(R^+) \mid w \not\geqslant s_\gamma\}$ and thus we obtain

$$\dim T(w,\tau) = \#\{\gamma \in \tau(R^+) \mid w \geqslant s_\gamma \tau\}.$$

□

THEOREM 6.1.3. *Let w, τ be as in Theorem 6.1.2. Then*

$$\{X_\beta,\ \beta \in \tau(R^-) \mid w \geqslant s_\beta \tau\}$$

is a basis for $T(w,\tau)$.

PROOF. Let $\beta \in \tau(R^-)$ be such that $w \geqslant s_\beta \tau$. We have (by Lemma 5.1.1), $X_\beta \in T(w,\tau)$. On the other hand, by Theorem 6.1.2,
$\dim T(w,\tau) = \#\{\gamma \in \tau(R^+) \mid w \geqslant s_\beta \tau\}$. The result follows from this. □

COROLLARY 6.1.4. *$X(w)$ is smooth at e_τ if and only if*

$$l(w) = \#\{\alpha \in R^+ \mid w \geqslant \tau s_\alpha\}.$$

PROOF. We have, $X(w)$ is smooth at e_τ if and only if $\dim T(w,\tau) = \dim X(w)$, and the result follows in view of Theorem 6.1.2 (note that if $\beta = \tau(\alpha)$, then $s_\beta\tau = \tau s_\alpha$). □

COROLLARY 6.1.5. *$X(w)$ is smooth if and only if*

$$l(w) = \#\{\alpha \in R^+ \mid w \geqslant s_\alpha\}.$$

PROOF. If Sing$X(w)$ (the singular locus of $X(w)$) is not empty, then it is a closed B-stable subset of $X(w)$, and hence is a union of Schubert subvarieties of $X(w)$. In particular, we obtain that Sing$X(w)$ is non-empty if and only if $e_{\mathrm{id}} \in$ Sing$X(w)$. Thus $X(\tau)$ is smooth if and only if it is smooth at e_{id}, and the result follows from Theorem 6.1.2. □

6.2. The G-module $V_k(\omega_d)$

Let $G = SL(n)$, and $V = k^n$. We denote the standard basis for k^n by $\{e_1, \cdots, e_n\}$. We follow **[27]** for indexing the roots. Given a root $\beta = \epsilon_j - \epsilon_k, 1 \leqslant j, k \leqslant n$, the element $X_{-\beta}$ of the Chevalley basis of $\mathfrak{g}$ is given by $X_{-\beta} = E_{kj}$, where E_{kj} is the elementary matrix with 1 at the (k, j)-th place, and 0's elsewhere. We have,

$$X_{-\beta}e_i = \begin{cases} 0, & \text{if } i \neq j \\ e_k, & \text{if } i = j. \end{cases}$$

For $1 \leqslant d \leqslant l(= n-1)$, we have, the Weyl module $V_k(\omega_d) = \wedge^d V$. For $w = (a_1 \cdots a_n) \in W$, we shall denote by $w^{(d)}$ the element in W^d representing the coset wW_d. For $w = (a_1 \cdots a_n) \in W$, we have, the extremal weight vector $q_{w^{(d)}}$ in $V_k(\omega_d)$ of weight $w(\omega_d)$ is given by $q_{w^{(d)}} = \pm e_{a_1} \wedge \cdots \wedge e_{a_d}$. Let $\beta \in \tau(R^+)$, say $\beta = \tau(\epsilon_j - \epsilon_k), 1 \leqslant j < k \leqslant n$. We have $X_{-\beta} = E_{a_k a_j}$, and it follows easily that for $1 \leqslant d \leqslant l$, $X_{-\beta}q_{w^{(d)}} \neq 0$ if and only if $a_j \in \{a_1, \cdots a_d\}$, $a_k \notin \{a_1, \cdots, a_d\}$, i.e., if and only if $j \leqslant d < k$, in which case
$X_{-\beta}q_{w^{(d)}} = \pm e_{a_1} \wedge \cdots \wedge e_{a_{j-1}} \wedge e_{a_k} \wedge e_{a_{j+1}} \wedge \cdots \wedge e_{a_d}$. Hence we obtain

$$X_{-\beta}q_{w^{(d)}} = \pm q_{(s_\beta\tau)^{(d)}}, \text{ when } X_{-\beta}q_{w^{(d)}} \neq 0.$$

Recall $N(w,\tau) = \{\beta \in \tau(R^+) \mid X_{-\beta} \in T(w,\tau)\}$. We have (in view of Theorem 6.1.3), $N(w,\tau) = \{\beta \in \tau(R^+) \mid w \geqslant s_\beta\tau\}$.

THEOREM 6.2.1. *Let $w, \tau \in W$, $w \geqslant \tau$, and $\beta \in \tau(R^+)$. Then $\beta \in N(w,\tau)$ if and only if $m_w(\tau(\omega_d) - \beta) = m(\tau(\omega_d) - \beta)$, for all $1 \leqslant d \leqslant l$, where $m(\tau(\omega_d) - \beta)$ (resp. $m_w(\tau(\omega_d) - \beta)$) denotes the multiplicity of $\tau(\omega_d) - \beta$ in $V_k(\omega_d)$ (resp. V_{w,ω_d} (cf. Chapter 4, §4.2)).*

PROOF. Given $d, 1 \leqslant d \leqslant l$, and $\beta = \tau(\epsilon_j - \epsilon_k), 1 \leqslant j < k \leqslant n$, from our discussion above, we see easily that

$$m(\tau(\omega_d) - \beta) = \begin{cases} 0, & \text{if } d < j \text{ or } d \geqslant k \\ 1, & \text{if } j \leqslant d < k. \end{cases}$$

Also note that if $m_w(\tau(\omega_d)-\beta) \neq 0$, then $q_{(s_\beta\tau)^{(d)}}$ is a generator of the corresponding (one dimensional) weight space. Hence we obtain that

$$m_w(\tau(\omega_d) - \beta) = m(\tau(\omega_d) - \beta)$$

for all $1 \leqslant d \leqslant l$ if and only if for all d, $j \leqslant d < k$, $w^{(d)} \geqslant (s_\beta\tau)^{(d)}$, i.e., if and only if $w \geqslant s_\beta\tau$. The result follows from this (in view of the definition of $N(w,\tau)$). □

THEOREM 6.2.2. *Let $G = SL(n)$. Let $w, \tau \in W$, $w \geqslant \tau$, $\beta \in \tau(R^+)$. Then the following are equivalent.*

(1) $w \geqslant s_\beta\tau$.
(2) $m_w(\tau(\omega_d) - \beta) = m(\tau(\omega_d) - \beta)$, *for all $1 \leqslant d \leqslant l$, l being the rank of G.*
(3) $\beta \in N(w,\tau)$.

PROOF. The result follows from Theorems 6.1.3 and 6.2.1. □

REMARK 6.2.3. The result in Theorem 6.2.2 need not hold for other classical groups (see **[93]**, **[94]** for details).

7. Multiplicity at a point on a variety X

7.1. Multiplicity of a local ring

Let A be a finitely generated k-algebra. Further let A be local with $\mathfrak{m}$ as the unique maximal ideal. For $l \geq 0$, let $\psi_A(l) = \text{length}(A/\mathfrak{m}^l)$ $(= \dim_k(A/\mathfrak{m}^l))$. ψ_A is called the *Hilbert-Samuel* function of A. Recall (see **[49]** for example) the following:

THEOREM 7.1.1. *There exists a polynomial $P_A(x) \in \mathbb{Q}[x]$, called the Hilbert-Samuel polynomial of A such that*

(1) $\psi_A(l) = P_A(l)$, $l \gg 0$.
(2) $\deg P_A(x) = \dim A$.
(3) *The leading coefficient of $P_A(x)$ is of the form $e_A/n!$, where $e_A \in \mathbb{Z}^+$ and* $n = \dim A$.

DEFINITION 7.1.2. With notations as in Theorem 7.1.1, the number e_A is called the *multiplicity* of A.

DEFINITION 7.1.3. For a point P on an algebraic variety X, the multiplicity of X at P is defined to be the number e_A, where $A = \mathcal{O}_{X,P}$, the stalk at P, and is denoted $\text{mult}_P X$.

PROPOSITION 7.1.4. *Let P be a point on an algebraic variety X. Then P is a smooth point of X if and only if $\text{mult}_P X = 1$.*

For a proof, see **[126]**.

Recall (see **[49]** for example) the following:

THEOREM 7.1.5. *Let $gr(A,\mathfrak{m}) = \bigoplus_{l \geqslant 0} \mathfrak{m}^l/\mathfrak{m}^{l+1}$. There exists a polynomial $Q_A(x) \in \mathbb{Q}[x]$, such that*

(1) $Q_A(l) = \dim_k \mathfrak{m}^l/\mathfrak{m}^{l+1}$, $l \gg 0$.

(2) $\deg Q_A(x) = \dim A - 1 (= n-1)$.

(3) *The leading coefficient of* $Q_A(x)$ *is of the form* $\frac{f_A}{(n-1)!}$, *where* $f_A \in \mathbb{Z}^+$.

REMARK 7.1.6. With notations as above, by considering the exact sequence

$$0 \to \mathfrak{m}^l/\mathfrak{m}^{l+1} \to A/\mathfrak{m}^{l+1} \to A/\mathfrak{m}^l \to 0,$$

we have for $l \gg 0$,

$$\begin{aligned} Q_A(l) &= P_A(l+1) - P_A(l) \\ &= \left[\frac{e_A}{n!}(l+1)^n + ()(l+1)^{n-1} + \cdots\right] - \left[\frac{e_A}{n!}(l)^n + ()(l)^{n-1} + \cdots\right] \\ &= \frac{e_A}{(n-1)!}(l)^{n-1} + \cdots . \end{aligned}$$

Hence we obtain

$$e_A = f_A.$$

Example. Let R be the polynomial algebra $k[x_1, \cdots x_n]$, $\mathfrak{a}$ the maximal ideal generated by $\{x_1, \cdots x_n\}$, and $A = R_{\mathfrak{a}}$ (the localization at $\mathfrak{a}$). Let $\mathfrak{m} = \mathfrak{a}R_{\mathfrak{a}}$. We have, $\mathfrak{m}^l/\mathfrak{m}^{l+1}$ is the span of monomials of total degree l and has dimension $\binom{l+n-1}{n-1}$. Hence $Q_A(l) = \binom{l+n-1}{n-1}$ and is a polynomial in l of degree $n-1$; further, the leading coefficient of $Q_A(l)$ is equal to $\frac{1}{(n-1)!}$. Therefore $e_A = 1$.

7.2. Multiplicity of a graded affine k-algebra

For this subsection we refer the reader to **[67]**, **[49]**. Let B be a graded, finitely generated k-algebra. The function $f_B(n) = \dim_k B_n, n \in \mathbb{Z}^+$ is called the *Hilbert function* of B, and $\sum_{n\in\mathbb{Z}^+} f_B(n)t^n$ is called the *Hilbert series* of B.

THEOREM 7.2.1. **[67]** *There exists a polynomial* $P_B(x) \in \mathbb{Q}[x]$ *of degree* $\dim(X)$, *where* $X = Proj(B)$ *such that* $f_B(n) = P_B(n)$, *for* $n \gg 0$. *Further, the leading coefficient of* $P_B(n)$ *is of the form* $c_B/r!$, *where* $c_B \in \mathbb{N}$ *and* $r = \deg P_B(x)$.

DEFINITION 7.2.2. Let $X \subset \mathbb{P}^n$ be a projective variety with B as the associated homogeneous coordinate ring. Then c_B is called the *degree* of X and is denoted $\deg X$.

We recall the following result (cf. **[67]**, Theorem 7.7) on $\deg X$:

THEOREM 7.2.3. *Let* X *be a variety of dimension* $\geqslant 1$ *in* $\mathbb{P}^n$, *and let* H *be a hyper surface not containing* X. *Let* $Z_1, \cdots, Z_s$ *be the irreducible components of* $X \cap H$. *Then*

$$\deg X \cdot \deg H = \sum_{j=1}^{s} i(X, H; Z_j) \deg Z_j,$$

where $i(X, H; Z_j)$ *denotes the intersection multiplicity of* Z_j *in* $X \cap H$, *and* $\deg H$ *is the degree of the homogeneous polynomial defining* H.
(refer to **[67]**, Chapter 1, §7, for the definition of intersection multiplicity)

REMARK 7.2.4. Let X, B be as in Definition 7.2.2. Consider $\widehat{X} = \mathrm{Spec}B$, the cone over X; denote its vertex by O. Then

(1) $$\mathrm{mult}_O \widehat{X} = \deg X.$$

REMARK 7.2.5. Let $P \in X$, X being an algebraic variety, and let $\mathcal{O}_{X,P} = A$ with $\mathfrak{m}$ as the unique maximal ideal of the local ring A. Let C_P be the *tangent cone* at P, namely $C_P = \mathrm{Spec}A(P)$, where $A(P) = \mathrm{gr}(A, \mathfrak{m})$. Then $\mathrm{mult}_P X = \mathrm{mult}_o C_P = \deg \mathrm{Proj}(A(P))$, where o is the vertex of C_P.

8. Degree of a Schubert variety

8.1. Chevalley multiplicity

Let $X_Q(w)$ be a Schubert variety in G/Q, Q being a parabolic subgroup. Let $L = L(\lambda)$ be a very ample line bundle on G/Q; note that for a simple root α, $(\lambda, \alpha^*) \neq 0$, if and only if α does not belong to S_Q. Consider the projective embedding

$$X_Q(w) \hookrightarrow G/Q \hookrightarrow \mathrm{Proj}\, H^0(G/Q, L).$$

Let $X_Q(w')$ be a Schubert divisor in $X_Q(w)$, say $w' = ws_\beta$, for some $\beta \in R^+$ (here, we suppose that $w, w' \in W_Q^{\min}$). The positive integer $m_\lambda(w, w') = (\lambda, \beta^*)$ is called the *Chevalley multiplicity* of $X_Q(w')$ in $X_Q(w)$.

Geometric interpretation of $m_\lambda(w, w')$. Let p_w be the extremal weight vector in $H^0(G/Q, L)$ of weight $-w(\lambda)$ (cf. Chapter 4, §4.1), and let H_w be the hyperplane in $\mathrm{Proj}\, H^0(G/Q, L)$ defined by p_w (note that $\deg H_w$=1). We have, $X_Q(w) \cap H_w$ is the union of all the Schubert divisors in $X_Q(w)$ (since $p_w \,|\, X_{Q(\tau)} \neq 0 \Leftrightarrow \tau \geqslant w$ (cf. Chapter 4, Lemma 4.3.1)). We have (cf. **[37]**), $m_\lambda(w, w') = i(X_Q(w), H_w; X_Q(w'))$, the intersection multiplicity of $X_Q(w')$ in $X_Q(w) \cap H_w$. Hence we obtain (in view of Theorem 7.2.3)

THEOREM 8.1.1.

$$\deg X_Q(w) = \sum m_\lambda(w, w') \cdot \deg X_Q(w'),$$

where the summation runs over all the Schubert divisors $X_Q(w')$ in $X_Q(w)$.

Combinatorial interpretation of $\deg X_Q(w)$**.** Consider the Hasse diagram given by the Chevalley-Bruhat order on $[id, w] := \{\tau \in W_Q^{\min} \,|\, \tau \leqslant w\}$ (note that $[id, w]$ is a ranked poset, i.e., all maximal chains have the same length, namely $l(w) + 1$; here, by a chain we mean a totally ordered subset of $[id, w]$). Consider an edge $\tau' \to \tau$, where $X(\tau')$ is a Schubert divisor in $X(\tau)$, and give it the weight $m(\tau, \tau')$. Then $\deg X_Q(w)$ is simply the number of maximal chains with the edges counted with the respective multiplicities; to be very precise, to a maximal chain $\underline{c} : \{\phi_0 = w > \phi_1 > \cdots > \phi_r = id\}$, where $r = l(w)$, the dimension of $X(w)$, assign the weight $n(\underline{c}) := \prod_{i=1}^r m(\phi_{i-1}, \phi_i)$. Then

$$\deg X_Q(w) = \sum_{\underline{c}} n(\underline{c}),$$

where the summation runs over all the maximal chains $\underline{c}$ in $[id, w]$.

In particular, taking $G = GL(n)$, $Q = P_d$, a maximal parabolic subgroup so that $G/Q = G_{d,n}$, the Grassmannian, and $\lambda = \omega_d$, we have $m(w, w') = 1$ for all w, w' such that $X_Q(w')$ is a Schubert divisor in $X_Q(w)$ (note that ω_d is minuscule for all $1 \leqslant d \leqslant n-1$, i.e., $(\omega_d, \beta^*) \leqslant 1$, $\forall \beta \in R^+$). Hence we obtain that the degree of a Schubert variety $X_{P_d}(w)$ in $G_{d,n}$ for the Plücker embedding $X_{P_d}(w) \hookrightarrow \mathbb{P}(\Lambda^d K^n)$ is simply the number of maximal chains in $[id, w]$.

Example: Let $G = SL(3)$, $\lambda = \omega_1 + \omega_2$, the adjoint representation. The corresponding weighted Hasse diagram is given by

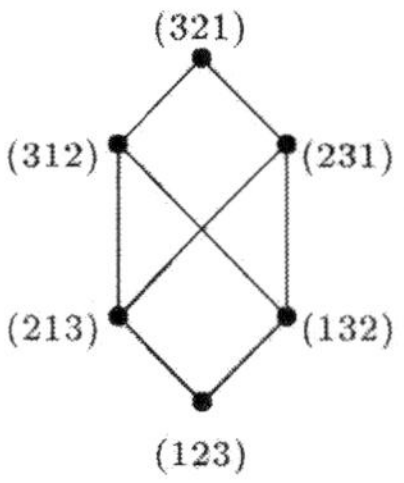

The edges $(213) \to (312)$, $(132) \to (231)$ have weights 2. All other edges have weight 1. For the embedding $X(w) \hookrightarrow G/B \hookrightarrow \mathrm{Proj}H^0(G/B, L(\lambda))$, we have, $\deg G/B = 6$, $\deg X(312) = \deg X(231) = 3$, $\deg X(213) = \deg X(132) = 1$.

Example: Consider $X(235) \subset G_{3,6}$ (= the Grassmannian of 3-planes in K^6). The corresponding weighted Hasse diagram of $[id, w]$ is given by

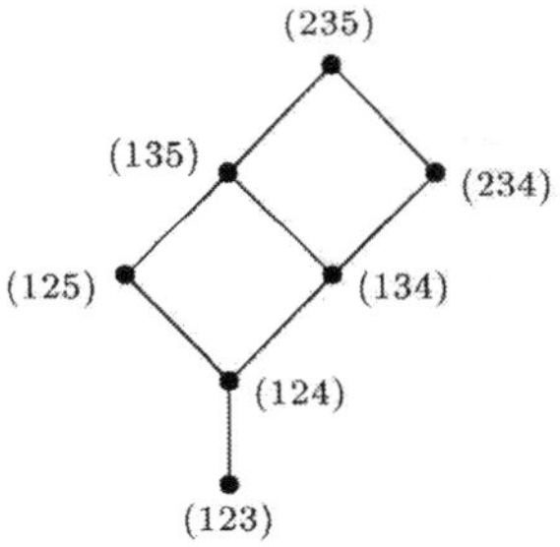

(The weight for any edge is just 1). For the corresponding Plücker embedding, we have, $\deg X(235) = 3$.

9. Multiplicity at a singular point

In this section, we give a recursive formula for the multiplicity at a singular point on a Schubert variety $X(w)$ in the Grassmannian.

Let $X = X(w)$ be a Schubert variety in the Grassmannian $G_{d,n}$ ($\cong SL(n)/P_d$). Let us denote P_d by just P. Given $x \in X$, we shall denote $\mathrm{mult}_x X$ by just $m_x X$. For $x \in B \cdot e_\tau$, we have $m_x(X) = m_{e_\tau}(X)$. Thus it suffices to determine $m_{e_\tau}(X)$, for

$\tau \leqslant w$. For $e_\tau \in X$, let us denote $m_{e_\tau}(X)$ by just $m_\tau(w)$. Consider a $\tau \in W^P$ such that $\tau \leqslant w$.

9.1. Evaluation of Plücker coordinates on $U_\tau^- e_\tau$

I. Let us first consider the case $\tau = \text{id}$. We identify $U^- e_{\text{id}}$ with

$$\left\{ \begin{pmatrix} & \mathrm{Id}_{d\times d} & \\ x_{d+1\,1} & \cdots & x_{d+1\,d} \\ \vdots & & \vdots \\ x_{n\,1} & \cdots & x_{n\,d} \end{pmatrix}, \quad x_{ij} \in k, \quad d+1 \leqslant i \leqslant n, \quad 1 \leqslant j \leqslant d \right\}.$$

Let A be the affine algebra of $U^- e_{\text{id}}$. Let us identify A with the polynomial algebra $k[x_{-\beta}, \beta \in R^+ \setminus R_P^+]$. To be very precise, we have

$$R^+ \setminus R_P^+ = \{\epsilon_j - \epsilon_i,\ 1 \leqslant j \leqslant d,\ d+1 \leqslant i \leqslant n\};$$

given $\beta \in R^+ \setminus R_P^+$, say $\beta = \epsilon_j - \epsilon_i$, we identify $x_{-\beta}$ with x_{ij}. Hence we obtain that the expression for $p_\theta|_{U^- e_{\text{id}}}$ in the local coordinates $x_{-\beta}$'s is homogeneous.

EXAMPLE 9.1.1. Consider $G_{2,4}$. Then

$$U^- e_{\text{id}} = \left\{ \begin{pmatrix} 1 & 0 \\ 0 & 1 \\ x_{31} & x_{32} \\ x_{41} & x_{42} \end{pmatrix}, \quad x_{ij} \in k \right\}.$$

On $U^- e_{\text{id}}$, we have $p_{12} = 1$, $p_{13} = x_{32}$, $p_{14} = x_{42}$, $p_{23} = x_{31}$, $p_{24} = x_{41}$, $p_{34} = x_{31}x_{42} - x_{41}x_{32}$.

Thus a Plücker coordinate is homogeneous in the local coordinates x_{ij}, $d+1 \leqslant i \leqslant n$, $1 \leqslant j \leqslant d$. This phenomenon does not hold for G/P, if P is not maximal as seen in the following example.

EXAMPLE 9.1.2. Consider $SL(4)/B$. We have

$$U^- e_{\text{id}} = \left\{ \begin{pmatrix} 1 & 0 & 0 & 0 \\ x_{21} & 1 & 0 & 0 \\ x_{31} & x_{32} & 1 & 0 \\ x_{41} & x_{42} & x_{43} & 1 \end{pmatrix}, \quad x_{ij} \in k \right\}.$$

Then, for example, on $U^- e_{\text{id}}$, $p_{23} = -x_{31} + x_{32}x_{21}$ is non-homogeneous!

II. Returning to $G_{d,n}$, let now τ be any other element in W^P, say $\tau = (a_1, \ldots, a_n)$. Then $U_\tau^- e_\tau$ consists of $\{N_{d,n}\}$, where $N_{d,n}$ is obtained from $\begin{pmatrix} \mathrm{Id} \\ X \end{pmatrix}_{n\times d}$ (with notations as above) by permuting the rows by τ^{-1}. (Note that $U_\tau^- e_\tau = \tau U^- e_{\text{id}}$.)

EXAMPLE 9.1.3. Consider $G_{2,4}$, and let $\tau = (2314)$. Then $\tau^{-1} = (3124)$, and

$$U_\tau^- e_\tau = \left\{ \begin{pmatrix} x_{31} & x_{32} \\ 1 & 0 \\ 0 & 1 \\ x_{41} & x_{42} \end{pmatrix}, \quad x_{ij} \in k \right\}.$$

We have on $U_\tau^- e_\tau$, $p_{12} = -x_{32}$, $p_{13} = x_{31}$, $p_{14} = x_{31}x_{42} - x_{41}x_{32}$, $p_{23} = 1$, $p_{24} = x_{42}$, $p_{34} = -x_{41}$.

As in the case $\tau = \mathrm{id}$, we find that for $\theta \in W^P$, $p_\theta|_{U_\tau^- e_\tau}$ is homogeneous in local coordinates. In fact we have

PROPOSITION 9.1.4. *Let $\theta \in W^P$. We have a natural isomorphism*

$$k[x_{-\beta}, \beta \in R^+ \setminus R_P^+] \cong k[x_{-\tau(\beta)}, \beta \in R^+ \setminus R_P^+],$$

given by

$$p_\theta \longmapsto p_{\tau\theta}$$

(here we have denoted $p_\theta|_{U^- \mathrm{id}}$, $p_{\tau\theta}|_{U_\tau^- e_\tau}$ by just p_θ, $p_{\tau\theta}$ respectively).

The proof is immediate from the above identifications of $U^- e_{\mathrm{id}}$ and $U_\tau^- e_\tau$.

As a consequence, we have

COROLLARY 9.1.5. *Let $\theta \in W^P$. Then the polynomial expression for $p_\theta|_{U_\tau^- e_\tau}$ in the local coordinates $\{x_{-\tau(\beta)}, \beta \in R^+ \setminus R_P^+\}$ is homogeneous.*

Given $y \in \mathfrak{g}$, we identify y with the corresponding right invariant vector field Dy on G. Considering $H^0(G/P, L^m)$, where $L = L_d$, and $m \geqslant 0$, as a $\mathfrak{g}$-module, as discussed in §5, we have $D_y f = yf$, $f \in H^0(G/P, L^m)$. Further, the evaluations of $\partial f/\partial x_{-\beta}$ and $X_{-\beta}$ at e_τ coincide for $\beta \in \tau(R^+ \setminus R_P^+)$.

9.2. The algebra $A_{w,\tau}$

Let

$$R(w,\tau) = \{\beta \in \tau(R^+ \setminus R_P^+) \mid w \geqslant s_\beta\tau\}.$$

We have (cf. Theorem 6.1.3), the Zariski tangent space $T(w,\tau)$ to $X(w)$ at e_τ is spanned by

$$\{X_{-\beta}, \beta \in R(w,\tau)\}.$$

For the rest of this section we fix a $\tau \in W^P$, $\tau \leqslant w$.

As above, we identify A_τ, the affine algebra of $U_\tau^- e_\tau$ with the polynomial algebra $k[x_{-\beta}, \beta \in \tau(R^+ \setminus R_P^+)]$. Let $A_{w,\tau} = A_\tau/I_w$, where I_w is the ideal of elements of A_τ that vanish on $X(w) \cap U_\tau^- e_\tau$.

Now $I(X(w))$, the ideal of $X(w)$ in G/P, is generated by $\{p_\theta, \theta \in W^P \mid w \not\geqslant \theta\}$ (cf. Chapter 6, §3.4). Hence we obtain (cf. Corollary 9.1.5) that I_w is homogeneous. Hence we get

$$(*) \qquad \mathrm{gr}\,(A_{w,\tau}, M_{w,\tau}) = A_{w,\tau},$$

where $M_{w,\tau}$ is the maximal ideal of $A_{w,\tau}$ corresponding to e_τ. In particular, denoting the image of $x_{-\beta}$ under the canonical map $A_\tau \to A_{w,\tau}$ by just $x_{-\beta}$, the set $\{x_{-\beta} \mid$

$\beta \in R(w,\tau)\}$ generates $A_{w,\tau}$. Let $R(w)$ be the homogeneous coordinate ring of $X(w)$ (for the Plücker embedding), $Y(w,\tau) = X(w) \cap U_\tau^- e_\tau$. Then $k[Y(w,\tau)] = A_{w,\tau}$ gets identified with the homogeneous localization $R(w)_{(p_\tau)}$, i.e. the subring of $R(w)_{p_\tau}$ (the localization of $R(w)$ with respect to p_τ) generated by the elements

$$\{\frac{p_\theta}{p_\tau},\ \theta \in W^P, \theta \leqslant w\}.$$

9.3. The integer $d_\tau(\theta)$

Let $\theta \in W^P$. Let $f_\tau(\theta)$ be the polynomial expressing $p_\theta|_{U_\tau^- e_\tau}$ in terms of the local coordinates $\{x_{-\beta}, \beta \in \tau(R^+ \setminus R_P^+)\}$. We define $d_\tau(\theta)$ (also denoted by $\deg_\tau(\theta)$) by

$$d_\tau(\theta) := \deg f_\tau(\theta)$$

(note that $f_\tau(\theta)$ is homogeneous, cf. Corollary 9.1.5). In fact, we have an explicit expression for $d_\tau(\theta)$, as follows:

PROPOSITION 9.3.1. *Let $\theta \in W^P$. Let $\tau = (a_1, \ldots, a_n)$, $\theta = (b_1, \ldots, b_n)$, τ, θ being considered as permutations. Let $r = \#\{a_1, \ldots, a_d\} \cap \{b_1, \ldots, b_d\}$. Then $d_\tau(\theta) = d - r$.*

PROOF. Let $s = d - r$, and

$$\{c_1, \ldots, c_s\} = \{a_1, \ldots, a_d\} \setminus \{a_1, \ldots, a_d\} \cap \{b_1, \ldots, b_d\}$$
$$\{f_1, \ldots, f_s\} = \{b_1, \ldots, b_d\} \setminus \{a_1, \ldots, a_d\} \cap \{b_1, \ldots, b_d\}.$$

For $1 \leqslant t \leqslant s$, let $\beta_t = \tau(\epsilon_i - \epsilon_j)(= \epsilon_{a_i} - \epsilon_{a_j})$, where i and j are given by $a_i = c_t$, $a_j = f_t$ (note that $i \leqslant d < j$, and hence $\beta_t \in \tau(R^+ \setminus R_P^+)$). For $\sigma = (\sigma_1, \ldots, \sigma_n) \in W^P$, let $q_\sigma = e_{\sigma_1} \wedge \ldots \wedge e_{\sigma_d}$, the extremal weight vector in $V_k(\omega_d)$ of weight $\sigma(\omega_d)$ (cf. Chapter 4, §4.1). As seen in §6.2, for $\beta \in R$, say $\beta = \epsilon_i - \epsilon_j$, we have

$$X_\beta q_\tau = \begin{cases} \pm q_{s_\beta \tau} & \text{if } j \in \{a_1, \ldots, a_d\},\ i \notin \{a_1, \ldots, a_d\} \\ 0 & \text{otherwise.} \end{cases}$$

(note that in the case when $j \in \{a_1, \ldots, a_d\}$ and $i \notin \{a_1, \ldots, a_d\}$, $(s_\beta\tau)^{(d)}$ is the d-tuple obtained from $(a_1, \ldots, a_d)$ by replacing j by i). Setting $D' = X_{-\beta_1} \cdots X_{-\beta_s}$, we have $D' q_\tau = \pm q_\theta$. This implies

$$\langle X_{-\beta_1} \ldots X_{-\beta_s} q_\tau, p_\theta \rangle = \pm 1,$$

where $\langle , \rangle$ is the canonical pairing on $\wedge^d V \times (\wedge^d V)^*$, $V = k^n$ (in view of the fact that the bases $\{p_\theta, \theta \in W^P\}$, $\{q_\theta, \theta \in W^P\}$ are dual to each other). Now using the $\mathfrak{g}$-invariance of $\langle , \rangle$, we obtain $\langle q_\tau, X_{-\beta_s} \ldots X_{-\beta_1} p_\theta \rangle = \pm 1$. Thus we obtain $X_{-\beta_s} \ldots X_{-\beta_1} p_\theta = \pm p_\tau$, and hence the monomial $x_{-\beta_s} \ldots x_{-\beta_1}$ appears in $f_\tau(\theta)$ (the polynomial expression for $p_\theta|_{U_\tau^- e_\tau}$ in terms of $x_{-\beta}$'s, $\beta \in \tau(R^+ \setminus R_P^+)$). The required result now follows. □

For $\theta \in W^P$, let $f_\tau(\theta)$ denote the polynomial expression for $p_\theta|_{U_\tau^- e_\tau}$ in terms of the local coordinates $\{x_{-\beta}, \beta \in \tau(R^+ \setminus R_P^+)\}$. We shall denote the restriction of $f_\tau(\theta)$ to $Y(w,\tau)$ $(= X(w) \cap U_\tau^- e_\tau)$ also by just $f_\tau(\theta)$. Let

$$Z_{w,\tau} = \{\theta \in W^P,\ \theta \leqslant w \mid \text{ either } \theta \geqslant \tau \text{ or } \tau \geqslant \theta\}.$$

9.4. A basis for the tangent cone

THEOREM 9.4.1. *With notations as above, given $r \in \mathbb{Z}^+$*

$$\{f_\tau(\theta_1)\dots f_\tau(\theta_m) \mid w \geqslant \theta_1 \geqslant \dots \geqslant \theta_m,\ \theta_i \in Z_{w,\tau},\ \sum_{i=1}^{m} d_\tau(\theta_i) = r\}$$

is a basis for $M_{w,\tau}^r / M_{w,\tau}^{r+1}$.

PROOF. For $F = p_{\theta_1} \cdots p_{\theta_m}$, let $\deg F$ denote the degree of $f_\tau(\theta_1) \cdots f_\tau(\theta_m)$. Let

$$A_r = \{F = p_{\theta_1} \cdots p_{\theta_m},\ w \geqslant \theta_i \mid \deg F = r\}.$$

Then in view of the relation (*) in §9.2, we have, A_r generates $M_{w,\tau}^r / M_{w,\tau}^{r+1}$. Let $F \in A_r$, say $F = p_{\tau_1} \cdots p_{\tau_m}$. From the results in Chapter 6, we know that $p_{\tau_1} \cdots p_{\tau_m}$ is a linear combination of standard monomials $p_{\theta_1} \cdots p_{\theta_m}$, $w \geqslant \theta_1$. We *claim* that in each $p_{\theta_1} \cdots p_{\theta_m}$, $\theta_i \in Z_{w,\tau}$, for all i. Suppose that for some $i, \theta_i \notin Z_{w,\tau}$. This means θ_i and τ are not comparable. Then using the fact that $f_\tau(\tau) = 1$, on $Y(w,\tau)$, we replace p_{θ_i} by $p_{\theta_i} p_\tau$ in $p_{\theta_1} \cdots p_{\theta_m}$. We now use the straightening relation (cf. Chapter 6, Proposition 3.3.3) $p_{\theta_i} p_\tau = \sum c_{\alpha,\beta} p_\alpha p_\beta$ on $X(w)$, where in each term $p_\alpha p_\beta$ on the right hand side, we have $\alpha < w$, and $\alpha >$ both θ_i and τ, and $\beta <$ both θ_i and τ; in particular, we have, in each term $p_\alpha p_\beta$ on the right hand side, α, β belong to $Z_{w,\tau}$. We now proceed as in the proof of Theorem 3.3.1, Chapter 6, to conclude that on $Y(w,\tau)$, F is a linear combination of standard monomials in p_θ's, $\theta \in Z_{w,\tau}$ which proves the Claim. The required result now follows from Claim. □

9.5. A recursive formula for $m_\tau(w)$

THEOREM 9.5.1. *(cf. [**108**]) With notations as above, we have*

$$m_\tau(w) \deg_\tau w = \sum m_\tau(w'),$$

where on the right we sum over all divisors $X(w')$ in $X(w)$ such that $e(\tau) \in X(w')$.

PROOF. Let H_τ be the union of all Schubert divisors $X(w')$ in $X(w)$ such that $e(\tau) \in X(w')$. Let $\varphi_w(r)$ (resp. $\varphi_{H_\tau}(r)$) be the Hilbert function for the tangent cone of $X(w)$ (resp. H_τ) at $e(\tau)$, i.e.

$$\varphi_w(r) = \dim(M_{\tau,w}^r / M_{\tau,w}^{r+1}).$$

Let

$$\mathcal{B}_{w,\tau}(r) = \left\{p_{\tau_1} \dots p_{\tau_m},\ \tau_i \in Z_{w,\tau} \mid (1)\, w \geqslant \tau_1 \geqslant \dots \geqslant \tau_m, (2) \sum d_\tau(\tau_i) = r\right\}.$$

Let

$$\mathcal{B}_1 = \{p_{\tau_1} \cdots p_{\tau_m} \in \mathcal{B}_{w,\tau}(r) \mid \tau_1 = w\},$$
$$\mathcal{B}_2 = \{p_{\tau_1} \cdots p_{\tau_m} \in \mathcal{B}_{w,\tau}(r) \mid \tau_1 < w\}.$$

We have $\mathcal{B}_{\tau,w}(r) = \mathcal{B}_1 \dot{\cup} \mathcal{B}_2$. Hence denoting $d_\tau(w)$ by d, we obtain

$$\varphi_w(r+d) = \varphi_w(r) + \varphi_H(r+d).$$

Taking $r \gg 0$ and comparing the coefficients of r^{u-1}, where $u = \dim X(w)$, we obtain the result. □

COROLLARY 9.5.2. *The Hilbert polynomial $F_{w,\tau}$ of $X_P(w)$ at e_τ is determined inductively by the formula*

$$F_{w,\tau}(r+d) = F_{w,\tau}(r) + F_{\tau,H}(r+d)$$

where H is the union of all Schubert divisors $X_P(w')$ in $X_P(w)$ such that $e_\tau \in X_P(w')$, $F_{\tau,H}$, the Hilbert polynomial of H at e_τ, and $d = \deg_\tau w$.

PROOF. We have, the leading coefficient in $F_{w,\tau}$ is $\dfrac{m_\tau(w)}{d!}$ (and $m_\tau(w)$ is determined as in Theorem 9.5.1). Also, the constant term in $F_{w,\tau}$ is 1; for, denoting by Y the tangent cone to $X(w)$ at e_τ, we have Y is Cohen-Macaulay (since $\mathrm{gr}\,(A_{w,\tau}, M_{w,\tau}) = A_{w,\tau}$, and $A_{w,\tau}$ is Cohen-Macaulay). Hence

$$\chi(\mathcal{O}_Y(r)) = h^0(Y, \mathcal{O}_Y(r)),$$

where χ is the Euler-Poincaré characteristic, and $h^0(Y, \mathcal{O}_Y(r))$ is the dimension of $H^0(Y, \mathcal{O}_Y(r))$. Now for $r \gg 0$, we have

$$F_{\tau,w}(r) = \chi(\mathcal{O}_Y(r)).$$

Hence the constant term in $F_{w,\tau}$ equals $h^0(Y, \mathcal{O}_Y(0))\ (= 1)$. The remaining coefficients in $F_{w,\tau}$ are inductively determined by the formula

$$F_{\tau,w}(r+d) = F_{\tau,w}(r) + F_{\tau,H}(r+d).$$

Here, $F_{\tau,H}$ is computed by induction on the number of components in H, say $H = X_P(w') \cup H'$. Then

$$F_{\tau,H}(r) = F_{\tau,H'}(r) + F_{\tau,w'}(r) - F_{\tau,\, H' \cap X_P(w')}(r),$$

where note that $H' \cap X_P(w')$ is a union of Schubert varieties of smaller dimension. □

We have more generally similar results for Schubert varieties in a minuscule G/P as well as Schubert varieties in the symplectic Grassmannian (cf. **[108]**). We have included below the statements of results for Schubert varieties in a minuscule G/P.

10. Results for Schubert varieties in a minuscule G/P

Let G be semisimple and P a maximal parabolic subgroup of minuscule type, i.e., the associated fundamental weight ω satisfies $(\omega, \beta^*) \leqslant 1$, for all $\beta \in R^+$; for example, any maximal parabolic in $SL(n)$ is minuscule, the corresponding G/P being a Grassmannian. Other examples of maximal parabolics which are minuscule are $G = Sp(2n)$, $P = P_1$, $G = SO(2n), P = P_{n-1}, P_n$, $G = SO(2n+1), P = P_n$ (note that for $G = Sp(2n)$, the corresponding G/P is a projective space, and for $G = SO(2n)$, $G = SO(2n+1)$, the corresponding G/P's are orthogonal Grassmannians consisting of maximal totally isotropic subspaces in $k^N, N = 2n, 2n+1$ equipped with a non-degenerate symmetric bilinear form). In addition, for $G = E_6, E_7$, P_1 is a minuscule maximal parabolic subgroup. The above list in fact exhausts all minuscule maximal parabolic subgroups for all simple G (note that there are no minuscule maximal parabolic subgroups for G of type E_8, F_4, G_2).

For $\theta \in W$, and $P = P_d$ (the maximal parabolic subgroup with $S \setminus \{\alpha_d\}$ as the associated set of simple roots), we shall denote by $\theta^{(d)}$, the element in W^{P_d} representing θW_{P_d}. For the rest of this section, we shall suppose that P is minuscule.

Let $P = P_d$ and $\tau, w \in W^P, w \geqslant \tau$. Set

$$R_P(w, \tau) = \{\beta \in \tau(R^+ - R_P^+) \mid w \geqslant (s_\beta \tau)^{(d)}\}.$$

Recall (cf. **[108]**)

THEOREM 10.0.1. *$T(w, \tau)$, the tangent space to $X_P(w)$ at e_τ is spanned by $\{X_{-\beta}, \beta \in R_P(w, \tau)\}$.*

Let U_τ^- be the unipotent subgroup of G generated by the root subgroups $U_{-\beta}$, $\beta \in \tau(R^+ - R_P^+)$. We have

$$U_{-\beta} \simeq \mathbb{G}_a, \; U_\tau^- \simeq \prod_{\beta \in \tau(R^+ - R_P^+)} U_{-\beta}.$$

We shall denote the coordinate system on $U_\tau^- e_\tau$ induced by the above identification by $\{x_{-\beta}, \beta \in \tau(R^+ - R_P^+)\}$. We shall denote by A_τ the polynomial algebra $k[x_{-\beta}, \beta \in \tau(R^+ - R_P^+)]$. Let $A_{\tau,w} = A_\tau / \mathfrak{I}_P(w)$, where $\mathfrak{I}_P(w)$ is the ideal of elements of A_τ that vanish on $X_P(w) \cap U_\tau^- e_\tau$. Then $A_{\tau,w}$ is generated as an algebra by $\{x_{-\beta}, \beta \in R_P(w, \tau)\}$ (in view of Theorem 10.0.1).

10.1. Homogeneity of $\mathfrak{I}_P(w)$

Let ω be the fundamental weight associated to P. We shall denote the extremal weight vectors in $H^0(G/P, L_\omega)$ by $\{p_\tau, \tau \in W^P\}$. Recall (cf. **[149]**) that $\{p_\tau, \tau \in W^P\}$ is a basis for $H^0(G/P, L_\omega)$. Given $y \in \mathfrak{g}$, we identify y with the corresponding right invariant vector field D_y on G. Considering $H^0(G/P, L_\omega)$ as a $\mathfrak{g}$-module, we have

$$D_y f = yf, \qquad f \in H^0(G/P, L_\omega).$$

Further, the evaluations of $\dfrac{\partial f}{\partial x_{-\beta}}$ and $X_{-\beta}f$ at e_τ coincide for $\beta \in \tau(R^+ - R_P^+)$ (recall that for a root γ, X_γ denotes the element in the Chevalley basis of $\mathfrak{g}$ associated to γ). Take an indexing of $\tau(R^+ - R_P^+)$.

Given $f \in H^0(G/P, L_\omega)$, let

$$\mathcal{D}_f = \{D = X_{-\beta_s} \cdots X_{-\beta_1} \mid Df = a_D p_\tau,\ a_D \in k^*\}$$

(here, $X_{-\beta_s} \cdots X_{-\beta_1}$ is considered as an element of $U(\mathfrak{g})$, the universal enveloping algebra of $\mathfrak{g}$). If $\mathcal{D}_f \neq \emptyset$, then f is a weight vector for the action of T, and in A_τ, f gets identified with $\sum_{D \in \mathcal{D}_f} a_D x_D$, where if $D = X_{-\beta_s} \cdots X_{-\beta_1}$, then x_D is the monomial $x_{-\beta_s} \cdots x_{-\beta_1}$.

LEMMA 10.1.1. *(cf.* **[108]***) Let $\tau, \theta \in W^P$. Suppose*

$$D_1 = X_{-\beta_s} \cdots X_{-\beta_1}, \qquad D_2 = X_{-\beta'_t} \cdots X_{-\beta'_1}$$

are such that $D_i p_\theta = a_{D_i} p_\tau$, $a_{D_i} \in k^*$, $i = 1, 2$. Then $s = t$, i.e., ord $D_1 =$ ord D_2.

Hence we obtain that for $f = p_\theta$, if $\mathcal{D}_f \neq \emptyset$, then for $D_1, D_2 \in \mathcal{D}_f$, ord $D_1 =$ ord D_2; thus in A_τ, p_θ gets identified with a homogeneous polynomial. Now $I_P(X(w))$, the ideal of $X_P(w)$ in G/P is generated by $\{p_\theta \mid w \not\geq \theta\}$ (cf. **[149]**). Hence we obtain that $I_P(w)$ (= the ideal of elements of A_τ that vanish on $X_P(w) \cap U_\tau^- e_\tau$) is homogeneous. Thus we get

$$\mathrm{gr}(A_{\tau,w}, M_{\tau,w}) = A_{\tau,w}$$

where $M_{\tau,w}$ is the maximal ideal in $A_{\tau,w}$ that corresponds to the point e_τ.

10.2. A basis for $(M_{\tau,w})^r/(M_{\tau,w})^{r+1}$

Let us fix $\tau, w \in W^P, w \geqslant \tau$. Let $\theta \in W^P$, $\theta \leqslant w$ be such that θ and τ are comparable i.e. either $\tau \geq \theta$ or $\theta \geq \tau$.

LEMMA 10.2.1. *(cf.* **[108]***) Let w, τ, θ be as above. Then there exists a $D = X_{-\beta_s} \cdots X_{-\beta_1}$, $\beta_i \in R(w, \tau)$, $1 \leq i \leq s$ such that $Dp_\theta = a_D p_\tau, a_D \in k^*$.*

Set

$$N_P(\theta) = \{D = X_{-\beta_s} \cdots X_{-\beta_1},\ \beta_i \in R_P(w, \tau),\ 1 \leqslant i \leqslant s \mid Dp_\theta = a_D p_\tau,\ a_D \in k^*\}.$$

Note that $N_P(\theta) \neq \emptyset$ (in view of the above Lemma).

DEFINITION 10.2.2. With notations as above, define $d_\theta :=$ ord D, where $D \in N_P(\theta)$ (note that the right hand side is independent of D (cf. Lemma 10.1.1), so the definition makes sense).

REMARK 10.2.3. If $\theta = w$, then we refer to d_w as the *degree* of $X_P(w)$ at e_τ and denote it by $\deg_\tau w$.

DEFINITION 10.2.4. With notations as above, define $x_\theta := \sum_{D \in N_P(\theta)} a_D x_D$, where for $D = X_{-\beta_s} \cdots X_{-\beta_1}$, we set $x_D = x_{-\beta_s} \cdots x_{-\beta_1}$.

THEOREM 10.2.5. *(cf. [108]) With notations as above, the set* $\{(x_{\tau_1} \cdots x_{\tau_m})\}$ *satisfying*

(1) $w \geqslant \tau_1 \geqslant \cdots \geqslant \tau_m$
(2) *for* $1 \leq i \leq m$, *either* $\tau_i \geqslant \tau$ *or* $\tau_i \leqslant \tau$
(3) $\sum d_{\tau_i} = r$

is a basis for $(M_{\tau,w})^r/(M_{\tau,w})^{r+1}$.

REMARK 10.2.6. With the above notations, we have, p_θ (as an element of $A_{\tau,w}$) gets identified with x_θ.

THEOREM 10.2.7. *(cf. [108]) Denoting* $\text{mult}_{e_\tau} X$ *by* $m_\tau(w)$, *we have*

$$m_\tau(w)\ \deg_\tau w = \sum m_\tau(w')$$

where the sum on the right hand side runs over all Schubert divisors $X_P(w')$ *in* $X_P(w)$ *such that* $e_\tau \in X_P(w')$.

COROLLARY 10.2.8. *(cf. [108]) The Hilbert polynomial* $F_{\tau,w}$ *of* $X_P(w)$ *at* e_τ *is determined inductively by the formula*

$$F_{\tau,w}(r+d) = F_{\tau,w}(r) + F_{\tau,H}(r+d)$$

where H *is the union of all Schubert divisors* $X_P(w')$ *in* $X_P(w)$ *such that* $e_\tau \in X_P(w')$ *and* $d = \deg_\tau w$.

(Here, $F_{\tau,H}$ is computed by induction on the number of components in H, say $H = X_P(w') \cup H'$. Then

$$F_{\tau,H}(r) = F_{\tau,H'}(r) + F_{\tau,w'}(r) - F_{\tau,\, H' \cap X_P(w')}(r),$$

where note that $H' \cap X_P(w')$ is a union of Schubert varieties of smaller dimension.)

11. The symplectic Grassmannian

Let $V = k^{2n}$ together with a nondegenerate, skew-symmetric bilinear form $(\,,)$. Let $H = SL(V)$ and

$$G = Sp(V) = \{A \in SL(V) \mid A \quad \text{leaves the form } (\ ,\ ) \text{ invariant}\}.$$

Taking the matrix of the form (with respect to the standard basis $\{e_1, ..., e_{2n}\}$ of V) to be

$$E = \begin{pmatrix} 0 & J \\ -J & 0 \end{pmatrix}$$

where J is the anti-diagonal $(1, \cdots, 1)$ of size $n \times n$, we may realize $Sp(V)$ as the fixed point set of a certain involution σ on $SL(V)$, namely $G = H^\sigma$, where $\sigma : H \longrightarrow H$ is given by $\sigma(A) = E({}^tA)^{-1}E^{-1}$ as follows:

$$\begin{aligned} G = Sp(2n) &= \{A \in SL(2n) \mid {}^tAEA = E\} \\ &= \{A \in SL(2n) \mid E^{-1}({}^tA)^{-1}E = A\} \\ &= \{A \in SL(2n) \mid E({}^tA)^{-1}E^{-1} = A\} \\ &= H^\sigma. \end{aligned}$$

(note that $E^{-1} = -E$). Let P_n be the maximal parabolic subgroup with $S \setminus \{\alpha_n\}$ as the associated set of simple roots. We shall denote P_n by P. Then W^P may be identified (cf. **[90]**) as

$$W^P = \left\{ (a_1, \cdots, a_n) \;\middle|\; \begin{array}{l} (1)\ 1 \leqslant a_1 < \cdots < a_n \leqslant 2n \\ (2)\ \text{for } 1 \leqslant i \leqslant n, \text{ precisely one of } \{i, i'\} \text{ is in } \{a_1, \cdots, a_n\} \end{array} \right\}$$

(here, $i' = 2n + 1 - i$).

11.1. Description of $T(w, \tau)$

Let $w, \tau \in W^P$, $w \geqslant \tau$. Let $S_P(w, \tau) = \{\beta \in \tau(R^+ - R_P^+) \mid X_{-\beta} \in T(w, \tau)\}$. As above, for $\theta \in W$, we shall denote by $\theta^{(n)}$, the element in W^P representing θW_P. We have (cf. **[92]**, **[93]**)

THEOREM 11.1.1. *Let $\beta = \tau(\alpha)$, where $\alpha \in R^+$.*

(1) *Let $\alpha = \epsilon_j - \epsilon_n, 1 \leqslant j < n$, or $2\epsilon_j, 1 \leqslant j \leqslant n$. Then $\beta \in S_P(w, \tau)$ if and only if $w \geqslant (s_\beta \tau)^{(n)}$.*

(2) *Let $\alpha = \epsilon_j + \epsilon_k, 1 \leqslant j < k \leqslant n$.*
   (a) *Let $\tau > (\tau s_\alpha)^{(n)} (= (s_\beta \tau)^{(n)})$. Then $\beta \in S_P(w, \tau)$ necessarily.*
   (b) *Let $\tau < (\tau s_\alpha)^{(n)}$.*
      (i) *Let $\tau > (\tau s_{2\epsilon_j})^{(n)}$ or $(\tau s_{2\epsilon_k})^{(n)}$. Then $\beta \in S_P(w, \tau)$ if and only if $w \geqslant (s_\beta \tau)^{(n)}$.*
      (ii) *Let $\tau < (\tau s_{2\epsilon_j})^{(n)}$ and $(\tau s_{2\epsilon_k})^{(n)}$.*
         (A) *If $\tau < (\tau s_{\epsilon_j - \epsilon_k})^{(n)}$, then $\beta \in S_P(w, \tau)$ if and only if $w \geqslant (s_\beta \tau)^{(n)} (= (\tau s_{\epsilon_j + \epsilon_k})^{(n)})$ or $(\tau s_{2\epsilon_j})^{(n)}$.*
         (B) *If $\tau > (\tau s_{\epsilon_j - \epsilon_k})^{(n)}$, then $\beta \in S_P(w, \tau)$ if and only if $w \geqslant (\tau s_{\epsilon_j - \epsilon_k} s_{2\epsilon_j})^{(n)}$.*

We have a canonical inclusion

$$G/P \hookrightarrow G_{n,2n}$$

$G_{n,2n}$ being the Grassmannian of n-dimensional subspaces in $V = k^{2n}$. Let us consider the Plucker embedding of $G_{n,2n}$:

$$G/P \hookrightarrow G_{n,2n} \hookrightarrow \mathbb{P}(\Lambda^n V).$$

Let L be the tautological line bundle on $\mathbb{P}(\Lambda^n V)$. We shall denote the restrictions of L to $G_{n,2n}$ and G/P also by just L. We have a canonical surjective map (induced by restriction)

$$H^0(G_{n,2n}, L) \longrightarrow H^0(G/P, L).$$

For $\theta = (\theta_1, \cdots, \theta_n)$, $1 \leqslant \theta_1 < \cdots < \theta_n \leqslant 2n$, let us denote the associated Plücker coordinate on $G_{n,2n}$ by f_θ and the restriction of f_θ to G/P by f'_θ. Let $\theta = (a_1, \cdots, a_r, y_1, y'_1, y_2, y'_2, \cdots, y_s, y'_s)$, where $y_j \leqslant n$, $1 \leqslant j \leqslant s$ and $a'_i (= 2n+1-a_i) \notin \theta, 1 \leqslant i \leqslant r$. Let $\{x_1, \cdots, x_t\}$ be the complement of $\{|a_1|, \cdots, |a_r|, y_1, y_2, \cdots, y_s\}$ in $\{1, \cdots, n\}$ (where $|q| = \min\{q, 2n+1-q\}$, $1 \leqslant q \leqslant 2n$).

Let $\theta' = (a_1, \cdots, a_r, x_1, x_1', x_2, x_2', \cdots, x_t, x_t')$. Then we have (cf. [**104**])

(1) $f_\theta' = f_{\theta'}'$.

(2) $\{f_\theta' : \theta \in I_{n,2n}\}$ is a basis of $H^0(G/P, L)$.

Note that if $\theta = \theta'$ (in which case $s = 0$) then $f_\theta' = p_\theta$, the extremal weight vector in $H^0(G/B, L)$ of weight $\theta(\omega_n)$. If $\theta \neq \theta'$, then set

$$\rho = (a_1, \cdots, a_r, y_1, y_2, \cdots, y_s, x_1', \cdots, x_s'),$$
$$\xi = (a_1, \cdots, a_r, y_1', y_2', \cdots, y_s', x_1, \cdots, x_s).$$

We have, ρ and ξ are comparable. We set $\delta =$ the bigger of $\{\rho, \xi\}$, $\varphi =$ the smaller of $\{\rho, \xi\}$, $p_{\delta,\varphi} = f_\theta' (= f_{\theta'}')$. (A pair as (δ, ϕ) is called an *admissible pair* (cf. [**106**]).)

REMARK 11.1.2. With notations as above, we have for $\tau \in W^P$, $\tau \geqslant \theta$ (resp. $\tau \leqslant \theta$) if and only if $\tau \geqslant \theta'$ (resp. $\tau \leqslant \theta'$).

REMARK 11.1.3. It is shown in [**104**] that $\{p_{\delta,\varphi},\ (\delta, \varphi)$ an admissible pair$\}$ is a basis for $H^0(G/P, L)$.

In view of the above facts, we can derive the results for G/P from those for $G_{n,2n}$ (as discussed in §9).

Let $w, \tau \in W^P$, $w \geqslant \tau$. Let $U_\tau^-, A_\tau, A_{w,\tau}, M_{w,\tau}$ etc., be defined in a similar way as in §9.2. We have, $A_{\tau,w}$ is generated as an algebra by $\{x_{-\beta},\ \beta \in S_P(w, \tau)\}$ (in view of Theorem 11.1.1). For $f \in H^0(G/P, L)$, $\mathfrak{D}_f$ is defined in a similar way as in §10.1. If $\mathfrak{D}_f \neq \emptyset$, then in A_τ, f gets identified with $\sum_{D \in \mathfrak{D}_f} a_D x_D$, where if $D = X_{-\beta_s} \cdots X_{-\beta_1}$, then x_D is the monomial $x_{-\beta_s} \cdots x_{-\beta_1}$.

11.2. Homogeneity of $I(X_P(w))$

Let $I(X_P(w))$ be the ideal of $X_P(w)$ in G/P. Then $I(X_P(w))$ is generated by $\{p_{\delta,\varphi} \mid w \not\geqslant \delta\}$ (cf. [**106**]). It is easily seen through the identification $p_{\delta,\varphi} = f_\theta'$ that for $f = p_{\delta,\varphi}$, if $\mathfrak{D}_f \neq \emptyset$, then for $D_1, D_2 \in \mathfrak{D}_f$, ord$D_1 =$ ordD_2. Thus in A_τ, $p_{\delta,\varphi}$ gets identified with a homogeneous polynomial. In particular, we obtain that $I_P(w)$ (= the ideal of elements of A_τ that vanish on $X_P(w) \cap U_\tau^- e_\tau$) is homogeneous. Thus we get

$$\mathrm{gr}(A_{w,\tau}, M_{w,\tau}) = A_{w,\tau}$$

where $M_{w,\tau}$ is the maximal ideal in $A_{\tau,w}$ that corresponds to the point e_τ.

LEMMA 11.2.1. *(cf. [**108**]) Let $w, \tau \in W^P$, $w \geqslant \tau$. Let $p_{\delta,\phi}$ be such that $p_{\delta,\phi} = f_\theta'$. Further, let θ and τ be comparable as n-tuples. Then there exists an operator $D = X_{-\beta_s} \cdots X_{-\beta_1}$, $\beta_i \in R_P(w, \tau)$, $1 \leqslant i \leqslant s$, where*

$$R_P(w, \tau) = \{\beta \in \tau(R^+ \setminus R_P^+) \mid w \geqslant s_\beta \tau\}$$

such that $D p_{\delta,\phi} = a_D p_\tau$, $a_D \in k^*$.

LEMMA 11.2.2. *(cf. [**108**]) Let $w, \tau \in W^P$, $w \geqslant \tau$. Let $p_{\delta,\phi}$ be such that $p_{\delta,\phi} = f_\theta'$. Further, let θ and τ be comparable as n-tuples. Suppose*

$$D_1 = X_{-\beta_s} \cdots X_{-\beta_1}, \qquad D_2 = X_{-\beta_t'} \cdots X_{-\beta_1'}$$

are such that $D_i p_{\delta,\phi} = a_{D_i} p_\tau$, $a_{D_i} \in k^*$, $i = 1, 2$. *Then* $s = t$, *i.e., ord* $D_1 =$ *ord* D_2.

With notations as in Lemma 11.2.2, set

$$N_P(\delta,\phi) = \{D = X_{-\beta_s}\cdots X_{-\beta_1},\\ \beta_i \in S_P(w,\tau),\ 1 \leqslant i \leqslant s \mid Dp_{\delta,\phi} = a_D p_\tau,\ a_D \in k^*\}$$

. Lemma 11.2.1 together with the fact that $R_P(w,\tau) \subseteq S_P(w,\tau)$ implies that

$$N_P(\delta,\phi) \neq \emptyset.$$

DEFINITION 11.2.3. Notations being as above, define $d_{\delta,\phi} :=$ ord D where $D \in N_P(\delta,\phi)$ (note that the right hand side is independent of D (cf. Lemma 11.2.2), so the definition makes sense). If $\delta = \phi$, then we shall denote $d_{\delta,\delta}$ by just d_δ. For $\delta = \phi = w$, we refer to d_w as the *degree* of $X_P(w)$ at e_τ and denote it by $\deg_\tau w$.

DEFINITION 11.2.4. With notations as above, define
$x_{\delta,\phi} := \sum_{D\in N_P(\delta,\phi)} a_D x_D$, where for $D = X_{-\beta_s}\cdots X_{-\beta_1}$, we set

$$x_D = x_{-\beta_s}\cdots x_{-\beta_1}.$$

Note that from our discussion above, we have, in $A_{w,\tau}$, $p_{\delta,\phi}$ gets identified with $x_{\delta,\phi}$.

THEOREM 11.2.5. *(cf.* **[108]***) With notations as above,*

$$\left\{ x_{\delta_1,\phi_1}\cdots x_{\delta_m,\phi_m} \;\middle|\; \begin{array}{l} (1)\ w \geqslant \delta_1 \geqslant \phi_1 \geqslant \delta_2 \geqslant \cdots \geqslant \phi_m \\ (2)\ \text{for } 1 \leq i \leq m,\ \delta_i \text{ and } \tau \text{ are comparable} \\ (3)\ \sum d_{\delta_i,\phi_i} = r \end{array} \right\}$$

is a basis for $(M_{w,\tau})^r/(M_{w,\tau})^{r+1}$.

Let $X_P(w')$ be a Schubert divisor in $X_P(w)$, say $w' = ws_\beta$, for some $\beta \in R^+$. Denote $m(w,w') = (\omega_n, \beta^*)$, the Chevalley multiplicity of $X_P(w')$ in $X_P(w)$ (cf. Chapter 6, §8.1).

THEOREM 11.2.6. *(cf.* **[108]***) Denoting* $\mathrm{mult}_{e_\tau} X(w)$ *by* $m_\tau(w)$, *we have*

$$m_\tau(w)\ \deg_\tau w = \sum m_\tau(w')m(w,w'),$$

where the sum on the right hand side runs over all Schubert divisors $X_P(w')$ *in* $X_P(w)$ *such that* $e_\tau \in X_P(w')$.

COROLLARY 11.2.7. *(cf.* **[108]***) The Hilbert polynomials* $F_{w,\tau}$ *of* $X_P(w)$ *at* e_τ *are determined inductively by the formula*

$$F_{w,\tau}(r+d+d') = F_{w,\tau}(r+d') + \sum F_{\delta_i,\tau}(r+d+d_i') + F_{\tau,H}(r+d+d')$$

where $d = \deg_\tau w$, H *is the union of all Schubert divisors* $X_P(w')$ *in* $X_P(w)$ *such that* $e_\tau \in X_P(w')$, *and* $\{\delta_i, 1 \leqslant i \leqslant m\}$ *are all the elements of* W^P *such that*

(1) $w \geqslant \delta_i$ *and* (w,δ_i) *is an admissible pair.*
(2) τ *and* δ_i *are comparable.*

and $d' = \sum d_{w,\delta_i}, d'_i = d' - d_{w,\delta_i}$. *(Here, for* $\delta \in W^P$ *such that* δ *and* τ *are comparable,* $F_{\delta_i,\tau}(r)$ *denotes all standard monomials* $p_{\delta_1,\phi_1} \cdots p_{\delta_m,\phi_m}$ *such that* $\delta \geqslant \delta_1$ *and* $\deg x_{\delta_1,\phi_1} \cdots x_{\delta_m,\phi_m} = r$ *)*

12. Singular loci of Schubert varieties in the Grassmannian

We apply the results of Section 9 to determine the components of Sing$X(w)$, the singular locus of $X(w)$, $X(w)$ being a Schubert variety in $G_{d,n}$.

As seen in Chapter 7, §1.3,

$$W^{P_d} = \{(a_1, \ldots, a_d) \mid 1 \leqslant a_1 < \cdots < a_d \leqslant n\}.$$

In this section it is more convenient to use the language of partitions. To each $(a_1, \ldots, a_d) \in W^{P_d}$ we associate a partition $\mathbf{a} = (\mathbf{a}_1, \ldots, \mathbf{a}_d)$, where $\mathbf{a}_i = a_{d-i+1} - (d - i + 1)$. For any partition $\mathbf{a} = (\mathbf{a}_1, \ldots, \mathbf{a}_d)$, we will denote by $X_{\mathbf{a}}$ the Schubert variety corresponding to $(a_1, \ldots, a_d)$.

The dimension of $X_{\mathbf{a}}$ in this setup equals $|\mathbf{a}| := \mathbf{a}_1 + \ldots \mathbf{a}_d$. It is clear that $\mathbf{a}_i$ cannot be larger than $n - d$. We shall denote the original d-tuple $(a_1, \ldots, a_d)$ by $\mathbf{a}^{\#}$. In Theorem 9.5.1, we showed that for any $\mathbf{a}$ and $\mathbf{b}$, where $\mathbf{a} \geqslant \mathbf{b}$ we have

$$(*) \qquad \deg(\mathbf{b}, \mathbf{a}) m_{\mathbf{b}}(\mathbf{a}) = \sum m_{\mathbf{b}}(\mathbf{a}'),$$

where $m_{\mathbf{b}}(\mathbf{a})$ is the multiplicity of $X_{\mathbf{a}'}$ at $e_{\mathbf{b}}$ (or any other point of the cell $Be_{\mathbf{b}}$), $\deg(\mathbf{b}, \mathbf{a})$ denotes $\deg_{e_{\mathbf{b}}} X_{\mathbf{a}}$, and on the right hand side, we sum over all partitions $\mathbf{a}'$ such that $|\mathbf{a}'| = |\mathbf{a}| - 1$ and $\mathbf{b} \subseteq \mathbf{a}' \subset \mathbf{a}$. Then as seen in §9, $\deg(\mathbf{b}, \mathbf{a}) = d - \#(\mathbf{a}^{\#} \cap \mathbf{b}^{\#})$. The equation $(*)$ determines the functions $m_{\mathbf{b}}(-)$ uniquely, provided $m_{\mathbf{b}}(\mathbf{b}) = 1$ and $m_{\mathbf{b}}(\mathbf{c}) = 0$ unless $\mathbf{b} \subseteq \mathbf{c}$.

We want to use $(*)$ to determine the singular points in X. To do this, we first prove some easy properties of the function $m_{\mathbf{b}}(\mathbf{a})$.

Given two partitions $\mathbf{a}, \mathbf{b}$, where $\mathbf{a} \geqslant \mathbf{b}$, the *skew partition* $\mathbf{a}/\mathbf{b}$ is defined to be the set-theoretic difference $\mathbf{a} \setminus \mathbf{b}$.

Example. $\mathbf{a} = (5441)$, $\mathbf{b} = (432)$. Then $\mathbf{a}/\mathbf{b}$ consists of the starred boxes in the picture below:

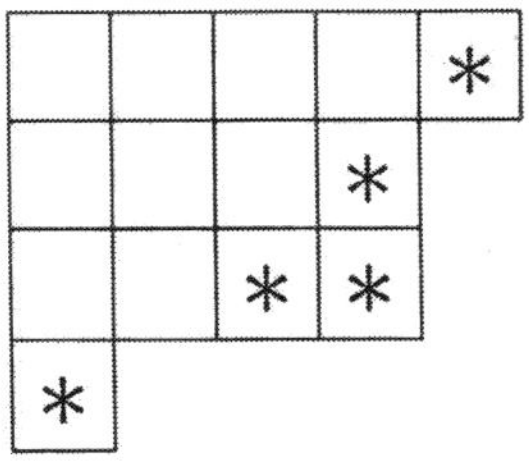

A Young diagram of the form $(p, 1, 1, \cdots, 1)$ is called a *hook*. For example $\lambda = (5, 1, 1, 1)$.

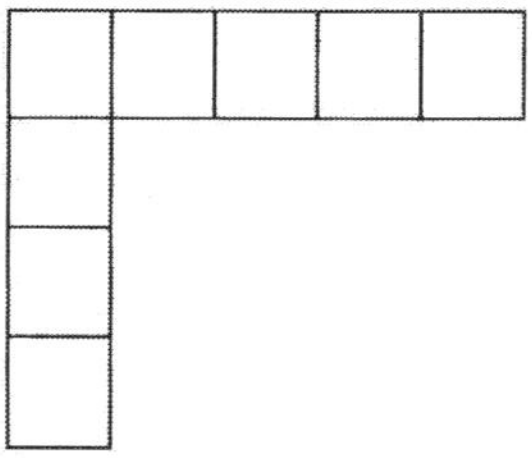

LEMMA 12.0.1. *Let* $\mathbf{a}, \mathbf{b}, \mathbf{a}', \mathbf{b}'$ *be four partitions such that* $\mathbf{a}/\mathbf{b} = \mathbf{a}'/\mathbf{b}'$. *Then* $m_{\mathbf{b}}(\mathbf{a}) = m_{\mathbf{b}'}(\mathbf{a}')$.

PROOF. We use induction on $r = |\mathbf{a}/\mathbf{b}|$. If $r = 0$, then the statement is obvious. To prove the inductive step, we use $(*)$ for $m_{\mathbf{b}}(\mathbf{a})$ and $m_{\mathbf{b}'}(\mathbf{a}')$. The right hand sides are the same, so it is enough to prove that $\deg(\mathbf{b}, \mathbf{a}) = \deg(\mathbf{b}', \mathbf{a}')$. This is however easily checked. □

LEMMA 12.0.2. *Let* $\mathbf{a}/\mathbf{b}$ *be a disconnected Young diagram* $\mathbf{a}_1/\mathbf{b}_1 \times \mathbf{a}_2/\mathbf{b}_2$. *Then* $m_{\mathbf{b}}(\mathbf{a}) = m_{\mathbf{b}_1}(\mathbf{a}_1) m_{\mathbf{b}_2}(\mathbf{a}_2)$.

PROOF. It is enough to check, using $(*)$, that

$$\deg(\mathbf{b}, \mathbf{a}) = \deg(\mathbf{b}_1, \mathbf{a}_1) + \deg(\mathbf{b}_2, \mathbf{a}_2),$$

which is again a straightforward verification. □

Now we describe $\mathrm{Sing}(X_{\mathbf{a}})$, the singular locus of $X_{\mathbf{a}}$. Let us recall that $\mathrm{Sing}(X_{\mathbf{a}})$ is a union of $X_{\mathbf{b}}$ such that $m_{\mathbf{b}}(\mathbf{a}) > 1$. To describe $\mathrm{Sing}(X_{\mathbf{a}})$, we use the "rectangle" notation. Let $\mathbf{a}$ be a partition. we write

$$\mathbf{a} = (\mathbf{p}_1^{q_1}, \ldots, \mathbf{p}_r^{q_r}) = (\underbrace{\mathbf{p}_1, \ldots, \mathbf{p}_1}_{q_1}, \underbrace{\mathbf{p}_2, \ldots, \mathbf{p}_2}_{q_2}, \cdots \underbrace{\mathbf{p}_r, \ldots, \mathbf{p}_r}_{q_r}),$$

where $\mathbf{p}_i$ denotes the Young diagram consisting of p_i boxes arranged in one row. With this notation, we say that $\mathbf{a}$ consists of r rectangles: $\mathbf{p}_1 \times q_1, \ldots, \mathbf{p}_r \times q_r$.

THEOREM 12.0.3. *Let* $\mathbf{a}$ *consist of* r *rectangles. Then*

(1) $\mathrm{Sing} X_{\mathbf{a}}$ *has* $r - 1$ *components* $X_{\mathbf{a}_1}, \ldots, X_{\mathbf{a}_{r-1}}$, *where*

$$\mathbf{a}_i = (\mathbf{p}_1^{q_1}, \ldots, \mathbf{p}_{i-1}^{q_{i-1}}, \mathbf{p}_i^{q_i - 1}, (\mathbf{p}_{i+1} - 1)^{q_{i+1}+1}, \mathbf{p}_{i+2}^{q_{i+2}}, \ldots, \mathbf{p}_r^{q_r}).$$

(2) *For a rectangular partition* $\mathbf{a}$, $X_{\mathbf{a}}$ *is nonsingular.*

PROOF. (2) is obvious, because the Schubert variety corresponding to a rectangle is just a smaller Grassmannian. To prove (1), let us first observe that $\mathbf{a}/\mathbf{a}_i$ is the hook $(\mathbf{p}_i - \mathbf{p}_{i+1} + 1, 1^{q_{i+1}})$, hence by Lemma 12.0.1, $m_{\mathbf{a}_i}(\mathbf{a}) > 1$. Moreover, for any $\mathbf{b} \underset{\neq}{\supset} \mathbf{a}_i$, $\mathbf{a}/\mathbf{b}$ is disconnected with two components. One of the components is a column, the other is a row, so $m_{\mathbf{b}}(\mathbf{a}) = 1$ by Lemma 12.0.2. This shows that $X_{\mathbf{a}_i}$ are components of $\mathrm{Sing} X_{\mathbf{a}}$. Let $X_{\mathbf{b}}$ be any other component of $\mathrm{Sing} X_{\mathbf{a}}$. Then we have that $\mathbf{b}$ is not contained in any $\mathbf{a}_i$ and $\mathbf{a}/\mathbf{b}$ is a disjoint union of r skew diagrams of the form $\mathbf{p}/\mathbf{q}$, where $\mathbf{p}$ is a rectangle, the corresponding multiplicities being 1, we obtain (by Lemma 12.0.2) that $m_{\mathbf{b}}(\mathbf{a}) = 1$. The result follows from this. □

REMARK 12.0.4. The result in Theorem 12.0.3 has been obtained by Svanes [**157**], and Lascoux [**109**] by other methods.

13. Singular loci of Schubert varieties in the symplectic and the orthogonal Grassmannians

TYPE C.

Let $G = \mathrm{Sp}(2n)$, and $P = P_n$, the maximal parabolic with associated set of simple roots being $S \setminus \{\alpha_n\}$ (notations being as in [**27**]). Then identifying Sp$(2n)$ as in §11, G/P can be identified with the isotropic Grassmannian of maximal totally isotropic n spaces in a $2n$-dimensional space with a non-degenerate skew-symmetric bilinear form (,). We can identify W^{P_n} (cf. [**90**]) as

$$W^{P_n} = \left\{ (a_1 \cdots a_n) \;\middle|\; \begin{array}{l} (1)\ 1 \leq a_1 < a_2 < \cdots < a_n \leq 2n,\ 1 \leqslant i \leqslant n \\ (2) \text{ for } \ 1 \leq i \leq 2n, \ \text{ if } \ i \in \{a_1, ..., a_n\} \\ \qquad \text{then } \ 2n + 1 - i \notin \{a_1, ..., a_n\} \end{array} \right\}.$$

To $(a_1, \cdots, a_n) \in W_{P_n}$, we associate the partition $\mathbf{a} := (\mathbf{a}_1, \cdots, \mathbf{a}_n)$, where $\mathbf{a}_{n+1-i} = a_i - i$. For example, the partition associated with $(13679) \in C_5$ is $(4, 3, 3, 1, 0)$. The conditions on the a_i's imply that the partition $\mathbf{a}$ is a *self-dual* partition contained in an $n \times n$ square. For a partition $\mathbf{a} = (\mathbf{a}_1, \cdots, \mathbf{a}_n)$, we shall denote by $X_{\mathbf{a}}$ the Schubert variety corresponding to $(a_1, \cdots, a_n)$. Thus Schubert varieties in G/P are indexed by self-dual partitions contained in n^n.

THEOREM 13.0.1. *(cf.* [**108**]*). Let* $\mathbf{a}$ *be a self-dual partition. Then* Sing $X_{\mathbf{a}} = \cup X_{\mathbf{b}}$*, where* $\mathbf{b} \subset \mathbf{a}$*, and either* $\mathbf{a}/\mathbf{b}$ *is a sum of two hooks that are dual to each other, or* $\mathbf{a}/\mathbf{b}$ *is a self-dual hook (different from a single box).*

TYPE B.

Let $V = k^{2n+1}$ together with a non degenerate symmetric bilinear form (,). Taking the matrix of the form (,) (with respect to the standard basis $\{e_1, \cdots, e_{2n+1}\}$ of V) to be E, the $2n+1 \times 2n+1$ anti-diagonal matrix with 1 all along the anti-diagonal except at the $n+1 \times n+1$-th place where the entry is 2 (note that the associated quadratic form Q on V is given by $Q(\sum_{i=1}^{2n+1} x_i e_i) = x_{n+1}^2 + \sum_{i=1}^{n} x_i x_{2n+2-i}$), we may realize $G = SO(V)$ as the fixed point set $SL(V)^\sigma$, where $\sigma : SL(V) \to SL(V)$ is given by $\sigma(A) = E^{-1}({}^tA)^{-1}E$.

Let $P = P_n$, the maximal parabolic with associated set of simple roots being $S \setminus \alpha_n$ (notations being as in [**27**]). Then G/P can be identified with the isotropic Grassmannian of maximal totally isotropic n spaces in V. We can identify W^{P_n} (cf.

[**90**]) as

$$W^{P_n} = \left\{ (a_1 \cdots a_n) \;\middle|\; \begin{array}{c} (1)\ 1 \leq a_1 < a_2 < \cdots < a_n \leq 2n, \\ a_i \neq n+1,\ 1 \leqslant i \leqslant n \\ (2) \text{ for } 1 \leq i \leq 2n, \text{ if } i \in \{a_1, ..., a_n\} \\ \text{then } 2n+2-i \notin \{a_1, ..., a_n\} \end{array} \right\}.$$

To $(a_1, \cdots, a_n) \in W_{P_n}$, we associate the partition $\mathbf{a} := (\mathbf{a}_1, \cdots, \mathbf{a}_n)$, where

(1)
$$\mathbf{a}_{n+1-i} = \begin{cases} a_i - i & \text{if } a_i \leq n \\ a_i - i - 1 & \text{if } a_i > n \end{cases}$$

For example, the partition associated with $(137810) \in B_5$ is $(4, 3, 3, 1, 0)$. The conditions on the a_i's imply that the partition $\mathbf{a}$ is a *self-dual* partition contained in an $n \times n$ square. For a partition $\mathbf{a} = (\mathbf{a}_1, \cdots, \mathbf{a}_n)$, we shall denote by $X_{\mathbf{a}}$ the Schubert variety corresponding to $(a_1, \cdots, a_n)$. Thus here again Schubert varieties in G/P are indexed by self-dual partitions contained in n^n.

THEOREM 13.0.2. *(cf.* [**108**]*). Let* $\mathbf{a}$ *be a self-dual partition. Then* $\operatorname{Sing} X_{\mathbf{a}} = \cup X_{\mathbf{b}}$*, where* $\mathbf{b} \subset \mathbf{a}$*, and either* $\mathbf{a}/\mathbf{b}$ *is a disjoint sum of two hooks that are dual to each other, or,* $\mathbf{a}/\mathbf{b} = (r+i, r^{r-1}, 1^i) \,/\, ((r-1)^{r-1})$ *for some* r, i *with* $i > 0$ *(the sum of two hooks dual to each other connected at one box), or* $\mathbf{a}/\mathbf{b} = (r^2, 2^{r-2}) \,/ (0^r)$ *for some* $r > 2$ *(self-dual double hook).*

TYPE D.

Let $V = k^{2n}$ together with a non-degenerate symmetric bilinear form (,). Taking the matrix of the form (,) (with respect to the standard basis $\{e_1, \cdots, e_{2n}\}$ of V) to be E, the anti-diagonal $(1, \cdots, 1)$ of size $2n \times 2n$, we may realize $G = SO(V)$ as the fixed point set $SL(V)^\sigma$, where $\sigma : SL(V) \to SL(V)$ is given by $\sigma(A) = E^{-1}({}^tA)^{-1}E$. Set $H = SL(V)$.

Let $P = P_n$, the maximal parabolic with associated set of simple roots being $S \setminus \{\alpha_n\}$ (notations being as in [**27**]). Then G/P can be identified with the isotropic Grassmannian of maximal totally isotropic n spaces in V. We can identify W^{P_n} (cf. [**90**]) as

$$W^{P_n} = \left\{ (a_1 \cdots a_n) \;\middle|\; \begin{array}{l} (1)\ 1 \leq a_1 < a_2 < \cdots < a_n \leq 2n,\ 1 \leqslant i \leqslant n \\ (2)\ \#\{i, 1 \leqslant i \leqslant n \mid a_i > n\} \text{ is even} \\ (3) \text{ for } 1 \leq i \leq 2n, \text{ if } i \in \{a_1, ..., a_n\} \\ \qquad \text{then } 2n+1-i \notin \{a_1, ..., a_n\} \end{array} \right\}.$$

For $w_1 = (a_1, \cdots, a_n), w_2 = (b_1, \cdots, b_n)$ in W^{P_n}, $w_1 \geqslant w_2$, if and only if $a_i \geqslant b_i$, $1 \leqslant i \leqslant n$ (cf. [**90**]). Let $P = P_n, Q = P_{n-1}$. Consider the map $\delta : W^P \to W^Q, \delta(a_1, \cdots, a_n) = (b_1, \cdots, b_{n-1})$, where $(b_1, \cdots, b_{n-1})$ is obtained from $(a_1, \cdots, a_{n-1})$ by replacing n by $n' (= n+1)$ (resp. n' by n) if n (resp. n') is present in $\{a_1, \cdots, a_{n-1}\}$. Note that if $a_n > n$, then precisely one of

$\{n, n'\}$ is present in $(a_1, \cdots, a_{n-1})$; if $a_n = n$, then $(a_1, \cdots, a_n) = (1, \cdots, n)$, and $\delta(a_1, \cdots, a_n) = (1, \cdots, n-1)$. It is easily seen that δ is a bijection preserving the Bruhat order. In fact δ is induced by the isomorphism of the varieties $G/P \to G/Q$.

Let us denote $W' = W(SO(2n-1))$, and define $\theta : W'^{P_{n-1}} \to W^{P_n}$ as $\theta(a_1, \cdots, a_{n-1}) = (a_1, \cdots, a_n)$, where $a_n = n$ or $n'(= n+1)$ and the choice is made so that $\#\{i, 1 \leqslant i \leqslant n \mid a_i > n\}$ is even (the i' in $(a_1, \cdots, a_{n-1})$ (resp. $\theta(a_1, \cdots, a_{n-1})$) should be understood as $2n - i$ (resp. $2n + 1 - i$)). Then it is easily seen that θ is a bijection preserving the Bruhat order.

In view of the isomorphisms θ, and δ, we have results for Schubert varieties in $G/P_{n-1}, G/P_n, G$ being $SO(2n)$ similar to Theorem 13.0.2.

14. A closed formula for multiplicity in the Grassmannian

In this section, we give a closed formula for $m_{\mathbf{b}}(\mathbf{a})$.

I. We first discuss the case $\mathbf{b} = \mathbf{0}$, the zero partition (note that the associated point is e_{id}). We denote by $m(\mathbf{a})$ the function $m_{\mathbf{0}}(\mathbf{a})$. First we observe (cf. §9) that

$$\deg(\mathbf{0}, \mathbf{a}) = \#\{\mathbf{a}_r \mid \mathbf{a}_r \geqslant r\}.$$

Now we introduce the Frobenius notation. Let $\mathbf{a}$ be a partition such that $\deg(\mathbf{0}, \mathbf{a}) = r$. Then there exist $c_1 > c_2 > \cdots > c_r$ and $d_1 > \cdots > d_r$ such that

$$\mathbf{a} = \big(c_1 + 1, c_2 + 2, \ldots, c_r + r, \\ (r)^{d_r}, (r-1)^{d_{r-1} - d_r - 1}, \ldots, (2)^{d_2 - d_3 - 1}, (1)^{d_1 - d_2 - 1}\big).$$

In this situation we shall denote $\mathbf{a} = (c_1, \ldots, c_r; d_1, \ldots, d_r)$. We call the number r the *Durfee square size of* $\mathbf{a}$. We have a nice description of $m(\mathbf{a})$ as a determinant, as given by the following.

THEOREM 14.0.1. *Let* $\mathbf{a}$ *be a partition with* $\deg(\mathbf{0}, \mathbf{a}) = r$, *and let* $c_1, \ldots, c_r$; $d_1, \ldots, d_r$ *be the numbers as defined above. Let* $[a, b] = (a+b)!/a!b!$ *be the binomial coefficient* $\binom{a+b}{b}$. *Then*

$$m(\mathbf{a}) = \det\big([c_i, d_j]\big)_{1 \leqslant i,j \leqslant r}. \qquad (**)$$

PROOF. We will show that the function defined by $(**)$ satisfies the inductive formula $(*)$ of §12. Let $\mathbf{a}$ be the partition of Durfee square size r. Let $\mathbf{a} = (c_1, \ldots, c_r; d_1, \ldots, d_r)$ (in the Frobenius notation). We want to prove formula $(*)$ for partition $\mathbf{a}$. We consider two cases.

Case 1. $c_r > 0$ or $d_r > 0$. In this case the partitions $\mathbf{a}'$ (appearing on the R.H.S. of (*) (of §12)) are also of Durfee square size r, and they are $(c_1, \ldots, c_i - 1, \ldots, c_r; d_1, \ldots, d_r)$ for all $c_i > c_{i+1}$, and $(c_1, \ldots, c_r; d_1, \ldots, d_j - 1, \ldots, d_r)$ for all $d_j > d_{j+1}$. We notice that we can sum over all i, j because for $c_i = c_{i+1}$ or $d_j = d_{j+1}$ the determinant $(**)$ is obviously zero. Now substitute the determinants $(**)$ on both sides of $(*)$. We compare the terms in all determinants corresponding to a fixed permutation σ. We notice that the contribution on both sides from such terms is the same. For example, for a diagonal term we get on the left side $r[c_1, d_1][c_2, d_2]\dots[c_r, d_r]$ and on the right side $\sum[c_1, d_1]\dots[c_i-1, d_i]\dots[c_r, d_r]+\sum[c_1, d_1]\dots[c_j, d_j-1]\dots[c_r, d_r]$. In view of the formula $[c, d] = [c-1, d] + [c, d-1]$ both contributions are equal, and thus $(*)$ is satisfied

Case 2. $c_r = d_r = 0$. The proof is the same in *Case 1*. The only difference is that while substituting the determinantal expression $(**)$ on both sides of $(*)$, on the right hand side we get one determinant of lower degree. The term corresponding to a permutation τ on $r-1$ letters in such a determinant has to be added to the terms corresponding to the permutation σ such that $\sigma(t) = \tau(t)$ for $1 \leqslant t \leqslant r-1, \sigma(r) = r$ in the determinants of order r. Then it is easily checked as above that contributions corresponding to every permutation cancel out. □

II. Let $w = (i_1, \cdots, i_d)$, $\tau = (j_1, \cdots, j_d)$, where we suppose that $\tau \leqslant w$. Then expressing $p_w \mid_{U_\tau^-}$ in terms of the coordinates $\{x_{-\beta},\ \beta \in \tau(R^+ \setminus R_P^+)\}$, we have (as seen in Section 9)

$$\deg_\tau w = d - \#\{i_k \mid i_k \in \{j_1, \dots, j_d\}\}.$$

Let us denote $\mathbf{i} = (i_1, \cdots, i_d)$, $\mathbf{j} = (j_1, \cdots, j_d)$

$$\deg(\mathbf{j}, \mathbf{i}) := d - \#\{i_k \mid i_k \in \{j_1, \dots, j_d\}\}.$$

Then by Theorem 9.5.1, we have,

$$(*) \qquad m_\tau(w)\ \deg(\mathbf{j}, \mathbf{i}) = \sum m_\tau(w')$$

where the sum on the right hand side runs over all Schubert divisors $X_P(w')$ in $X_P(w)$ such that $e_\tau \in X_P(w')$.

Similar to the interpretation for $\deg X(w)$ in terms of weighted chains in $[id, w]$ (cf. §8), we have an interpretation for $m_\tau(w)$ in terms of maximal weighted chains in the Hasse diagram of $[\tau, w]$, where the maximal chains are given weights as follows:

To each element $\phi \in [\tau, w]$, assign the number $\frac{1}{\deg_\tau \phi}$ or 1 according as $\phi >$ or $= \tau$. To a maximal chain $\underline{c} := \{\phi_0 = w > \phi_1 > \cdots > \phi_r = \tau\}$, where $r = l(w) - l(\tau)$, the codimension of $X(\tau)$ in $X(w)$, assign the weight

$$m(\underline{c}) := \frac{1}{\prod_{t=0}^{r} \deg_\tau \phi_t}.$$

Then $m_\tau(w)$ is simply $\sum_{\underline{c}} m(\underline{c})$, where the summation runs over all the maximal chains $\underline{c}$ in $[\tau, w]$.

Example Consider $X(356) \subset G_{3,6}$. Take $\tau = (235)$. We have the following Hasse diagram

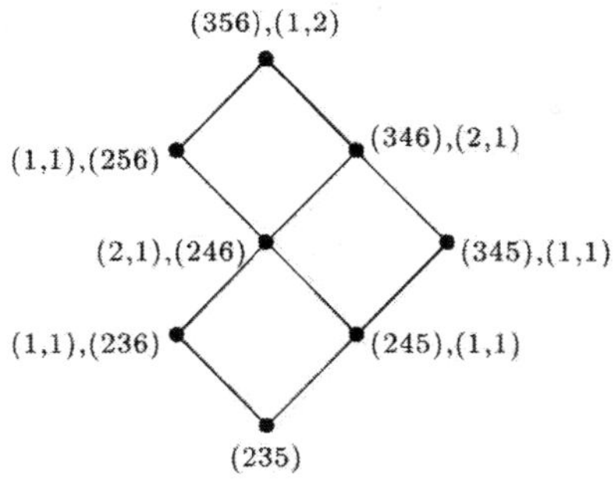

where at each vertex ϕ, the pair (a, b) stands for $(\deg \phi, m_\tau(\phi))$. We have the following five maximal chains:

$$\underline{c}_1 := \{(356) > (256) > (246) > (236) > (235)\},\ m(\underline{c}_1) = \frac{1}{2}$$

$$\underline{c}_2 := \{(356) > (256) > (246) > (245) > (235)\},\ m(\underline{c}_2) = \frac{1}{2}$$

$$\underline{c}_3 := \{(356) > (346) > (246) > (236) > (235)\},\ m(\underline{c}_3) = \frac{1}{4}$$

$$\underline{c}_4 := \{(356) > (346) > (246) > (245) > (235)\},\ m(\underline{c}_4) = \frac{1}{4}$$

$$\underline{c}_5 := \{(356) > (346) > (345) > (245) > (235)\},\ m(\underline{c}_5) = \frac{1}{2}.$$

Hence we obtain $\sum_{i=1}^{5} m(\underline{c}_i) = 2$ which is in fact equal to $m_\tau(w)$ (as computed from (*)).

Rosenthal (cf. [**139**]) has given a closed formula for $m_\tau(w)$ which we describe below.

Let

$$|\mathbf{i}| := i_1 + \cdots + i_d$$

and

$$V(\mathbf{x}) := V(x_1, \ldots, x_d) := \det \begin{bmatrix} 1 & \ldots & \ldots & 1 \\ x_1 & \ldots & \ldots & x_d \\ \vdots & & & \vdots \\ x_1^{d-1} & \ldots & \ldots & x_d^{d-1} \end{bmatrix} = \prod_{u>k} (x_u - x_k).$$

$V(\mathbf{x}) \in K[x_1, \ldots, x_d]$ is an alternating polynomial function (note that $V(\mathbf{x})$ is a Vandermonde determinant).

Let us denote $M_{\mathbf{j}}(\mathbf{i}) = m_\tau(w)$. Then by Theorem 9.5.1, $M_{\mathbf{j}}(\mathbf{i})$ is the unique solution of the partial difference equation

$$(\dagger) \qquad M_{\mathbf{j}}(\mathbf{i}) = \frac{1}{\deg(\mathbf{j},\mathbf{i})} \sum_{\substack{|\mathbf{k}|=|\mathbf{i}|-1 \\ \mathbf{j}\leq\mathbf{k}<\mathbf{i}}} M_{\mathbf{j}}(\mathbf{k})$$

subject to the initial condition $M_{\mathbf{j}}(\mathbf{j}) = 1$ and subject to the boundary condition $M_{\mathbf{j}}(\mathbf{k}) = 0$ for any index set $\mathbf{k}$ having the property that some of the indices $1 \leq k_1 \leq \cdots \leq k_d \leq n$ are equal.

THEOREM 14.0.2. *(cf. [139]) Consider index sets* $\mathbf{j} \leq \mathbf{i}$ *and let*

$$s_k := \#\{j_p \mid i_k < j_p\}.$$

Then the multiplicity $M_{\mathbf{j}}(\mathbf{i})$ *is given by*

$$(*) \qquad M_{\mathbf{j}}(\mathbf{i}) = \frac{1}{1!\cdots(d-1)!} \sum_{k_1=0}^{s_1} \cdots \sum_{k_d=0}^{s_d} (-1)^{|\mathbf{k}|} \binom{s_1}{k_1} \cdots \binom{s_d}{k_d} V(\mathbf{i}+\mathbf{k}).$$

The theorem is proved using the following three Lemmas.

LEMMA 14.0.3.

$$(**) \qquad \sum_{k=1}^{d} V(x_1,\ldots,x_{k-1},x_k-1,x_{k+1},\ldots,x_d) = d\,V(x_1,\ldots,x_d).$$

PROOF. For $\kappa = 1,\ldots,d$, view the left hand side of (**) as a polynomial $\varphi_\kappa(x_\kappa) \in K(x_1,\ldots,x_{\kappa-1},x_{\kappa+1},\ldots,x_d)[x_\kappa]$. Now $\varphi_\kappa(x_\kappa)$ is a polynomial of degree $d-1$ and the $d-1$ numbers $x_1,\ldots,x_{\kappa-1},x_{\kappa+1},\ldots,x_d$ are all roots of $\varphi_\kappa(x_\kappa)$. It follows that the left hand side of (**) when viewed as an element of $K[x_1,\ldots,x_d]$ must be a constant multiple of $V(x_1,\ldots,x_d)$. A comparison of the leading coefficients shows that this constant is d and this concludes the proof. □

LEMMA 14.0.4. *Let* $d \geq n$ *be positive integers and let* x *be an indeterminate. Then*

$$(**) \qquad \sum_{k=0}^{d} (-1)^k \binom{d}{k} (x+k)^n = \begin{cases} 0 & \text{if } n<d \\ (-1)^d d! & \text{if } n=d. \end{cases}$$

PROOF. We have,

$$\sum_{k=0}^{d} (-1)^k \binom{d}{k} \sum_{q=0}^{n} \binom{n}{q} x^{n-q} k^q = \sum_{q=0}^{n} \binom{n}{q} x^{n-q} \sum_{k=0}^{d} (-1)^k \binom{d}{k} k^q.$$

The proof now follows from the identity

$$\sum_{k=0}^{d} (-1)^k \binom{d}{k} k^q = \begin{cases} 0 & \text{if } q<d \\ (-1)^d d! & \text{if } q=d \end{cases}$$

which is verified by induction. □

LEMMA 14.0.5. *Assume $f(x) \in K[x]$ is a polynomial. Then*

$$\sum_{k=0}^{s+1}(-1)^k\binom{s+1}{k}f(x+k-1)=\sum_{k=0}^{s}(-1)^k\binom{s}{k}\left[f(x+k-1)-f(x+k)\right].$$

PROOF. We have,

$$\begin{aligned}&\sum_{k=0}^{s+1}(-1)^k\binom{s+1}{k}f(x+k-1)\\&=f(x-1)+\sum_{k=1}^{s}(-1)^k\left[\binom{s}{k}+\binom{s}{k-1}\right]f(x+k-1)+(-1)^{s+1}f(x+s)\\&\qquad=\sum_{k=0}^{s}(-1)^k\binom{s}{k}f(x+k-1)-\sum_{k=0}^{s}(-1)^k\binom{s}{k}f(x+k)\end{aligned}$$

□

PROOF OF THEOREM 14.0.2. (cf. [**139**]) We first show that the expression (*) in Theorem 14.0.2 satisfies the initial condition $M_{\mathbf{j}}(\mathbf{j})=1$. If $\mathbf{j}=\mathbf{i}$, then $s_1,\ldots,s_d$ attain their maximum possible value: $s_1=d-1$, $s_2=d-2$, …, $s_d=0$. It follows that

$$\begin{aligned}&M_{\mathbf{j}}(\mathbf{j})=\\&\frac{1}{1!\cdots(d-1)!}\sum_{k_1=0}^{d-1}\cdots\sum_{k_{d-1}=0}^{1}\sum_{k_d=0}^{0}(-1)^{|\mathbf{k}|}\binom{d-1}{k_1}\cdots\binom{1}{k_{d-1}}\binom{0}{k_d}V(\mathbf{j}+\mathbf{k})\\&=\frac{1}{1!\cdots(d-1)!}\det\left[\sum_{k_1=0}^{d-1}(-1)^{k_1}\binom{d-1}{k_1}\begin{bmatrix}1\\ j_1+k_1\\ \vdots\\ (j_1+k_1)^{d-1}\end{bmatrix},\ldots,\begin{bmatrix}1\\ j_d\\ \vdots\\ (j_d)^{d-1}\end{bmatrix}\right]\end{aligned}$$

$$\overset{Lemma\ 14.0.4}{=}\ \frac{1}{1!\cdots(d-1)!}\det\begin{bmatrix}0&\ldots&0&1\\ \vdots&&-1&*\\ 0&\ldots&\ldots&\vdots\\ (-1)^{d-1}(d-1)!&*&\ldots&*\end{bmatrix}=1.$$

Next we show that (*) in Theorem 14.0.2 satisfies the boundary condition. For this assume that $1\le i_1\le\cdots\le i_t=i_{t+1}\le\cdots\le i_d\le n$. It follows that $s_t=s_{t+1}$. Since $V(x_1,\ldots,x_d)$ is alternating in the indeterminates $x_1,\ldots,x_d$ it follows that the summands in (*) (of Theorem 14.0.2) cancel pairwise, i.e. $M_{\mathbf{j}}(\mathbf{i})=0$ as soon as some of the indices $1\le i_1\le\cdots\le i_d\le n$ are equal.

Finally we show that (*) in Theorem 14.0.2 satisfies the recurrence relation (†). For this let

$$l := \#\{i_k \mid i_k \in \{j_1, \dots, j_d\}\},$$

i.e. $l = d - \deg(\mathbf{j}, \mathbf{i})$. Without loss of generality we can assume that $i_1, \dots, i_l \in \{j_1, \dots, j_d\}$. (The proof for the other situations is identical).

In order to show that the expression (*) (of Theorem 14.0.2) for $M_{\mathbf{j}}(\mathbf{i})$ satisfies the recurrence relation we compute:

$$(***) \qquad \frac{1}{(d-l)} \left\{ M_{\mathbf{j}}(i_1 - 1, i_2, \dots, i_d) + \dots + M_{\mathbf{j}}(i_1, \dots, i_d - 1) \right\}.$$

For each of the d summands in (***) we will apply (*) (of Theorem 14.0.2). Note that by our assumption the summation bound s_k increases by one in the kth term

$$M_{\mathbf{j}}(i_1, \dots, i_k - 1, \dots, i_d)$$

if $1 \leq k \leq l$ and it does not increase for $l + 1 \leq k \leq d$. With this in mind (***) is equal to:

$$\begin{aligned}
&\frac{1}{(d-l)} \frac{1}{1! \cdots (d-1)!} \Bigg\{ \\
&\sum_{k_1=0}^{s_1+1} \cdots \sum_{k_d=0}^{s_d} (-1)^{|\mathbf{k}|} \binom{s_1+1}{k_1} \binom{s_2}{k_2} \cdots \binom{s_d}{k_d} V(i_1+k_1-1, i_2+k_2, \dots, i_d+k_d) \\
&+ \dots + \sum_{k_1=0}^{s_1} \cdots \sum_{k_d=0}^{s_d} (-1)^{|\mathbf{k}|} \binom{s_1}{k_1} \cdots \binom{s_l+1}{k_l} \cdots \binom{s_d}{k_d} V(i_1+k_1, \dots, i_d+k_d) \\
&+ \sum_{k_1=0}^{s_1} \cdots \sum_{k_d=0}^{s_d} (-1)^{|\mathbf{k}|} \binom{s_1}{k_1} \cdots \binom{s_l}{k_l} \binom{s_{l+1}}{k_{l+1}} \cdots \binom{s_d}{k_d} V(i_1+k_1, \dots, i_d+k_d) \\
&+ \dots + \sum_{k_1=0}^{s_1} \cdots \sum_{k_d=0}^{s_d} (-1)^{|\mathbf{k}|} \binom{s_1}{k_1} \cdots \binom{s_d}{k_d} V(i_1+k_1, \dots, i_d+k_d-1) \Bigg\} \\
&\overset{\text{Lemma 14.0.5}}{=} \frac{1}{(d-l)} \frac{1}{1! \cdots (d-1)!} \Bigg\{ \sum_{k_1=0}^{s_1} \cdots \sum_{k_d=0}^{s_d} (-1)^{|\mathbf{k}|} \binom{s_1}{k_1} \cdots \binom{s_d}{k_d} \Bigg[\\
&\qquad V(i_1+k_1-1, i_2+k_2, \dots, i_d+k_d) - V(\mathbf{i}+\mathbf{k}) \\
&\qquad + \dots + V(i_1+k_1, \dots, i_l+k_l-1, \dots, i_d+k_d) - V(\mathbf{i}+\mathbf{k}) \\
&\qquad + V(i_1+k_1, \dots, i_{l+1}+k_{l+1}-1, \dots, i_d+k_d) \\
&\qquad + \dots + V(i_1+k_1, \dots, i_{d-1}+k_{d-1}, i_d+k_d-1) \Bigg] \Bigg\}
\end{aligned}$$

Lemma 14.0.3
$$\overset{\text{Lemma 14.0.3}}{=} \frac{1}{(d-l)}\frac{1}{1!\cdots(d-1)!}\left\{(d-l)\sum_{k_1=0}^{s_1}\cdots\sum_{k_d=0}^{s_d}(-1)^{|\mathbf{k}|}\binom{s_1}{k_1}\cdots\binom{s_d}{k_d}V(\mathbf{i}+\mathbf{k})\right\}.$$

This last expression represents exactly the formula (*) (of Theorem 14.0.2) for $M_{\mathbf{j}}(\mathbf{i})$ and this completes the proof. □

In many situations it is possible to simplify the formula (*) (of Theorem 14.0.2) considerably. One such instance is described in the following example:

Example: Assume the indices $\mathbf{i},\mathbf{j}$ satisfy $j_d \leq i_1$. In this situation the numbers $s_1,\ldots,s_d$ attain the smallest possible value, i.e. $s_1 = \ldots = s_d = 0$. Formula (*) (of Theorem 14.0.2) simplifies in this case to:

$$M_{\mathbf{j}}(\mathbf{i}) = \frac{1}{1!\cdots(d-1)!}V(\mathbf{i}) = \frac{1}{1!\cdots(d-1)!}\prod_{u>k}(i_u - i_k).$$

It follows in particular that in this case $M_{\mathbf{j}}(\mathbf{i}) = 1$ if and only if $\mathbf{i} = (n-d+1, n-d+2,\ldots,n)$, i.e. $X(w) = G_{d,n}$.

The closed formula for $M_{\mathbf{j}}(\mathbf{i})$ (as given by Theorem 14.0.2) has recently been improved upon by Rosenthal and Zelevinsky ([**140**]):

For $1 \leqslant k \leqslant d$, let $s_k = \#\{j_p \mid i_k < j_p\}$. Then

$$M_{\mathbf{j}}(\mathbf{i}) = (-1)^{s_1+\cdots+s_d}\det\begin{bmatrix} \binom{i_1}{-s_1} & \cdots & \cdots & \binom{i_d}{-s_d} \\ \binom{i_1}{1-s_1} & \cdots & \cdots & \binom{i_d}{1-s_d} \\ \vdots & & & \vdots \\ \binom{i_1}{d-1-s_1} & \cdots & \cdots & \binom{i_d}{d-1-s_d} \end{bmatrix}$$

where the binomial coefficient $\binom{a}{b}$ is supposed to be 0 if $b < 0$.

15. A criterion for smoothness

The problem of the explicit determination of the irreducible components of Sing$X(w)$ is difficult even for Type A. There is a conjecture in Type A which originally appeared in [**103**]. First observe that in $G = SL(4)$, $X(3412)$, $X(4231)$ are the only two singular Schubert varieties. In [**103**], it is shown that the situation for a general $SL(n)$ is quite analogous to that of $SL(4)$ as given by the following Theorem:

THEOREM 15.0.1. *(cf. [**103**]) Let $\eta = (a_1 \ldots a_n) \in \mathfrak{S}_n$. Then $X(\eta)$ is singular if and only if there exist i, j, k, l, $1 \leqslant i < j < k < l \leqslant n$ such that either (1) or (2) below holds.*

(1) $a_k < a_l < a_i < a_j$

(2) $a_l < a_j < a_k < a_i$.

PROOF. We shall index the simple roots as in [**27**]. For $1 \leqslant i \leqslant n-1$, we shall denote the reflection with respect to the simple root α_i by s_i. For $w \in W$, we shall denote by $X(w)_r$ (resp. $X(w)_l$) the Schubert variety in G/B (resp. $B\backslash G$) associated to w. When we write $X(w)$ in some statement, it is to be understood that the statement holds for both $X(w)_r$ and $X(w)_l$.

Towards proving Theorem 15.0.1, we first prove some preparatory results (cf. Lemmas 15.0.2-15.0.6, Theorem 15.0.7 below).

LEMMA 15.0.2. *(cf.* **[96]***) Let* $w \in W$.

(1) *w has a reduced expression $w = u'_{n-1} \cdots u'_1$, where u'_i is either id or a left-end segment of $s_i \cdots s_{n-1}$.*
(2) *w has a reduced expression $w = v'_1 \cdots v'_{n-1}$, where v'_i is either id or a right-end segment of $s_{n-1} \cdots s_i$.*

Let $R_w = \{\alpha \in R^+ \mid w \geqslant s_\alpha\}$. Note that $\#R_w(= \dim T(w, \mathrm{id})) \geqslant l(w)$. Let $S_w = \{\alpha \in R_w \mid \alpha$ does not involve $\alpha_1\}$, $N_w = \{\alpha \in R_w \mid \alpha$ involves $\alpha_1\}$, where $\alpha_1 = \epsilon_1 - \epsilon_2$.

LEMMA 15.0.3. *Let $w = (a_1 \cdots a_n)$. Let $w = w's_1 \cdots s_{m-1}$, where $w' = u'_{n-1} \cdots u'_2$ (cf. Lemma 15.0.2). Let $a_1 = t$. Then $\#N_w = min\{t, m\} - 1$.*

PROOF. First observe that $a_m = 1$. The required result follows from the fact that $w \geqslant$ the transposition (i, j) (where $i < j$), if and only if $a_k \geqslant j$ for some $k \leqslant i$ and $a_l \leqslant i$ for some $l \geqslant j$. □

LEMMA 15.0.4. *Let $w = w's_1 \cdots s_{m-1}$, where $w' = u'_{n-1} \cdots u'_2$. Then $\#\{R_w \setminus R_{w'}\} \geqslant m - 1$.*

PROOF. Fix j, $2 \leqslant j \leqslant m - 1$. We shall exhibit a root $\beta_j \in R_w \setminus R_{w'}$. If $s_j \not\leqslant w'$ (equivalently s_j does not occur in w'), then $\beta_j = \alpha_j \in R_w \setminus R_{w'}$. Let then $s_j \leqslant w'$.
Let $2 \leqslant i \leqslant j$ be such that $s_j s_{j-1} \cdots s_i \leqslant w'$, but $s_j s_{j-1} \cdots s_i s_{i-1} \not\leqslant w'$ (note that such an i exists since $s_1 \not\leqslant w'$). The facts that $s_j s_{j-1} \cdots s_i \leqslant w'$, $s_{i-1} s_i \cdots s_j \leqslant s_1 \cdots s_{m-1}$, and $w's_1 \cdots s_{m-1}$ is a reduced expression for w, imply that

$$s_j s_{j-1} \cdots s_i s_{i-1} s_i \cdots s_j \leqslant w.$$

Thus the root $\beta_j = \epsilon_{i-1} - \epsilon_{j+1} \in R_w \setminus R_{w'}$. Also it is clear that for $j \neq j'$, $\beta_j \neq \beta_{j'}$. Thus $\{\alpha_1, \beta_j, 2 \leqslant j \leqslant m - 1\} \subseteq R_w \setminus R_{w'}$. The required result now follows. □

LEMMA 15.0.5. *Let $w = s_{t-1} \cdots s_1 w''$. Then $\#\{R_w \setminus R_{w'}\} \geqslant t - 1$.*

The proof is similar to that of Lemma 15.0.4.

LEMMA 15.0.6. *Let $\lambda, \mu \in W$. Let $\lambda = s_1 \ldots s_m$ be a reduced expression for λ. Suppose that $\mu s_i < \mu$ for all $1 \leqslant i \leqslant m$. Then $\mu'\lambda' \leqslant \mu$ for all $\mu' \leqslant \mu$ and $\lambda' \leqslant \lambda$. In particular, $\mu\lambda \leqslant \mu$*

See **[96]** for a proof.

Next we prove a theorem which gives a criterion for smoothness in terms of some equidimensionality properties.

For $1 \leqslant i \leqslant n - 1$, let P_i be the maximal parabolic subgroup with $S \setminus \{\alpha_i\}$ as the associated set of simple roots. Let Q_i, $1 \leqslant i \leqslant n - 1$ be the parabolic subgroups defined inductively as

$$Q_1 = P_1$$
$$Q_2 = Q_1 \cap P_2$$
$$Q_3 = Q_2 \cap P_3$$
$$\vdots$$
$$Q_{i+1} = Q_i \cap P_{i+1}\,.$$

For $0 \leqslant i \leqslant n-1$, let $\pi_{i,r}$ (resp. $\pi_{i,l}$) be the projection $Q_i/B \to Q_i/Q_{i+1}$ (resp. $B\backslash Q_i \to Q_{i+1}\backslash Q_i$), where $Q_0 = G$, $Q_n = B$.

Note that in view of Lemma 3.1.1, Chapter 7 for any $\tau \in W$, $X(\tau)_l \cap B\backslash Q_i$, $X(\tau)_r \cap Q_i/B$ are irreducible.

THEOREM 15.0.7. *Let $w \in W$. Then $X(w)$ is smooth if and only if there exist $\tau_0, \tau_1, \cdots, \tau_n$, $\tau_i \in W_{Q_i}$ with $\tau_0 = w$ such that for $0 \leqslant i \leqslant n-1$ the following two conditions hold:*

(1) *Either $\pi_{i,l}|_{X(\tau_i)_l}$ or $\pi_{i,r}|_{X(\tau_i)_r}$ is equi-dimensional, i.e., the fibers have the same dimension.*
(2) *τ_{i+1} is given by $X(\tau_i)_l \cap B\backslash Q_{i+1}$ or $X(\tau_i)_r \cap Q_{i+1}/B$ according as $\pi_{i+1,l}|_{X(\tau_i)_l}$ or $\pi_{i+1,r}|_{X(\tau_i)_r}$ is equi-dimensional.*

PROOF. Let $X(w)$ be smooth. We have (cf. Lemma 15.0.3)

$$\#N_w = \min\{t, m\} - 1 = m - 1, \text{ say.}$$

Let T_w be the tangent space to $X(w)$ at e_{id}. Then the smoothness hypothesis implies that

(*) $$\#R_w = \dim T_w = l(w)$$

Using the reduced expression $w = w's_1 \cdots s_{m-1}$, where $w' = u'_{n-1} \cdots u'_2$, we have (cf. Lemma 15.0.4)

$$\#R_w \geqslant \#R_{w'} + m - 1 \geqslant l(w') + m - 1 = l(w).$$

Hence (*) implies equality everywhere. In particular, we have, $\#R_{w'} = l(w')$ which implies the smoothness of $X(w')$. Let $\tau_1 = w'$, and τ_i, $2 \leqslant i \leqslant n$ be the elements corresponding to w' satisfying (1) and (2) (by induction hypothesis applied to w'). We shall now show that $X(w)_l \to P_1\backslash G$ is equidimensional. We have (in view of smoothness), $\#R_w = l(w)$. Also,

$$\#R_w = \#S_w + \#N_w = \#S_w + m - 1$$

($\#N_w = m - 1$, by our assumption) and

$$l(w) = l(w') + m - 1.$$

Hence we obtain $l(w') = \#S_w$ which in turn implies that $\#R_{w'} = \#S_w$ (by smoothness of $X(w')$). Thus we get $S_w = R_{w'}$ (note that $S_w \supseteq R_{w'}$). This now implies $w's_i < w'$, $2 \leqslant i \leqslant m-1$ (for, $w's_i > w'$ for some $i, 2 \leqslant i \leqslant m-1$ would imply

$l(w's_i) > l(w') = \#S_w \geqslant \#R_{w's_i} \geqslant l(w's_i)$, which is a contradiction). From this we obtain (in view of Lemma 15.0.6)

$$X(w)_l \cap B\backslash P_1 = X(w')_l$$

which implies that $X(w)_l \to P_1\backslash G$ is equi-dimensional. Thus $X(w)_l \to P_1\backslash G$ is a fibration with fibers isomorphic to $X(w')_l$, and hence the smoothness of $X(w)_l$ implies the smoothness of $X(w')_l$. We take $\tau_1 = w'$; τ_i, $i \geqslant 2$ are obtained by applying induction hypothesis to the (smooth) Schubert variety $X(w')_l$ in $B\backslash P_1$.

A similar proof holds if $\#N_w = t - 1$; in this case, we have $X(w)_r \to G/P_1$ is equidimensional We work with the expression $w = s_{t-1} \cdots s_1 w''$, and proceed as above.

Let now w satisfy conditions (1) and (2). We shall show that $X(w)$ is smooth. Let us suppose that $X(w)_l \to P_1\backslash G$ is equidimensional, and that $X(w)_l \cap B\backslash P_1 = X(\tau_1)_l$. As above, if $w = w's_1 \cdots s_{m-1}$, where $w' = u'_{n-1} \cdots u'_2$, then we obtain (in view of equi-dimensionality)

(**) $$X(w)_l \cap B\backslash P_1 = X(w')_l, \text{ and } w' = \tau_1$$

Now (**) implies $S_w = R_{w'}$ (note that $S_w \supseteq R_{w'}$). Also, by induction we have, $X(w')$ is smooth (note that $w' = \tau_1$) and hence, $\#R_{w'} = l(w')$. Now

$$m - 1 \geqslant \#N_w = \#R_w - \#S_w = \#R_w - \#R_{w'} \geqslant m - 1$$

(cf. Lemma 15.0.4). Hence equality holds everywhere. In particular we obtain $\#R_w = \#S_w + m - 1 = R_{w'} + m - 1 = l(w') + m - 1 = l(w)$. Thus we obtain $\dim T_w = l(w)$, which proves that $X(w)$ is smooth. □

Proof of Theorem 15.0.1. Let $w = (a_1 \cdots a_n)$ satisfy either condition (1) or (2) of Theorem 15.0.1. We shall now show that $X(w)$ does not satisfy either condition (1) or (2) of Theorem 15.0.7 from which it will follow that $X(w)$ is singular. If possible, let us assume that w satisfies the conditions (1) and (2) of Theorem 15.0.7. Let us suppose that $X(w)_l \to P_1\backslash G$ is equidimensional, and that

(A) $$X(w)_l \cap B\backslash P_1 = X(\tau_1)_l.$$

Writing $w = w's_1 \cdots s_{m-1}$, we have, $w' = \tau_1$. Hence in view of (A), we have, $w's_i < w'$, $2 \leqslant i \leqslant m - 1$. Let $w' = (1b_2 \cdots b_n)$. Then $w = w'(23 \cdots m1m + 1 \cdots) = (b_2 b_3 \cdots b_m 1 \cdots b_n)$, where

(B) $$b_2 > b_3 > \cdots > b_n$$

(since $w's_i < w'$; note that $w's_i < w' \iff w'(\epsilon_i - \epsilon_{i+1}) (= \epsilon_{b_i} - \epsilon_{b_{i+1}}) > 0$). Our assumption implies that $X(w')$ is smooth and hence by induction (on $\dim X(w)$) we may suppose that w' does not satisfy the conditions of Theorem 15.0.1. This (together with our assumption) implies that $1 \in \{a_i, a_j, a_k, a_l\}$ (notations being as in Theorem 15.0.1). This implies $a_k = 1$ if (1) of Theorem 15.0.1 holds, and $a_l = 1$ if (2) of Theorem 15.0.1 holds. This implies $a_i < a_j$ or $a_j < a_k$ according as (1) or (2) of Theorem 15.0.1 holds. We have, $a_i, a_j \in \{b_2, \cdots, b_m\}$ in the former case, and

$a_j, a_k \in \{b_2, \cdots, b_m\}$ in the latter case; this is not possible (in view of (B) above; note that for $a_t \in \{b_2, \cdots, b_m\}$, $a_t = b_{t+1}$). Hence our assumption is wrong and the result follows.

Let now $X(w)$ be singular, equivalently $X(w)$ does not satisfy either (1) or (2) of Theorem 15.0.7. We shall now show that w satisfies either condition (1) or (2) of Theorem 15.0.1. Writing $w = w's_1 \cdots s_{m-1}$, $w = (a_1 \cdots a_n)$, $w' = (1b_2 \cdots b_n)$, we have $w = (b_2 b_3 \cdots b_m 1 \cdots b_n)$. If w' satisfies either condition (1) or (2) of Theorem 15.0.1, then clearly so does w. Let us then suppose that w' does not satisfy the conditions (1) & (2) of Theorem 15.0.1. This implies by induction (on dim$X(w)$) that $X(w')$ satisfies the conditions (1) and (2) of Theorem 15.0.7. This together with the hypothesis that $X(w)$ is singular implies (in view of Theorem 15.0.7) that $\pi_{i,l}|_{X(\tau_i)_l}$, $\pi_{i,r}|_{X(\tau_i)_r}$ are not equidimensional. This implies (cf. Lemma 15.0.6) $w's_i > w'$ for some $2 \leqslant i \leqslant m-1$, and $s_j w'' > w''$ for some $2 \leqslant j \leqslant t-1$, where $w = w's_1 \cdots s_{m-1}$, $w = s_{t-1} \cdots s_1 w''$. This implies (cf. Theorem 16.0.2) the existence of integers p, q such that

$$w = (\cdots a_q a_{q+1} \cdots) \text{ (resp. } w = (\cdots p \cdots p+1 \cdots)),$$

where $p+1 < t$, $q+1 < m$, and $a_q < a_{q+1}$. Let $a_q = b$, $a_{q+1} = c$. We now distinguish the following two cases.

Case 1. 1 occurs after $p+1$ in w.

In this case, condition(2) of Theorem 15.0.1 holds for w with $a_i = t, a_j = p, a_k = p+1, a_l = 1$.

Case 2. 1 occurs before $p+1$ in w.

We first note that $p+1 \neq b, c$ (since b and c occur before 1 in w while $p+1$ occurs after 1). We further divide this case into the following three subcases.

Sub case 2(a). $b > p+1$.

In this case, condition (1) of Theorem 15.0.1 holds for w with $a_i = b, a_j = c, a_k = 1, a_l = p+1$.

Sub case 2(b). $b < p+1$, $t > c$.

In this case, condition (2) of Theorem 15.0.1 holds for w with $a_i = t, a_j = b, a_k = c, a_l = 1$.

Sub case 2(c). $b < p+1$, $t < c$.

In this case, condition (1) of Theorem 15.0.1 holds for w with $a_i = t, a_j = c, a_k = 1, a_l = p+1$.

This completes the proof of Theorem 15.0.1. □

16. A conjecture

In $G = SL(4)$, Sing $X(3412) = X(1324)$ and Sing $X(4231) = X(2143)$. The conjecture below states that for a general n, the situation is "nothing more than this".

Let then $X(w) \subset G/B$ be singular (G being $SL(n)$). Let $w = (a_1 \cdots a_n)$. Since $X(w)$ is singular, there exist i, j, k, l, $1 \leqslant i < j < k < l \leqslant n$ such that

(1) either $a_k < a_l < a_i < a_j$ or $a_l < a_j < a_k < a_i$.

It is shown in [**103**] that in the former case, if w' is obtained from w by replacing a_i, a_j, a_k, a_l respectively by a_k, a_i, a_l, a_j, then $e_{w'} \in \operatorname{Sing} X(w)$; and in the latter case, if w' is obtained from w by replacing a_i, a_j, a_k, a_l respectively by a_j, a_l, a_i, a_k, then $e_{w'} \in \operatorname{Sing} X(w)$.

For $w \in W$, let P_w (resp. Q_w) be the maximal element of the set of parabolic subgroups which leave $\overline{BwB}$ (in G) stable under multiplication on the left (resp. right).

DEFINITION 16.0.1. Given parabolic subgroups P, Q, we say that $\overline{BwB}$ is P-Q stable if $P \subset P_w$ and $Q \subset Q_w$.

THEOREM 16.0.2. *Let* $w = (a_1 \dots a_n)$. *Then*

$$S_{Q_w} = \{\epsilon_i - \epsilon_{i+1} | a_i > a_{i+1}\}$$

$$S_{P_w} = \{\epsilon_i - \epsilon_{i+1} | i + 1 \textit{ appears before } i \textit{ in } \{a_1, a_2, \dots, a_n\}\ \}$$

PROOF. This follows immediately from Chapter 3, Proposition 7.0.1. Let $\alpha = \epsilon_i - \epsilon_{i+1}$. Then $\alpha \in P_w$ (resp. Q_w) if and only if $w^{-1}(\alpha)$ (resp. $w(\alpha)$) is < 0 (cf. Chapter 3, Proposition 7.0.1). The result now follows from this (note that for $1 \leqslant t \leqslant n$, $w(t) = a_t$, $w^{-1}(t) = m$, where m is given by $a_m = t$). □

Remark. In combinatorial terms, S_{Q_w} is indexed by the descents of w, and S_{P_w} is indexed by the descents of w^{-1}.

Remark. If $X(\tau)$ is an irreducible component of $\operatorname{Sing} X(w)$, then $\overline{B\tau B}$ is P_w-Q_w stable.

16.1. Conjecture for type A

The set F_η

Let $\eta = (a_1 \dots a_n) \in \mathfrak{S}_n$. Let E_η be the set of all $\tau' \leqslant \eta$ such that either 1) or 2) below holds.

1) There exist i, j, k, l, $1 \leqslant i < j < k < l \leqslant n$, such that

(a) $a_k < a_l < a_i < a_j$

(b) if $\tau' = (b_1 \dots b_n)$, then there exist i', j', k', l', $1 \leqslant i' < j' < k' < l' \leqslant n$ such that $b_{i'} = a_k$, $b_{j'} = a_i$, $b_{k'} = a_l$, $b_{l'} = a_j$

(c) if τ (resp. η') is the element obtained from η (resp. τ') by replacing a_i, a_j, a_k, a_l resp. by a_k, a_i, a_l, a_j (respectively $b_{i'}, b_{j'}, b_{k'}, b_{l'}$ resp. by $b_{j'}, b_{l'}, b_{i'}, b_{k'}$), then $\tau' \geqslant \tau$ and $\eta' \leqslant \eta$.

2) There exist i, j, k, l, $1 \leqslant i < j < k < l \leqslant n$, such that

(a) $a_l < a_j < a_k < a_i$

(b) if $\tau' = (b_1 \dots b_n)$, then there exist i', j', k', l', $1 \leqslant i' < j' < k' < l' \leqslant n$ such that $b_{i'} = a_j$, $b_{j'} = a_l$, $b_{k'} = a_i$, $b_{l'} = a_k$

(c) if τ (resp. η') is the element obtained from η (resp. τ') by replacing a_i, a_j, a_k, a_l resp. by a_j, a_l, a_i, a_k (respectively $b_{i'}, b_{j'}, b_{k'}, b_{l'}$ resp. by $b_{k'}, b_{i'}, b_{l'}, b_{j'}$), then $\tau' \geqslant \tau$ and $\eta' \leqslant \eta$.

Let $F_\eta = \{\tau \in E_\eta \mid \overline{B\tau B} \text{ is } P_\eta\text{-}Q_\eta \text{ stable}\}$.

CONJECTURE . *(cf.* **[103]***) For* $\eta \in \mathcal{S}_n$, *The singular locus of* $X(\eta)$ *is equal to* $\cup_\lambda X(\lambda)$, *where* λ *runs over the maximal (under the Bruhat order) elements of* F_η.

For example, the conjecture says that Sing $X(35142)$ is $X(13254)$ (which is confirmed by the Tables in **[16]**). Note, that the bad pattern in (35142) is 3512. These numbers are rearranged to get 1325 but they are not in exactly the same positions in (13254).

The Examples (1) and (2) below show the necessity of the two conditions $\tau \le \tau'$ and $w' \le w$.

Example 1. Let $w = (52431) \in S_5$. We have from the Tables in **[16]**, Sing $X(w) = X(21543)$. A 4231 pattern occurs in the subsequence (5231), and the corresponding τ is given by $\tau = (21453)$. Set $\tau' = (42153)$ and consider the corresponding $w' = (45231)$. We have, $\tau \leqslant \tau'$, but $w' \nleq w$. Further, $\tau' \notin$ Sing $X(w)$ thus confirming the conjecture.

Example 2. Let $w = (456312) \in S_6$. We have from the Tables in **[16]**, Sing $X(w) = X(415326) \cup X(154326) \cup X(146325)$. A 3412 pattern occurs in the subsequence (4512), and the corresponding τ is given by $\tau = (146325)$. Set $\tau' = (146253)$ and consider the corresponding $w' = (456123)$. We have, $w' \le w$, but $\tau \nleqslant \tau'$. Further, $\tau' \notin$ Sing $X(w)$ thus confirming the conjecture.

Example 3. Let $w = (35142) \in S_5$. We have from the Tables in **[16]**, Sing $X(w) = X(13254)$. A 3412 pattern occurs in the subsequence (3512), and the corresponding τ is given by $\tau = (13245)$. Set $\tau' = (13254)$ and consider the corresponding $w' = (35124)$. We have, $\tau \leqslant \tau'$, $w' \le w$. Further, $\tau' \in$ Sing $X(w)$ thus confirming the conjecture.

Remark. This conjecture has been verified for all elements of $\mathcal{S}_7$. See the Tables in **[16]**.

Remark. This conjecture has been verified to be true for a certain class of Schubert varieties, namely the class of Schubert varieties related to Ladder determinantal varieties (cf. **[62]** and Chapter 12).

Remark. Recently, Ghasharov (cf. **[58]**) has proved the sufficiency part of the Conjecture, i.e., if $\theta \in F_\eta$, then $e_\theta \in$ Sing$X(\eta)$. In fact, if τ is as in 1(c) or 2(c) of §16.1, then it is shown in **[103]** that $e_\theta \in$ Sing$X(\eta)$.

CHAPTER 9

Rational smoothness and Kazhdan-Lusztig theory

In this chapter, we give a brief review of rational smoothness and Kazhdan-Lusztig Theory. Even though rational smoothness is a weaker notion than smoothness, for $G = SL(n)$, the two notions coincide. We have included the results due to Brenti ([**28**],[**29**]), Carrell-Peterson ([**33**], [**34**]) Deodhar ([**43**], [**44**], [**45**]), Kazhdan-Lusztig ([**80**], [**81**]), Kumar ([**87**]), Lascoux-Schützenberger([**110**], [**111**]), Sankaran-Vanchinathan ([**141**], [**142**]), Zelevinsky ([**164**]). We define rational smoothness in §1. In §2, Kazhdan-Lusztig polynomials are introduced and results of Kazhdan-Lusztig, Peterson, relating Kazhdan-Lusztig polynomials and rational smoothness are described. In §3, we describe Carrell-Peterson's results on rational smoothness. In §4, we present Kumar's Criterion for smoothness and rational smoothness. §5 is a summary of combinatorial formulas for Kazhdan-Lusztig polynomials. In §6, we describe results on small resolutions for Schubert varieties in the Grassmannian, and the symplectic and orthogonal Grassmannians. In §7, we present results on the factorization of the Poincaré polynomial of a Schubert variety. In §8, we describe a generating function for the number of smooth Schubert varieties in the flag variety. This chapter contains statements of the various results mentioned. We have not included proofs for any of them. We have, nevertheless, pointed the appropriate references.

1. Definition of rational smoothness

Definition. A variety X of dimension d is *rationally smooth* if for all $x \in X$, the singular cohomology $H^i(X, X \setminus \{x\}, \mathbb{Q}) = 0$, $i \neq 2d$, and $H^{2d}(X, X \setminus \{x\}, \mathbb{Q})$ is one-dimensional. A point $x_0 \in X$ is said to be *rationally smooth* if there exists a Zariski-open rationally smooth neighborhood of $x_0 \in X$.

Remark 1. If X is a complex projective variety, then a result of McCrory (cf. [M]) implies that X is rationally smooth if and only if $H^*(X, \mathbb{C})$ (the singular cohomology of X; see [**153**] for the discussion on singular cohomology of a variety) admits Poincaré duality. This is the same thing as saying $H^*(X, \mathbb{C})$ and $IH^*(X)$ coincide, where $IH^*(X)$ denotes the intersection cohomology groups of X (cf. [**63**]); note that this is stronger than saying Poincaré polynomial of X $(= \sum \dim H^i(X, \mathbb{C})q^i)$ is symmetric.

Remark 2. Smoothness implies rational smoothness in Schubert varieties, but the converse does not hold in general. Nevertheless, the two notions are equivalent for type **A, D, E**, as given by the following Theorem:

THEOREM 1.0.1. *Let G be of type* **A, D, E**. *Then $X(w)$ is smooth if and only if it is rationally smooth.*

Theorem 1.0.1 was proven in Type A by Deodhar (cf. [**43**]). Peterson announced a proof for all three types the details of which appear in [**35**].

2. Kazhdan-Lusztig polynomials

Let A be the ring $\mathbb{Z}[q, q^{-1}]$ of Laurent polynomials over $\mathbb{Z}$ in the indeterminate q. The *Hecke algebra* $\mathcal{H}$ of W is the algebra over A with basis elements $\{T_w : w \in W\}$ and the following relations:

$$T_{s_i}T_w = T_{s_iw} \qquad \text{if } l(s_iw) > l(w)$$
$$T_{s_i}T_{s_i} = (q-1)T_{s_i} + qT_1.$$

Note that the Hecke algebra reduces to the group algebra of W if q is set equal to 1. One feature of $\mathcal{H}$ is that the basis elements T_w are invertible over A.

EXAMPLE 2.0.1. For any simple reflection $s \in S$,

$$T_s^{-1} = q^{-1}T_s - (1 - q^{-1})T_{id}.$$

DEFINITION 2.0.2. The expansion of $(T_{w^{-1}})^{-1}$ in the basis $\{T_w\}$

$$(T_{w^{-1}})^{-1} = (-1)^{l(w)}q^{-l(w)} \sum_{v \leq w} (-1)^{l(x)} R_{v,w}(q) T_v$$

defines the polynomials $R_{v,w}(q) \in \mathbb{Z}[q]$. Set $R_{v,w} = 0$ if $v \not\leq w$.

The R-polynomials play an important role in the computation of Kazhdan-Lusztig polynomials and are often easier to work with. The Remark below gives a recursive formula for computing R_{uv} and Theorem 5.1.2 below states a combinatorial formula for computing a related family of polynomials.

REMARK 2.0.3. The following facts are easily checked.

(1) $R_{w,w} = 1$.
(2) For any $s \in S$ such that $ws < w$,

$$R_{v,w} = \begin{cases} R_{vs,ws} & vs < v \\ (q-1)R_{v,ws} + qR_{vs,ws} & vs > v. \end{cases}$$

(3) $R_{v,w}$ is a polynomial of degree $l(w) - l(v)$.

One can define an involution i on $\mathcal{H}$ by interchanging q and q^{-1} and setting $i(T_w) = (T_{w^{-1}})^{-1}$. Using this involution one defines the *Kazhdan-Lusztig polynomials*, $P_{x,w}$, $x, w \in W$, $x \leqslant w$:

THEOREM 2.0.4. *(cf.* [**80**]*) For each $w \in W$, there exists a unique element $C'_w \in \mathcal{H}$ having the following properties:*

(1) $i(C'_w) = C'_w$.
(2) *Expanding* C'_w, *one has*

$$C'_w = (q^{-1/2})^{l(w)} \sum_{x \leq w} P_{x,w} T_x$$

where $P_{w,w} = 1$, $P_{x,w}(q) \in \mathbb{Z}[q]$ *has degree* $\leq \frac{1}{2}(l(w) - l(x) - 1)$ *if* $x < w$, *and* $P_{x,w} = 0$ *if* $x \not\leq w$.

From Kazhdan-Lusztig's proof of the uniqueness of C'_w they also obtain several equivalent formulas which can be used to compute $P_{v,w}$ in terms of the $R_{x,y}$ polynomials. One version of this formula is

$$q^{l(w)-l(x)}\overline{P}_{v,w} - P_{v,w} = \sum_{v<y\leq w} R_{v,y} P_{y,w}$$

where $\overline{P}_{v,w} = P_{v,w}(q^{-1})$. Since the degree of $P_{v,w}$ is at most $\frac{1}{2}(l(w) - l(x) - 1)$, $P_{v,w}$ is obtained from the right hand side by ignoring all terms of higher degree.

THEOREM 2.0.5. **[80]** *The following are equivalent for any* $v \leq w$ *in* W,

(1) $X(w)$ *is rationally smooth at* e_v.
(2) $P_{x,w} = 1$ *for all* $v \leq x \leq w$.
(3) *For all* $v \leq x \leq w$,

$$\sum_{x \leq y \leq w} R_{y,w} = q^{l(w)-l(x)}.$$

In fact, Condition 2 of Theorem 2.0.5 can often be replaced by a weaker condition:

THEOREM 2.0.6. *Let* W *be any Weyl group such that the Kazhdan-Lusztig polynomials are known to have nonnegative coefficients. Then* $X(w)$ *is rationally smooth at* e_τ *if and only if* $P_{w,\tau} = 1$.

Theorem 2.0.6 was first proved by Deodhar in **[43]** for $W = S_n$. It was extended by Carrell and Peterson to all Weyl groups with the Kazhdan-Lusztig positivity condition in **[34]** (cf. Theorem 3.0.3).

THEOREM 2.0.7. *(D. Peterson). Let* G *be of type* **A, D** *or* **E**. *Then* $X(w)$ *is smooth at* e_τ *if and only if* $P_{w,\tau} = 1$.

THEOREM 2.0.8. **[81]** *Let* G *be a semisimple, simply connected algebraic group. Let* W *be the Weyl group of* G. *For* $w \in W$, *let* $\mathcal{H}^{\cdot}(w)$ *be the intersection cohomology sheaf (with respect to middle perversity) of the Schubert variety* $X(w)$. *Then for* $v \in W$, $v \leqslant w$, *we have*

$$P_{v,w}(q) = \sum_{j \geq 0} \dim \mathcal{H}^{2j}(w)_v \, q^j$$

where $\mathcal{H}^{2j}(w)_v$ *denotes the stalk of* $\mathcal{H}^{2j}(w)$ *at the point* e_v.

For background on intersection cohomology, we refer the reader to **[23]**, **[63]**, **[154]**.

COROLLARY 2.0.9. *The coefficients of the Kazhdan-Lusztig polynomial $P_{v,w}(q)$ are all non-negative if W is the Weyl group of a semisimple algebraic group.*

A long standing open problem is to find a combinatorial description of the coefficients of $P_{v,w}$. Such a combinatorial formula would also be of interest in terms of the representation theory of Verma modules (see **[75]** for background). Let $M(\lambda)$ be the Verma module with highest weight $\lambda \in \mathfrak{h}^*$ and let $L(\lambda)$ be the unique irreducible quotient of $M(\lambda)$. For any $\mathfrak{g}$-module M, let $[M : L(\lambda)]$ be the multiplicity of $L(\lambda)$ in M. The formal character $\mathrm{ch}(M)$ is given by

$$\mathrm{ch}(M) = \sum [M : L(\lambda)] \,\mathrm{ch}L(\lambda).$$

Define a "shifted" action of W on $\mathfrak{h}^*$ by $w \cdot \lambda = w(\lambda + \rho) - \rho$ for $w \in W$, $\lambda \in \mathfrak{h}^*$, and $\rho = \frac{1}{2}\sum_{\alpha \in R^+} \alpha$. Combining results of Verma **[159]**, Bernstein-Gelfand-Gelfand **[11]**, van den Hombergh **[158]**, one has $[M(v\cdot\lambda_0) : L(\mu)] \neq 0$ if and only if $\mu = w\cdot\lambda_0$ for some w with $v \leq w$. Furthermore, $[M(v \cdot \lambda_0) : L(y \cdot \lambda_0)]$ is independent of λ_0. Hence, we can define $m(v, y)$ to be $[M(v \cdot \lambda_0) : L(y \cdot \lambda_0)]$ for any $v \leq w$ in W.

THEOREM 2.0.10. *The multiplicity $m(v, w)$ is equal to $P_{v,w}(1)$.*

Theorem 2.0.10 was originally conjectured in **[80]** and was proved independently by Beilinson-Bernstein **[10]** and Brylinski-Kashiwara **[32]**. This theorem has since been extended in several ways. We refer the reader to **[45]** for a more complete history and generalizations.

3. Carrell-Peterson's results on rational smoothness

Let $a(x, w)$ denote the average value of $l(y)$ for $y \in [x, w]$. If $x = \mathrm{id}$, we shall denote $a(x, w)$ by a_w. For $y \in [x, w]$, let $p(y, w) = \#\{r \in \mathcal{R} \mid y < ry \leqslant w\}$, and $q(x, y) = \#\{r \in \mathcal{R} \mid x \leqslant ry < y\}$. Note that Deodhar's Inequality (cf. Chapter 8, §5.1) implies that $p(y, w) \geqslant l(w) - l(y)$, and $q(x, y) \geqslant l(y) - l(x)$.

THEOREM 3.0.1. *(cf.* **[34]***) Let $x \leqslant w$. Then*

(1) $q(x, y) = l(y) - l(x), \forall y \in [x, w]$ *implies* $a(x, w) \geqslant 1/2(l(x) + l(w))$ *with equality if and only if* $p(y, w) = l(w) - l(y), \forall y \in [x, w]$.
(2) $p(y, w) = l(w) - l(y), \forall y \in [x, w]$ *implies* $a(x, w) \leqslant 1/2(l(x) + l(w))$ *with equality if and only if* $q(x, y) = l(y) - l(x), \forall y \in [x, w]$.

DEFINITION 3.0.2. Define the *Poincaré polynomial* of $X(w)$ to be

$$p_w(t) = \sum_{x \leqslant w} t^{l(x)}.$$

Note that the true Poincaré polynomial for the cohomology ring of $X(w)$ is $p_w(t) = \sum_{x \leqslant w} t^{2l(x)}$, however, we prefer to work in complex dimensions.

THEOREM 3.0.3. *(cf.* **[34]***) Assume that the coefficients of all Kazhdan-Lusztig polynomials $P_{u,v}$ are non-negative for all $u, v \in W$. Let $x \leqslant w \in W$. Then the following are equivalent:*

(1) $P_{x,w} = 1$.

(2) $P_{y,w} = 1$, $\forall y \in [x, w]$.
(3) $p(y, w) = l(w) - l(y)$, $\forall y \in [x, w]$.

3.0.4. T-stable curves. Recall (cf. Chapter 8, §5.1) the following:

For a root $\alpha > 0$, let Z_α denote the $SL(2)$-copy in G corresponding to α (namely, Z_α is the subgroup of G generated by U_α and $U_{-\alpha}$). Then $Z_\alpha \cdot e_x$ is a T-stable curve in G/B, and conversely any T-stable curve in G/B is of this form. Given $y \leqslant w$, let $\mathcal{R}(y, w) = \{\alpha \in R^+ \mid s_\alpha y \leqslant w\}$. Then there are precisely $r(y, w)$ T-stable curves in $X(w)$ passing through e_y, where $r(y, w) = \#\mathcal{R}(y, w)$. In view of Deodhar's Inequality, we have $r(y, w) \geqslant l(w)$ (taking $x = e$ in Deodhar's Inequality (cf. Chapter 8, §5.1)). Thus there are at least $l(w)$ T-stable curves in $X(w)$ passing through e_y.

THEOREM 3.0.5. *(cf. [**34**]) For a Schubert variety $X(w)$ the following are equivalent:*

(1) *$X(w)$ is rationally smooth at e_x.*
(2) *$P_{x,w} = 1$.*
(3) *$P_{y,w} = 1$ for all $y \in [x, w]$.*
(4) *For $x \leqslant y \leqslant w$, $p(y, w) = l(w) - l(y)$.*
(5) *For $x \leqslant y \leqslant w$, $r(y, w) = l(w)$, i.e., e_y lies on exactly $l(w)$ T-invariant curves on G/B.*

DEFINITION 3.0.6. Define the *Bruhat graph of w* to be the graph with vertices in the set $\{v \in W : v < w\}$ and edges between u and v if $u = v\sigma$ for some reflection σ (not necessarily a simple reflection). Note, this is not the same as the graph of the Bruhat poset.

THEOREM 3.0.7. *(cf. [**34**]) For a Schubert variety $X(w)$ the following are equivalent:*

(1) *$X(w)$ is rationally smooth.*
(2) *$P_{e,w} = 1$.*
(3) *$P_{y,w} = 1$ for all $y \leq w$.*
(4) *$p_w(t)$ is symmetric.*
(5) *$p(y, w) = l(w) - l(y), \forall y \leqslant w$.*
(6) *For $y \leqslant w$, $r(y, w) = l(w)$, i.e., e_y lies on exactly $l(w)$ T - invariant curves on G/B.*
(7) *$a_w = \frac{1}{2}l(w)$.*
(8) *The Bruhat graph (defined above) is regular, i.e. every vertex has the same number of edges.*

Remark. Note that in the above Theorem, (2), (4), and (7) involve only w (and id), while (3), (5), (6), and (8) involve *all $y \leqslant w$*.

THEOREM 3.0.8. *(cf. [**34**]) The set of e_τ which are not rationally smooth on $X(w)$ is a B-stable closed subvariety of $X(w)$ codimension at least* 3.

4. Kumar's Criterion for smoothness and rational smoothness

The nil Hecke Ring. Throughout this section we assume W is the Weyl group of a semi-simple algebraic group. Let $S(\mathfrak{h}^*)$ be the symmetric algebra of $\mathfrak{h}^*$ (the dual of the Cartan subalgebra $\mathfrak{h}$), and let $Q(\mathfrak{h}^*)$ be the quotient field of $S(\mathfrak{h}^*)$ viewed as a W-field. Let Q_W denote the $Q(\mathfrak{h}^*)$-module with basis $\{\delta_w,\ w \in W\}$ endowed with a ring structure defined as follows:

$$f\delta_v \cdot g\delta_w = fv(g)\delta_{vw}$$

where $v, w \in W,\ f, g \in Q(\mathfrak{h}^*)$.

DEFINITION 4.0.1. Recall, $s_1, \ldots, s_n$ are the simple reflections in W. For each simple reflection s_i, let x_i be the element of Q_W defined as

$$x_i = -(\delta_e + \delta_{s_i})\frac{1}{\alpha_i},$$

where e is the identity element in W. For any $w \in W$ and any reduced expression $s_{i_1} \cdots s_{i_r} = w$, define $x_w = x_{i_1} \cdots x_{i_r}$ (it can be seen that x_w depends only on w and not on the reduced expression chosen). Expanding x_w in the basis of $\{\delta_v\}$ uniquely defines the coefficients $c_{w,v} \in Q(\mathfrak{h}^*)$, namely

$$x_w = \sum_v c_{w,v}\delta_v.$$

Note that what we call $c_{w,v}$, Kumar calls $c_{w^{-1},v^{-1}}$. Kostant and Kumar originally introduced the elements x_w in **[86]** and show they are well defined and closely related to the Schubert cycles in the cohomology ring of G/B. The x_w play a role analogous to the divided difference operators in the theory of Bernstein-Gelfand-Gelfand **[13]** and Demazure **[42]**. The coefficients $c_{w,v}$ play an important role analogous to the Kazhdan-Lusztig polynomials.

An explicit expression for $c_{w,v}$ can be derived directly from the definition of x_w.

THEOREM 4.0.2. **[8, 87]** *Fix a reduced expression for w say $s_{a_1}s_{a_2}\ldots s_{a_p}$, then*

$$c_{w,v} = (-1)^{l(w)} \sum s_{a_1}^{\epsilon_1}\alpha_{a_1}^{-1} s_{a_2}^{\epsilon_2}\alpha_{a_2}^{-1} \ldots s_{a_p}^{\epsilon_p}\alpha_{a_p}^{-1}$$

where the sum is over all sequences $(\epsilon_1\epsilon_2\ldots\epsilon_p)$ such that $\epsilon_i = 0$ or 1 for all $1 \le i \le p$, $v = s_{a_1}^{\epsilon_1}s_{a_2}^{\epsilon_2}\ldots s_{a_p}^{\epsilon_p}$ (not necessarily reduced), and each s_i acts on all α_j^{-1} to the right.

For example, take $w = s_1s_2s_1 \in \mathfrak{S}_n$ and $v = \text{id}$. Say we want to compute the coefficient of δ_{id} in $x_w = x_1x_2x_1$. Then (4.0.2) gives

$$\begin{aligned} c_{w,v} &= (-1)^3\left[\frac{1}{\alpha_1}\cdot\frac{1}{\alpha_2}\cdot\frac{1}{\alpha_1} + \frac{1}{s_1\alpha_1}\cdot\frac{1}{s_1\alpha_2}\cdot\frac{1}{s_1^2\alpha_1}\right] \\ &= -\left(\frac{1}{\alpha_1^2\alpha_2} - \frac{1}{\alpha_1^2(\alpha_1+\alpha_2)}\right) \\ &= \text{coefficient of } \delta_{\text{id}} \text{ in } x_w. \end{aligned}$$

Criteria for smoothness and rational smoothness

THEOREM 4.0.3. [**87**, Theorem 5.5] *Let $v, w \in W, v \leqslant w$. Then $X(w)$ is smooth at e_v if and only if*

$$c_{w,v} = (-1)^{l(w)-l(v)} \prod_{\beta \in S(w,v)} \beta^{-1},$$

where $S(w,v) = \{\beta \in R^+ \mid vs_\beta \leq w\}$. Furthermore, $X(w)$ is rationally smooth at e_v if and only if for each $v \leq x \leq w$ there exists a positive integer $d_{w,x}$ such that

$$c_{w,x} = d_{w,x}(-1)^{l(w)-l(x)} \prod_{\beta \in S(w,x)} \beta^{-1}.$$

Theorem 4.0.3 holds in arbitrary characteristic (see [**30**] or [**87**, Remark 8.10]). Brion's proof uses equivariant Chow group theory for torus actions.

Variation on Kumar's Theorem. Explicit computations using Kumar's theorem are not efficient due to the fact that $c_{w,v}$ is a rational function. Below we state a more efficient algorithm for applying Kumar's theorem in the case G is semisimple using the inverse matrix, given by

$$\delta_v = \sum_v k_{v,w} x_w.$$

It is known that $k_{v,w}$ belongs to $S(\mathfrak{h}^*)$ and $k_{v,w} = 0$ unless $v \leq w$ [**86, 14**] (see Theorem 4.0.6 below). This algorithm is still however exponential in the length of the Weyl group element.

THEOREM 4.0.4. *(cf. [**87**]) Given any $v, w \in W$ such that $v \leq w$, we have*

$$c_{w,v} \prod_{\alpha > 0} \alpha = (-1)^{l(w)-l(v)} k_{ww_0, vw_0},$$

where w_0 is the unique element of W of longest length.

COROLLARY 4.0.5. *Given any $v, w \in W$ such that $v \leq w$, the Schubert variety X_{vw_0} is smooth at e_{ww_0} if and only if*

$$k_{v,w} = \prod_{\gamma \in T(v,w)} \gamma$$

where $T(v,w) = \{\gamma \in R_+ \,:\, v \not\leq w\sigma_\gamma\} = \{\beta \in R_+ \,:\, w_0 w \sigma_\beta \not\leq w_0 v\}$.

Furthermore, there is an explicit formula for $k_{v,w}$. Let $R(w)$ be the set of all sequences $\mathbf{b} = b_1 \dots b_p$ such that $s_{b_1} s_{b_2} \dots s_{b_p}$ is a reduced expression for w.

THEOREM 4.0.6. *(cf. [**14**]) Let $v, w \in W$ and fix a reduced word $\mathbf{b} = b_1 b_2 \dots b_p \in R(w)$. Then*

$$k_{v,w} = \sum_{b_{i_1} b_{i_2} \dots b_{i_k} \in R(v)} r_{\mathbf{b}}(i_1) r_{\mathbf{b}}(i_2) \cdots r_{\mathbf{b}}(i_k)$$

where $r_{\mathbf{b}}(j)$ is defined by

$$r_{\mathbf{b}}(j) = r_{b_1 b_2 \dots b_p}(j) = s_{b_1} s_{b_2} \dots s_{b_{j-1}}(\alpha_{b_j}),$$

and the sum is over all sequences $1 \leq i_1 < i_2 < \ldots < i_k \leq p$ *such that* $b_{i_1} b_{i_2} \ldots b_{i_k} \in R(v)$. *Furthermore, the sum in* (4.0.6) *is independent of the choice of* $\mathbf{b} \in R(w)$.

Representation Theoretic results on the Tangent Cone

Notation:

- $S(w,v) = \{\alpha \in R^+ \mid vs_\alpha \leq w\}$ *i.e* edges adjacent to v in Bruhat graph of w.
- T maximal torus
- $\mathfrak{h}$ = LieT = Cartan subalgebra of Lie$G$
- $\mathfrak{h}^*_{\mathbb{Z}} = \{\lambda \in \mathfrak{h}^* \mid \lambda(\widehat{\alpha_i}) \in \mathbb{Z}\, \forall i\}$ = integral weights
- $R(T)$ group algebra of the character group $X(T)$ with basis $\{e^\lambda\}_{\lambda \in \mathfrak{h}^*_{\mathbb{Z}}}$
- $Q(T)$ quotient field of $R(T)$
- $*$ involution on $Q(T)$ taking $e^\lambda \to e^{-\lambda}$
- $gr(\mathcal{O}(e_v, X(w)))$ graded local ring at $e_v \in X(w)$
- $T_v(X(w)) = \mathrm{Spec}(gr(\mathcal{O}(e_v, X(w))))$ (scheme theoretic) tangent cone
- $S^l(\mathfrak{h}^*)$ homogeneous polynomials on $\mathfrak{h}$ of degree l
- $Q(\mathfrak{h}^*)$ quotient field of $S(\mathfrak{h}^*) = \bigoplus S^l(\mathfrak{h}^*)$

With this notation one defines the special elements y_w in $Q(T)$.

DEFINITION 4.0.7. For each simple reflection s_i, let y_i be the element of $Q_W(T)$ (smash product of $\mathbb{Z}[W]$ with $Q(T)$) defined as

$$y_i = (\delta_e + \delta_{s_i}) \frac{1}{1 - e^{-\alpha_i}}.$$

For any $w \in W$ and any reduced expression $s_{i_1} \cdots s_{i_r} = w$, define $y_w = y_{i_1} \cdots y_{i_r}$. Expanding y_w in the basis of $\{\delta_v\}$ uniquely defines the coefficients $b_{w,v} \in Q(T)$, namely

$$y_w = \sum_v b_{w,v} \delta_v.$$

Again, note that what we call $b_{w,v}$, Kumar calls $b_{w^{-1},v^{-1}}$.

The relationship between the coefficients $b_{w,v}$ and $c_{w,v}$ is dependent on the following bracket operation.

DEFINITION 4.0.8. If $a = \sum n_\lambda e^\lambda \in R(T)$ and k is a non-negative integer then define $(a)_k \in S^k(\mathfrak{h}^*)$ to be the homogeneous component of a of degree k (each λ being given degree 1), *i.e.*

$$(a)_k = \sum n_\lambda \frac{\lambda^k}{k!}.$$

Define $[a]$ as

$$[a] = (a)_{k_0}$$

where k_0 is the smallest non-negative integer such that $(a)_{k_0} \neq 0$. If a is a scalar, then $[a] = a$. If $q = \frac{a}{b} \in Q(T)$ for some $a, b \in R(T)$ then

$$[q] = \frac{[a]}{[b]} \in Q(\mathfrak{h}^*)$$

THEOREM 4.0.9. *(cf. [87], Theorem 2.2) For any $v \leq w$ in W, $gr(\mathcal{O}(e_v, X(w)))$ is a weight module under the T action such that each component is finite dimensional and*

$$ch\, gr(\mathcal{O}(e_v, X(w))) = *b_{w,v} \in Q(T)$$

Further, $[ch\, gr(\mathcal{O}(e_v, X(w)))] = [*b_{w,v}] = c_{w,v}$ as elements of $Q(\mathfrak{h}^*)$.

5. Summary of combinatorial formulas for Kazhdan-Lusztig polynomials

As was shown in Section 2, the Kazhdan-Lusztig polynomials can be used to determine whether or not a Schubert variety is rationally smooth at a point. Due to the enormous number of formulas related to Kazhdan-Lusztig theory in the literature, we will not attempt to give this subject complete coverage but instead we will summarize a selection of combinatorial algorithms and give references to the appropriate literature.

5.1. Brenti's results

In this section we describe some of the algorithms for computing $R_{u,v}$ and $P_{u,v}$ formulated by Francesco Brenti. These algorithms have been developed in several papers. We refer the reader to [29] for a more complete overview of the history, consequences, and conjectures related to the formulas below. These results are stated here for S_n only. See Remark 5.1.8 for extensions to other Weyl groups.

Brenti's algorithm for computing $R_{u,v}$ is stated in terms of a related family of polynomials $\tilde{R}_{u,v}$. Define $\tilde{R}_{u,v}$ to be the unique polynomial satisfying

$$R_{u,v} = q^{\frac{l(v)-l(u)}{2}} \tilde{R}_{u,v}(q^{1/2} - q^{-1/2}).$$

DEFINITION 5.1.1. Given $u \in S_n$, let $C_{i,j}(u)$ be the set of all sequences of the form $(u(i_1), \ldots, u(i_k))$ where $1 \leq k \leq n$, $i = i_1 < \cdots < i_k = j$, and, $u(i_1) < \cdots < u(i_k)$. If $u < v \in S_n$, let $C(u,v)$ be $C_{v^{-1}(d), u^{-1}(d)}(u)$ where $d = \max\{1 \leq i \leq n : u^{-1}(i) \neq v^{-1}(i)\}$, i.e. $v^{-1}(d)$ is the position in v of the largest number which does not appear in the same position in both u and v.

For example, $C_{1,5}(1427635) = \{(1,6), (1,4,6), (1,2,6)\}$. We think of the elements in $C_{i,j}(u)$ as permutations in cycle notation consisting of single cycles. Multiplying a permutation on the right by a cycle rotates the numbers to the left, e.g. $(1,4,6)1427635 = 4627135$.

THEOREM 5.1.2. *Let $u, v \in S_n$, then*

$$\tilde{R}_{u,v} = \begin{cases} \sum_{w \in C(u,v)} q^{k(w)-1} \tilde{R}_{w^{-1}u,v}(q) & u < v \\ 1 & u = v \\ 0 & u \not\leq v. \end{cases}$$

where $k(w)$ is the number of elements in the cycle w.

To illustrate the algorithm we compute $\tilde{R}_{u,v}$ for $u = 1427635$, $v = 6457213$. First, $d = 6$ since 7 appears in the same position in both u and v. Then, $v^{-1}(6) = 1$ and $u^{-1}(6) = 5$ so $C(u,v) = C_{1,5}(1427635) = \{(1,6), (1,4,6), (1,2,6)\}$. Hence,

$$\begin{aligned} \tilde{R}_{1427635,6457213} &= q\tilde{R}_{6427135,6457213} + q^2\tilde{R}_{6127435,6457213} + q^2\tilde{R}_{6417235,6457213} \\ &= q(q^3) + q^2(2q^4 + q^6) + q^2(q^2 + q^4) \\ &= 2q^4 + 3q^6 + q^8 \end{aligned}$$

REMARK 5.1.3. $\tilde{R}_{u,v}(q)$ is clearly a polynomial in q with nonnegative integer coefficients. The recursion in Theorem 5.1.2 will terminate after at most n steps.

Next, we begin describing Brenti's algorithm for the Kazhdan-Lusztig polynomials. This requires the introduction of polynomials $\gamma_{\beta_1,\ldots,\beta_s}(q)$ and operators U_j.

DEFINITION 5.1.4. Let U_j be the operator acting on the space of all Laurent polynomials in q with real coefficients given by

$$U_j\left(\sum_{i \in \mathbb{Z}} a_i q^i\right) = \sum_{i \geq j} a_i q^i$$

for any $j \in \mathbb{Q}$. For any composition $\beta = \beta_1, \ldots, \beta_s \in \{1, 2, 3, \ldots\}$, define $\gamma_{\beta_1,\ldots,\beta_s}$ recursively by

$$\gamma_{\beta_1,\ldots,\beta_s} = \begin{cases} -1 & \beta_1 = 1, s = 1 \\ (q-1)U_{\frac{|\beta|}{2}}\left(q^{|\beta|-1}\gamma_{\beta_2,\ldots,\beta_s}\left(\frac{1}{q}\right)\right) & \beta_1 = 1, s \geq 2 \\ (q-1)\gamma_{\beta'}(q) + U_{\frac{|\beta|+1}{2}}\left(q^{|\beta|-1}(1-q)\gamma_{\beta'}\left(\frac{1}{q}\right)\right) & \beta_1 > 1 \end{cases}$$

where $\beta' = \beta_1 - 1, \beta_2, \ldots, \beta_s$ and $|\beta| = \beta_1 + \beta_2 + \cdots + \beta_s$.

EXAMPLE 5.1.5. We have

$$\begin{aligned} \gamma_1 &= -1 \\ \gamma_{1,1} &= -q^2 + q \\ \gamma_2 &= q^2 + q + 1 \\ \gamma_{2,1} &= -q^3 + q^2 + -q \end{aligned}$$

DEFINITION 5.1.6. Define $B(u,v)$ to be a labeled directed graph with vertex set $\{x \in S_n : u \leq x \leq v\}$ and directed edges from x to y labeled by $\lambda(x,y) = j - i + \sum_{k=1}^{i-1}(n-k)$ if and only if $y = x(i,j)$ for some $1 \leq i < j \leq n$ and

$l(y) > l(x)$. For any directed path $\Gamma = (x_0, x_1, \ldots, x_k)$, say i is a *descent* in Γ if $\lambda(x_{i-1}, x_i) > \lambda(x_i, x_{i+1})$. Assign Γ a composition

$$D(\Gamma) = (\beta_1, \ldots, \beta_s)$$

if Γ has $s - 1$ descents, $|\beta| = k$ and for each i β_i is the length of the path between the $(i-1)^{st}$ and i^{th} descents.

For example, take $u = (15243)$ and $v = (45231)$. There are 22 paths in $B(u, v)$ below from u to v.

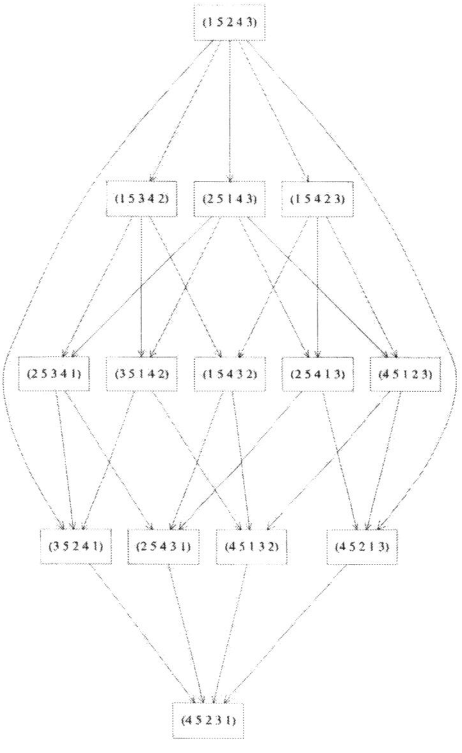

Among these paths is $\Gamma = 15243 \xrightarrow{9} 15342 \xrightarrow{2} 35142 \xrightarrow{9} 35241 \xrightarrow{3} 45231$, hence $D(\Gamma) = (1, 2, 1)$. The 22 paths give rise to a multiset of 22 compositions, namely

$$\{(2), (1,1), (4), (1,3)^2, (2,2)^4, (3,1)^3, (1,1,2)^3, (1,2,1)^4, (2,1,1)^2, (1,1,1,1))\}$$

THEOREM 5.1.7. **[28]** *Let $u < v \in S_n$, then*

$$P_{u,v}(q) = (-1)^{l(v)-l(u)} \sum_{\Gamma} q^{\frac{1}{2}(l(v)-l(u)-|D(\Gamma)|)} \gamma_{D(\Gamma)}(q),$$

where the sum is over all directed paths in $B(u,v)$ from u to v.

Note, that $\frac{1}{2}(l(v) - l(u) - |D(\Gamma)|)$ is always a non-negative integer since $l(v) - l(u)$ and $|D(\Gamma)|$ must have the same parity.

Continuing with the example above and applying Theorem 5.1.7 we obtain

$$P_{u,v} = (-1)^{8-4}(q\,\gamma_{(2)} + q\,\gamma_{(1,1)} + \gamma_{(4)} + 2\gamma_{(1,3)} + 4\gamma_{(2,2)} + 3\gamma_{(3,1)} \\ + 3\gamma_{(1,1,2)} + 4\gamma_{(1,2,1)} + 2\gamma_{(2,1,1)} + \gamma_{(1,1,1,1)})$$

$$\begin{aligned}
\gamma_{(11)} &= -q^2 + q^1 \\
\gamma_{(2)} &= q^2 - q^1 + 1 \\
\gamma_{(4)} &= q^4 - 3q^3 + 2q^2 - 3q^1 + 1 \\
\gamma_{(13)} &= -q^4 + 3q^3 - 2q^2 \\
\gamma_{(22)} &= q^4 - 2q^3 + q^2 \\
\gamma_{(31)} &= -q^4 + 3q^3 - 2q^2 + q^1 \\
\gamma_{(112)} &= 0 \\
\gamma_{(121)} &= -q^3 + q^2 \\
\gamma_{(211)} &= 0 \\
\gamma_{(1111)} &= 0
\end{aligned}$$

Therefore,

$$P_{u,v} = q + 1.$$

REMARK 5.1.8. Brenti's algorithm for Kazhdan-Lusztig polynomials can be extended to other finite Weyl groups as well. Simply replace the edge labels in $B(u,v)$ with the reflection s_β if $x = ys_\beta$. Fix an order on the set of all reflections by choosing a reduced word $s_{i_1}s_{i_2}\dots s_{i_p}$ for the unique longest element of the Weyl group, then every reflection can be written as $s_{i_1}\dots s_{i_{j-1}}s_{i_j}s_{i_{j-1}}\dots s_{i_1}$ for some $1 \le j \le p$. This naturally gives rise to an ordering on reflections where s_{i_1} is the smallest and $s_{i_1}\dots s_{i_{p-1}}s_{i_p}s_{i_{p-1}}\dots s_{i_1}$ is the largest. Now, any chain in $B(u,v)$ can be partitioned into increasing segments by breaking the chain whenever two adjacent edges are labeled by reflections where the first is bigger than the second.

5.2. Deodhar's and Lascoux-Schützenberger's algorithms

Deodhar's algorithm

Deodhar has defined a statistic on subexpressions of reduced expressions call the *defect*. This statistic naturally arises in the Hecke algebra and in a formula for Kazhdan-Lusztig polynomials.

DEFINITION 5.2.1. Fix a reduced expression $\mathbf{r} = r_1r_2\dots r_p = s_{j_1}s_{j_2}\dots s_{j_p}$ for $w \in W$. Let $\sigma = (\sigma_0, \sigma_1, \dots, \sigma_p)$ be a sequence of Weyl group elements such that $\sigma_{j-1}\sigma_j \in \{id, r_j\}$ for all $1 \le j \le k$, i.e. σ corresponds with some subexpression $r_{i_1}r_{i_2}\dots r_{i_k}$. Let $\mathcal{S}$ be the set of all such sequences for the fixed reduced expression $\mathbf{r}$. Define the *defect* of $\sigma \in \mathcal{S}$ by $d(\sigma) = \#\{j : \sigma_{j-1}r_j \le \sigma_{j-1}\}$. Let $\pi(\sigma) = \sigma_p$.

For example, in type A, take $\mathbf{r} = s_1s_2s_3s_2s_1s_2$, then

σ	$d(\sigma)$	$\pi(\sigma)$
$(id, s_1, s_1s_2, s_1s_2s_3, s_1s_2s_3, s_1s_2s_3, s_1s_2s_3)$	0	$s_1s_2s_3$
$(id, s_1, s_1s_2, s_1s_2, s_1s_2, s_1s_2s_1, s_1s_2s_1, s_1s_2s_1)$	1	$s_1s_2s_1$
$(id, id, s_2, s_2, s_2, s_2, s_2)$	2	s_2

THEOREM 5.2.2. **[45]** *Assume the coefficients of all Kazhdan-Lusztig polynomials for W are non-negative. Then for each pair $x, y \in W$, there exists a set $E_{min} \subset \mathcal{S}$ such that*

$$P_{x,y} = \sum_{\sigma \in E_{min} s.t.\ \pi(\sigma)=x} q^{d(\sigma)}.$$

In **[45]**, Deodhar gives a recursive algorithm for computing the set $E_{\min}$ starting from $\mathcal{S}$.

THEOREM 5.2.3. **[45]** *Given a reduced expression $y = r_1 \ldots r_p$, let $D'_y = C'_{r_1}C'_{r_2}\ldots C'_{r_p}$. Then $\{D'_y : y \in W\}$ forms a basis for the Hecke algebra and*

$$D'_y = q^{-l(y)/2} \sum_{\sigma \in \mathcal{S}} q^{d(\sigma)} T_{\pi(\sigma)}.$$

5.2.4. Lascoux and Schützenberger's algorithm.
Lascoux and Schützenberger **[111]** have given an algorithm for computing the Kazhdan-Lusztig polynomials in terms of paths along trees for the "Grassmannian" permutations in the symmetric group, i.e. permutations with at most one descent in one-line notation. Lascoux **[110]** has extended this algorithm to the set of vexillary permutations i.e. 2143-avoiding permutations.

5.3. Bona's results

Recall from Chapter 8 that we say a permutation $w = (w_1 \ldots w_n) \in S_n$ *avoids* a pattern $(a_1 \ldots a_k) \in S_k$ if no subsequence of w has the same relative order as $(a_1 \ldots a_k)$. As seen in Chapter 8, permutations indexing smooth Schubert varieties in type A are precisely the permutations which avoid the pair of patterns 3412 and 4231 (Chapter 8, Theorem 15.0.1). Let $S_n(u, v)$ be the number of $w \in S_n$ such that w avoids both u and v. M. Bona has completely classified all pairs of permutations in S_n such that $S_n(u, v) = S_n(3412, 4231)$ i.e. equinumerous to the smooth permutations.

Two pairs of patterns are *equivalent* if one can be transformed into the other by some combination of reversing, complementing, and inverse-taking.

THEOREM 5.3.1. **[21]** *The following five inequivalent pairs of permutation classes are equinumerous:*

$$S_n(3412, 4231) = S_n(1342, 2431) = S_n(1342, 3241)$$
$$= S_n(1342, 2314) = S_n(1324, 2413).$$

There are no other inequivalent pairs (u, v) for which the values of $S_n(u, v)$ equals $S_n(3412, 4231)$ for all n.

6. Small resolutions

Let X be an irreducible complex variety.

DEFINITION 6.0.1. A proper, birational morphism $p : \tilde{X} \to X$, with $\tilde{X}$ smooth is called a *resolution of* X. Such a resolution is said to be *small* if

$$\operatorname{codim}_X\{x \in X \mid \dim p^{-1}(x) \geqslant i\} > 2i,\ i > 0.$$

THEOREM 6.0.2. *(cf.* **[63]***) Let $p : \tilde{X} \to X$ be a small resolution. Then for any $i \geqslant 0$, for the intersection cohomology sheaf $\mathcal{H}(X)$ (with respect to the middle perversity), the stalk $\mathcal{H}^i(X)_x$ is isomorphic to the singular cohomology group $H^I(p^{-1}(x);\mathbb{C})$.*

Combining Theorem 6.0.2 with Theorem 2.0.8, we obtain

THEOREM 6.0.3. *Let $p : \tilde{X} \to X(w)$ of a Schubert variety $X(w)$ in G/B, G being semisimple. Let $X(\tau) \subset X(w)$. Then the Poincaré polynomial $P_t(p^{-1}(e_\tau))$ is equal to the Kazhdan-Lusztig polynomial $P_{\tau,w}(q)$, with $q = t^2$.*

REMARK 6.0.4. Thus, given a Schubert variety $X(w) \subset G/B$, if we can find a small resolution $p : \tilde{X} \to X(w)$, then for $\tau \leqslant w$, $P_{\tau,w}(q)$ may be computed as the Poincaré polynomial $P_t(p^{-1}(e_\tau))$. Zelevinsky (cf. **[164]**) constructed small resolutions for Schubert varieties in the Grassmannian. Towards generalizing Zelevinsky's results, Sankaran and Vanchinathan (cf. **[141]**, **[142]**) describe a class of Schubert varieties in the symplectic and orthogonal Grassmannians for which they construct small resolutions. They also describe a class of Schubert varieties in the symplectic Grassmannian not admitting any small resolutions! We recall these results below.

6.1. Bott-Samelson resolution

(cf. **[25]**, **[66]**, **[42]**) Let Q be a parabolic subgroup, $Q \supset B$. For a simple root α, we shall denote the associated rank one parabolic by P_α. If $\alpha = \alpha_i$, we shall denote P_α by just P_i. Let $X(\lambda) \subset G/Q$ be any Schubert variety, and P_λ the largest subgroup of G which leaves $X(\lambda)$ stable for the action of G on G/Q by left multiplication. Clearly, P_λ is a parabolic subgroup containing B. We shall refer to P_λ as the *stabilizer* of $X(\lambda)$. Note that it is possible to find a parabolic subgroup $P \subset P_\lambda$ and a Schubert variety $X(\lambda')$, $\lambda' < \lambda$, such that $P_\lambda X(\lambda') = PX(\lambda') = X(\lambda)$; for example, suppose P_λ is a rank one parabolic with α as the associated simple root, then we may take λ' to be $s_\alpha\lambda$. Let $R_0 = P_\lambda \cap P_{\lambda'}$, $P_\lambda \times^{R_0} X(\lambda') = P_\lambda \times X(\lambda')/\sim$, where $\sim$ is the equivalence relation $(gr, rx) \sim (g, x)$, $g \in P_\lambda$, $r \in R_0, x \in X(\lambda)$. Then the map $\pi_0 : P_\lambda \times^{R_0} X(\lambda') \to X(\lambda)$, $(g, x) \mapsto gx$ is surjective and P_λ-equivariant (for the P_λ-action $p \cdot (g, x) = (pg, x)$, $p \in P_\lambda$), but not birational in general. However it is possible to choose P and $\lambda' < \lambda$ such that if $R_1 = P \cap P'_\lambda$, then dim P/R_1 equals the codimension of $X(\lambda')$ in $X(\lambda)$, and so $\pi_1 : P \times^{R_1} X(\lambda') \to X(\lambda)$ is P_λ-equivariant and birational; for example, if α is a simple root such that the rank one parabolic $P_\alpha \subseteq P_\lambda$, then we can choose $\lambda' = s_\alpha\lambda$ and $P = P_\alpha$. Since any 1-dimensional Schubert variety is smooth, iterating this construction we obtain a $P_{(1)}(=$

P_λ)-equivariant resolution

$$p : P_{(1)} \times^{R_1} P_{(2)} \times \cdots \times^{R_{r-2}} P_{(r-1)} \times^{R_{r-1}} X(\lambda^r) = \tilde{X}(\lambda) \to X(\lambda).$$

where $X(\lambda^r)$ is smooth, and

$$\begin{aligned} &\lambda^1 = \lambda,\ \lambda^2 = \lambda', \cdots \\ &P_{(i)} \subset P_{\lambda^i}, \qquad 1 \leqslant i \leqslant r, \\ &P_{(r)} = P_{\lambda^r}, \\ &R_i = P_{(i)} \cap P_{(i+1)}, \qquad 1 \leqslant i \leqslant r-1. \end{aligned}$$

In fact, at each step, $P_{(i)}$ may be chosen to be the stabilizer of $X(\lambda^i)$.

The above resolution is usually referred to as a *Bott-Samelson resolution*. When $X(\lambda) \subset G/B$, such resolutions were obtained by H. Hansen (cf. [**66**]) and M. Demazure (cf. [**42**]).

An Example of a Bott-Samelson resolution. Take $G = SL(4)$, $\lambda = (2341)$. We have $\lambda = s_1 s_2 s_3$, $P_\lambda = P_1$. We may take $\lambda^2 = s_2 s_3$, $P_{(2)} = P_2$, $\lambda^3 = s_3$, $P_{(3)} = P_3$, $\lambda^4 = id$.

6.2. Small resolutions and Schubert varieties in the Grassmannian

THEOREM 6.2.1. *(cf.* [**164**]*) Zelevinsky's small resolution: Let $G = SL(n)$, and P a maximal parabolic subgroup so that G/P is the Grassmannian. Let X be a Schubert variety in G/P. There exist Schubert varieties*

$$X = X(\lambda^1) \supset X(\lambda^2) \supset \cdots \supset X(\lambda^r),$$

such that

$$P_{(1)} \times^{R_1} P_{(2)} \times \cdots \times^{R_{r-2}} P_{(r-1)} \times^{R_{r-1}} X(\lambda^r) = \tilde{X}(\lambda^1) \to X(\lambda^1)$$

is a $P_{(1)}$-equivariant small resolution of $X(\lambda^1)$, where $P_{(i)}$ is the stabilizer of $X(\lambda^i)$, $1 \leqslant i \leqslant r$ and $R_i = P_{(i)} \cap P_{(i+1)}$, $1 \leqslant i \leqslant r-1$.

6.3. Small resolutions and Schubert varieties in the symplectic Grassmannian

Let $G = Sp(2n)$ and $P = P_n$, the maximal parabolic subgroup associated to the simple root α_n (the simple roots being indexed as in [**27**]). As seen in Chapter 8, §13, G/P can be identified with the isotropic Grassmannian of maximal totally isotropic n spaces in a $2n$-dimensional space with a non-degenerate skew-symmetric bilinear form (,). The set W^P of minimal representatives of W/W_P may be identified with

$$\big\{(a_1, \cdots, a_n),\ a_i \in \{1, 2, \cdots, 2n\} \mid \text{if } k \in (a_1, \cdots, a_n), \\ \text{then } 2n+1-k \notin (a_1, \cdots, a_n),\ 1 \leqslant k \leqslant 2n\big\}.$$

Note that for $1 \leqslant k \leqslant 2n$, precisely one of $\{k, 2n+1-k\}$ belongs to $(a_1, \cdots, a_n)$. Thus, if r is such that $a_r \leqslant n$, $a_{r+1} > n$, then $(a_1, \cdots, a_n)$ is completely known by knowing $(a_1, \cdots, a_r)$. Hence W^P may be identified with $\bigcup_{0 \leqslant r \leqslant n} I_{r,n}$, where $I_{r,n} = \big\{(i_1, \cdots, i_r),\ 1 \leqslant i_1 < i_2 < \cdots < i_r \leqslant n\big\}$.

THEOREM 6.3.1. *(cf. [**141**]) Let $X(\lambda) \subset G/P_n$, $\lambda \in I_{r,n}$. Let $\lambda_r < n-r$. There exist Schubert varieties*

$$X(\lambda) = X(\lambda^1) \supset X(\lambda^2) \supset \cdots \supset X(\lambda^s),$$

such that

$$P_{(1)} \times^{R_1} P_{(2)} \times \cdots \times^{R_{s-2}} P_{(s-1)} \times^{R_{s-1}} X(\lambda^s) = \tilde{X}(\lambda^1) \to X(\lambda^1)$$

is a $P_{(1)}$-equivariant small resolution of $X(\lambda^1)$, where $P_{(i)}$ is the stabilizer of $X(\lambda^i)$, $1 \leqslant i \leqslant s$ and $R_i = P_{(i)} \cap P_{(i+1)}$, $1 \leqslant i \leqslant s-1$.

6.4. Small resolutions and Schubert varieties in the orthogonal Grassmannian

Let $G = SO(2n)$ and $P = P_n$, the maximal parabolic subgroup associated to the simple root α_n. As seen in Chapter 8, §13, G/P can be identified with the isotropic Grassmannian of maximal totally isotropic n spaces in a $2n$-dimensional space with a non-degenerate symmetric bilinear form (,). Then as above, W^P may be identified with $\bigcup_{0 \leqslant r \leqslant n,\ n-r \text{ even}} I_{r,n}$.

THEOREM 6.4.1. *(cf. [**141**]) Let $X(\lambda) \subset G/P_n$, $\lambda \in I_{r,n}$, where $n-r$ is even. Further, let either $\lambda_r < n-r$ or $r \geqslant 2$, $\lambda_r = n$, $\lambda_{r-1} \leqslant n-r$. There exist Schubert varieties*

$$X(\lambda) = X(\lambda^1) \supset X(\lambda^2) \supset \cdots \supset X(\lambda^s),$$

such that

$$P_{(1)} \times^{R_1} P_{(2)} \times \cdots \times^{R_{s-2}} P_{(s-1)} \times^{R_{s-1}} X(\lambda^r) = \tilde{X}(\lambda^1) \to X(\lambda^1)$$

is a $P_{(1)}$-equivariant small resolution of $X(\lambda^1)$, where $P_{(i)}$ is the stabilizer of $X(\lambda^i)$, $1 \leqslant i \leqslant s$ and $R_i = P_{(i)} \cap P_{(i+1)}$, $1 \leqslant i \leqslant s-1$.

6.5. Schubert varieties not admitting any small resolutions

Suppose $p : \tilde{X} \to X$ is a resolution of a normal irreducible variety. Then by Zariski's Main Theorem (cf. [**67**]), the fiber $p^{-1}(x)$ over any singular point x of X has positive dimension. Hence, any normal irreducible variety X with codimension 2 singular locus cannot have any small resolution.

Using this fact, Sankaran-Vanchinathan construct a class of Schubert varieties in $Sp(2n)/B$ which do not admit any small resolution (by showing that these Schubert varieties have codimension 2 singular loci):

THEOREM 6.5.1. *(cf. [**141**]) Let $G = Sp(2n)$. Let $\lambda = (n)$ $(\in I_{1,n})$, $n \geqslant 3$, $I_{d,n}$ being as in Chapter 6. Let Q be any parabolic subgroup contained in P_n, the maximal parabolic subgroup with $S \setminus \{\alpha_n\}$ as the associated set of simple roots. Let $X(\Lambda)$ be the inverse image of $X(\lambda)$ $(\subset G/P_n)$ under the canonical projection $G/Q \to G/P_n$. Then $X(\Lambda)$ does not admit any small resolution.*

6.6. The $2 \times m$ matrix notation for λ

Let $\lambda \in I_{r,n}$. A maximal subsequence of consecutive integers in λ will be referred to as a *block* of λ. Clearly, λ is the concatenation of its blocks. Let a_i be the length, k_i the last term of the i-th block in λ. Then λ determines a $2 \times m$ matrix $\begin{pmatrix} k_1 & \cdots & k_m \\ a_1 & \cdots & a_m \end{pmatrix}$, where $0 < k_1 < \cdots < k_m \leqslant n$, $0 \leqslant a_i \leqslant k_i - k_{i-1}$, $k_0 = 0$, $\sum a_i = r$, and conversely such a $2 \times m$ matrix determines an unique element of $I_{r,n}$. In the sequel, we shall denote λ by the associated $2 \times m$ matrix $\begin{pmatrix} k_1 & \cdots & k_m \\ a_1 & \cdots & a_m \end{pmatrix}$.

REMARK 6.6.1. (cf. [**141**]) Note that G being $SL(n)$, and $\lambda = \begin{pmatrix} k_1 & \cdots & k_m \\ a_1 & \cdots & a_m \end{pmatrix}$, P_λ is simply the parabolic subgroup with $S \setminus \{\alpha_{k_1}, \cdots, \alpha_{k_m}\}$ as the associated set of simple roots.
Set $b_i = k_{i+1} - k_i - a_{i+1}$, $0 \leqslant i < m$, $k_0 = 0$.

6.7. Description of the small resolution in terms of the $2 \times m$ matrix

Schubert varieties in the Grassmannian
Let $\lambda \in I_{r,n}$ so that $X(\lambda)$ is a Schubert variety in the Grassmannian $G_{r,n}$. As above, let

$$\lambda = \begin{pmatrix} k_1 & \cdots & k_m \\ a_1 & \cdots & a_m \end{pmatrix}.$$

Set $a_0 = b_m = \infty$. Choose $0 \leqslant i < m$ so that $b_i \leqslant a_i$, and $a_{i+1} \leqslant b_{i+1}$; note that such an i exists since $b_0 < a_0 \,(= \infty)$ and $b_m \,(= \infty) > a_m$. Let

$$\lambda' = \begin{pmatrix} k_1 & \cdots & k_{i-1} & k_i + a_{i+1} & k_{i+2} & \cdots & k_m \\ a_1 & \cdots & a_{i-1} & a_i + a_{i+1} & a_{i+2} & \cdots & a_m \end{pmatrix}.$$

Then the number of blocks in λ' is $m - 1$ except the case when $m = 1$, in which case $X(\lambda')$ is a smooth variety. This choice of λ' corresponds to the first step in the Bott-Samelson type resolution as given by Theorem 6.2.1. Iterating this procedure, we obtain the Bott-Samelson type resolution as given by Theorem 6.2.1.
Example. Let $\lambda = (3, 4, 6, 9, 10,) \in I_{5,10}$ so that it determines a Schubert variety in the Grassmannian $G_{5,10}$, of 5-planes in 10-space. In the matrix notation,

$$\lambda = \begin{pmatrix} 4 & 6 & 10 \\ 2 & 1 & 2 \end{pmatrix}.$$

We have, $(a_1, a_2, a_3) = (2, 1, 2)$ and $(b_0, b_1, b_2) = (2, 1, 2)$ (as above, we set $a_0 = b_3 = \infty$). Let $i = 1$ so that $a_i \geq b_i$ and $a_{i+1} \leq b_{i+1}$. This leads to

$$\lambda' = (3, 4, 5, 9, 10) = \begin{pmatrix} 5 & 10 \\ 3 & 2 \end{pmatrix}.$$

Then $X(\lambda')$ is a subvariety of $X(\lambda)$ such that $P_\lambda \cdot X(\lambda') = X(\lambda)$. Here P_λ denotes the stabilizer of $X(\lambda)$. Again, for this λ', one has $(a'_1, a'_2) = (3, 2)$, $(b'_0, b'_1) = (2, 3)$.

Choosing $i = 1$ the conditions $a'_i \geq b'_i$ and $a'_{i+1} \leq b'_{i+1}$ are satisfied. This choice of i leads to $\lambda'' = (3,4,5,6,7) = \begin{pmatrix} 7 \\ 5 \end{pmatrix}$. Note that $X(\lambda'')$ is a smooth subvariety of $X(\lambda')$ and $P_{\lambda'} \cdot X(\lambda'') = X(\lambda')$.

We have, P_λ is the parabolic with $S \setminus \{\alpha_4, \alpha_6\}$ as the associated set of simple roots. Similarly, $P_{\lambda'}$ (resp. $P_{\lambda''}$) is the parabolic with $S \setminus \{\alpha_5\}$ (resp. $S \setminus \{\alpha_7\}$) as the associated set of simple roots. Write $R = P_\lambda \cap P_{\lambda'}, R' = P_{\lambda'} \cap P_{\lambda''}$. Then

$$P_\lambda \times_R P_{\lambda'} \times_{R'} X(\lambda'') \longrightarrow X(\lambda)$$

is a small resolution of the Schubert variety $X(\lambda)$.

Schubert varieties in the symplectic and the orthogonal Grassmannians

Let $G = Sp(2n)$ or $SO(2n)$. Let $X(\lambda) \subset G/P_n$. Let $\lambda = \begin{pmatrix} k_1 & \cdots & k_m \\ a_1 & \cdots & a_m \end{pmatrix}$.

REMARK 6.7.1. (cf. **[141]**) As in Remark 6.6.1, we have, P_λ is simply the parabolic subgroup with $S \setminus \{\alpha_{k_1}, \cdots, \alpha_{k_m}\}$ as the associated set of simple roots, if either $G = Sp(2n)$ or $G = SO(2n)$ and $a_m \geqslant 2$ if $k_m = n$; if $G = SO(2n)$, $a_m = 1$, $k_m = n$, then P_λ is the parabolic sub group with $S \setminus \{\alpha_{k_1}, \cdots, \alpha_{k_{m-1}}\}$ as the associated set of simple roots.

Let

$$\lambda = \begin{cases} \begin{pmatrix} k_1 & \cdots & k_m \\ a_1 & \cdots & a_m \end{pmatrix}, & G = Sp(2n),\ SO(2n), \\ \begin{pmatrix} k_1 & \cdots & k_m & n \\ a_1 & \cdots & a_m & 1 \end{pmatrix}, & G = SO(2n). \end{cases}$$

Let M denote the number of blocks in λ; note that $M = m$, or $m + 1$.

Set $b_i = k_{i+1} - k_i - a_{i+1}, 0 \leqslant i < m, k_0 = 0, k_{m+1} = n, a_{m+1} = 1, b_m = N - k_m$, $a_0 = \infty$, where $N = n + 1$ or n, according as $G = Sp(2n)$ or $G = SO(2n)$.

Choose i, $0 \leqslant i < m$, such that $b_i \leqslant a_i$, $a_{i+1} \leqslant b_{i+1}$. Note that such an i exists, since $a_0 = \infty$, and $b_m > a_m$. Let

$$\lambda' = \begin{pmatrix} k_1 & \cdots & k_{i-1} & k_i + a_{i+1} & k_{i+2} & \cdots & k_M \\ a_1 & \cdots & a_{i-1} & a_i + a_{i+1} & a_{i+2} & \cdots & a_M \end{pmatrix}.$$

6.8. The depth sequence $c(\tau, \lambda)$ (cf. [112], [164])

Assume that the following conditions on λ hold:

(1) $k_m < N - a_m$.

(2) $(a_m + a_{m-1} + \cdots + a_i) - (b_{m-1} + \cdots + b_i) < N - k_m$, $i \geqslant 1$.

Under the above assumptions on λ, $X(\lambda)$ admits a small resolution; indeed choose i, $0 \leqslant i < m$ such that $b_i \leqslant a_i$, $a_0 = \infty$, $a_{i+1} \leqslant b_{i+1}$; note that such an i exists since $b_0 < a_0 (= \infty)$ and $b_m > a_m$; with λ' as above, this corresponds to the first step in the Bott-Samelson type resolution as given by Theorems 6.3.1, 6.4.1.

Now the conditions (1) and (2) hold on λ' also, and iterating this procedure, we obtain the Bott-Samelson type resolution as given by Theorems 6.3.1, 6.4.1.

Recall (cf. [**141**]) the following:

PROPOSITION 6.8.1. *Let $\lambda \in I_{r,n}$. Let $X(\tau)$ be a Schubert subvariety of $X(\lambda)$.*

(1) *Let $G = SL(n)$ or $Sp(2n)$. Then $X(\tau)$ is P_λ-stable if and only if there exists a sequence $c(\tau, \lambda) := (c_1, \cdots, c_m)$ of non-negative integers such that $0 \leqslant a_i + c_i - c_{i-1} \leqslant k_i - k_{i-1}$, where $c_m = 0$ if $G = SL(n)$, and*

$$\tau = \begin{pmatrix} k_1 & k_2 & \cdots & k_m \\ a_1 + c_1 & a_2 + c_2 - c_1 & \cdots & a_m + c_m - c_{m-1} \end{pmatrix}.$$

(2) *Let $G = SO(2n)$. Then $X(\tau)$ is P_λ-stable if and only if there exists a sequence $c(\tau, \lambda) := (c_1, \cdots, c_m)$ of non-negative integers such that*

$$0 \leqslant a_i + c_i - c_{i-1} \leqslant k_i - k_{i-1},$$

where $c_m \equiv 0 \,(mod\ 2)$ if $k_m = n$, and $0 \leqslant a_m + c_m - c_{m-1} \leqslant 1$ if $(k_m, a_m) = (n, 1)$ such that

(a) *If $k_m \leqslant n - 2$, then*

$$\tau = \begin{pmatrix} k_1 & k_2 & \cdots & k_m & n \\ a_1 + c_1 & a_2 + c_2 - c_1 & \cdots & a_m + c_m - c_{m-1} & \epsilon \end{pmatrix},$$

where $\epsilon = 1/2(1 - (-1)^{c_m})$.

(b) *If $k_m = n - 1$, then*

$$\tau = \begin{pmatrix} k_1 & k_2 & \cdots & k_m & n \\ a_1 + c_1 & a_2 + c_2 - c_1 & \cdots & a_m + c_m - c_{m-1} & \epsilon \end{pmatrix},$$

where $\epsilon = 1/2(1 - (-1)^{c_m})$, and $a_m + c_m - c_{m-1} = 0$ if $\epsilon = 1$.

(c) *If $k_m = n$, then*

$$\tau = \begin{pmatrix} k_1 & k_2 & \cdots & k_m \\ a_1 + c_1 & a_2 + c_2 - c_1 & \cdots & a_m + c_m - c_{m-1} \end{pmatrix}.$$

With τ, λ as in Theorem 6.8.1, the sequence $c(\tau, \lambda) := (c_1, \cdots, c_m)$ is called the *depth* of τ in λ.

For $a, b \in \mathbb{Z}$, let $\left[{a \atop b}\right]$ denote the Gaussian binomial coefficient $(q^a - 1) \cdots (q^{a-b+1} - 1)/(q^b - 1) \cdots (q - 1)$, where $\left[{a \atop b}\right]$ is to be understood as 0 if either $b < 0$ or $a < b$, and $\left[{0 \atop 0}\right] = 1$.

THEOREM 6.8.2. *(cf. [**164**], [**142**])*

(1) *Let $X(\lambda)$ be a Schubert variety in the Grassmannian $G_{r,n}$. For the small resolution $p : \tilde{X}(\lambda) \to X(\lambda)$ (cf. 6.2.1), we have the following inductive formula for the Poincaré polynomial of the fiber over e_ϕ:*

$$P_{\phi,\lambda}(q) = \sum_d q^{(c_i - d)(c_{i+1} - d)} \left[{a_{i+1} - c_i + c_{i+1} \atop c_{i+1} - d}\right] \left[{b_i + c_i - c_{i+1} \atop c_i - d}\right] P_{\phi(d),\lambda'}(q)$$

where λ' is as in §6.7, and $X(\phi(d))$ is the $P_{\lambda'}$-stable subvariety of $X(\lambda')$ with depth sequence $c(\phi(d),\lambda') = (c_1,\cdots,c_{i-1},d,c_{i+2},\cdots,c_m)$.

(2) *Let $X(\lambda)$ be a Schubert variety in G/P_n, $G = Sp(2n), SO(2n)$. Further, let λ satisfy the conditions in 6.8. For the small resolution $p : \tilde{X}(\lambda) \to X(\lambda)$ (cf. 6.3.1, 6.4.1), we have the following inductive formula for the Poincaré polynomial of the fiber over e_ϕ:*

$$P_{\phi,\lambda}(q) = \sum_d q^{(c_i-d)(c_{i+1}-d)} \begin{bmatrix} a_{i+1}-c_i+c_{i+1} \\ c_{i+1}-d \end{bmatrix} \begin{bmatrix} b_i+c_i-c_{i+1} \\ c_i-d \end{bmatrix} P_{\phi(d),\lambda'}(q)$$

where λ' is as in §6.7, and $X(\phi(d))$ is the $P_{\lambda'}$-stable subvariety of $X(\lambda')$ with depth sequence $c(\phi(d),\lambda') = (c_1,\cdots,c_{i-1},d,c_{i+2},\cdots,c_m)$.

6.9. Reduction of the computation of $P_{\tau,\lambda}(q)$

Let $G = Sp(2n), SO(2n)$. Let $\lambda,\phi \in W/W_{P_n}$ be as in Theorem 6.8.2. Let λ_0,ϕ_0 in W be the representatives of maximal length of λ,ϕ respectively. Let $\Lambda = \begin{pmatrix} k_1 & \cdots & k_m & N+c_m \\ a_1 & \cdots & \cdots a_m & c_m \end{pmatrix}$, where $N = n$ or $n+1$ according as $G = SO(2n)$ or $Sp(2n)$. Now Λ defines a Grassmannian Schubert variety $X^{Gr}(\Lambda) \subset SL(N+c_m)/P_{r+c_m}$, where $r = \sum_{1\leqslant i\leqslant m} a_i$. Let $X^{Gr}(\Phi)$ be the P_Λ-stable Schubert variety of $X^{Gr}(\Lambda)$ with depth sequence $(c_1,c_2,\cdots,c_m,0)$. Let $\Lambda_0,\Phi_0 \in W(SL(N+c_m))$ denote the representatives of maximal length of Λ,Φ respectively.

We denote by $P^W_{y,w}$ the Kazhdan-Lusztig polynomial of the Weyl group W defined by the pair $y,w \in W$, $y \leqslant w$.

THEOREM 6.9.1. *(cf. [**142**]) Let $G = Sp(2n), SO(2n)$, and W the Weyl group of G. Let λ be as in Theorem 6.8.2. Let $\theta \in W$ be any element such that $\theta \leqslant \lambda_0$, λ_0 being as above. Let $\overline{\theta} \in W/W_{P_n}$ denote its projection. Let $X(\phi) = P_\lambda \cdot X(\overline{\theta}) \subset X(\lambda)$. With notations as in 6.9, we have*

$$P^W_{\theta,\lambda_0} = P^W_{\phi_0,\lambda_0} = P^{W'}_{\Phi_0,\Lambda_0}$$

where $W' = W(SL(N+c_m))$, the Weyl group of $SL(N+c_m)$.

Example. (cf. [**142**]) Let $G = SO(24)$. Then $W = W(SO(24))$ is a subgroup of S_{24}. Let $\lambda = \begin{pmatrix} 6 & 8 \\ 3 & 1 \end{pmatrix} \in W/W_{P_{12}}$. Note that λ satisfies conditions (1) and (2) of 6.8 and hence admits a small resolution. Then

$$\lambda_0 = (24,23,22,18,16,15,14,13,8,6,5,4) \in W$$

(note that it suffices to give the first twelve entries in λ_0). Take

$$\theta = (5,6,8,13,21,16,14,7,1,10,3,2).$$

Then $\theta \leqslant \lambda_0$. We have, $\overline{\theta} = (1,2,3,5,6,7,8,10,13,14,16,21)$. Now P_λ, the stabilizer of $X(\lambda)(\subset SO(24)/P_{12})$ is the parabolic subgroup with $S \setminus \{\alpha_6,\alpha_8\}$ as the associated set of simple roots. The P_λ-saturation of $X(\theta)$ is $X(\phi)$, where

$\phi = \begin{pmatrix} 6 & 8 & 12 \\ 5 & 2 & 1 \end{pmatrix}$. Hence by Theorem 6.8.2, $P_{\theta,\lambda_0}(q)$, the Kazhdan-Lusztig polynomial $= P_t(p^{-1}(e_\phi))$, the Poincaré polynomial of the fiber over e_ϕ for the small resolution $p : \tilde{X}(\lambda) \to X(\lambda)$ (with $q = t^2$).

Now for $\lambda = \begin{pmatrix} 6 & 8 \\ 3 & 1 \end{pmatrix}$, one has $(k_1, k_2) = (6,8)$, $(a_1, a_2) = (3,1)$, $(b_0, b_1) = (3,1)$. Hence, the choice $i = 1$ leads to the first step of a small resolution of $X(\lambda)$. When $i = 1$, we have $\lambda' = \begin{pmatrix} 7 \\ 4 \end{pmatrix}$. One then takes $\lambda'' = \begin{pmatrix} 4 \\ 4 \end{pmatrix}$. We have, $X(\lambda'') \simeq SO(16)/P_8$ and hence is smooth.

Now $C(\phi, \lambda) = (c_1, c_2) = (2,3)$. Since $i = 1$, the possible non-zero terms in the formula for $P_t(p^{-1}(e_\phi))$ (as given by Theorem 6.8.2) are the terms corresponding to $d = 0, 1$, or 2. Also, we have, $b_2 - c_2 + c_1 = 0$, and hence the term corresponding to $d = c_1 = 2$ is the only non-zero term. Thus we obtain

$P_{\phi,\lambda}(q) = q^0 \left[\begin{smallmatrix} 1+3-2 \\ 1 \end{smallmatrix}\right] \cdot \left[\begin{smallmatrix} 0 \\ 0 \end{smallmatrix}\right] \cdot P_{\phi(2),\lambda'}(q)$. Now $\phi(2) = \begin{pmatrix} 7 \\ 6 \end{pmatrix}$, and as above (using the formula given by Theorem 6.8.2), we obtain

$P_{\phi(2),\lambda'}(q) = q^0 \left[\begin{smallmatrix} 4+2-0 \\ 2 \end{smallmatrix}\right] \cdot 1 = \left[\begin{smallmatrix} 6 \\ 2 \end{smallmatrix}\right] = (1 + q^2 + q^4)(1 + q + q^2 + q^3 + q^4)$.

Hence we obtain

$P_{\theta,\lambda_0}(q) = P_{\phi,\lambda}(q) = (1 + q)(1 + q^2 + q^4)(1 + q + q^2 + q^3 + q^4)$.

REMARK 6.9.2. The Kazhdan-Lusztig polynomials for Schubert varieties in a minuscule G/P have been worked out in [**20, 31**] also.

7. Factoring the Poincaré polynomial of a Schubert variety

Recall that if the Poincaré polynomial of a Schubert variety is symmetric then $X(w)$ is rationally smooth by Theorem 3.0.7. It turns out that these Poincaré polynomials are not just symmetric, but in fact have nice factorizations as a product of symmetric factors. The first theorem below is a generalization of the Kostant-Macdonald formula (originally proved by Kostant and then Macdonald [**119**]). The factoring rules given in Theorem 7.0.2 can be used as a recursive criteria for rational smoothness. While other factoring rules also hold for the Poincaré polynomials, these are sufficient to completely factor the Poincaré polynomials of all rationally smooth Schubert varieties of classical type.

THEOREM 7.0.1. [**33**] *Let W be any Weyl group. If $X(w)$ is smooth then the Poincaré polynomial factors as*

$$p_w(t) = \prod_{\{\alpha>0,\ s_\alpha \leq w\}} \frac{1 - t^{\mathrm{ht}\alpha+1}}{1 - t^{\mathrm{ht}\alpha}}.$$

Remark. Note that the converse to Theorem 7.0.1 need not be true. For example, in type C, $X(263415)$, $X(513462)$, and $X(153426)$ are not smooth but their Poincaré polynomials do factor as in (7.0.1). Thus, the existence of such a factorization is only a necessary condition for smoothness.

Remark. The theorem does not extend to all rationally smooth Schubert varieties. The following Schubert varieties are rationally smooth in type C, but (7.0.1) does not hold: $X(362514)$, $X(531642)$, and $X(632541)$. Nevertheless, the existence of such a factorization is a sufficient condition for rational smoothness **[33]**.

Gasharov has given another factorization of the Poincaré polynomials.

THEOREM 7.0.2. **[57]**

(1) *Let $w \in \mathcal{S}_n$, say $w = (w_1, \dots, w_n)$, and assume $w_d = n$ and $w_n = e$. The Poincaré polynomial of w factors in the form*

$$p_w(t) = (1 + t^1 + \cdots + t^\mu)p_{w'}$$

under the following circumstances:

  (a) *If $n = w_d > w_{d+1} > \cdots > w_n$, then p_w factors with $w' = ws_d \cdots s_{n-1}$ and $\mu = n - d$.*

  (b) *If w contains a consecutive sequence ending in $w_n = e$, then p_w factors with $w' = s_{n-1} \cdots s_{e+1}s_e w$ and $\mu = n - e$.*

(2) *$p_w(t)$ factors completely using the above algorithm if and only if $X(w)$ is smooth.*

For example, apply Theorem 7.0.2 to the element $w_0 = n \dots 1$ (in one-line notation) to get the well-known formula for the Poincaré polynomial of $\mathcal{S}_n$ (see **[74]**, 3.7), namely

$$p_{w_0}(t) = \prod_{k=1}^{n-1}(1 + t + \dots + t^k).$$

8. Generating function for the number of smooth Schubert varieties

There is a surprising generating function for the number of permutations with smooth Schubert varieties. This formula was first proved by M. Haiman (preprint dated 1992) and also appears in the work of M. Bona **[21]** using results of Stankova **[155]**.

THEOREM 8.0.1. *[H] Let v_n be the number of $w \in \mathcal{S}_n$ for which $X(w)$ is smooth. Then the generating function $V(t) = \sum_n v_n t^n$ is given by*

$$\begin{aligned} V(t) =& \frac{1 - 5t + 3t^2 + t^2\sqrt{1-4t}}{1 - 6^t + 8^{t^2 - 4t^3}} \\ =& 1 + t + 2t^2 + 6t^3 + 22t^4 + 88t^5 + 366t^6 \\ &+ 1552t^7 + 6652t^8 + 28696t^9 + O(t^{10}). \end{aligned}$$

The following data shows the number of smooth Schubert varieties in types A, B, C, and D:

	A	B	C	D
$n = 1$	1	2	2	1
$n = 2$	2	7	7	4
$n = 3$	6	28	28	22
$n = 4$	22	116	114	108
$n = 5$	88	490	472	490
$n = 6$	366	2094	1988	2164
$n = 7$	1552	9014	8480	9474

REMARK 8.0.2. The problem of finding a generating function for the number of (rationally) smooth Schubert varieties is still open for all types except type A.

CHAPTER 10

Degenerations of Schubert varieties to schemes defined by monomials

In this chapter we prove the degenerations of Schubert varieties (in the Grassmannian as well as $SL(n)/B$) to schemes defined by monomials (using the Standard Monomial Theory). We then use the combinatorial properties of these schemes to conclude their Cohen-Macaulayness, and then deduce the Cohen-Macaulayness of Schubert varieties thus providing yet another proof for the Cohen-Macaulayness of Schubert varieties. In §1, we briefly review flat families. In §2, §3, the degenerations to schemes defined by monomials are carried out for Schubert varieties in the Grassmannian, Flag variety respectively. Using the results of §2, §3, we provide another proof of the arithmetic Cohen-Macaulayness of Schubert varieties.

1. Flat family (or Flat deformation)

DEFINITION 1.0.1. Let A be a ring and M an A-module. We say that M is *flat* over A if an exact sequence of A modules remains exact after tensoring with M.

DEFINITION 1.0.2. Let $f : X \to Y$ be a morphism of schemes, and $\mathcal{F}$ an $\mathcal{O}_X$-module. We say that $\mathcal{F}$ is flat over Y at a point $x \in X$ if the stalk $\mathcal{F}_x$ is a flat $\mathcal{O}_{Y,f(x)}$-module. We say that $\mathcal{F}$ is flat over Y if it is flat over Y at every $x \in X$. The morphism f is said to be *flat* if $\mathcal{O}_X$ is flat over Y.

DEFINITION 1.0.3. Let $f : X \to Y$ be a morphism of schemes. The collection $\{f^{-1}(y), y \in Y\}$ consisting of the fibers of f is called a *family* of schemes parametrized by Y.

Example. To give a family of closed subschemes of a scheme X parametrized by a scheme Y is simply to give a closed subscheme Z of $X \times Y$; for, the fiber $p_2^{-1}(y)$ under $p_2|_Z : Z \to Y$ gets identified with a closed subscheme of X.

In general, the fibers of a morphism can vary in a highly discontinuous manner. Flatness prevents this pathological behaviour. Thus, if $f : X \to Y$ is flat, then all fibers of f are equidimensional (i.e., all the irreducible components of $f^{-1}(y), y \in Y$, have the same dimension) and are of the same dimension (see **[67]** for a proof of this fact).

In the above example, let us take $X = \mathbb{P}_k^n$. Let Y be an irreducible variety and let Z be a closed subscheme of $\mathbb{P}_Y^n := \mathbb{P}_k^n \times Y$. This gives a family $\{X_y(=$

$f^{-1}(y)), y \in Y\}$ parametrized by Y. Then $p_2|_Z : Z \to Y$ is flat if and only if the Hilbert polynomial P_y of X_y (X_y being considered as a closed subvariety of $\mathbb{P}^n_k$) is independent of y; in particular, all the fibers have the same dimension when $p_2|_Z$ is flat (see **[67]**, Theorem 9.9).

When the parametrizing space Y is taken to be $\mathbb{A}^1$ ($=$ Spec $k[t]$), then the family becomes a 1-parameter family parametrized by k. For $u \in k$, the fiber at $u = 0$ is called the *special fiber*, and the fiber at u invertible is called the *general fiber*. In this situation many geometric properties for the general fiber may be inferred by knowing them for the special fiber; for example, Cohen-Macaulayness and normality are two such properties (see **[65]** Theorem 12.2.1). In the above discussion, one may equally work with Spec $k[[t]]$ as the parametrizing space (where $k[[t]]$ is the ring of formal power series).

In this chapter, we construct a flat 1-parameter family whose general fiber is $\widehat{X}$, the cone over a Schubert variety X in the flag variety and whose special fiber is a scheme Y defined by monomials. We refer to this as the *degeneration* of the Schubert variety to the monomial scheme Y. In §2, we carry out the degeneration of a Schubert variety in the Grassmannian. In §3, we carry out the degeneration of a Schubert variety in the flag variety. Using the results of §3, we prove in §4 the Cohen-Macaulayness and normality of the multicone over a Schubert variety in the flag variety.

2. Degenerations of Schubert varieties in the Grassmannian

Let $G = SL(n)$. Fix r, $1 \leqslant r \leqslant n-1$. Consider the polynomial algebra $P = k[x_\beta, \beta \in T_\tau]$, where $\tau \in I_{r,n}$ and $T_\tau = \{\phi \in I_{r,n} \mid \phi \leqslant \tau\}$ *(a left half space)*.

DEFINITION 2.0.1. For $\tau \in I_{r,n}$, let I_τ be the ideal of P generated by the set of monomials

$$\{x_\gamma x_\delta \mid \gamma \text{ and } \delta \text{ are noncomparable elements of } T_\tau\}.$$

We define the *discrete algebra associated to* τ (or *to the Schubert variety* $X(\tau)$ (in the Grassmannian $G_{r,n}$)) to be the k-algebra P/I_τ, and denote it by D_τ.

The main result of this section is that Spec D_τ is a degeneration of $\widehat{X(\tau)}$ (the cone over the Schubert variety $X(\tau)$), i.e. there exists a one parameter flat family with generic fiber (t invertible) $\widehat{X(\tau)}$ and special fiber ($t = 0$) Spec D_τ. We keep the notations of Chapter 6. We first collect some preliminary results.

If γ and δ are noncomparable elements of T_t, then by Chapter 6, Proposition 3.3.3, we have a quadratic relation in R_τ (the homogeneous coordinate ring of $X(\tau)$ for the Plücker embedding) of the form

$$(*) \qquad p_\gamma p_\delta = \sum_{\lambda,\mu} a_{\lambda\mu} p_\lambda p_\mu, \quad a_{\lambda\mu} \in k^*,$$

where the monomials in the right hand side are standard, i.e. $\lambda \geqslant \mu$; further, $\lambda >$ both γ and δ, and $\mu <$ both γ and δ. We shall denote by Q the set of all such quadratic relations for various pairs of noncomparable elements γ, δ in T_τ.

LEMMA 2.0.2. *There exists a function $\theta : T_\tau \to \mathbb{Z}_+$ such that given any quadratic relation $p_\gamma p_\delta = \sum_{\lambda,\mu} a_{\lambda\mu} p_\lambda p_\mu$ in Q, we have $\theta(\lambda) + \theta(\mu) > \theta(\gamma) + \theta(\delta)$ for all λ, μ appearing in the sum.*

PROOF. Since T_τ is a finite partially ordered set, we can define an order function $\operatorname{ord} : T_\tau \to \mathbb{Z}_+$ such that $\beta > \beta'$ implies $\operatorname{ord}\beta > \operatorname{ord}\beta'$. For $\beta \in T_\tau$ set

$$\theta(\beta) = N^{\operatorname{ord}\beta},$$

where N is a positive integer. We claim that for a sufficiently large N, θ is the required function. Indeed, using the relation (*) above, we have $\operatorname{ord}\gamma < \operatorname{ord}\lambda$ and $\operatorname{ord}\delta < \operatorname{ord}\lambda$; therefore, choosing a sufficiently large N, we can achieve the inequality

$$N^{\operatorname{ord}\lambda} + N^{\operatorname{ord}\mu} > N^{\operatorname{ord}\gamma} + N^{\operatorname{ord}\delta}$$

for all quadratic relations in Q. □

We recall the following well known Lemma from commutative algebra.

LEMMA 2.0.3. *Let M be a finitely generated module over the power series ring $A = k[[t]]$ and let K be the quotient field of A. Then M is a free module if and only if*

$$\dim_k M \otimes_A k \leqslant \dim_K M \otimes_A K.$$

PROOF. If M is free, then obviously the inequality holds.
Conversely, suppose that the inequality holds. If $\dim_k M \otimes_A k = m$, then by Nakayama's lemma M can be generated by m elements and therefore we have an exact sequence

$$0 \to N \to A^m \to M \to 0$$

of A-modules. Tensoring the sequence by K and using the inequality, we conclude that $N \otimes_A K = 0$. Since N is torsion free, $N = 0$, and hence M is isomorphic to A^m. □

THEOREM 2.0.4. *Let $\tau \in I_{r,n}$. There exists a flat family over* Spec $k[[t]]$ *whose generic fiber (t invertible) is $\widehat{X(\tau)}$ and the special fiber ($t = 0$) is* Spec D_τ, *where D_τ is the discrete algebra associated to τ.*

PROOF. Let R_τ be the homogeneous coordinate ring of $X(\tau)$ (cf. Chapter 6, §3.1) and let Q be as above (namely, the set of quadratic relations of the form (*)). For each quadratic relation $p_\gamma p_\delta = \sum_{\lambda,\mu} a_{\lambda\mu} p_\lambda p_\mu$ in Q, consider the polynomial

$$F_{\gamma,\delta} = x_\gamma x_\delta - \sum_{\lambda,\mu} a_{\lambda\mu} x_\lambda x_\mu$$

in $P = k[x_\phi, \phi \in T_\tau]$, and let J be the ideal of P generated by all such polynomials. If $\overline{x}_\phi$ denotes the canonical image of x_ϕ in P/J, then it is clear that P/J is generated by standard monomials in $\overline{x}_\phi$, $\phi \in T_\tau$. Moreover, we have a canonical surjective map $f_\tau : P/J \to R_\tau$ which sends $\overline{x}_\phi$ to p_ϕ. Since R_τ has a basis consisting of standard monomials in p_ϕ's, it follows that f_τ is injective. Hence P/J is isomorphic to R_τ.

Now using Lemma 2.0.2, choose a function $\theta : T_\tau \to \mathbb{Z}_+$ such that for each quadratic relation $p_\gamma p_\delta = \sum_{\lambda,\mu} a_{\lambda\mu} p_\lambda p_\mu$ in Q, $\theta(\lambda) + \theta(\mu) - \theta(\gamma) - \theta(\delta) > 0$. Further, let $A = k[[t]]$. Corresponding to each quadratic relation in Q as above, consider the polynomial

$$F_{\gamma,\delta,t} = x_\gamma x_\delta - \sum_{\lambda,\mu} a_{\lambda\mu} t^{\theta(\lambda)+\theta(\mu)-\theta(\gamma)-\theta(\delta)} x_\lambda x_\mu$$

in $P_A := A[x_\phi, \phi \in T_\tau]$. Denoting by $\mathcal{J}$ the ideal generated by the polynomials $F_{\gamma,\delta,t}$ in P_A, we set $\mathcal{R}_\tau = P_A/\mathcal{J}$. Let K be the quotient field of A and let $P_K = K[x_\phi, \phi \in T_\tau]$. If $\tilde{J}$ is the ideal of P_K generated by the polynomials $F_{\gamma,\delta}$ and $\tilde{\mathcal{J}}$ is the ideal generated by the polynomials $F_{\gamma,\delta,t}$, then it is easy to see that $P_K/\tilde{\mathcal{J}} \cong P_K/\tilde{J}$, the isomorphism being induced by the K-automorphism of P_K which sends x_β to $t^{-\theta(\beta)} x_\beta$. Thus we have

$$\mathcal{R}_\alpha \otimes_A K = P_A/\mathcal{J} \otimes_A K \cong P_K/\tilde{\mathcal{J}} \cong P_K/\tilde{J} \cong P/J \otimes_k K \cong R_\tau \otimes_k K$$

and $\mathcal{R}_\tau \otimes_A k \cong \mathcal{R}_\tau / t\mathcal{R}_\tau \cong D_\tau$. This shows that the generic fiber (t invertible) of $\mathcal{R}_\tau$ is R_τ and the special fiber ($t = 0$) is D_τ.

Now we show that $\mathcal{R}_\tau$ is flat over A. From the definition of D_τ it is clear that D_τ is generated by standard monomials in the canonical images $\overline{x}_\phi$ of x_ϕ, $\phi \in T_\tau$. Therefore, using the above isomorphisms, it follows that for every $m \geqslant 0$, we have,

$$\begin{aligned} \dim_k(\mathcal{R}_\tau)_m \otimes_A k &= \dim_k(D_\tau)_m \leqslant \dim_k(R_\tau)_m \\ &= \dim_K(\mathcal{R}_\tau)_m \otimes_k K = \dim_K(\mathcal{R}_\tau)_m \otimes_A K, \end{aligned}$$

where $(\mathcal{R}_\tau)_m$, $(D_\tau)_m$, $(R_\tau)_m$ denote the m-th graded components of the respective rings. Thus, in view of Lemma 2.0.3, each graded component of $\mathcal{R}_\tau$ is a free A-module, and hence $\mathcal{R}_\tau$ is a flat A-algebra. □

COROLLARY 2.0.5. *(i) D_τ has a basis consisting of standard monomials in $\overline{x}_\phi$, $\phi \in T_\tau$.*

(ii) D_τ is reduced.

(iii) If $\mathbf{P} = \mathrm{Proj} P$, where $P = k[x_\phi, \phi \in T_\tau]$, then the degree of $X(\tau)$ in the Plücker embedding is the same as the degree of Proj D_τ *in* $\mathbf{P}$.

PROOF. (i) Since for each $m \geqslant 0$, $(\mathcal{R}_\tau)_m$ is a free A-module, we have

$$\dim_k(\mathcal{R}_\tau)_m \otimes_A k = \dim_K(\mathcal{R}_\tau)_m \otimes_A K,$$

and therefore the inequalities $(*)$ in the proof of the above theorem imply that

$$(**) \qquad \dim_k(D_\tau)_m = \dim_k(R_\tau)_m, \text{ for each } m \geqslant 0.$$

Since distinct standard monomials on $X(\tau)$ of degree m form a basis of $(R_\tau)_m$, we obtain $\dim_k(D_\tau)_m = \#\{\text{standard monomials on } X(\tau) \text{ of degree } m\}$. This together with the fact that $(D_\tau)_m$ is generated by standard monomials (of degree m) implies that $(D_\tau)_m$ has a basis consisting of standard monomials of degree m.

(ii) Let $f = \sum a_i f_i$, where $a_i \in k$, $a_i \neq 0$ and f_i are standard monomials in D_τ. From the definition of D_τ it is clear that any product of standard monomials in D_τ is either zero or a standard monomial. Therefore, using (i) it follows that $f^m \neq 0$ for every $m \geqslant 0$. This proves that D_τ is reduced.

(iii) The equality $(**)$ in the proof of (i) implies that the Hilbert polynomial of $X(\tau)$ is the same as that of D_τ, which proves the assertion. □

Let T be a subset of $I_{r,n}$. Recall that a *chain of length m in T* is a sequence $\Lambda = (\lambda(0), \lambda(1), \dots, \lambda(m))$ of elements of T such that $\lambda(i) > \lambda(i+1)$, $0 \leqslant i \leqslant m-1$. We say that Λ is *maximal* if it is not a subchain of any chain in T different from Λ.

For a chain Λ in T_τ, $\tau \in I_{r,n}$ let $L(\Lambda)$ denote the linear space in $\mathbf{P} = \operatorname{Proj} P$ defined by the equations $x_\phi = 0$, $\phi \in T_\tau \setminus \Lambda$, where $T_\tau \setminus \Lambda$ denotes the set of elements in T_τ not appearing in Λ.

PROPOSITION 2.0.6. *If $M(T_\tau)$ denotes the set of maximal chains in T_τ, then*

$$\operatorname{Proj} D_\tau = \bigcup_{\Lambda \in M(T_\tau)} L(\Lambda) \quad \text{scheme theoretically,}$$

i.e. the linear spaces $L(\Lambda)$, $\Lambda \in M(T_\tau)$, are the irreducible components of $\operatorname{Proj} D_\tau$.

PROOF. The ideal of the scheme theoretic union on the right hand side is

$$I = \bigcap_{\Lambda \in M(T_\tau)} I_\Lambda,$$

where I_Λ is the ideal of $L(\Lambda)$ (note that I_Λ is generated by the set $\{x_\phi \mid \phi \in T_\tau \setminus \Lambda\}$). We shall show that I is the ideal defining $\operatorname{Proj} D_\tau$, i.e. I is generated by the set of monomials $S = \{x_\gamma x_\delta \mid \gamma, \delta \text{ are noncomparable elements of } T_\tau\}$. In fact, if γ, δ are noncomparable in T_τ, then both of them together cannot belong to a maximal chain in T_τ. This shows that $x_\gamma x_\delta \in I_\Lambda$ for all $\Lambda \in M(T_\tau)$, and hence $S \subset I$. Now, if $F \in I$, it is easy to see that every monomial of F is in I. Thus, to complete the proof it suffices to show that every monomial $F \in I$ is divisible by an element of S. In fact, since $F \in I_\Lambda$ for all $\Lambda \in M(T_\tau)$, it is clear that F is not a standard monomial, and hence it is divisible by an element of S. □

COROLLARY 2.0.7. *The degree of $X(\tau)$ in the Plücker embedding is equal to the number of maximal chains in T_τ.*

The proof is immediate from Theorem 2.0.4 and the above Proposition.

2.1. Stanley-Reisner ring ([138])

Given a finite partially ordered set H, let $A = k[x_\alpha,\ \alpha \in H]$, I_H the ideal generated by $\{x_\gamma x_\delta \mid \gamma, \delta \text{ are noncomparable elements of } H\}$. Then the ring $R_H := A/I_H$ is called the *Stanley-Reisner ring* associated to H.

THEOREM 2.1.1. Spec D_τ *is Cohen-Macaulay.*

PROOF. We have, for $H = T_\tau, R_H = D_\tau$, and the Cohen-Macaulayness of Stanley-Reisner ring R_{T_τ} follows from [**18**]. □

Combining Theorem 2.0.4 with the above Theorem, we obtain

THEOREM 2.1.2. $\widehat{X(\tau)}(= \operatorname{Spec} R_\tau)$ *is Cohen-Macaulay.*

3. Degenerations of Schubert varieties in the flag variety

Let $G = SL(n)$. We preserve the notations of Chapter 7. Let Q be a parabolic subgroup, say $Q = \bigcap_{i=1}^{r} P_i$, where $P_i, 1 \leq i \leq r$ are the maximal parabolic subgroups containing Q. Let $w \in W^Q$, and let

$$R(\tau) = \bigoplus_a H^0(X(\tau), L_a)$$

where $a = (a_1, \cdots, a_r)$, $a_i \in \mathbb{Z}^+$, $L_a = \bigotimes_{i=1}^{r} L_i, 1 \leq i \leq r$.

3.1. The element λ^a

Let $A = S_{m_1} \times \cdots \times S_{m_r}$, the product of the symmetric groups. If $\lambda = (\lambda_{ij})$ is a $(m_1 + \cdots + m_r)$-tuple of elements where $\lambda_{ij} \in W^i, 1 \leq j \leq m_i, 1 \leq i \leq r$, and $a \in A$, $a = (a_1, \cdots, a_r)$, then by λ^a we denote the $(m_1 + \cdots + m_r)$-tuple obtained from λ by permuting the first m_1 entries according to a_1, the next m_2 entries according to a_2, and so on.

3.2. The partial order $\overset{\mathrm{L}}{\geq}$

Let $M = \{\lambda = (\lambda_{ij}), \lambda_{ij} \in W^i, 1 \leq j \leq m_i, 1 \leq i \leq r\}$. We define a partial order $\geq^L$ on M as follows: Let $\lambda = (\lambda_{ij}), \mu = (\mu_{ij})$ be two elements of M. Let us write $\lambda = (\lambda_1, \cdots, \lambda_m), \mu = (\mu_1, \cdots, \mu_m)$, where $m = m_1 + \cdots + m_r$. We say $\lambda \overset{\mathrm{L}}{\geq} \mu$, if either $\lambda = \mu$ or there exists a $t, 1 \leq t \leq m$ such that $\lambda_i = \mu_i, 1 \leq i < t$, and $\lambda_t > \mu_t$ (note that for any i, $1 \leq i \leq m, \lambda_i, \mu_i$ are both elements of W^P for some maximal parabolic P, and $\geq$ on W^P denotes the Bruhat order).

PROPOSITION 3.2.1. *Let $Z = \bigcup_{i=1}^{l} X(\varphi_i)$ be a union of Schubert varieties in $SL(n)/Q$. Let $p(\lambda)$ be a non-zero, non-standard monomial on Z of type $\underline{m} = (m_1, \cdots, m_r)$. Let*

$$(*) \qquad p(\lambda) = \sum_{i=1}^{N} a_i p(\tau_i),\ a_i \in k^*$$

be the expression for $p(\lambda)$ as a sum of standard monomials on Z (cf. Chapter 7). Then for every $i, \tau_i \overset{\mathrm{L}}{>} \lambda^a, a \in A$ (A being as above).

PROOF. We may assume $m_1 \neq 0$ (if $m_1 = 0, m_2 \neq 0$, then we may work with $Q' = \bigcap_{i=2}^{r} P_i$ etc.). Let us write $\tau_i = (\tau_{jk}^{(i)})$, where $\tau_{jk}^{(i)} \in W^j(= W^{P_j})$, $1 \leq k \leq m_j$, $1 \leq j \leq r$. Among $\{\tau_{11}^{(i)}, 1 \leq i \leq N\}$, choose a minimal one, which we may suppose to be $\tau_{11}^{(1)}$ (by reindexing the τ_i's if necessary). Let us denote $\tau_{11}^{(1)}$ by just τ_{11}. Let us index the τ_i's so that $\tau_1, \cdots, \tau_s$ are all the elements such that $\{\tau_{11}^{(i)} = \tau_{11}, 1 \leq i \leq s\}$ (note in particular that $\tau_{11}^{(i)} \neq \tau_{11}, s < i \leq N$). For each $\tau_i, 1 \leq i \leq s$, let us fix a defining sequence $\theta = (\theta_{jk}^{(i)})$ on Z (note that τ_i being standard on Z, is standard on some $X(\varphi_t)$ and we fix a defining sequence (Chapter 7, Definition 2.3.4) for τ_i on $X(\varphi_t)$; note that $X(\theta_{11}^{(i)}) \subseteq X(\varphi_t)$).

Let $Z_1 = \bigcup_{i=1}^{s} X(\theta_{11}^{(i)})$. Let us now restrict (*) to Z_1. Then $p(\tau_j)|_{Z_1} \equiv 0$, $j > s$ (note that under $\pi : G/Q \to G/P_1$, $\pi(X(\theta_{11}^{(i)})) = X(\tau_{11})$, for $1 \leq i \leq s$, and $\pi(X(\theta_{11}^{(i)})) \not\subseteq X(\tau_{11})$, for $s < i \leq N$. Thus $p(\tau_j)|_{Z_1} \equiv 0$, for $j > s$). Further, $p(\tau_j)$, for $1 \leq j \leq s$, is standard on Z_1. By linear independence of standard monomials on Z_1, we have $p(\lambda)|_{Z_1} \not\equiv 0$ (note that $\sum_{i=1}^{s} a_i p(\tau_i)$ is a non-zero sum of standard monomials on Z_1). This implies in particular that $\lambda_{1j} \leq \tau_{11}, 1 \leq j \leq m_1$ (note that $\pi(Z_1) = X(\tau_{11}), \pi$ being as above).

If $\lambda_{ij} = \tau_{11}$ for some j, $1 \leq j \leq m_1$, then we cancel $p_{\tau_{11}}$, and use induction on m_1 and r, the reasoning being the same for the case of a maximal parabolic, i.e., the case $r = 1$. (Note that $I(Z_1)$, the ideal of $Z_1 = \bigcap_{i=1}^{s} I(X(\theta_{11}^{(i)}))$, since Z_1 is reduced. Hence if $p_{\tau_{11}} F \in I(Z_1)$, then $F \in I(Z_1)$, since $I(X(\theta_{11}^{(i)}))$ is prime, and $p_{\tau_{11}} \notin I(X(\theta_{11}^{(i)})), 1 \leq i \leq s$.) Thus we may assume $\lambda_{ij} < \tau_{11}$, for all $1 \leq j \leq m_1$. From this we obtain that for every τ_i such that $\tau_{11}^{(i)}$ is minimal in $\{\tau_{11}^{(j)}, 1 \leq j \leq N\}$, $\tau_i \overset{\mathrm{L}}{>} \lambda^a, a \in A$. If for some i, $\tau_{11}^{(i)}$ is not minimal, then there exists a j such that $\tau_{11}^{(i)} > \tau_{11}^{(j)}$, and $\tau_{11}^{(j)}$ is a minimal element in $\{\tau_{11}^{(l)}, 1 \leq l \leq N\}$. But then from the discussion above, we have $\tau_{11}^{(j)} > \lambda_{1t}, 1 \leq t \leq m_1$. Hence it follows that $\tau_{11}^{(i)} > \lambda_{1t}, 1 \leq t \leq m_1$. From this we obtain that $\tau^{(i)} \overset{\mathrm{L}}{>} \lambda^a, a \in A$. □

REMARK 3.2.2. We shall refer to a relation as in Proposition 3.2.1 as a *straightening relation on Z.*

3.3. Degeneration

We preserve the above notations and introduce a few more notations. Given $X(w)$ in G/Q, let us denote its projection under $\pi_i : G/Q \to G/P_i$ by $X(w^{(i)}), 1 \leq i \leq r$. We shall denote W^{P_i} by just W^i. We set Y_w or just Y as

$$Y = \{\beta \in W^i, \beta \leq w^{(i)}, 1 \leq i \leq r\}$$

and

$$Y_i = \{\beta \in W^i, \beta \leq w^{(i)}\}.$$

Note that $Y = \cup_{i=1}^{r} Y_i$. For a monomial T of type $(m_1, \cdots, m_r)$, and $\alpha \in Y_i$, we set

$$T_\alpha = \{\beta \in Y_i \,|\, p_\beta \text{ occurs in } T, \text{ and } \beta \geq \alpha\}$$

$$T^\alpha = \{\beta \in T_\alpha \mid \beta > \alpha\}.$$

We have (from Chapter 7)

THEOREM 3.3.1. *Let* $R = \bigoplus_m H^0(X(w), L_m)$, *where* $m = (m_1, \cdots, m_r)$, $m_i \in \mathbb{Z}^+$. *Then*

(1) *R is a k-algebra with algebra generators consisting of*

$$\{p_\beta, \beta \in W^i, \beta \leq w^{(i)}, 1 \leq i \leq r\}.$$

(2) *Monomials in p_β's standard on $X(w)$ form a k-basis of R.*

(3) *In any straightening relation* $p(\lambda) = \sum_{i=1}^N a_i p(\tau_i)$, $a_i \in k^*$, *for every* i, $\tau_i \overset{\mathrm{L}}{>} \lambda^a$, $a \in A$ *(notations being as in Proposition 3.2.1).*

Set $J_R^1 = \Big\{\beta \in W^1 \mid$ there exists a straightening relation in R, and a monomial F on the right hand side of this relation such that p_β occurs more times in F than it does on the left hand side of this relation $\Big\}$.

LEMMA 3.3.2. *Fix a maximal element α in J_R^1. Let $I = (p_\alpha)$. Let $I_j = I^j, j \geq 1$. Then I_j has a basis consisting of standard monomials on $X(w)$ involving p_α at least j times.*

PROOF. Let T be a non-zero monomial in I^j. Further let T be non-standard on $X(w)$. Let

$$T = \sum a_i T_i,\ a_i \in k^*$$

be the expression for T as a sum of standard monomials on $X(w)$. For each i, we have (by the maximality assumption on α), $T^\alpha \supseteq T_i^\alpha$. This together with the lexicographic order condition (cf. Proposition 3.2.1) implies that $T_\alpha \subseteq (T_i)_\alpha$ (and $T^\alpha = T_i^\alpha$). Hence we obtain that for every i, $(T_i)_\alpha \setminus (T_i)^\alpha \supseteq T_\alpha \setminus T^\alpha$, and hence

$$\#\big((T_i)_\alpha \setminus (T_i)^\alpha\big) \geq \#(T_\alpha \setminus T^\alpha) \geq j$$

(since $T \in I^j$). Thus, each T_i involves p_α at least j times. □

PROPOSITION 3.3.3. *Let α be as in Lemma 3.3.2. Let*

$$\mathcal{R} = \cdots Rt^2 \oplus Rt \oplus R \oplus I_1 t^{-1} \oplus I_2 t^{-2} \oplus \cdots,$$

where t is an indeterminate and $I_j = I^j$. Then

(1) *$\mathcal{R}$ is a $k[t]$-algebra with algebra generators given by $\{P_\beta, \beta \in Y\}$, where* $P_\beta = p_\beta, \beta \neq \alpha$, and $P_\alpha = p_\alpha t^{-1}$.

(2) *Standard monomials give a $k[t]$-basis for $\mathcal{R}$ (here by a standard monomial in $\mathcal{R}$, we mean a monomial of the form $P(\Lambda) = \pi p_{\lambda_{ij}}$, where $\Lambda = (\lambda_{ij})$ is a standard Young diagram on $X(w)$).*

(3) *In any straightening relation* $P(\Lambda) = \sum a_i P(\tau_i)$, $a_i \neq 0$, *where* Λ *is a nonstandard Young diagram on* $X(w)$ *of type* $(m_1, \cdots, m_r)$, *we have for all* i $\tau_i \overset{\mathrm{L}}{>} \Lambda^a, a \in A$.

PROOF. Assertion (1) is clear.

(2) *Linear Independence:* The $k[t, t^{-1}]$-algebra $R[t, t^{-1}] = \mathcal{R}[t^{-1}]$ has a $k[t, t^{-1}]$-basis given by standard monomials (by change of base). Since t is a unit in $k[t, t^{-1}]$, the standard monomials in the P_β's are also linearly independent over $k[t, t^{-1}]$ (note that $P_\beta = p_\beta$, or $p_\alpha t^{-1}$ according as $\beta \neq \alpha$, or $= \alpha$) and hence are linearly independent over $k[t]$.

Generation: For $a \in R$, let $n(a)$ be the integer r such that $a \in I_r$, $a \notin I_{r+1}$ (we call $n(a)$ the *order* of a, and denote it also by ord a). Now given a monomial f in the p_β's, f determines an unique monomial F in the P_β's given by $F = f t^{-n(f)}$ (and conversely). Further F is standard if and only if f is. Let now F be a nonstandard monomial in $\mathcal{R}$. This implies that f is nonstandard. Let

$$f = \sum a_i f_i,\ a_i \in k^*$$

be the straightening relation for f. This implies

$$F t^{n(f)} = \sum a_i F_i t^{n(f_i)},\ a_i \in k^*$$

where we have (by Lemma 3.3.2) $n(f_i) \geq n(f)$. Hence

$$F = \sum a_i F_i t^{m_i},\ a_i \in k^*$$

where $m_i \geq 0$. This proves the generation.

The assertion (3) follows from Proposition 3.2.1. □

PROPOSITION 3.3.4. *Let* α *be as in Lemma 3.3.2. Let* $I = (p_\alpha)$, *and let*

$$\mathcal{R}_\alpha = R/I \oplus I/I^2 \oplus \cdots \oplus I^j/I^{j+1} \oplus \cdots\ (= \mathrm{gr}_I R).$$

Then

(1) $\mathcal{R}_\alpha \cong \mathcal{R}/(t)$.

(2) $\mathcal{R}_\alpha$ *is a graded* k*-algebra with algebra generators consisting of*

$$\{\overline{p}_\beta, \beta \in Y\},$$

where $\overline{p}_\beta = p_\beta + I, \beta \neq \alpha$, and $\overline{p}_\alpha = p_\alpha + I^2$.

(3) *Standard monomials give a* k*-basis for* $\mathcal{R}_\alpha$ *(here, by a standard monomial in* $\mathcal{R}_\alpha$, *we mean a monomial of the form* $\overline{p(\Lambda)} = \pi \overline{p_{\lambda_{ij}}}$, *where* $\Lambda = (\lambda_{ij})$ *is a standard Young diagram on* $X(w)$*).*

(4) *In any straightening relation* $\overline{p(\Lambda)} = \sum a_i \overline{p(\tau_i)}$, $a_i \neq 0$, *where* Λ *is a nonstandard Young diagram on* $X(w)$ *of type* $(m_1, \cdots, m_r)$, *we have for all* i $\tau_i \overset{\mathrm{L}}{>} \Lambda^a, a \in A$.

(5) $J^1_{\mathcal{R}_\alpha} \subseteq J^1_R \setminus \{\alpha\}$.

PROOF. (1) The map

$$\theta : \mathcal{R}_\alpha \longrightarrow \mathcal{R}/(t) \ (= k[t]/(t) \otimes \mathcal{R}),$$
$$\overline{p}_\beta) \longmapsto 1 \otimes P_\beta$$

defines an isomorphism. From this assertion (1) follows.

Assertions (2) and (3) follow from (1) and Proposition 3.3.3.

(4) For any $a \in R$, as above let $n(a)$ = the integer l such that $a \in I^l, a \notin I^{l+1}$. Let f be a monomial, and let $j = n(f)$. Further, let f be not standard. Let $f = \sum a_i f_i$, $a_i \neq 0$ be the expression for f as a sum of standard monomials. We have, $n(f_i) \geq j$ (in view of Lemma 3.3.2). Hence we obtain, $n(f) \leq \min\{n(f_i)\}$.

If possible, let $n(f) < n(f_i)$, for all i. Then reading $f = \sum a_i f_i$ modulo I^{j+1}, we obtain $f \equiv 0 \pmod{I^{j+1}}$. Hence, $n(f) \geq j+1$, which is not possible, since $n(f) = j$. Thus $n(f) = \min\{n(f_i)\}$, and hence $\overline{f} = \sum a_i \overline{f_i}$ (in I^j/I^{j+1}), where the summation on the right hand side runs over $\{i \mid n(f_i) = \text{order} f\}$. Thus, if we write $f_i = p(\tau_i)$, and $f = p(\Lambda)$, then we obtain $\overline{p(\Lambda)} = \sum a_i \overline{p(\tau_i)}$, where the summation runs over $\{i \mid n(p(\tau_i)) = n(p(\Lambda))\}$. The assertion now follows from Proposition 3.2.1.

(5). From our discussion above, we obtain that I^j/I^{j+1} has a basis consisting of $\{\overline{p(\Lambda)} \mid \Lambda \text{ is standard, and } n(p(\Lambda)) = j\}$. Thus in any straightening relation

$$\overline{p(\Lambda)} = \sum a_i \overline{p(\tau_i)} \ (\text{in } I^j/I^{j+1})$$

p_α occurs the same number of times in *all* of the terms (including $\overline{p(\Lambda)}$). From this assertion (5) follows. This completes the proof of the Proposition. □

PROPOSITION 3.3.5. *The $k[t]$-algebra $\mathcal{R}$ is a flat deformation whose special fiber ($t = 0$) is $\mathcal{R}_\alpha$ and whose general fiber (t invertible) is R.*

PROOF. That $\mathcal{R}$ is a deformation with general fiber R is clear. That $\mathcal{R}_\alpha$ is the special fiber of $\mathcal{R}$ follows from Proposition 3.3.4, (1). That $\mathcal{R}$ is a flat deformation (i.e., $\mathcal{R}$ is $k[t]$-free) follows from Proposition 3.3.3, (2). □

3.4. The algebra $R^{\text{def}}(w)$

We first observe that in carrying out the above one-step deformation for R, we have used only those properties of R as given by Theorem 3.3.1. Now the properties as given by Theorem 3.3.1 hold for $\mathcal{R}_\alpha$ also (in view of Proposition 3.3.4). Hence we carry out a similar one step deformation for $\mathcal{R}_\alpha$ also. Thus proceeding we obtain

THEOREM 3.4.1. *There exists a sequence of graded k-algebras $R = R_{11}, R_{12} \cdots, R_{1t_1}$ such that*

1. *There is a flat deformation $\mathcal{R}_{1s}$ over $k[t]$ whose special fiber is R_{1s+1} and whose general fiber is $R_{1s}, 1 \leq s \leq t_1 - 1$.*
2. *R_{1t_1} has algebra generators $\{x_\gamma \mid \gamma \in W^j, \gamma \leq w^{(j)}, j = 1, \cdots, r\}$.*

(3) *Standard monomials in x_γ's give a k-basis for R_{1t_1}.*

(4) *In any straightening relation*

$$(*) \qquad x(\Lambda) = \sum_l b_l x(\tau_l), \; b_l \in k^*$$

(where Λ is a non-standard Young diagram on $X(w)$ of type $(m_1, \cdots, m_r)$), we have, $\tau_l \overset{\mathrm{L}}{>} \Lambda^a, a \in A$, for all l

(here, for a Young diagram $\theta = (\theta_{ij}), 1 \leq j \leq m_i, 1 \leq i \leq r$, on $X(w)$ of type $(m_1, \cdots, m_r), x(\theta) = \prod_i \prod_j x_{\theta_{ij}}$).

(5) $J^1(R_{1t_1}) = \emptyset$.

REMARK 3.4.2. In (*) above, let $\Lambda = \Lambda_{ij}, \tau_l = (\alpha_{ij}^{(l)}), 1 \leq j \leq m_i, 1 \leq i \leq r$. Then (4) and (5) of Theorem 3.4.1 imply that $\prod_j \Lambda_{1j} = \prod_j \alpha_{1j}^{(l)}$, for all l. We next work with J_S^2 (we have denoted R_{1t_1} by S for simplicity of notations), where $J_S^2 = \{\beta \in W^2 \mid$ there exists a straightening relation in S, and a monomial F on the right hand side of this relation such that x_β occurs more times in F than it does on the left hand side of this relation $\}$. We have (similar to Lemma 3.3.2)

LEMMA 3.4.3. *Fix a maximal element β in J_S^2. Let $I = (p_\beta)$, and $I_j = I^j, j \geq 1$. Then I_j has a basis consisting of standard monomials (in the x_γ's) involving x_β at least j times.*

PROOF. Let T be a nonzero monomial in I^j. Further, let

$$(*) \qquad T = \sum a_i T_i, \; a_i \in k^*$$

be the straightening relation for T. For each i, we have (by the maximality assumption on β), $T^\beta \supseteq T_i^\beta$ (notations as in the proof of Lemma 3.3.2). This together with the lexicographic condition (cf. Theorem 3.4.1, (4)) implies (in view of Remark 3.4.2) that $T_\beta \subseteq (T_i)_\beta$ (and $T^\beta = T_i^\beta$). From this we conclude as in Lemma 3.3.2 that I_j has a basis consisting of standard monomials in x_γ's involving x_β at least j times. □

Proceeding thus, we arrive at $R_{21} = R_{1t_1}, R_{22} \cdots, R_{2t_2}$ such that

(1) There is a flat deformation over $k[t]$ $\mathcal{R}_{2s}$ whose special fiber is R_{2s+1} and whose general fiber is $R_{2s}, 1 \leq s \leq t_2 - 1$.

(2) R_{2t_2} has algebra generators $\{y_\gamma \mid \gamma \in W^j, \gamma \leq w^{(j)}, j = 1, \cdots, r\}$.

(3) Standard monomials in y_γ's give a k-basis for R_{2t_2}.

(4) In any straightening relation

$$(*) \qquad y(\Lambda) = \sum_l c_l y(\varphi_l), \; c_l \in k^*$$

(where Λ is a non-standard Young diagram on $X(w)$ of type $(m_1, \cdots, m_r)$), we have, $\varphi_l \overset{\mathrm{L}}{>} \Lambda^a, a \in A$, for all l.

(5) $J^i(R_{2t_2}) = \emptyset$, $i = 1, 2$.

Proceeding thus (working with $J^3(R_{2t_2})$ etc.), we obtain

THEOREM 3.4.4. *There exists a sequence of graded k-algebras $R = R_1, \cdots, R_t$ satisfying*

(1) *There is a flat deformation $\mathcal{R}_s$ over $k[t]$ whose special fiber is R_{s+1} and whose general fiber is R_s, $1 \leq s \leq t-1$.*
(2) *R_t has algebra generators $\{q_\gamma \mid \gamma \in W^j, \gamma \leq w^{(j)}, j = 1, \cdots, r\}$.*
(3) *Standard monomials in q_γ's give a k-basis for R_t.*
(4) *$J^i(R_t) = \emptyset$, $i = 1, \cdots, r$.*

3.4.5. Summary. With notations as in Theorem 3.4.4, let us denote R_t by $R^{\mathrm{def}}(w)$. We have shown
(a) $R^{\mathrm{def}}(w)$ has a basis consisting of standard monomials. Further, any nonstandard monomial is zero (in view of Theorem 3.4.4, (4)).
(b) If $R^{\mathrm{def}}(w)$ is Cohen-Macaulay, then $R = R(w)$ is Cohen-Macaulay.

As a consequence of 3.4.5, (a), we obtain

PROPOSITION 3.4.6. *$R^{\mathrm{def}}(w)$ is defined by square-free monomials.*

PROOF. Let x_α be an indeterminate over k, one for each $\alpha \in W^j, \alpha \leq w^{(j)}, j = 1, \cdots, r$. Map $k[x_\alpha](= S$, say) onto $R^{\mathrm{def}}(w)$ by sending x_α to q_α. Let I be the kernel. Denote the ideal generated by the set of square-free nonstandard monomials by J. Clearly $J \subseteq I$. On the other hand, if $F = p(X_\lambda)$ is a monomial which maps to zero, then λ is not standard, and hence $p(X_\lambda)$ is divisible by some $p(X_\tau)$, where X_τ is a square-free non-standard monomial. □

REMARK 3.4.7. Note that in particular that $R^{\mathrm{def}}(w)$ is an algebra with straightening law (cf. [**39**]), a fact we shall make use of in §4. This is a significant improvement, as $R(w)$ itself is not an algebra with straightening law in any obvious way.

3.5. Concluding remarks

The above procedure suggests that the one-step deformation can be carried out in a more general context as described below:

Let R be a k-algebra with algebra generators $\{x_\alpha, \alpha \in H\}$, ($H$ being a finite indexing set), such that certain monomials in the x_α's give a k-basis for R. Let us call these as *standard monomials*. Let F be a non-standard monomial. Let $F = \sum a_i F_i$, $a_i \in k^*$ be the expression for F as a sum of standard monomials. We shall refer to such a relation as a *straightening relation*. Let $J_R = \{x_\alpha \mid$ there exists a straightening relation $F = \sum a_i F_i$ such that x_α is present in some $F_i\}$.

Suppose x_α in J_R is such that the ideal I^r, where $I = (x_\alpha)$ has a basis consisting of standard monomials involving x_α at least r times.

Call this property (*). Then it is easily seen that the above one-step deformation can be carried out. Thus if R is such that the property (*) above holds at each step, then the above one-step deformation can be carried out successively, and we arrive (by successive flat deformations) at the "discrete algebra" R^{Disc}, where R^{Disc} has algebra generators $\{y_\alpha, \alpha \in H\}$ such that

(1) Certain monomials (to be called *standard monomials*) in the y_α's form a k-basis for R.
(2) Any nonstandard monomial is $\equiv 0$.

The simpler algebra R^{Disc} may then be analyzed by studying either the combinatorics of H or the ideal theory of R^{Disc}. Examples of such R are given by Hodge algebras (cf. [**39**]), the rings $R(w)$ as discussed above, etc. In the previous section, we used the combinatorics of T_τ (cf. [**18**]) to conclude the Cohen-Macaulayness of R_{T_τ}. In this section, we shall prove the Cohen-Macaulayness of $R^{\text{def}}(w)$ by a "principal radical system" (cf. [**47**]).

3.6. Some preliminary results on the Schubert divisors in $X(w)$

Given a subset J of S (the set of simple roots), if Q is the parabolic subgroup such that $S_Q = J$, then we denote W_Q, W^Q also by W_J, W^J. If w is in W, then w can be factored uniquely as $w = uv, u \in W^J$, $v \in W_J$; in such a factorization, we have $l(w) = l(u) + l(v)$. In particular, if $u \in W^J$, then u is the unique member of its coset uW_J having minimal length. We note the following well known facts:

LEMMA 3.6.1. *Let $w \in W^J, w' \in W, w' > w, l(w') = l(w) + 1$. Then either $w' \in W^J$ or $w' = ws$ for some $s \in J$.*

LEMMA 3.6.2. *Let $w, w' \in W^J, w < w'$. Then all maximal chains $w' = u_0 > u_1 > \cdots > u_r = w$ in W^J have the same length $r = l(w') - l(w)$.*

REMARK 3.6.3. Let $\mathcal{R} = \{$the set of all reflections$\}$. Let $s, t \in \mathcal{R}$. Then $(st)^p = 1, 2,$ or 3. Let s (resp. t) be the transposition (i, j) (resp. (k, l)). Then $p = 1 \Leftrightarrow k = i,\ j = l$, and $p = 2 \Leftrightarrow \{i, j\} \cap \{k, l\} = \emptyset$.

LEMMA 3.6.4. *Let $s_1, s_2, t_1, t_2 \in \mathcal{R}, s_1 \neq s_2$,, and*

$$s_1 s_2 = t_1 t_2 . \tag{*}$$

If any distinct two of these reflections commute, then all of them commute and $\langle s_1, s_2\rangle = \langle t_1, t_2\rangle$.

PROOF. Since $s_1^2 = t_1^2 = 1$, the equation (*) is equivalent to either one of the following two equations

$$s_1 t_1 = s_2 t_2, \tag{**}$$

$$t_1 s_1 = t_2 s_2. \tag{***}$$

If $s_1 s_2 = s_2 s_1$, then $(s_1 s_2)^2 = 1$, and so $(t_1 t_2)^2 = 1$. Thus $t_1 t_2 = t_2 t_1$ also. If in addition $(s_1 t_j)^2 = 1$, then it is clear that $(s_i t_j)^2 = 1$, for all i, j and so all of these reflections commute.

Therefore we may assume $(s_1 t_1)^2 \neq 1$, in which case $(s_1 t_1)^3 = 1$, i.e., $s_1 t_1 s_1 t_1 = t_1 s_1$. Using (***) we find $s_1 t_2 s_2 t_1 = t_1 s_1$, while from (**) we obtain $s_1 t_2 s_1 t_2 = t_1 s_1$. As we are assuming $(s_1 t_2)^2 \neq 1$, we have $(s_1 t_2)^3 = 1$, and this together with the latter equation shows $t_1 s_1 = t_2 s_1$, or $t_1 = t_2$. This contradicts the fact that $s_1 \neq s_2$. Therefore if $s_1 s_2 = s_2 s_1$, then all the reflections commute.

Since (*) is equivalent to (**) or (***), the above proof also proves the case when $s_it_i = t_is_i$. Finally, let us assume that $s_1s_2 \neq s_2s_1$, $s_it_i \neq t_is_i$, and $s_1t_2 = t_2s_1$, say. We have $(s_1s_2)^3 = 1$, i.e. $1 = s_1s_2s_1s_2s_1s_2 = t_1t_2s_1s_2s_1s_2$ (using (*)) $= s_1t_2$ (note that $t_1s_1t_1 = s_1t_1s_1$). Hence we obtain that $s_1 = t_2$. This together with the hypothesis that two distinct elements of $\{s_i, t_i, i = 1, 2\}$ commute, implies that $s_2t_1 = t_1s_2$. Proceeding as above, we have
$1 = s_2s_1s_2s_1s_2s_1 = s_2t_1t_2s_1s_2s_1$ (using (*)) $= t_1s_2t_2s_1s_2s_1$ (since $s_2t_1 = t_1s_2$) $= t_1s_1$ (since $s_1 = t_2$). Hence we obtain $s_1 = t_1$. This implies $s_2 = t_2 = s_1 = t_1$. This completes the proof of the first statement.

Write $s_1 = (i_1, i_2)$, $s_2 = (j_1, j_2)$, $t_1 = (k_1, k_2)$ and $t_2 = (l_1, l_2)$. By Remark 3.6.3, we see that $\{i_1, i_2, j_1, j_2, k_1, k_2, l_1, l_2\}$ are all distinct, unless some of s_1, s_2, t_1, t_2 are equal. However if these integers are all distinct, it is clear that $s_1s_2 \neq t_1t_2$. Since $s_1 \neq s_2$, s_1 must equal either t_1 or t_2. This implies (in view of the fact that $s_1s_2 = t_1t_2$), $\{s_1, s_2\} = \{t_1, t_2\}$. The second assertion now follows. □

LEMMA 3.6.5. *Suppose $s_1 \neq s_2$ are in T, $t_1, t_2 \in T$ and $s_1s_2 = t_1t_2$. If $s_1s_2 = s_2s_1$, then $\langle s_1, s_2\rangle = \langle t_1, t_2\rangle$ and is a Klein four group. If $s_1s_2 \neq s_2s_1$, then $\langle s_1, s_2\rangle$ is isomorphic to $\mathbb{S}_3$ and $\langle s_1, s_2\rangle = \langle t_1, t_2\rangle$. In particular, at most three of $\{s_1, s_2, t_1, t_2\}$ are distinct.*

PROOF. The first statement is immediate from Lemma 3.6.4. Let us assume that $s_1s_2 \neq s_2s_1$. Our assumption implies (in view of Lemma 3.6.4) that no two of $\{s_1, s_2, t_1, t_2\}$ commute. That $\langle s_1, s_2\rangle \equiv \mathbb{S}_3$ follows from the fact that two non-commuting reflections in $\mathbb{S}_n$ generate $\mathbb{S}_3$. It remains to show that $\langle s_1, s_2\rangle = \langle t_1, t_2\rangle$. If two of $\{s_1, s_2, t_1, t_2\}$ coincide, then this result follows from the hypothesis that $s_1s_2 = t_1t_2$. We *claim* that all the four cannot be distinct. If possible, let us assume that all the four are distinct. Since no two of these four distinct reflections commute, we obtain that these four reflections are precisely of the form (i, j), (i, k), (i, l), (i, m). But then the hypothesis that $s_1s_2 = t_1t_2$ cannot hold. Hence the claim follows from this, and so also the required result. □

Suppose $w = (a_1, \ldots, a_n)$ and $w_1 = (b_1, \ldots, b_n)$. If $w > w_1$ and $l(w) = l(w_1) + 1$, then $w_1 = ws_\alpha$, where s_α is the reflection corresponding to some root $\alpha = e_i - e_j$ $(i < j)$. The condition $l(w_1) = l(w) - 1$ translates into the following two conditions:
1. $a_i > a_j$
2. there does not exist a k, $i < k < j$, such that $a_j < a_k < a_i$.

The main result of this section is the following.

PROPOSITION 3.6.6. *Suppose w, w_1, w_2 are in W^Q, $w_1 \neq w_2$ and $l(w) - 1 = l(w_1) = l(w_2)$, $w > w_1$, $w > w_2$. Also suppose that both w_1 and w_2 have strictly smaller projection on W/W_{i_1} than does w, where P_{i_1} is a maximal parabolic subgroup containing Q. Then the following statements hold.*

(1) There exists a $w' \in W$ such that $w_1, w_2 > w'$ and $l(w') = l(w) - 2$.

(2) $w' \in W^Q$.

(3) If $\theta \in W$ *and* $\theta < w_1$, $\theta < w_2$, *then* $\theta \leqslant w'$.

REMARK 3.6.7. The assumption that both w_1 and w_2 have strictly smaller projection than w is crucial. For instance if $w = (3412)$, $w_1 = (1432)$, $w_2 = (3214)$, then there is *no* w' of codimension two in w which is less than both w_1 and w_2. Note that w_2 does not have a strictly less projection than w on W/W_1. On the other hand both w_1 and w_2 have strictly less projection on W/W_2 than w does; however, if $Q = P_2 \cap P_3$, then $w_2 \notin W^Q$.

For another slightly different example, consider $Q = P_1 \cap P_3$. Then $w = (3142)$ is in W^Q, while both $w_1 = (1342)$ and $w_2 = (3124)$ are of codimension one in w and are both in W^Q. In addition, $w' = (1324)$ is the unique codimension two subvariety of w such that $w_1, w_2 \geqslant w'$. However, $w' \notin W^Q$. In this case both w_2 and w have the same projection on W/W_1.

PROOF. Write $w = (a_1, \ldots, a_n)$, $w_1 = (b_1, \ldots, b_n)$, $w_2 = (c_1, \ldots, c_n)$. As $l(w_1) = l(w) - 1 = l(w_2)$, we may write $w_1 = ws_\alpha$, $w_2 = ws_\beta$, where s_α, s_β are reflections corresponding to the roots $\alpha = \epsilon_i - \epsilon_j$, and $\beta = \epsilon_k - \epsilon_l$. Without loss of generality, we may assume $i \leqslant k$. Since both w_1 and w_2 have strictly less projection on W/W_{i_1} than w does, we see that $i, k \leqslant i_1$, while $j, l > i_1$. Note that

$$b_r = \begin{cases} a_r & \text{if } r \neq i, j \\ a_j & \text{if } r = i \\ a_i & \text{if } r = j \end{cases} \qquad \text{and} \qquad c_r = \begin{cases} a_r & \text{if } r \neq k, l \\ a_k & \text{if } r = l \\ a_l & \text{if } r = k. \end{cases}$$

There are three cases to consider. We list them here.

Case 1. $i < k, j < l$

Case 2. $i < k, j > l$

Case 3. $i = k, j < l$

Note that the case $i < k$, $j = l$ cannot occur; for, in this case we would have $a_j < a_i < a_k$. But then, $w_2 \notin W^Q$, since $c_k = a_j > c_i = a_i$.

Each case will require a separate argument.

Case 1. $i < k$, $j < l$. Then $s_\alpha s_\beta = s_\beta s_\alpha$, and we define $w' = ws_\alpha s_\beta = ws_\beta s_\alpha$. If $w' = (d_1, \ldots, d_n)$, then

$$d_r = \begin{cases} a_r & \text{if } r \neq i, j, k, l \\ a_j & \text{if } r = i \\ a_i & \text{if } r = j \\ a_k & \text{if } r = l \\ a_l & \text{if } r = k. \end{cases}$$

We first show that $l(w') = l(w) - 2$ and that $w' < w_1$, $w' < w_2$. Since $w' = w_1 s_\beta$, we need to show that there is no r, $k < r < l$ such that $b_l < b_r < b_k$. (Notice that $b_k > b_l$, since $b_k = a_k$, $b_l = a_l$ and $a_k > a_l$ as $w_2 = ws_\beta$ is of codimension one in w.

Also, there is no r, $k < r < l$ such that $a_l < a_r < a_k$.) Since $b_r = a_r$ for $k < r < l$, $r \neq j$, the only possible bad case is if $b_l < b_j < b_k$, i.e. if $a_l < a_i < a_k$.

However, $w_2 \in W^Q$ so that $c_i < c_k$. Since $c_i = a_i$, $c_k = a_l$, $a_i < a_l$, the case $a_l < a_i$ cannot occur. Thus $l(w') = l(w_1) - 1 = l(w) - 2$. It also easily follows that $w' < w_2$.

Next we must prove that $w' \in W^Q$.

Since $w_1, w_2 \in W^Q$, we see that $a_{i-1} < a_j < a_i$ and $a_{k-1} < a_l < a_k$. Choose s, r such that $i_s + 1 \leqslant j \leqslant i_{s+1}$ and $i_r + 1 \leqslant l \leqslant i_{r+1}$. Then $a_j < a_i < a_{j+1}$ and$a_l < a_k < a_{l+1}$. (If no such r or s exists, then there are no conditions to be checked.)

Since $i < k$, $a_j < a_i \leqslant a_{k-1} < a_l < a_k$, so that $d_1 < \cdots < d_{i_1}$. If $s \neq r$ the observations above prove that

$$d_{i_s+1} < \cdots < d_{i_{s+1}}$$

while

$$d_{i_r+1} < \cdots < d_{i_{r+1}}.$$

If $s = r$, then we need to prove that $d_{i_s+1} < \cdots < d_{i_{s+1}}$, i.e. that

$$a_{i_j+1} < \cdots < a_{j-1} < a_i < a_{j+1} < \cdots < a_{l-1} < a_k < a_{l+1} < \cdots < a_{i_s+1}.$$

This follows immediately from the above inequalities and from the fact that $w \in W^Q$. It follows that $w' \in W^Q$.

Lastly, we must show that w' has the property (3). If $(a_1, \ldots, a_k)$ are distinct integers, by $(a_1, \ldots, a_k)^{\text{inc}}$ we denote the unique sequence $(a_{\pi 1}, \ldots, a_{\pi k})$ such that $a_{\pi 1} < \cdots < a_{\pi k}$. If $\alpha = (e_1, \ldots, e_n)$ and $\beta = (f_1, \ldots, f_n)$ are in W, recall that $\beta \leqslant \alpha$ if and only if for all i, $1 \leqslant i \leqslant n$, $(f_1, \ldots, f_i)^{\text{inc}} \leqslant (e_1, \ldots, e_i)^{\text{inc}}$.

Now let θ be as in (3). Set $\theta = (e_1, \ldots, e_n)$. If $\theta \not\leqslant w'$, then we may choose m minimal such that

$$(e_1, \ldots, e_m)^{\text{inc}} \not\leqslant (d_1, \ldots, d_m)^{\text{inc}}.$$

By assumption,

$$(e_1, \ldots, e_m)^{\text{inc}} \leqslant (c_1, \ldots, c_m)^{\text{inc}}$$

and

$$(e_1, \ldots, e_m)^{\text{inc}} \leqslant (b_1, \ldots, b_m)^{\text{inc}}.$$

If $(x_1, \ldots, x_m)$ and $(y_1, \ldots, y_m)$ are two sets of increasing integers, then we set $z = (z_1, \ldots, z_m) = \min\{(x_1, \ldots, x_m), (y_1, \ldots, y_m)\}$, where $z_i = \min\{x_i, y_i\}$. Clearly then,

$$(e_1, \ldots, e_m)^{\text{inc}} \leqslant \min\{(c_1, \ldots, c_m)^{\text{inc}}, (b_1, \ldots, b_m)^{\text{inc}}\}.$$

We *claim* that

$$\min\{(c_1, \ldots, c_m)^{\text{inc}}, (b_1, \ldots, b_m)^{\text{inc}}\} = (d_1, \ldots, d_m)^{\text{inc}}.$$

This clearly contradicts our minimality assumption on m and hence we obtain that $\theta \leqslant w'$.

If $m > l$, then $\{c_1, \ldots, c_m\} = \{b_1, \ldots, b_m\} = \{d_1, \ldots, d_m\}$ and our claim is obvious. If $m \leqslant i_1$, then since $w_1, w_2 \in W^Q$, and $c_1 < \cdots < c_m$, it is clear that

$$d_i = \min\{b_i, c_i\}.$$

If $i_1 < m < j$, then

$$\begin{aligned}(b_1, \ldots, b_m) &= (a_1, \ldots, a_{i-1}, a_j, a_i, \ldots, a_m),\\ (c_1, \ldots, c_m) &= (a_1, \ldots, a_{k-1}, a_l, a_k, \ldots, a_m),\\ (d_1, \cdots, d_m) &= (a_1, a_{i-1}, a_j, a_{i+1}, a_{k-1}, a_l, a_k, \ldots, a_m).\end{aligned}$$

To show $(d_1, \ldots, d_m)^{\text{inc}} = \min\{(b_1, \ldots, b_m), (c_1, \ldots, c_m)\}$, it is enough to show that the places of $b_i = a_j$ and $b_k = a_k$ in $(b_, \ldots, b_m)^{\text{inc}}$ are the same respectively as those of $c_i = a_i$ and $c_k = a_l$ in $(c_1, \ldots, c_m)^{\text{inc}}$. For, in this case since $a_l < a_k$ and $a_j < a_i$, $\min\{(b_1, \ldots, b_m), (c_1, \ldots, c_m)\} = (d_1, \ldots, d_m)$. The place of b_i is equal to one plus the number of $b_r < b_i$ with $r \leqslant m$. Thus to show that b_i has the same place in $(b_1, \ldots, b_m)^{\text{inc}}$ as does c_i in $(c_1, \ldots, c_m)^{\text{inc}}$, it is enough to show that the number of $b_r < a_j = b_i$ with $r \leqslant m$ is the same as the number of $c_r < a_i = c_i$ with $r \leqslant m$. We note that if $a_r < a_i$ and $r \leqslant m < j$, then $a_r < a_j$. If not, there exists an $r < j$ such that $a_j < a_r < a_i$. Since if $r < i$, $a_r < a_j$ (as $w_1 \in W^Q$), we see that $i < r < j$. This contradicts the fact that $l(w_1) = l(w) - 1$. However, $\{b_r \mid r \leqslant m, r \neq i\} = \{a_r \mid r \leqslant m, r \neq i\} = \{c_r \mid r \leqslant m, r \neq i, k\} \cup \{a_k\}$. Since $a_j < a_i < a_l < a_k$, neither a_l nor a_k have any effect upon the number of $a_r < a_j$ or $< a_i$. The proof for places of b_k and c_k is similar. This proves our assertion.

If $j \leqslant m < l$, then $\{b_1, \ldots, b_m\} = \{a_1, \ldots, a_m\}$, and

$$(c_1, \ldots, c_m) = (a_1, \ldots, a_i, \ldots, a_{k-1}, a_l, a_{k+1}, \ldots, a_m).$$

Since $w_2 < w$, we have

$$(c_1, \ldots, c_m)^{\text{inc}} \leqslant (a_1, \ldots, a_m)^{\text{inc}} = (b_1, \ldots, b_m)^{\text{inc}},$$

so that

$$\min\{(c_1, \ldots, c_m)^{\text{inc}}, (b_1, \ldots, b_m)^{\text{inc}}\} = (c_1, \ldots, c_m)^{\text{inc}}.$$

However, $\{d_1, \ldots, d_m\} = \{c_1, \ldots, c_m\}$, and the assertion is clear. This finishes the proof of Case 1.

Case 2. $i < k$, $j < l$. Again $s_\alpha s_\beta = s_\beta s_\alpha$, and we set $w' = w s_\alpha s_\beta = w s_\beta s_\alpha$. The proof of (1) and (2) in this case is similar to that of Case 1 and we leave the details to the reader. Write $w' = (d_1, \ldots, d_n)$.

Let $\theta = (e_1, \ldots, e_n)$ be as in (3). We may assume $\theta \not\leqslant w'$. Choose m to be the least such that

$$(e_1, \ldots, e_m)^{\text{inc}} \not\leqslant (d_1, \ldots, d_m)^{\text{inc}}.$$

Again, we make the *claim* that

$$(z_1, \ldots, z_m) = \min\{(c_1, \ldots, c_m)^{\text{inc}}, (b_1, \ldots, b_m)^{\text{inc}}\} = (d_1, \ldots, d_m)^{\text{inc}}.$$

If $m \leqslant i_1$, this is again easy to see using the fact that $w_1, w_2 \in W^Q$.

If $i_1 < m < l$, then $\{c_1, \ldots, c_m\} = \{a_1, \ldots, a_{k-1}, a_l, a_k, \ldots, a_m\}$, while $\{b_1, \ldots, b_m\} = \{a_1, \ldots, a_{i-1}, a_j, a_i, \ldots, a_m\}$ and

$$\{d_1, \ldots, d_m\} = \{a_1, \ldots, a_{i-1}, a_j, a_i, \ldots, a_{k-1}, a_l, a_k, \ldots, a_m\}.$$

As above, it is enough to show that the places b_i and b_k in $(b_1, \ldots, b_m)^{\text{inc}}$ are the same respectively as the places of c_i and c_k in $(c_1, \ldots, c_m)^{\text{inc}}$. Again, this follows as above.

If $l \leqslant m < j$, then we have $\{c_1, \ldots, c_m\} = \{a_1, \ldots, a_m\}$, and so $(z_1, \ldots, z_m) = (b_1, \ldots, b_m)^{\text{inc}}$ as $w_1 < w$. However, $\{d_1, \ldots, d_m\} = \{b_1, \ldots, b_m\}$ in this case. Thus $(d_1, \ldots, d_m)^{\text{inc}} = (z_1, \ldots, z_m)$ as required. Finally, $m > j$ implies $\{d_1, \ldots, d_m\} = \{a_1, \ldots, a_m\} = \{b_1, \ldots, b_m\} = \{c_1, \ldots, c_m\}$, and our claim is obvious.

Case 3. $i = k, j < l$. Since $l(w_1) = l(w_2) = l(w) - 1$, we have, $a_j < a_i$ and $a_l < a_i$. On the other hand, if $a_l < a_j$, then $a_l < a_j < a_i$, and then $ws_\beta = w_2$ would not have length $l(w) - 1$. Hence

$$a_j < a_l < a_i.$$

We define $w' = (d_1, \ldots, d_n)$ by

$$d_r = \begin{cases} a_r & \text{if } r \neq i, j, l \\ a_j & \text{if } r = i \\ a_l & \text{if } r = j \\ a_i & \text{if } r = l. \end{cases}$$

$w' = w_2 s_\alpha$, $w' = w_1(s_\alpha s_\beta s_\alpha)$. We note that $s_\alpha s_\beta \neq s_\beta s_\alpha$; consequently $(s_\alpha s_\beta)^3 = 1$, and $s_\alpha s_\beta s_\alpha = s_\beta s_\alpha s_\beta$.

We will show that $l(w') = l(w_2) - 1$ and $w' < w_2$. For this, it suffices to show that $c_j < c_i$, and if $i < m < j$, then c_m does not lie between c_i and c_j. However, $c_j = a_j$, $c_m = a_m$ for $i < m < j$ and $c_i = a_l$. By the discussion above, $a_j < a_l$, so that $c_j < c_i$. If $a_j < a_m < a_l$, then $a_j < a_m < a_i$, which contradicts the assumption that w_1 is not of codimension one in w. Thus $w' < w_2$ and $l(w') < l(w) - 2$. Next we observe that $w' < w_1$ (since $b_l < b_j$). However, $b_i = a_i$ and $b_l = a_l$; since $a_l < a_i$, we see that $b_l < b_j$.

Next, we need to prove that $w' \in W^Q$. If possible, let us assume that there is an m with $i_m + 1 \leqslant j < l \leqslant i_{m+1}$. Assumption implies that w_1 is not in W^Q as $b_j < b_l$, since $b_j = a_i$, $b_l = a_l$. Thus such an m does not exist; it now easily follows that $w' \in W^Q$ (from the fact that w_1 and w_2 are in W^Q).

Lastly, we prove (3). Let $\theta = (e_1, \ldots, e_n)$ be as in (3). We have assumed that for any $1 \leqslant m \leqslant n$,

$$\begin{cases} (e_1, \ldots, e_m)^{\text{inc}} \leqslant (c_1, \ldots, c_m)^{\text{inc}} & \text{and} \\ (e_1, \ldots, e_m)^{\text{inc}} \leqslant (b_1, \ldots, b_m)^{\text{inc}}. \end{cases}$$

However, it is trivial to check that

$$(d_1, \ldots, d_m)^{\text{inc}} = \begin{cases} (b_1, \cdots, b_m)^{\text{inc}} & \text{if } m < j \\ (c_1, \cdots, c_m)^{\text{inc}} & \text{if } j \leqslant m < l \\ (a_1, \cdots, a_m)^{\text{inc}} & \text{if } m > l \end{cases}.$$

Thus in all cases,

$$(e_1, \ldots, e_m)^{\text{inc}} \leqslant (d_1, \ldots, d_m)^{\text{inc}},$$

which shows that $\theta \leqslant w'$. This completes the proof. □

3.7. Ideal theoretic unions and intersections in $R^{\text{def}}(w)$

In this section we apply our calculations of the previous section to understand the ideal theoretic union and intersection in the deformed ring $R^{\text{def}}(G/Q)$ of §3.4.5. Accordingly, for the remainder of this section, we fix a parabolic $Q = P_{i_1} \cap \cdots \cap P_{i_k}$ with $i_1 < \cdots < i_k$. If $w \in W^Q$, we let $R^{\text{def}}(w)$ (or $R^{\text{def}}(X(w))$) be the ring R_t given by Theorem 3.4.4. The k-algebra $R^{\text{def}}(w)$ has a basis consisting of standard monomials on $X(w)$, and any nonstandard monomial is zero. For the variety $X(w) = G/Q$ itself, we set $R := R^{\text{def}}(G/Q)$. There is a surjection from R onto $R^{\text{def}}(w)$; we denote the kernel of this surjection by $I(w)$, where $I(w)$ is generated by the set of all square-free standard monomials on G/Q which are *not* standard on $X(w)$. We denote this set of square-free monomials by $J(w)$. We let H be the set of all square-free standard monomials on G/Q. We define a partial order on H as follows: if m and m' are in H, we say that $m \geqslant m'$ if and only if m' contains m as a submonomial. Of course, any monomial can be written uniquely in the form

$$m = {m_1}^{n_1} \ldots {m_k}^{n_k},$$

where

$$m_1 > \cdots > m_k.$$

With this partial order, R becomes an algebra with straightening law on H (cf. [**39**]). Indeed, any ring defined by square free monomials is such an algebra. We will use only the most basic properties of such an algebra in the discussion of this section. Although R has algebra generators indexed by the partially ordered set $\bigcup_{t=1}^{r} W^{i_t}$, $R^{\text{def}}(w)$ is not an algebra with straightening law on the set of all $\alpha \in \bigcup_{t=1}^{r} W^{i_t}$ such that $p_\alpha|_{X(w)} \neq 0$. It is possible to have two standard monomials $p_\tau p_\alpha$ and $p_\alpha p_\beta$ on $X(w)$ without $p_\tau p_\beta$ being standard. In fact if $G = SL(3)$, and $\tau = (2)$, $\alpha = (1)$, $\beta = (13)$ with $w = (312)$, then this is the case.

DEFINITION 3.7.1. A subset $J \subset H$ is said to be an *ideal* of H if whenever $\alpha \in J$ and $\beta \leqslant \alpha$, then $\beta \in J$. If J is an ideal of H, we let $\overline{J}$ denote the ideal of R generated by the monomials in J.

REMARK 3.7.2. The set $J(w)$ is an ideal of H. For, $m \in J(w)$ if and only if m is nonstandard on $X(w)$; in particular, if $m' \leqslant m$, then m' is also nonstandard on $X(w)$.

LEMMA 3.7.3. *Suppose R is an algebra with straightening law on H over A. If I and J are two ideals of H, then*

(1) $\overline{I} + \overline{J} = \overline{I + J}$,
(2) $\overline{I \cap J} = \overline{I} \cap \overline{J}$.

For a proof, see [**50**].

LEMMA 3.7.4. *Let $w_1, \dots, w_k \in W^Q$.*

(1) $\left[I(w_1 \cap \cdots \cap I(w_{k-1})\right] + I(w_k) =$

$$\left[I(w_1) + I(w_k)\right] \cap \cdots \cap \left[I(w_{k-1}) + I(w_k)\right]$$

Assume that $w, w_1, w_2 \in W^Q$ and w_1 and w_2 have strictly smaller projections on W/W_{i_1} than does w. Let w' be an element of W^Q as in Proposition 3.6.6. Then

(2) $I(w_1) + I(w_2) = I(w')$.

PROOF. Since $\overline{J(w_i)} = I(w_i)$ by Lemma 3.7.3, to prove (1) it is enough to prove that

$$\left[J(w_1) \cap \cdots \cap J(w_{k-1})\right] \cup J(w_k) = \left[J(w_1) \cup J(w_k)\right] \cap \cdots \cap \left[J(w_{k-1}) \cup J(w_k)\right].$$

However this equality is simply DeMorgan's law.

To prove (2) it suffices to show that

$$J(w_1) \cup J(w_2) = J(w').$$

Clearly $J(w_1) \cup J(w_2) \subset J(w')$, as any nonstandard monomial on $X(w_i)$, $i = 1, 2$ remains nonstandard on $X(w')$.

Conversely, suppose that m is a monomial which is standard on both $X(w_1)$ and $X(w_2)$. We must show that m is standard on $X(w')$. Let λ be the Young diagram to which m corresponds. Let θ be the (unique) minimal defining sequence for λ (cf. Chapter 7, Corollary 3.1.4) and let θ_1 be the first member of θ. Since m is standard on both $X(w_1)$ and $X(w_2)$, we have $w_1 \geqslant \theta_1$ and $w_2 \geqslant \theta_1$. Hence Proposition 3.6.6, (3) implies $\theta_1 \leqslant w'$. It follows that λ is a standard Young diagram on $X(w')$, i.e. m is standard on $X(w')$, and the required result follows. □

In proving that $R^{\text{def}}(w)$ is Cohen-Macaulay, it will be crucial to understand the representation of a hyperplane section of $X(w)$ as a union of Schubert varieties. Of course we actually do this in R.

Let $w \in W^Q$ and let $\tau = w^{(i_1)}$ (notation as in the previous section).

DEFINITION 3.7.5. Define

$$B(w) = \{w' \in W^Q \mid w \geqslant w', l(w) - 1 = l(w') \text{ and } p_\tau|_{X(w')} = 0\}.$$

To say $p_\tau|_{X(w')} = 0$ is simply to say that the projection τ' of w' on W^{i_1} is strictly smaller than τ. Define $J_\tau = \langle p_\alpha \in W^{i_1},\ \alpha \leqslant \tau\rangle$, the ideal generated by $\{p_\alpha \in W^{i_1},\ \alpha \leqslant \tau\}$.

LEMMA 3.7.6. *Fix $w \in W^Q$. Let λ be a standard Young diagram on $X(w)$ and suppose $\lambda_1 \in W^{i_1}$ (λ_1 being the first element in λ) and λ_1 is strictly less than the projection τ of w on W/W_{i_1}. Then λ is a standard Young diagram on some $X(w')$, where $w' \in B(w)$.*

PROOF. Since λ is standard on $X(w)$, there is a defining sequence θ for λ such that $w \geqslant \theta_1$. (Notice that $\theta_1 \in W^Q$.) Since θ_1 projects to λ_1 and $\lambda_1 < \tau$, we see that $\theta_1 < w$. We induct on $l(w) - l(\theta_1)$ to prove the Lemma.

If $l(w) - l(\theta_1) = 1$, then $\theta_1 \in W^Q$ and λ is standard on $X(\theta_1)$.

If $l(w) - l(\theta_1) = 2$, by **[18]** there are precisely two elements w_1 and w_2 in W such that $w > w_1, w_2 > \theta_1$. Write $w_1 = ws_1$, $w_2 = ws_2$, $\theta_1 = w_1t_1$, $\theta_1 = w_2t_2$. Then $s_1t_1 = s_2t_2$, and by Lemma 3.6.5, $\langle s_1, s_2\rangle = \langle t_1, t_2\rangle = \langle s_1, t_1\rangle = \langle s_2, t_2\rangle$. The facts that $w, \theta_1 \in W^Q$ imply that at least one of w_1, w_2 is in W^Q.

Case 1. $w_1, w_2 \in W^Q$. We need to show that at least one of w_1, w_2 has a different projection on W/W_{i_1} than does w. If not, we have $s_1, s_2 \in W_{i_1}$. In this case $\langle t_1, t_2\rangle = \langle s_1, s_2\rangle \subset W_{i_1}$, and so $s_1t_1 \in W_{i_1}$. Then $\theta_1 = ws_1t_1$ has the same projection on W/W_{i_1} as w does. This is a contradiction.

Case 2. $w_1 \in W^Q$, $w_2 \notin W^Q$. Then by Lemma 3.6.1, $t_2 \in W_Q$. If $s_1 \notin W_{i_1}$, then $w_1 \in W^Q$ and has strictly smaller projection on W/W_{i_1} than does w. So it is enough to prove $s_1 \notin W_{i_1}$. If possible, let $s_1 \in W_{i_1}$. If $s_1s_2 = s_2s_1$, then $\langle s_1, s_2\rangle = \langle t_1, t_2\rangle$ by Lemma 3.6.4, and so $s_1 = t_2 \in W_Q$. But then $w_1 \notin W^Q$. Let then $s_1s_2 \neq s_2s_1$. Since $t_2 \in W_Q$, $t_1 = s_1t_2s_1 \in W_{i_1}$. But then $s_1t_1 \in W_{i_1}$, and this is again a contradiction.

Finally, assume that $l(w) - l(\theta_1) > 2$. By Proposition 3.6.6, there is a $w_1 \in W^Q$ such that $w > w_1 > \theta_1$ and $l(w_1) = l(w) - 1$. If $w_1 \in B(w)$, then there is nothing to prove. If not, then w_1 projects to τ. By induction, we may assume that there is a w', $w' \in W^Q$, $w' \in B(w_1)$, $w_1 > w' > \theta_1$, $l(w') = l(w_1) - 1$. In this case $l(w) = l(w') + 2$ and $w' \in W^Q$, and the result follows from the case $l(w) - l(w') = 2$ discussed above. □

THEOREM 3.7.7. *With notations as above,*
(1) *We have*

$$(*) \qquad (p_\tau, I(w)) = \bigcap_{w' \in B(w)} I(w') \cap (J_\tau + I(w))$$

(2) *Let $w = (a_1, \ldots, a_n) \in W^Q$. If $a_{i_1+1} < a_{i_1}$, then $J_\tau + I(w) \supset I(w')$ for some $w' \in B(w)$.*

(Here, J_τ is as in Definition 3.7.5)

PROOF. (1) If $w' < w$, then $I(w') \supset I(w)$. Since $p_\tau|_{X(w')} = 0$ if $w' \in B(w)$, $p_\tau \in I(w')$. Thus $(p_\tau, I(w)) \subseteq \bigcap_{w' \in B(w)} I(w') \cap \big(J_\tau + I(w)\big)$. To prove the reverse inclusion, let F be a monomial in the right-hand intersection of $(*)$. Since we may read $(*)$ modulo $I(w)$, we may assume F is standard on $X(w)$, and we are working in $R^{\text{def}}(w) = R/I(w)$. As $F \in \big(J_\tau + I(w)\big)/I(w)$, we may assume

$F = \sum_{i=1}^{m} p_{\alpha_i} F_i$, where $\alpha_i \in W^{i_1}$, $\tau \geqslant \alpha_i$, and $p_{\alpha_i} F_i$ is standard on $X(w)$. We wish to show that $\alpha_i = \tau$. Choose α_i minimal among $\{\alpha_1, \ldots, \alpha_m\}$, and rearrange them so that $\alpha_i = \alpha_1$. It suffices to show $\alpha_1 = \tau$. Now $p_{\alpha_1} F_1$ being standard on $X(w)$, it corresponds to a standard Young diagram λ on $X(w)$ with $\lambda_1 = \alpha_1$. If $\alpha_1 < \tau$, then Lemma 3.7.6 implies that λ remains standard on some $X(w')$ with $w' \in B(w)$. In particular, $p_{\alpha_1} F_1$ $(= F)$ is nonzero on $X(w')$. However by assumption, $F|_{X(w')} = 0$ for all such w'. From this it follows that $\alpha_1 = \tau$ which in turn implies the reverse inclusion.

We now prove (2). Let d be the largest integer $i_1 + 1 \leqslant d \leqslant i_2$ such that $a_d < \max\{a_1, \ldots, a_{i_1}\}$. Then there is a k with $a_{k-1} < a_d < a_k$ (possibly $k = 1$, in which case we set $a_{k-1} = 0$). Consider the element

$$w' = (a_1, \ldots, a_{k-1}, a_d, a_{k+1}, \ldots, a_{i_1}, \ldots, a_{d-1}, a_k, a_{d+1}, \ldots, a_n).$$

Note that $w \geqslant w'$, and $l(w) - 1 = l(w')$ (since if $k + 1 \leqslant r \leqslant i_1$, $a_r > a_k$ and $a_r > a_d$, and if $i_1 + 1 \leqslant r \leqslant d - 1$, then by choice $a_r < a_d < a_k$). Further, the projection of w' on W/W_{i_1} is strictly less than τ. Thus $w' \in B(w)$. We *claim* that $I(w) + J_\tau \supset I(w')$. Let m be a monomial of multidegree $(m_{i_1}, \ldots, m_{i_k})$ in $I(w')$. Let $m \notin I(w)$. We shall show that $m \in J_\tau$. First we observe that $m_{i_1} \neq 0$. For, if $m_{i_1} = 0$, then we set $Q' = P_{i_2} \cap \cdots \cap P_{i_k}$, and consider the projection $X(\theta)$ of $X(w')$ under $\pi : G/Q \to G/Q'$. Then m is not standard on $X(\theta)$, while m is standard on $X(w)$. But this is a contradiction, since $\pi(X(w)) = \pi(X(w')) = X(\theta)$ (and $m_{i_1} = 0$ by assumption). Thus we obtain $m_{i_1} \neq 0$. Let λ_1 be the first element in the Young diagram λ representing m. The fact that $m \notin I(w)$ implies that $\lambda_1 \leqslant \tau$. Hence $m \in J_\tau$, as asserted. This completes the proof of (2). □

4. Arithmetic Cohen-Macaulayness of multi-cones over Schubert varieties

In this section we will prove the following

THEOREM 4.0.1. *Let $G = SL(n)$. Let $Q = P_{i_1} \cap \cdots \cap P_{i_r}$ be any parabolic subgroup of G containing B, and let $w \in W^Q$. Let $\mathbf{a} = (a_1, \ldots, a_r)$, $a_i \in \mathbb{Z}^+$, and $R(w) = \oplus_{\mathbf{a}} H^0(X(w), L_{i_1}^{a_1} \otimes \cdots \otimes L_{i_r}^{a_r})$. Then $R(w)$ is Cohen-Macaulay.*

As we observed in §3.4.5, to prove the above theorem it suffices to prove

THEOREM 4.0.2. *Let $Q = P_{i_1} \cap \cdots \cap P_{i_r}$, and let $w \in W^Q$. Then the ring $R^{\mathrm{def}}(w)$ is Cohen-Macaulay.*

We will prove Theorem 4.0.2 by induction on the number r of maximal parabolics containing Q. We will also induct on $\dim R^{\mathrm{def}}(w)$; to descend to the Cohen-Macaulay property along the class of Schubert varieties we will use Theorem 3.7.7 and several elementary lemmas from homological algebra.

LEMMA 4.0.3. *Let $R = \bigoplus_{n \geqslant 0} R_n$ be a Noetherian graded ring with $R_0 = k$ and let I and J be two homogeneous ideals of R such that $\dim R/I = \dim R/J = d = \dim R/(I + J) + 1$. If R/I, R/J and $R/(I + J)$ are Cohen-Macaulay, then $R/I \cap J$ is Cohen-Macaulay of dimension d.*

This is just Lemma 6.2.5, Chapter 6 (where it has been proved).

REMARK 4.0.4. If R is as above and x is a homogeneous element of R which is not a zero-divisor, then R is Cohen-Macaulay if and only if R/xR is Cohen-Macaulay.

LEMMA 4.0.5. *Let* $Q = P_{i_1} \cap \cdots \cap P_{i_r}$, *and let* $w \in W^Q$. *Let* $w = (a_1, \ldots, a_n)$.

(1) *If* $w' \in B(w)$, *then* $\dim R^{\mathrm{def}}(w') = \dim R^{\mathrm{def}} - 1$.
(2) *If* $w_1, w_2 \in B(w)$ *and* w' *is the element as defined in Proposition 3.6.6, then*
$$\dim R^{\mathrm{def}}(w') = \dim R^{\mathrm{def}}(w) - 2.$$
(3) *If* $a_{i_1+1} < a_{i_1}$, *then*
$$\dim R^{\mathrm{def}}(w)/J_\tau = \dim R^{\mathrm{def}}(w) - 1,$$
where recall that $J_\tau = \langle p_\alpha \mid \alpha \in W^{i_1},\ \alpha \leqslant \tau \rangle$.

PROOF. To compute the dimension of $R^{\mathrm{def}}(w)$, we note that $\dim R^{\mathrm{def}}(w) = \dim R(w) = r + l(w)$.

(1) This is clear, since $l(w') = l(w) - 1$.

(2) As $w' \in W^Q$, and $l(w') = l(w) - 2$, (2) follows.

(3) Observe that $R/(J_\tau + I(w)) = R^{\mathrm{def}}(w)/J_\tau$ is isomorphic to $S^{\mathrm{def}}(w)$, where R is the deformed ring of G/Q and S is the deformed ring of G/Q', $Q' = P_{i_2} \cap \cdots \cap P_{i_r}$. Hence
$$\dim R^{\mathrm{def}}(w)/J = \dim S^{\mathrm{def}}(w) = l(w) + r - 1 = \dim R^{\mathrm{def}}(w) - 1.$$
□

To prove Theorem 4.0.2, we prove a more general result which will allow us to use induction.

THEOREM 4.0.6. *Let* $Q = P_{i_1} \cap \cdots \cap P_{i_r}$, *and* $w \in W^Q$. *Let* $w_1, \ldots, w_k \in B(w)$. *Let* $\tau = w^{(i_1)}$, *and let* $J = I(w) + J_\tau$. *Then*

(1) $R^{\mathrm{def}}(w)$ *is Cohen-Macaulay.*
(2) $R/I(w_1) \cap \cdots \cap I(w_k)$ *is Cohen-Macaulay.*
(3) *If* $w = (a_1, \ldots, a_n)$ *and* $a_{i_1} < a_{i_1+1}$, *then* $R/\big(I(w_1) \cap \cdots \cap I(w_k) \cap J\big)$ *is Cohen-Macaulay, where* R *is the deformed ring of* G/Q.

PROOF. We will prove all three statements by induction on r and on $\dim R^{\mathrm{def}}(w)$. By this induction we may assume all three statements hold for any $w' \in W^{Q'}$, $Q' = P_{i_2} \cap \cdots \cap P_{i_r}$.

We prove (2) first, by induction on k. If $k = 1$, $R/I(w_1) \equiv R^{\mathrm{def}}(w_1)$, and $\dim R^{\mathrm{def}}(w_1) < \dim R^{\mathrm{def}}(w)$. Hence by induction $R/I(w_1)$ $(= R^{\mathrm{def}}(w_1))$ is Cohen-Macaulay. Suppose $k > 1$. Let $K = I(w_2) \cap \cdots \cap I(w_k)$, and $d = \dim R^{\mathrm{def}}(w)$. By induction, R/K is Cohen-Macaulay of dimension $d - 1$. Also, $\dim R/I(w_1) = d - 1$, and $R/I(w_1)$ is Cohen-Macaulay. By Lemma 4.0.3, to show that $R/K \cap I(w_1)$ is Cohen-Macaulay it suffices to show that $R/\big(K + I(w_1)\big)$ is Cohen-Macaulay of dimension $d-2$. By Lemma 3.7.4,

$$K + I(w_1) = \big(I(w_1) + I(w_2)\big) \cap \cdots \cap \big(I(w_1) + I(w_k)\big).$$

Since each pair w_1, w_j is in $B(w)$, Proposition 3.6.6 shows that there is a unique $t_j \in W^Q$ such that $w_1 > t_j$, $w_j > t_j$, and $l(t_j) = l(w_1) - 1 = l(w_j) - 1$; further by Lemma 3.7.4, $I(w_1) + I(w_j) = I(t_j)$. Therefore

$$K + I(w_1) = I(t_2) \cap \cdots \cap I(t_r),$$

and $w_1 > t_j$, with $l(w_1) - 1 = l(t_j)$. If we can choose w_1 in such a way that each $t_j \in B(w_1)$, then by induction we may conclude that

$$R/K + I(w_1) = R/I(t_2) \cap \cdots \cap I(t_r)$$

is Cohen-Macaulay of dimension $d-2$. That such a w_1 can be chosen is proved in the following Lemma.

LEMMA 4.0.7. *Let $w \in W^Q$, $w_1, \ldots, w_k \in B(w)$. Then we may choose w_1 (after rearranging $w_1, \ldots, w_k$) in such a way that $t_j \in B(w_1)$, where the notations are as above.*

PROOF. Let $w = (a_1 \cdots a_n)$. Write $w_j = ws_j$ and let $\beta_j = \epsilon_{k_j} - \epsilon_{l_j}$ (here, β_j is the positive root associated to the reflection s_j). Since $w_j \in B(w)$, we have $k_j \leqslant i_1$ and $l_j > i_1$. Among the set $\{l_1, \ldots, l_k\}$ choose a maximal element. This maximal element, say l_1, is unique. For, if $l_1 = l_2$, then $\beta_1 = \epsilon_{k_1} - \epsilon_{l_1}$, $\beta_2 = \epsilon_{k_2} - \epsilon_{l_1}$. Since w covers w_1 and w_2 (i.e., $X(w_i)$ are Schubert divisors in $X(w)$, $i = 1, 2$), $a_{l_1} < a_{k_1}$ and $a_{l_1} < a_{k_2}$. Suppose (without loss of generality) that $k_1 < k_2$. Since $w \in W^Q$, we have, $a_{k_1} < a_{k_2}$. Then $a_{l_1} < a_{k_1} < a_{k_2}$, which implies $l(ws_2) < l(w) - 1$. Thus l_1 is unique. We *claim* that $t_m \in B(w_1)$, $m \geqslant 2$ for w_1 chosen in this manner. We shall prove the Claim for t_2, the proof being the same for any t_m. Write $\beta_1 = \epsilon_i - \epsilon_j$, $\beta_2 = \epsilon_k - \epsilon_l$. By assumption $j > l$.

Case 1. $i < k$. Since $j \neq l$, $s_1 s_2 = s_2 s_1$ and $t_2 = ws_1s_2 = w_1 s_2$. Thus $t_2 = (d_1, \ldots, d_n)$, where

$$d_r = \begin{cases} a_r & \text{if } r \neq i, j, k, l \\ a_j & \text{if } r = i \\ a_i & \text{if } r = j \\ a_l & \text{if } r = k \\ a_k & \text{if } r = l. \end{cases}$$

The projection of t_2 on W/W_{i_1} is $(d_1, \ldots, d_{i_1})$ and this is strictly less than the projection of w_1. For, if $w_1 = (b_1, \ldots, b_n)$, then

$$b_r = \begin{cases} a_r & \text{if } r \neq i, j \\ a_j & \text{if } r = i \\ a_i & \text{if } r = j. \end{cases}$$

Thus $d_s = b_s$ for $s \leqslant i_1$ unless $s = k$, in which case $a_l = d_k < b_k = a_k$.

Case 2. $i > k$. Again $s_1 s_2 = s_2 s_1$ and the same argument shows that t_2 has strictly less projection on W/W_{i_1}.

Thus for Case 1 and Case 2 any choice of w_1 would suffice.

Case 3. $i = k$. Recall that $j > l$. Then by Proposition 3.6.6, if $t_2 = (d_1, \ldots, d_n)$, then

$$d_r = \begin{cases} a_r & \text{if } r \neq i, j, l \\ a_l & \text{if } r = i \\ a_j & \text{if } r = j \\ a_i & \text{if } r = l. \end{cases}$$

Also, $a_l < a_j < a_i$. By choice of w_1, $w_1 = (b_1, \ldots, b_n)$, where

$$b_r = \begin{cases} a_r & \text{if } r \neq i, j \\ a_j & \text{if } r = i \\ a_i & \text{if } r = j. \end{cases}$$

Thus $d_r = b_r$ for $r \leqslant i_1$ unless $r = i$; in this case $a_l = d_i < b_i = a_j$, and so $t_2 \in B(w_1)$. This proves Lemma 4.0.7 and also finishes the proof of Theorem 4.0.6, part (2).

We now prove part (3) of Theorem 4.0.6. By part (2), if $K = I(w_1) \cap \cdots \cap I(w_k)$, then R/K is Cohen-Macaulay of dimension $d-1$. By Lemma 4.0.5, R/J has dimension $d-1$, and is Cohen-Macaulay, since $R/J = R'/I'(w)$, where $'$ means that we are working in G/Q' (recall that $Q' = P_{i_2} \cap \cdots \cap P_{i_k}$). Thus we apply our induction on the number of maximal parabolics containing Q. By Lemma 4.0.3, to prove that $R/K \cap J$ is Cohen-Macaulay it suffices to prove that $R/K + J$ is Cohen-Macaulay of dimension $d-1$. However

$$J + \big(I(w_1) \cap \cdots \cap I(w_k)\big) = \big(J + I(w_1)\big) \cap \cdots \cap \big(J + I(w_k)\big),$$

and

$$R/\big(J + I(w_l)\big) = R'/I'(w_l).$$

Since $a_{i_1} < a_{i_1+1}$, we have, $a_1 < \cdots < a_{i_1} < a_{i_1+1} < \cdots < a_{i_2}$, and hence $w \in W^{Q'}$. As each w_l has strictly smaller projection in G/P_{i_1} than does w, we see that if $w_l = ws$, $s = e_i - e_j$, then $i \leqslant i_1$ and $j > i_2$. For, if $j < i_2$, then $l(ws) \neq l(w) - 1$. When we apply s, we do not change the relative order of any of the segments $a_{i_m+1}, \ldots, a_{i_{m+1}}$. Thus $w_l \in W^{Q'}$ also. In addition, since $j > i_2$ and w_l has strictly smaller projection on W/W_{i_1} than w, it easily follows that w_l has strictly smaller projection on W/W_{i_2} than does w. Thus $w_l \in B'(w)$, $1 \leqslant l \leqslant k$. The induction hypothesis now implies that $R'/I'(w_1) \cap \cdots \cap I'(w_k) = R/(J + K)$ is Cohen-Macaulay of dimension $d-2$. This proves part (3).

Finally, we prove part (1). By Theorem 3.7.7,

$$(I(w), p_\tau) = \begin{cases} \bigcap_{w' \in B(w)} I(w'), & \text{if } a_{i_1} > a_{i_1+1} \\ \bigcap_{w' \in B(w)} I(w') \cap J, & \text{if } a_{i_1} < a_{i_1+1} \end{cases}$$

In either case, the proof of parts (2) and (3) show that $R/(I(w), p_\tau)$ is Cohen-Macaulay. If p_τ is not a zero-divisor in $R/I(w)$, then clearly $R/I(w) = R^{\mathrm{def}}(w)$ is Cohen-Macaulay. However it is clear that since w projects to τ, if F is any standard monomial on $X(w)$, with Young diagram λ, then $p_\tau F$ is also standard with diagram (τ, λ). This shows that p_τ is not a zero-divisor in $R^{\mathrm{def}}(w)$. □

□

COROLLARY 4.0.8. *Let $P_1, \ldots, P_l$ (where $l =$ rank(G)) be any ordering of the maximal parabolics, $Q_i = P_i \cap \cdots \cap P_l$ and $w \in W^{Q_i}$. Then $R(w)$ is normal.*

PROOF. Serre's criterion for normality (cf. [**148**]) implies that if a variety X is Cohen-Macaulay, and is regular in codimension 1 (i.e., $\operatorname{codim} \operatorname{Sing} X \geqslant 2$), then X is normal. For the case of $X = X(w)$, Chevalley (cf. [**37**]) has shown that X is regular in codimension 1. This, together with Theorem 4.0.1 implies that $X(w)$ is normal. □

CHAPTER 11

Degenerations of Schubert varieties to toric varieties

In this chapter we construct Gröbner bases for Schubert varieties in $SL(n)/B$. We further show that all Schubert varieties in the Grassmannian as well as a certain class of Schubert varieties which includes $SL(n)/B$ (namely, the class of Kempf varieties) in $SL(n)/B$ degenerate to toric varieties. These results are proved using Standard monomial bases for Schubert varieties in $SL(n)/B$. To be more precise, using the standard monomial bases, we construct Gröbner bases for Schubert varieties. We then use the Gröbner bases and construct a flat 1-parameter family whose general fiber is the cone over a Schubert variety and special fiber is a toric variety. After reviewing the generalities on Gröbner bases in §1, we construct Gröbner bases for Schubert varieties in §2. We then review the generalities on distributive lattices and toric varieties in §3, §4 respectively. We then prove some results relating distributive lattices and flat deformations in §5, §6. We then carry out the degenerations of Schubert varieties to toric varieties in §7-9.

1. Generalities on Gröbner bases

Let k be a field, and consider the ring $k[x_1, \ldots, x_n]$ of polynomials in n variables $x_1, \ldots, x_n$. We recall below some generalities on Gröbner bases; for more details one may refer to **[38]**, **[49]**.

DEFINITION 1.0.1. A total order $\preceq$ on the set of monomials in $k[x_1, \ldots, x_n]$ is called a *monomial order* if for given monomials $\underline{m}, \underline{m}_1, \underline{m}_2$, with $\underline{m} \neq 1$, $\underline{m}_1 \prec \underline{m}_2$ implies $\underline{m}_1 \prec \underline{m} \cdot \underline{m}_1 \prec \underline{m} \cdot \underline{m}_2$.

For the rest of this section, a fixed monomial order $\preceq$ is considered.

If f is a nonzero polynomial in $k[x_1, \ldots, x_n]$, then the greatest monomial (with respect to $\preceq$) occurring in f is called the *initial monomial of* f, and we denote it by $\text{in}(f)$; the coefficient of $\text{in}(f)$ is called the initial coefficient of f. For a family of polynomials $\mathcal{F} \subset k[x_1, \ldots, x_n]$, the ideal generated by its elements will be denoted by $\langle \mathcal{F} \rangle$, and the set of the initial monomials of all polynomials in $\mathcal{F}$ will be denoted by $\text{in}(\mathcal{F})$.

DEFINITION 1.0.2. Let $I \subset k[x_1, \ldots, x_n]$ be an ideal. A finite set of polynomials $\mathcal{F} \subset I$ is called a *Gröbner basis for* I *with respect to* $\preceq$ if $\langle \text{in}(\mathcal{F}) \rangle = \langle \text{in}(I) \rangle$.

DEFINITION 1.0.3. A *minimal Gröbner basis for I with respect to $\preceq$* is a Gröbner basis $\mathcal{F}$ for I with respect to $\preceq$ such that the initial coefficients of the elements in $\mathcal{F}$ are all 1, and for any $f \in \mathcal{F}$, $\mathrm{in}(f) \notin \langle \mathrm{in}(\mathcal{F} \setminus \{f\})\rangle$.

DEFINITION 1.0.4. A *reduced Gröbner basis for I with respect to $\preceq$* is a Gröbner basis $\mathcal{F}$ for I with respect to $\preceq$ such that the initial coefficients of the elements in $\mathcal{F}$ are all 1, and for any $f \in \mathcal{F}$, none of the monomials present in f lies in $\langle \mathrm{in}(\mathcal{F} \setminus \{f\})\rangle$.

PROPOSITION 1.0.5. *Any Gröbner basis for I generates I as an ideal.*

In the case when I is the defining ideal of an algebraic variety X, a Gröbner basis for I will be also called a *Gröbner basis for X*.

PROPOSITION 1.0.6. *A nonzero ideal $I \subset k[x_1, \dots, x_n]$ has a unique reduced Gröbner basis (with respect to a given monomial order).*

1.1. Lexicographic order

Assume that the variables $x_1, \dots, x_n$ are totally ordered as follows: $x_1 < \cdots < x_n$. A monomial $\underline{m}$ of degree r in the polynomial ring $k[x_1, \dots, x_n]$ will be written in the form $\underline{m} = x_{i_1} \dots x_{i_r}$, with $1 \leqslant i_1 \leqslant \dots \leqslant i_r \leqslant n$. The *lexicographic order* on the set of monomials $\underline{m} \in k[x_1, \dots, x_n]$ is denoted by $\preceq_{lex}$, and defined as follows: $x_{i_1} \dots x_{i_r} \prec_{lex} x_{j_1} \dots x_{j_s}$ if and only if either $r < s$, or $r = s$ and there exists an $l < r$ such that $i_1 = j_1, \dots, i_l = j_l, i_{l+1} < j_{l+1}$. It is easy to check that $\preceq_{lex}$ is a monomial order.

2. Gröbner bases for unions of Schubert varieties

Let Q be a parabolic subgroup, say $Q = P_{k_1} \cap \cdots \cap P_{k_r}$, P_{k_i}, $1 \leqslant i \leqslant r$ being the maximal parabolic subgroups containing Q. Let $X(w)$ be a Schubert variety in G/Q. For $\mathbf{a} = (a_1, \dots, a_r) \in \mathbb{Z}_+^r$, let $|\mathbf{a}| = a_1 + \cdots + a_r$ and $R(w)_{\mathbf{a}} = H^0(X(w), L^{\mathbf{a}})$, where $L^{\mathbf{a}} = L_{k_1}^{a_1} \otimes \cdots \otimes L_{k_r}^{a_r}$, L_{k_i} being the ample generator of $\mathrm{Pic}(G/P_{k_i})$. Now define

$$R(w) = \bigoplus_{\mathbf{a} \in \mathbb{Z}_+^r} H^0(X(w), L^{\mathbf{a}}).$$

For $X(w) = G/Q$, we shall denote $R(w)$(resp. $R(w)_{\mathbf{a}}$) by just R (resp. $R_{\mathbf{a}}$). Recall the following (cf. Chapter 3, §9):

THEOREM 2.0.1. (1) *For each $\mathbf{a} \in \mathbb{Z}_+{}^r$ the canonical map*

$$\theta_{\mathbf{a}} : \bigotimes_{i=1}^{r} S^{a_i}(H^0(G/Q, L_{k_i})) \longrightarrow R_{\mathbf{a}}$$

is surjective. Let $I_{\mathbf{a}}$ be its kernel.

(2) *Let I be the kernel of the canonical map*

$$\theta : \bigoplus_{\mathbf{a} \in \mathbb{Z}_+{}^r} \bigotimes_{i=1}^{r} S^{a_i}(H^0(G/Q, L_{k_i})) \longrightarrow R,$$

Then I is multigraded, generated by $I_2 = \bigcup_{|\mathbf{a}|=2} I_{\mathbf{a}}$.

If $\underline{n}$ is a nonstandard monomial of degree $\mathbf{a}$, then, by Theorem 4.3.1, Chapter 7, $\underline{n}$ can be written in a unique way as a linear combination of standard monomials of degree $\mathbf{a}$, modulo the ideal I:

$$(*) \qquad \underline{n} = \sum_{i=1}^{t} c_i \underline{s}_i \pmod{I}, \qquad c_i \in k^*.$$

We refer to (*) as a *straightening relation*. Denote

$$(\dagger) \qquad f_{\underline{n}} = \underline{n} - \sum_{i=1}^{t} c_i \underline{s}_i,$$

$$\mathcal{F}_{\mathbf{a}} = \{f_{\underline{n}} \mid \underline{n} \text{ is a nonstandard monomial of degree } \mathbf{a}\},$$

$$\mathcal{F} = \bigcup_{\mathbf{a}} \mathcal{F}_{\mathbf{a}}, \qquad \mathcal{F}_2 = \bigcup_{|\mathbf{a}|=2} \mathcal{F}_{\mathbf{a}}.$$

Clearly, $\mathcal{F}_{\mathbf{a}} \subset I_{\mathbf{a}}$ and $\mathcal{F} \subset I$.

2.1. Let $H = \bigcup_{i=1}^{r} W^{k_i}$ We define a partial order $\succeq$ on the H as follows: Given $\tau_1 = (a_1, \dots, a_r)$, $\tau_2 = (b_1, \dots, b_s)$, where $r, s \in \{k_1, \dots, k_r\}$,

$$\tau_1 \succeq \tau_2 \iff r \leqslant s \text{ and } a_t \geqslant b_t \text{ for } 1 \leqslant t \leqslant r.$$

It is easily seen that $(H, \succeq)$ is a distributive lattice (see §3 below for the definition of a distributive lattice).

We now extend $\succeq$ to a total order on H, also denoted by $\succeq$. This induces a total order on the set $\{p_\tau \mid \tau \in H\}$: $p_{\tau_1} \prec p_{\tau_2} \iff \tau_1 \succ \tau_2$ (we have taken the order this way for a specific purpose). Thus, according to our convention, a monomial of degree r in the polynomial ring $k[p_\tau, \tau \in H]$ will be written in the form $p_{\tau_1} \cdots p_{\tau_r}$, with $\tau_1 \succeq \cdots \succeq \tau_r$. We take the lexicographic order on monomials induced by the above total order.

In view of Proposition 3.2.1, Chapter 10, we have

THEOREM 2.1.1. *Let*

$$f_{\underline{n}} = \underline{n} - \sum_{i=1}^{t} c_i \underline{s}_i, \qquad c_i \in k^*$$

as in (†), be a typical element in $\mathcal{F}$. Then $\mathrm{in}(f_{\underline{n}}) = \underline{n}$, *i.e.* $\underline{s}_i \prec_{lex} \underline{n}$, *for all* i.

THEOREM 2.1.2. *We have* $\mathrm{in}(\mathcal{F}) = \mathrm{in}(I)$.

PROOF. Since $\mathrm{in}(\mathcal{F}) \subset \mathrm{in}(I)$, and $\mathrm{in}(\mathcal{F})$ consists of all the nonstandard monomials, it is enough to prove that the leading monomial of any element $f \in I$ is nonstandard. Assume this is not true, and let f be an element of I such that $\mathrm{in}(f)$ is a standard monomial. Let $\underline{s}_0, \underline{s}_1, \ldots, \underline{s}_t$ be all the standard monomials, including $\mathrm{in}(f) = \underline{s}_0$, and $\underline{n}_1, \ldots, \underline{n}_l$ all the nonstandard monomials appearing in f, so that f is written as

$$f = a_0\underline{s}_0 + \sum_{i=1}^{t} a_i\underline{s}_i + \sum_{j=1}^{l} b_j\underline{n}_j, \quad a_0, a_i, b_j \in k^*.$$

Consider the polynomial $f' = f - \sum_{j=1}^{l} b_j f_{\underline{n}_j}$ ($f_{\underline{n}_j}$ being as in (†)). Then $\mathrm{in}(f') = \underline{s}_0$, since $\mathrm{in}(f) = \underline{s}_0$, $\mathrm{in}(f_{\underline{n}_j}) = \underline{n}_j$, and $\underline{n}_j \prec_{lex} \underline{s}_0$, for all j. Therefore f' is a nontrivial linear combination of standard monomials, since the coefficient of $\underline{s}_0$ in its expression is $a_0 \neq 0$. On the other hand, $f' \equiv 0 \pmod{I}$, contradicting the linear independence of standard monomials. Hence our assumption is wrong, and $\mathrm{in}(f)$ is nonstandard. □

THEOREM 2.1.3. *$\mathcal{F}_2$ is the reduced Gröbner basis for I with respect to the lexicographic order.*

PROOF. We have to show that $\langle \mathrm{in}(\mathcal{F}_2)\rangle = \langle \mathrm{in}(I)\rangle$. In view of Theorem 2.1.2, in order to show that $\mathcal{F}_2$ is a Gröbner basis, it is enough to show that $\langle \mathrm{in}(\mathcal{F}_2)\rangle = \langle \mathrm{in}(\mathcal{F})\rangle$. Since $\langle \mathrm{in}(\mathcal{F}_2)\rangle \subset \langle \mathrm{in}(\mathcal{F})\rangle$, it suffices to show that for any nonstandard monomial $\underline{n}$, $\mathrm{in}(f_{\underline{n}}) \in \langle \mathrm{in}(\mathcal{F}_2)\rangle$, i.e. $\underline{n} \in \langle \mathrm{in}(\mathcal{F}_2)\rangle$.

Let $\underline{n} = p_{\tau_1} \cdots p_{\tau_r}$ be nonstandard of degree r. Then there exists an i such that $p_{\tau_i}p_{\tau_{i+1}}$ is nonstandard. Since $\underline{n} \in \langle p_{\tau_i}p_{\tau_{i+1}}\rangle$ and $p_{\tau_i}p_{\tau_{i+1}} = \mathrm{in}(f_{p_{\tau_i}p_{\tau_{i+1}}})$, we conclude that $\underline{n} \in \langle \mathrm{in}(\mathcal{F}_2)\rangle$. Hence $\mathcal{F}_2$ is a Gröbner basis for I. The fact that $\mathcal{F}_2$ is reduced can be easily seen from the form of its elements. □

2.2. Let $w \in W^Q$, and $X(w)$ the Schubert variety in G/Q corresponding to w. Under $\pi_i : G/Q \to G/P_{k_i}$, let $\pi_i(X(w)) = X(w^{(i)})$, where $w^{(i)} \in W^{k_i}$. For $\tau \in W^{k_i}$, $1 \leqslant i \leqslant r$, we have $p_\tau|_{X(w^{(i)})} \neq 0$ if and only if $w^{(i)} \geqslant \tau$.

We have (cf. Chapter 3, §9):

THEOREM 2.2.1. (1) *The restriction maps $H^0(G/Q, L^{\mathbf{a}}) \to H^0(X(w), L^{\mathbf{a}})$, $\mathbf{a} \in \mathbb{Z}_+^r$, are surjective.*

(2) *The kernel $\mathfrak{a}(w)$ of the epimorphism $R \to R(w)$ is multigraded say $\mathfrak{a}(w) = \bigoplus_{\mathbf{a}} \mathfrak{a}(w)_{\mathbf{a}}$, and generated by $\{p_\tau \mid \tau \in W^{k_i}, 1 \leqslant i \leqslant r, w^{(i)} \not\geqslant \tau\}$, i.e. $\mathfrak{a}(w)$ is generated by $\bigoplus_{|\mathbf{a}|=1} \mathfrak{a}(w)_{\mathbf{a}}$.*

Denote $H_w = \{\tau \in W^{k_i} \mid 1 \leqslant i \leqslant r,\ w^{(i)} \geqslant \tau\}$, and $A(w) = k[p_\tau, \tau \in H_w]$. Consider the canonical epimorphism

$$A \to A(w), \qquad \Big(A \text{ being } \bigoplus_{\mathbf{a}\in\mathbb{Z}_+{}^r} \bigotimes_{i=1}^{r} S^{a_i}(H^0(G/Q, L_{k_i}))\Big),$$

and denote its kernel by $J(w)$; then $J(w)$ is generated by $\{p_\tau, \tau \in H \setminus H_w\}$. By Theorem 2.2.1 we obtain an epimorphism $A(w) \to R(w)$, whose kernel is $I + J(w) \pmod{J(w)}$, where, recall that I is the kernel of the epimorphism $A \to R$, R

being $\bigoplus_{\mathbf{a}\in\mathbb{Z}_+^r} H^0(G/Q, L^{\mathbf{a}})$. We shall denote this kernel by $I(w)$; thus $R(w) = A(w)/I(w)$.

For an element $f \in A(w)$, we shall denote its image in $R(w)$ by f^w. For a monomial $\underline{n}$ which is nonstandard on $X(w)$, let $\underline{n}^w = \sum c_i \underline{s}_i^w$ be the expression for $\underline{n}^w$ as a sum of standard monomials on $X(w)$. Let us denote $f_{\underline{n}^w} = \underline{n}^w - \sum c_i \underline{s}_i^w \in A(w)$, and set

$$\mathcal{F}_{\mathbf{a}}^w = \{f_{\underline{n}^w} \mid \underline{n}^w \text{ is a nonstandard monomial on } X(w) \text{ of degree } \mathbf{a}\},$$

$$\mathcal{F}^w = \bigcup_{\mathbf{a}} \mathcal{F}_{\mathbf{a}}^w, \qquad \mathcal{F}_2^w = \bigcup_{|\mathbf{a}|=2} \mathcal{F}_{\mathbf{a}}^w.$$

Clearly, $\mathcal{F}_{\mathbf{a}}^w \subset I(w)_{\mathbf{a}}$ and $\mathcal{F}^w \subset I(w)$.

In view of Theorems 2.0.1, part (2), and 2.2.1, part (2), we have

THEOREM 2.2.2. *The multigraded ideal $I(w)$ is generated by $\mathcal{F}_2^w$.*

Using Proposition 3.2.1, we obtain:

THEOREM 2.2.3. *Let*

$$f_{\underline{n}^w} = \underline{n}^w - \sum_{i=1}^{t} c_i \underline{s}_i^w, \qquad c_i \in k^*$$

be a typical element in $\mathcal{F}_{\mathbf{a}}^w$, where $\underline{n}^w$ is a nonstandard monomial and $\underline{s}_i^w$ are standard monomials on $X(w)$, all of degree $\mathbf{a}$. Then $\operatorname{in}(f_{\underline{n}}^w) = \underline{n}$, *i.e. $\underline{s}_i^w \prec_{lex} \underline{n}^w$, for all i.*

Proceeding as in Theorem 2.1.2, we obtain:

THEOREM 2.2.4. *We have* $\operatorname{in}(\mathcal{F}^w) = \operatorname{in}(I(w))$.

THEOREM 2.2.5. *$\mathcal{F}_2^w$ is the reduced Gröbner basis for $I(w)$ with respect to the lexicographic order.*

PROOF. We need to show that $\langle \operatorname{in}(\mathcal{F}_2^w\rangle = \langle \operatorname{in}(I(w))\rangle$. In view of Theorem 2.2.4, in order to prove that $\mathcal{F}_2^w$ is a Gröbner basis, it is enough to prove that $\langle \operatorname{in}(\mathcal{F}_2^w)\rangle = \langle \operatorname{in}(\mathcal{F}^w)\rangle$. Since $\langle \operatorname{in}(\mathcal{F}_2^w)\rangle \subset \langle \operatorname{in}(\mathcal{F}^w)\rangle$, it suffices to show that for any nonstandard monomial $\underline{n}^w$, $\operatorname{in}(f_{\underline{n}^w}) \in \langle \operatorname{in}(\mathcal{F}_2^w)\rangle$, i.e. $\underline{n}^w \in \langle \operatorname{in}(\mathcal{F}_2^w)\rangle$. Let $\underline{n}^w = x_{\tau_1}^w \cdots x_{\tau_r}^w$, where $x_{\tau_i}^w$ is $p_{\tau_i}|_{X(w)}$. Then, in view of Theorem 2.2.1, there exists a pair (i, j) such that (τ_i, τ_j) is not standard on $X(w)$. Hence $\underline{n}^w \in \langle x_{\tau_1}^w x_{\tau_j}^w\rangle$; now, $x_{\tau_i}^w x_{\tau_j}^w = \operatorname{in}(f_{p_{\tau_i}p_{\tau_j}}^w)$ (note that $f_{p_{\tau_i}p_{\tau_j}}^w = p_{\tau_i}p_{\tau_j} - \sum c_{\alpha\beta}p_\alpha p_\beta$, where $p_{\tau_i}p_{\tau_j} = \sum c_{\alpha\beta}p_\alpha p_\beta$ is the straightening relation for $p_{\tau_i}p_{\tau_j}$ on $X(w)$). Thus we obtain $\underline{n}^w \in \langle \operatorname{in}(\mathcal{F}_2^w)\rangle$. □

Let $Z = \bigcup_{i=1}^r X(w_i)$ be a union of Schubert varieties in G/Q. Let $H_Z = \bigcup_{i=1}^r H_{w_i}$ and $A_Z = k[p_\tau, \tau \in H_Z]$. Consider the canonical epimorphism $A \to A_Z$ and denote its kernel by J_Z; then J_Z is generated by $\{p_\tau \mid \tau \in H \setminus H_Z\}$. Using the surjective map $R \to R_Z = \bigoplus_{\mathbf{a}} H^0(Z, L^{\mathbf{a}})$, we obtain an epimorphism $A_Z \to R_Z$ whose kernel is $I + J_Z$, which we denote by I_Z (as above, $R = \bigoplus_{\mathbf{a}\in\mathbb{Z}_+^r} H^0(G/Q, L^{\mathbf{a}})$). For an element $f \in A_Z$, we shall denote its image in R_Z by f^Z. For a monomial $\underline{n}$ which is nonstandard on Z, let $\underline{n}^Z = \sum c_i \underline{s}_i^Z$

be the expression for $\underline{n}^Z$ as a sum of standard monomials on Z. Let us denote $f_{\underline{n}^Z} = \underline{n}^Z - \sum c_i \underline{s}_i^Z \in A_Z$, and set

$$\mathcal{F}_{\mathbf{a}}^Z = \{f_{\underline{n}^Z} \mid \underline{n}^Z \text{ is a nonstandard monomial on } Z \text{ of degree } \mathbf{a}\},$$

$$\mathcal{F}^Z = \bigcup_{\mathbf{a}} \mathcal{F}_{\mathbf{a}}^Z, \qquad \mathcal{F}_2^Z = \bigcup_{|\mathbf{a}|=2} \mathcal{F}_{\mathbf{a}}^Z.$$

Noting that Theorem 2.2.1 holds when $X(w)$ is replaced by Z (cf. **[137]**, **[85]**), and proceeding as above, we obtain:

THEOREM 2.2.6. (1) $\text{in}(\mathcal{F}^Z) = \text{in}(I_Z)$.
(2) $\mathcal{F}_2^Z$ *is the reduced Gröbner basis for* I_Z *with respect to the lexicographic order.*

3. Generalities on distributive lattices

DEFINITION 3.0.1. A *lattice* is a partially ordered set $(\mathcal{L}, \leqslant)$ such that, for every pair of elements $x, y \in \mathcal{L}$, there exist elements $x \vee y$ and $x \wedge y$, called the *join*, respectively the *meet* of x and y, defined by:

$$x \vee y \geqslant x,\ x \vee y \geqslant y, \text{ and if } z \geqslant x \text{ and } z \geqslant y, \text{ then } z \geqslant x \vee y,$$
$$x \wedge y \leqslant x,\ x \wedge y \leqslant y, \text{ and if } z \leqslant x \text{ and } z \leqslant y, \text{ then } z \leqslant x \wedge y.$$

It is easy to check that the operations $\vee$ and $\wedge$ are commutative and associative.

DEFINITION 3.0.2. An element $z \in \mathcal{L}$ is called the *zero* of $\mathcal{L}$, denoted by 0, if $z \leqslant x$ for all x in $\mathcal{L}$. An element $z \in \mathcal{L}$ is called the *one* of $\mathcal{L}$, denoted by 1, if $z \geqslant x$ for all x in $\mathcal{L}$.

DEFINITION 3.0.3. Given a lattice $\mathcal{L}$, a subset $\mathcal{L}' \subset \mathcal{L}$ is called a *sublattice* of $\mathcal{L}$ if $x, y \in \mathcal{L}'$ implies $x \wedge y \in \mathcal{L}'$, $x \vee y \in \mathcal{L}'$.

DEFINITION 3.0.4. Two lattices $\mathcal{L}_1$ and $\mathcal{L}_2$ are *isomorphic* if there exists a bijection $\varphi : \mathcal{L}_1 \to \mathcal{L}_2$ such that, for all $x, y \in \mathcal{L}_1$,

$$\varphi(x \vee y) = \varphi(x) \vee \varphi(y) \text{ and } \varphi(x \wedge y) = \varphi(x) \wedge \varphi(y).$$

DEFINITION 3.0.5. A lattice is called *distributive* if the following identities hold:

$$x \wedge (y \vee z) = (x \wedge y) \vee (x \wedge z)$$
$$x \vee (y \wedge z) = (x \vee y) \wedge (x \vee z).$$

EXAMPLE 3.0.6. The lattice of all subsets of the set $\{1, 2, \ldots, n\}$ is denoted by $\mathcal{B}(n)$, and called the *Boolean algebra of rank* n.

One has the following (see **[5]**):

THEOREM 3.0.7. *Any finite distributive lattice is isomorphic to a sublattice of a Boolean algebra of finite rank, and hence, in particular, to a sublattice of a finite chain product.*

3.1. An important example

Given an integer $n \geqslant 1$, $\mathcal{C}(n)$ will denote the chain $\{1 < \cdots < n\}$, and for $n_1, \ldots, n_d > 1$, $\mathcal{C}(n_1, \ldots, n_d)$ will denote the chain product lattice $\mathcal{C}(n_1) \times \cdots \times \mathcal{C}(n_d)$ consisting of all d-tuples $(i_1, \ldots, i_d)$, with $1 \leqslant i_1 \leqslant n_1, \ldots, 1 \leqslant i_d \leqslant n_d$. For $(i_1, \ldots, i_d)$, $(j_1, \ldots, j_d)$ in $\mathcal{C}(n_1, \ldots, n_d)$, we define

$$(i_1, \ldots, i_d) \leqslant (j_1, \ldots, j_d) \iff i_1 \leqslant j_1, \ldots, i_d \leqslant j_d \,.$$

We have

$$(i_1, \ldots, i_d) \vee (j_1, \ldots, j_d) = (\max\{i_1, j_1\}, \ldots, \max\{i_d, j_d\})$$
$$(i_1, \ldots, i_d) \wedge (j_1, \ldots, j_d) = (\min\{i_1, j_1\}, \ldots, \min\{i_d, j_d\}).$$

$\mathcal{C}(n_1, \ldots, n_d)$ is a finite distributive lattice, and its zero and one are $(1, \ldots, 1)$, $(n_1, \ldots, n_d)$ respectively.

Note that there is a total order $\lhd$ on $\mathcal{C}(n_1, \ldots, n_d)$ extending $<$, namely the lexicographic order, defined by $(i_1, \ldots, i_d) \lhd (j_1, \ldots, j_d)$ if and only if there exists $l < d$ such that $i_1 = j_1, \ldots, i_l = j_l, i_{l+1} < j_{l+1}$. Also note that two elements $(i_1, \ldots, i_d) \lhd (j_1, \ldots, j_d)$ are non-comparable with respect to $\leqslant$ if and only if there exists $1 < h \leqslant d$ such that $i_h > j_h$.

Sometimes we denote the elements of $\mathcal{C}(n_1, n_2, \ldots, n_d)$ by $x_{i_1 \ldots i_d}$, with $1 \leqslant i_1 \leqslant n_1, \ldots, 1 \leqslant i_d \leqslant n_d$.

4. Generalities on toric varieties

DEFINITION 4.0.1. (cf. [**53**], [**84**]) An *equivariant affine embedding* of a torus T is an affine variety X containing T as an open subset and equipped with a T-action $T \times X \to X$ extending the action $T \times T \to T$ given by multiplication. If in addition X is normal, then X is called an *affine toric variety*.

For our purpose, we also need the definition of a toric ideal (cf. [**156**])

Let $T = (k^*)^m$ be the m-dimensional torus. Let M be the character group (=$\mathrm{Hom}_{\mathrm{alg.gp.}}(T, \mathbb{G}_m)$) of T. Then M can be identified with $\mathbb{Z}^m$. Let $\mathcal{A} = \{\mathbf{a}_1, \ldots, \mathbf{a}_n\}$ be a subset of $\mathbb{Z}^m$. Consider the map

$$\pi_{\mathcal{A}} : \mathbb{Z}_+^n \to \mathbb{Z}^m, \quad \mathbf{u} = (u_1, \ldots, u_n) \mapsto u_1\mathbf{a}_1 + \cdots + u_n\mathbf{a}_n.$$

Let $k[\mathbf{x}] = k[x_1, \ldots, x_n]$, $k[\mathbf{t}^{\pm 1}] = k[t_1, \ldots, t_m, t_1^{-1}, \ldots, t_m^{-1}]$.

The map $\pi_{\mathcal{A}}$ induces a homomorphism of semigroup algebras

$$\widehat{\pi_{\mathcal{A}}} : k[\mathbf{x}] \longrightarrow k[\mathbf{t}^{\pm 1}],$$
$$x_i \longmapsto \mathbf{t}^{\mathbf{a}_i}.$$

DEFINITION 4.0.2. The kernel of $\widehat{\pi}_{\mathcal{A}}$ is denoted by $I_{\mathcal{A}}$ and called the *toric ideal* associated to $\mathcal{A}$. A variety of the form $\mathcal{V}(I_{\mathcal{A}})$, the affine variety of the zeroes in k^n of $I_{\mathcal{A}}$, is called an *affine toric variety*.

Note that a toric ideal is prime.

REMARK 4.0.3. Consider the action of T on k^n given by $\mathbf{t}e_i = \mathbf{t}^{\mathbf{a}_i}e_i$ (here, e_i, $1 \leq i \leq n$ are the standard basis vectors of k^n). Then $\mathcal{V}(I_{\mathcal{A}})$ is simply the Zariski closure of the T-orbit through $(1, 1, \ldots, 1)$. In particular, $\mathcal{V}(I_{\mathcal{A}})$ is an equivariant affine embedding of $T = (k^*)^m$.

REMARK 4.0.4. In the above definition, we do not require $\mathcal{V}(I_{\mathcal{A}})$ to be normal. Using [**84**], we have that $\mathcal{V}(I_{\mathcal{A}})$ is normal if and only if the semi subgroup S of M generated by $\mathcal{A}$ is saturated (here, S is said to be saturated, if for $\mathbf{a} \in M, r\mathbf{a} \in S \implies \mathbf{a} \in S$ (where r is an integer > 1) (cf. [**84**])).

Recall the following (see [**156**]).

PROPOSITION 4.0.5. *The toric ideal $I_{\mathcal{A}}$ is spanned as a k-vector space by the set of binomials*

$$(*) \qquad \{\mathbf{x}^{\mathbf{u}} - \mathbf{x}^{\mathbf{v}} \mid \mathbf{u}, \mathbf{v} \in \mathbb{Z}_+^n \text{ with } \pi_{\mathcal{A}}(\mathbf{u}) = \pi_{\mathcal{A}}(\mathbf{v})\}.$$

(Here, a *binomial* is a polynomial with at most two terms.)

4.1. An example

Let us fix the integers $n_1, \ldots, n_d > 1$, and let $n = \prod_{i=1}^d n_i$, $m = \sum_{i=1}^d n_i$. Let $\mathbf{e}_1^l, \ldots, \mathbf{e}_{n_l}^l$ be the unit vectors in $\mathbb{Z}^{n_l}$, for $1 \leqslant l \leqslant d$.

For $1 \leqslant \xi_1 \leqslant n_1, \ldots, 1 \leqslant \xi_d \leqslant n_d$, define

$$\mathbf{a}_{\xi_1 \ldots \xi_d} = \mathbf{e}_{\xi_1}^1 + \cdots + \mathbf{e}_{\xi_d}^d \in \mathbb{Z}^{n_1} \oplus \cdots \oplus \mathbb{Z}^{n_d}$$

and let

$$\mathcal{A}_{n_1, \ldots, n_d} = \{\mathbf{a}_{\xi_1 \ldots \xi_d} \mid 1 \leqslant \xi_1 \leqslant n_1, \ldots, 1 \leqslant \xi_d \leqslant n_d\}.$$

The corresponding map

$$\pi_{\mathcal{A}} : \mathbb{Z}_+^{n_1 \cdot \cdots \cdot n_d} \longrightarrow \mathbb{Z}^{n_1 + \cdots + n_d}$$

is defined as follows: for $1 \leqslant l \leqslant d$ and $1 \leqslant i_l \leqslant n_l$ fixed, the $(n_1 + \cdots + n_{l-1} + i_l)$-th coordinate of $\pi_{\mathcal{A}}(\mathbf{u})$ is given by $\sum u_{\xi_1 \ldots \xi_{l-1} \xi_l \xi_{l+1} \ldots \xi_d}$, the sum being taken over the elements $(\xi_1, \ldots, \xi_{l-1}, \xi_l, \xi_{l+1}, \ldots, \xi_d)$ of $\mathcal{C}(n_1, \ldots, n_d)$ with $\xi_l = i_l$. We call this subset the l-th *slice* of $\mathcal{C}(n_1, \ldots, n_d)$ defined by i_l, and denote it by $\{\xi_l = i_l\}$. The components (or *entries*) of an element $\mathbf{u} \in \mathbb{Z}^{n_1 \cdot \cdots \cdot n_d}$ are indexed by the elements $(i_1, \ldots, i_d)$ of $\mathcal{C}(n_1, \ldots, n_d)$. If $(j_1, \ldots, j_d) \in \{\xi_l = i_l\}$, sometimes we also say that $u_{j_1 \ldots j_d}$ itself belongs to the slice $\{\xi_l = i_l\}$.

The map $\pi_{\mathcal{A}}$ induces the map

$$\widehat{\pi_{\mathcal{A}}} : k[x_{11 \ldots 1}, \ldots, x_{\xi_1 \xi_2 \ldots \xi_d}, \ldots, x_{n_1 n_2 \ldots n_d}] \to k[t_{11}, \ldots, t_{1n_1}, \ldots, t_{d1}, \ldots, t_{dn_d}]$$

given by

$$x_{\xi_1 \ldots \xi_d} \longmapsto t_{1\xi_1} \cdots t_{d\xi_d}, \qquad \text{for } 1 \leqslant \xi_1 \leqslant n_1, \ldots, 1 \leqslant \xi_d \leqslant n_d.$$

5. The algebra associated to a distributive lattice

DEFINITION 5.0.1. Given a finite lattice $\mathcal{L}$, the *ideal associated to* $\mathcal{L}$, denoted by $I(\mathcal{L})$, is the ideal of the polynomial ring $k[\mathcal{L}]$ generated by the set of binomials

$$\mathcal{G}_{\mathcal{L}} = \{xy - (x \wedge y)(x \vee y) \mid x, y \in \mathcal{L} \text{ non-comparable}\}.$$

By Theorem 3.0.7, a finite distributive lattice $\mathcal{L}$ may be identified with a sublattice of a finite chain product lattice . Hence it inherits a total order extending the given partial order. In turn, this total order induces the lexicographic order on the monomials in $k[\mathcal{L}]$, as in §1.1.

The following theorem shows that the ideal associated to a chain product lattice is toric.

THEOREM 5.0.2. (1) *We have* $I(\mathcal{C}(n_1, \dots, n_d)) = I_{\mathcal{A}_{n_1,\dots,n_d}}$.

(2) *The set of binomials*

$$\mathcal{G} = \{xy - (x \wedge y)(x \vee y) \mid x, y \in \mathcal{C}(n_1, \dots, n_d) \textit{ non-comparable}\}$$

is a Gröbner basis for $I(\mathcal{C}(n_1, \dots, n_d))$ *with respect to the lexicographic order.*

PROOF. Let $\mathcal{C} = \mathcal{C}(n_1, \dots, n_d)$ and $\mathcal{A} = \mathcal{A}_{n_1,\dots,n_d}$. Let $f \in I_{\mathcal{A}}$; by Proposition 4.0.5, there exist $\mathbf{u}_i, \mathbf{v}_i \in \mathbb{Z}_+^n$ with $\pi_{\mathcal{A}}(\mathbf{u}_i) = \pi_{\mathcal{A}}(\mathbf{v}_i)$, and $c_i \in k^*$, $1 \leqslant i \leqslant s$ such that

$$f = \sum_{i=1}^{s} c_i(\mathbf{x}^{\mathbf{u}_i} - \mathbf{x}^{\mathbf{v}_i}) \tag{a}$$

for some $s \geqslant 1$, with the property that s is the smallest integer $\geqslant 1$ such that f can be expressed as a linear combination of s binomials in the set (*) of Proposition 4.0.5. Now we rewrite f as

$$f = \sum_{i=1}^{s} a_i \mathbf{x}^{\mathbf{u}_i} + \sum_{i=1}^{s} b_i \mathbf{x}^{\mathbf{v}_i}, \quad a_i, b_i \in k.$$

Then none of the coefficients $a_1, \dots, a_s, b_1, \dots, b_s$ is zero. Indeed, suppose that $a_i = 0$ for some $1 \leqslant i \leqslant s$. This implies that there exists $j \in \{1, \dots, s\}$, $j \neq i$ such that either $c_j = c_i$ and $\mathbf{v}_j = \mathbf{u}_i$, or $c_j = -c_i$ and $\mathbf{u}_j = \mathbf{u}_i$.

In the first case we have

$$c_i(\mathbf{x}^{\mathbf{u}_i} - \mathbf{x}^{\mathbf{v}_i}) + c_j(\mathbf{x}^{\mathbf{u}_j} - \mathbf{x}^{\mathbf{v}_j}) = c_i(\mathbf{x}^{\mathbf{u}_j} - \mathbf{x}^{\mathbf{v}_i}), \quad \pi_{\mathcal{A}}(\mathbf{u}_j) = \pi_{\mathcal{A}}(\mathbf{v}_i). \tag{b}$$

In the second case we have

$$c_i(\mathbf{x}^{\mathbf{u}_i} - \mathbf{x}^{\mathbf{v}_i}) + c_j(\mathbf{x}^{\mathbf{u}_j} - \mathbf{x}^{\mathbf{v}_j}) = c_i(\mathbf{x}^{\mathbf{v}_j} - \mathbf{x}^{\mathbf{v}_i}), \quad \pi_{\mathcal{A}}(\mathbf{v}_j) = \pi_{\mathcal{A}}(\mathbf{v}_i). \tag{c}$$

But (a), (b) and (c) imply that f can be written as a linear combination of $s - 1$ binomials in the set (*) of Proposition 4.0.5, contradicting the choice of s. Thus $a_i \neq 0$, $1 \leqslant i \leqslant s$, and similarly $b_i \neq 0$, $1 \leqslant i \leqslant s$. This shows that $\text{in}(f) = \text{in}(\mathbf{x}^{\mathbf{u}_i} - \mathbf{x}^{\mathbf{v}_i})$ for some $1 \leqslant i \leqslant s$.

Let us write

$$\mathbf{x}^{\mathbf{u}_i} - \mathbf{x}^{\mathbf{v}_i} = \mathbf{x}^{\mathbf{w}}(\mathbf{x}^{\mathbf{u}} - \mathbf{x}^{\mathbf{v}}),$$

where $\mathbf{u}, \mathbf{v}, \mathbf{w} \in \mathbb{Z}_+^n$, with $\pi_A(\mathbf{u}) = \pi_A(\mathbf{v})$ and $\operatorname{supp}(\mathbf{u}) \cap \operatorname{supp}(\mathbf{v}) = \emptyset$. Let us suppose that $\mathbf{x}^\mathbf{u} \succeq_{lex} \mathbf{x}^\mathbf{v}$, i.e. $\operatorname{in}(\mathbf{x}^\mathbf{u} - \mathbf{x}^\mathbf{v}) = \mathbf{x}^\mathbf{u}$ and $\operatorname{in}(f) = \operatorname{in}(\mathbf{x}^{\mathbf{u}_i} - \mathbf{x}^{\mathbf{v}_i}) = \mathbf{x}^{\mathbf{u}_i}$. Let $x_{i_1 \dots i_d}$ be the smallest variable appearing in $\mathbf{x}^\mathbf{u}$, i.e. $(i_1, \dots, i_d)$ is the smallest element of $\operatorname{supp}(\mathbf{u})$ with respect to $\triangleleft$. Then $\mathbf{x}^\mathbf{v}$ contains a variable $x_{k_1 \dots k_d}$, with $(k_1, \dots, k_d) \triangleleft (i_1, \dots, i_d)$. Since $\pi_A(\mathbf{u}) = \pi_A(\mathbf{v})$, the sum of the entries in every slice is the same for both $\mathbf{u}$ and $\mathbf{v}$. In particular, since all the entries of $\mathbf{u}$ in the slices $\{\xi_1 = i\}$, with $1 \leqslant i < i_1$, are 0 (by the choice of $(i_1, \dots, i_d)$), all the entries of $\mathbf{v}$ in these slices must also be 0. This implies that $(k_1, \dots, k_d) \in \{\xi_1 = i_1\}$.

Let $1 < h \leqslant n_1$ such that $k_1 = i_1, \dots, k_{h-1} = i_{h-1}, k_h < i_h$. Then the sum of the elements of $\mathbf{v}$ in the slice $\{\xi_h = k_h\}$ is nonzero, which implies that

$$\{\xi_h = k_h\} \cap \operatorname{supp}(\mathbf{u}) \neq \emptyset.$$

Let $(j_1, \dots, j_d)$ be an element in this intersection. We have $(i_1, \dots i_d) \triangleleft (j_1, \dots, j_d)$ (by the definition of $(i_1, \dots i_d)$), and since $i_h > k_h = j_h$, we conclude that $(i_1, i_2, \dots, i_d)$ and $(j_1, \dots, j_d)$ are non-comparable. Thus we obtain that $\mathbf{x}^\mathbf{u}$ is divisible by $x_{i_1 \dots i_d} x_{j_1 \dots j_d}$. Hence $\mathbf{x}^{\mathbf{u}_i}$ is also divisible by $x_{i_1 \dots i_d} x_{j_1 \dots j_d}$. Therefore $\operatorname{in}(f)$ is divisible by the initial term of an element of the set

$$\mathcal{G} = \{xy - (x \wedge y)(x \vee y) \mid x, y \in \mathcal{C} \text{ non-comparable}\}$$

of generators of the ideal $I(\mathcal{C})$. Since $\mathcal{G} \subset I_A$, it follows that $\mathcal{G}$ is a Gröbner basis for I_A. In particular it is a set of generators for this ideal. Thus $\mathcal{G}$ generates both $I(\mathcal{C})$ and I_A, which implies the equality of the two ideals. □

THEOREM 5.0.3. *Let $\mathcal{L}$ be a finite distributive lattice. Then*

(1) *The ideal $I(\mathcal{L})$ is toric.*
(2) *The set of binomials*

$$\mathcal{G}_\mathcal{L} = \{xy - (x \wedge y)(x \vee y) \mid x, y \in \mathcal{L} \text{ non-comparable}\}$$

is a Gröbner basis for $I(\mathcal{L})$ with respect to the lexicographic order.

PROOF. By Theorem 3.0.7, we may assume that for some $n_1, \dots, n_d \geqslant 1$, $\mathcal{L}$ is a sublattice of $\mathcal{C}(n_1, \dots, n_d)$. Let us denote $\mathcal{C} = \mathcal{C}(n_1, \dots, n_d)$, $A = A_{n_1, \dots, n_d}$ and $\mathcal{G} = \mathcal{G}_\mathcal{C}$. Note that $\mathcal{G}_\mathcal{L}$ is the subset of $\mathcal{G}$ consisting of all binomials in $\mathcal{G}$ involving only the variables from $\mathcal{L}$. Let us denote

$$\mathcal{G}_\mathcal{L} = \{f_1, \dots, f_r\}, \quad \mathcal{G} \setminus \mathcal{G}_\mathcal{L} = \{g_1, \dots, g_s\}.$$

Let $g_i = xy - (x \wedge y)(x \vee y)$, with $x, y \in \mathcal{C}$ non-comparable, $1 \leqslant i \leqslant s$; then at least one of x and y does not belong to $\mathcal{L}$ ($\mathcal{L}$ being a sublattice of $\mathcal{C}$, $x, y \in \mathcal{L}$ would imply $x \wedge y, x \vee y \in \mathcal{L}$, so g_i would involve only variables from $\mathcal{L}$).

Let $A_\mathcal{L} \subset A$ be given by the elements in A indexed by the elements of $\mathcal{L}$, and let f be an element of

$$I_{A_\mathcal{L}} = \operatorname{Ker}\left(\widehat{\pi_A}\big|_{k[\mathcal{L}]}\right) = (\operatorname{Ker}\widehat{\pi_A}) \cap k[\mathcal{L}] = I_A \cap k[\mathcal{L}].$$

In the course of the proof of Theorem 5.0.2, we saw that $\mathrm{in}(f)$ is divisible by the initial term of a binomial in $\mathcal{G}$, and since $f \in k[\mathcal{L}]$, this binomial must be one of the f_i's, i.e. an element of $\mathcal{G}_{\mathcal{L}}$. Since $\mathcal{G}_{\mathcal{L}} \subset I_{A_{\mathcal{L}}}$, it follows that $\mathcal{G}_{\mathcal{L}}$ is a Gröbner basis for $I_{A_{\mathcal{L}}}$, hence $\mathcal{G}_{\mathcal{L}}$ generates $I_{A_{\mathcal{L}}}$. Therefore $I(\mathcal{L}) = I_{A_{\mathcal{L}}}$, and the stated assertions follow now. □

REMARK 5.0.4. (1) In **[69]**, it is shown that $\mathcal{R}_{\mathcal{L}} = k[\mathcal{L}]/I(\mathcal{L})$ ($\mathcal{L}$ being a finite distributive lattice) is a normal domain; on the other hand, one knows (see **[51]** for example) that a prime binomial ideal is toric, in the sense of Definition 4.0.2 (here, by a binomial, we mean a polynomial with at most two terms). Thus the result that $I(\mathcal{L})$ is toric ($\mathcal{L}$ being a finite distributive lattice) may also be concluded using **[69]** and **[51]**.

(2) In fact, given a finite lattice $\mathcal{L}$, $I(\mathcal{L})$ is toric if and only if $\mathcal{L}$ is distributive.

(3) We have included below a short geometric proof of the result that a prime binomial ideal is toric.

5.1. Varieties defined by binomials

Let $N \geqslant 1$, and let X be an affine variety in $\mathbb{A}^N$, not contained in any of the coordinate hyperplanes $\{x_i = 0\}$. Further, let X be irreducible, and let its defining prime ideal $I(X)$ be generated by l binomials

$$(*) \qquad x_1^{a_{i1}} \dots x_N^{a_{iN}} - \lambda_i x_1^{b_{i1}} \dots x_N^{b_{iN}}, \qquad 1 \leqslant i \leqslant l \ .$$

Consider the natural action of the torus $T_N = (k^*)^N$ on $\mathbb{A}^N$,

$$(t_1, \dots, t_N) \cdot (a_1, \dots, a_N) = (t_1 a_1, \dots, t_N a_N).$$

Let $X(T_N) = \mathrm{Hom}(T_N, \mathbb{G}_m)$ be the character group of T_N, and let $\epsilon_i \in X(T_N)$ be the character

$$\epsilon_i(t_1, \dots, t_N) = t_i, \qquad 1 \leqslant i \leqslant N.$$

For $1 \leqslant i \leqslant l$, let

$$\varphi_i = \sum_{t=1}^{N} (a_{it} - b_{it}) \epsilon_t \, .$$

Set $T = \bigcap_{i=1}^{l} \mathrm{Ker}\, \varphi_i$, and $X^\circ = \{(x_1, \dots, x_N) \in X \mid x_i \neq 0 \text{ for all } 1 \leqslant i \leqslant N\}$.

PROPOSITION 5.1.1. *Let notations be as above.*

(1) *There is a canonical action of T on X.*
(2) *X° is T-stable. Further, the action of T on X° is simple and transitive.*
(3) *T is a subtorus of T_N, and X is an equivariant affine embedding of T.*

PROOF. (1) We consider the (obvious) action of T on $\mathbb{A}^N$. Let

$$\begin{aligned}(x_1, \ldots, x_N) &\in X\\ \mathbf{t} = (t_1, \ldots, t_N) &\in T\\ (y_1, \ldots, y_N) &= \mathbf{t} \cdot (x_1, \ldots, x_N)\\ &= (t_1 x_1, \ldots, t_N x_N).\end{aligned}$$

Using the fact that $(x_1, \ldots, x_N)$ satisfies $(*)$, we obtain

$$\begin{aligned}y_1^{a_{i1}} \ldots y_N^{a_{iN}} &= t_1^{a_{i1}} \ldots t_N^{a_{iN}} x_1^{a_{i1}} \ldots x_N^{a_{iN}}\\ &= \lambda_i t_1^{b_{i1}} \ldots t_N^{b_{iN}} x_1^{b_{i1}} \ldots x_N^{b_{iN}} = \lambda_i y_1^{b_{i1}} \ldots y_N^{b_{iN}},\end{aligned}$$

for all $1 \leqslant i \leqslant l$, i.e. $(y_1, \ldots, y_N) \in X$. Hence $\mathbf{t} \cdot (a_1, \ldots, a_N) \in X$ for all $\mathbf{t} \in T$.

(2) Let $x = (x_1, \ldots, x_N) \in X^\circ$, and $\mathbf{t} = (t_1, \ldots, t_N) \in T$. Then, clearly $\mathbf{t} \cdot (x_1, \ldots, x_N) \in X^\circ$. Considering x as a point in $\mathbb{A}^N$, the isotropy subgroup in T_N at x is $\{\mathrm{id}\}$. Hence the isotropy subgroup in T at x is also $\{\mathrm{id}\}$. Thus the action of T on X° is simple.

Let $(x_1, \ldots, x_N), (x'_1, \ldots, x'_N) \in X^\circ$. Set $\mathbf{t} = (t_1, \ldots, t_N)$, where $t_i = x_i / x'_i$. Then, clearly $\mathbf{t} \in T$. Thus $(x_1, \ldots, x_N) = \mathbf{t} \cdot (x'_1, \ldots, x'_N)$. Hence the action of T on X° is simple and transitive.

(3) Now, fixing a point $x \in X^\circ$, we obtain from (2) that the orbit map $\mathbf{t} \mapsto \mathbf{t} \cdot x$ is in fact an isomorphism of T onto X°. Also, since X is not contained in any of the coordinate hyperplanes, the open set $X_i = \{(x_1, \ldots, x_N) \in X \mid x_i \neq 0\}$ is nonempty for all $1 \leqslant i \leqslant N$. The irreducibility of X implies that the sets X_i, $1 \leqslant i \leqslant N$, are open dense in X, and hence their intersection

$$X^\circ = \bigcap_{i=1}^{N} X_i = \{(x_1, \ldots, x_N) \in X \mid x_i \neq 0 \text{ for any } i\}$$

is an open dense set in X, and thus X° is irreducible. This implies that T is irreducible (and hence connected). Thus T is a subtorus of T_N. The assertion that X is an equivariant affine embedding of T follows from (1) and (2). □

REMARK 5.1.2. One can see that the ideal $I(X)$ is a toric ideal in the sense of Definition 4.0.2; more precisely, $I(X) = I_{\mathcal{A}}$, with $m = \dim T$, $n = N$, $\mathcal{A} = \{\rho_i, 1 \leqslant i \leqslant N\}$, where $\rho_i = \epsilon_i\big|_T$ (here $k[T]$ is identified with $k[t_1^{\pm 1}, \ldots, t_m^{\pm 1}]$, and the character group $X(T)$ with $\mathbb{Z}^m$).

6. Lattices and flat deformations

Let H be a finite distributive lattice. Let R be a k-algebra with generators

$$\{p_\alpha \mid \alpha \in H\}.$$

DEFINITION 6.0.1. A monomial $p_{\alpha_1} \ldots p_{\alpha_r}$ is said to be standard if $\alpha_1 \geqslant \ldots \geqslant \alpha_r$.

Suppose that the standard monomials form a k-basis for R. Given any nonstandard monomial $\underline{n}$, the expression

(1) $$\underline{n} = \sum c_i \underline{s}_i, \qquad c_i \in k^*$$

for $\underline{n}$ as a sum of standard monomials will be referred as a *straightening relation*. Let $P = k[x_\alpha, \alpha \in H]$, and consider the surjective map

$$\pi : P \to R, \qquad x_\alpha \mapsto p_\alpha.$$

Let us denote Ker π by I.

For $\alpha, \beta \in H$ with $\alpha > \beta$, we set

$$]\beta, \alpha[= \{\gamma \in H \mid \alpha > \gamma > \beta\}.$$

THEOREM 6.0.2. *Let H be a finite distributive lattice, and assume that there exists a lattice embedding $H \hookrightarrow \mathcal{C}$, where $\mathcal{C} = \mathcal{C}(n_1, \dots, n_d)$ for some $n_1, \dots, n_d \geqslant 1$, such that the entries of the d-tuple $(\theta_1, \dots, \theta_d)$ representing an element θ of H form a non-decreasing sequence, i.e. $\theta_1 \leqslant \dots \leqslant \theta_d$. Let $S = P/I(H)$, where $I(H)$ is as in Definition 5.0.1. Suppose that I is generated as an ideal by elements of the form $x_\tau x_\phi - \sum c_{\alpha\beta} x_\alpha x_\beta$ (where τ, ϕ are non-comparable, and $\alpha \geq \beta$). Further suppose that in the straightening relation*

(*) $$p_\tau p_\phi = \sum c_{\alpha\beta} p_\alpha p_\beta,$$

the following hold:

(1) *$p_{\tau\vee\phi} p_{\tau\wedge\phi}$ occurs on the right-hand side of (*) with coefficient* 1.
(2) *$\tau, \phi \in]\beta, \alpha[$, for every pair (α, β) appearing on the right-hand side of (*).*
(3) *Under the embedding $H \hookrightarrow \mathcal{C}$, we have $\tau \dot{\cup} \phi = \alpha \dot{\cup} \beta$, for every (α, β) on the right-hand side of (*).*

Then there exists a flat family (cf. Chapter 10) over Speck$[t]$ whose special fiber $(t = 0)$ is S and general fiber (t invertible) is R.

PROOF. Let $\theta \in H$. Let $\theta = (a_1, \dots, a_d) \in \mathcal{C}$ under the identification of H with a sublattice of $\mathcal{C}$, given by Theorem 3.0.7. We fix an integer $N \gg 0$, and let

$$N_\theta = \sum_{r=1}^{d} N^{d-r} a_r$$

be the integer $a_1 \dots a_d$ in the N-ary representation.

For τ, ϕ non-comparable, let

$$f_{\tau,\phi} = x_\tau x_\phi - \sum c_{\alpha\beta} x_\alpha x_\beta.$$

Then hypothesis implies that I is generated by $\{f_{\tau,\phi} \mid (\tau, \phi) \in Q\}$, where

$$Q = \{(\tau, \phi) \mid \tau, \phi \in H \text{ non-comparable}\}.$$

Let $A = k[t]$, and $P_A = A[x_\alpha, \alpha \in H]$. For $(\tau, \phi) \in Q$, we define the element $f_{\tau,\phi,t}$ in P_A as

$$f_{\tau,\phi,t} = x_\tau x_\phi - \sum c_{\alpha\beta} x_\alpha x_\beta t^{N_\alpha + N_\beta - N_\tau - N_\phi}.$$

Note that $N_{\tau\vee\phi} + N_{\tau\wedge\phi} = N_\tau + N_\phi$; for, if $\tau = (i_1, \ldots, i_d)$, $\phi = (j_1, \ldots, j_d)$, we have $\tau \vee \phi = (k_1, \ldots, k_d)$, $\tau \wedge \phi = (l_1, \ldots, l_d)$, with $k_r = \max\{i_r, j_r\}$ and $l_r = \min\{i_r, j_r\}$ for $1 \leqslant r \leqslant d$.

Also note that for any other (α, β) on the right-hand side of (*), by hypothesis $\alpha = (\alpha_1, \ldots, \alpha_d), \beta = (\beta_1, \ldots, \beta_d) \in \mathcal{C}$ and $\alpha > \tau \vee \phi$, and $\beta < \tau \wedge \phi$. Let s be the smallest integer $\leqslant d$ such that $\alpha_s > k_s$. Then the hypotheses that $\beta < \tau \wedge \phi$, and $\alpha \dot{\cup} \beta = \tau \dot{\cup} \phi$ imply that $\beta_r = l_r$, $1 \leqslant r \leqslant s$. Hence $\alpha_s + \beta_s > k_s + l_s = i_s + j_s$. Thus for $(\alpha, \beta) \neq (\tau \vee \phi, \tau \wedge \phi)$, we have $N_\alpha + N_\beta > N_\tau + N_\phi$.

Let $\mathcal{J}$ be the ideal in P_A generated by $\{f_{\tau,\phi,t} \mid (\tau, \phi) \in Q\}$, and $\mathcal{R} = P_A/\mathcal{J}$.

CLAIM. *(a)* $\mathcal{R}$ *is* $k[t]$*-free.*

(b) $\mathcal{R} \otimes_{k[t]} k[t, t^{-1}] \simeq R[t, t^{-1}]$.

(c) $\mathcal{R} \otimes_{k[t]} k[t]/(t) \simeq S$.

PROOF. We have

$$\mathcal{R} \otimes_{k[t]} k[t]/(t) = P_A/\langle \mathcal{J} + (t) \rangle = S.$$

This proves (c). Let $B = k[t, t^{-1}]$, and $P_B = B[x_\alpha, \alpha \in H]$. Let $\tilde{I}$(resp. $\tilde{\mathcal{J}}$) be the ideal in P_B generated by $\{f_{\tau,\phi} \mid (\tau, \phi) \in Q\}$(resp. $\{f_{\tau,\phi,t} \mid (\tau, \phi) \in Q\}$. We have

(A) $$P_B/\tilde{I} \simeq R[t, t^{-1}].$$

(B) $$P_B/\tilde{\mathcal{J}} \simeq \mathcal{R} \otimes_{k[t]} k[t, t^{-1}].$$

The automorphism

$$P_B \simeq P_B,$$
$$x_\alpha \mapsto t^{N_\alpha} x_\alpha$$

induces an isomorphism

(C) $$P_B/\tilde{I} \simeq P_B/\tilde{\mathcal{J}}.$$

From (A), (B), (C), we obtain (b). Finally, it remains to show (a). Let $X_\alpha = \overline{x}_\alpha$ (in $\mathcal{R} = P_A/\mathcal{J}$), $P_\alpha = t^{N_\alpha} X_\alpha$ and

$$\mathcal{M} = \{P_{\alpha_1} \ldots P_{\alpha_r} \mid \alpha_1 \geqslant \ldots \geqslant \alpha_r, \; r \in \mathbb{Z}_+\}.$$

We shall now show that $\mathcal{M}$ is a $k[t]$-basis for $\mathcal{R}$.

First we prove the linear independence. By base change,

$$\{P_{\alpha_1} \ldots P_{\alpha_r} \mid \alpha_1 \geqslant \ldots \geqslant \alpha_r, r \in \mathbb{Z}_+\}$$

is a $k[t, t^{-1}]$-basis for $R[t, t^{-1}]$. Denoting the isomorphism $P_B/\tilde{I} \simeq R[t, t^{-1}]$ by φ, we have $\{\varphi^{-1}(p_{\alpha_1} \cdots p_{\alpha_r}) \mid \alpha_1 \geqslant \ldots \geqslant \alpha_r, r \in \mathbb{Z}_+\}$ is a $k[t, t^{-1}]$-basis for $\mathcal{R}[t^{-1}]$. For a monomial $\underline{m} = p_{\tau_1} \ldots p_{\tau_r}$ (in $R[t, t^{-1}]$), we have $\varphi^{-1}(\underline{m}) = t^{-N_{\underline{m}}} X_{\tau_1} \ldots X_{\tau_r}$ (where $N_{\underline{m}} = \sum_{i=1}^r N_{\tau_i}$)$= u X_{\tau_1} \ldots X_{\tau_r}$, where $u = t^{-N_{\underline{m}}}$ is a unit in $k[t, t^{-1}]$.

Thus we obtain $\{X_{\alpha_1}\dots X_{\alpha_r} \mid \alpha_1 \geqslant \dots \geqslant \alpha_r, r \in \mathbb{Z}_+\}$ is a $k[t,t^{-1}]$-basis for $\mathcal{R}[t^{-1}]$. Hence $\{P_{\alpha_1}\dots P_{\alpha_r} \mid \alpha_1 \geqslant \dots \geqslant \alpha_r, r \in \mathbb{Z}_+\}$ is a $k[t,t^{-1}]$-basis for $\mathcal{R}[t^{-1}]$. In particular, we obtain that

$$\{P_{\alpha_1}\dots P_{\alpha_r} \mid \alpha_1 \geqslant \dots \geqslant \alpha_r, r \in \mathbb{Z}_+\}$$

is linearly independent over $k[t,t^{-1}]$, and hence over $k[t]$.

Next we prove the generation. Let $F = X_{\tau_1}\dots X_{\tau_r}$ be any monomial in $\mathcal{R}$, where $\tau_r \trianglelefteq \dots \trianglelefteq \tau_1$. Further, we suppose that there exists an i such that $\tau_i \not\geqslant \tau_{i+1}$. Let us denote τ_i, τ_{i+1} by τ, ϕ respectively. Then using the relation

$$P_\tau P_\phi = \sum c_{\alpha\beta} P_\alpha P_\beta\,,$$

we obtain

$$F = \sum a_i F_i\,,$$

where $F_i \preceq_{lex} F$ for each i. Hence, by induction, we obtain that F_i is a $k[t]$-linear combination of elements of $\mathcal{M}$. Now (a) follows. This completes the proof of the claim. □

Now claim implies that $\mathcal{R}$ is a flat family over $k[t]$, whose fiber over 0 is S and whose fiber over any $t-u$, $u \in k^*$ is R. □

In view of the above Theorem, Theorem 5.0.3 and **[69]** (cf. Remark 5.0.4), we obtain the following

THEOREM 6.0.3. *Let hypotheses be as in Theorem 6.0.2. Then there exists a flat family over Speck$[t]$ whose general fiber (t invertible) is R and special fiber ($t=0$) is a toric algebra (by a toric algebra we mean a quotient of a polynomial algebra by a toric ideal).*

7. Degenerations of Schubert varieties in the Grassmannian to toric varieties

Let $G = SL(n)$. Fix d, $1 \leqslant d \leqslant n-1$. Let $G_{d,n}, I_{d,n}$ etc., be as in Chapter 6.

PROPOSITION 7.0.1. *$(I_{d,n}, \geqslant)$ is a distributive lattice.*

PROOF. Let $\tau = (i_1,\cdots,i_d)$, $\phi = (j_1,\cdots,j_d)$ be two non-comparable elements in $I_{d,n}$. Set

$$\lambda = (k_1,\cdots,k_d), \qquad \mu = (l_1,\cdots,l_d)$$

where $k_t = \max\{i_t, j_t\}$, $l_t = \min\{i_t, j_t\}$, $1 \leqslant t \leqslant d$.

Note that $(k_{t-1}, k_t) = (i_{t-1}, i_t)$ or (i_{t-1}, j_t) or (j_{t-1}, i_t) or (j_{t-1}, j_t), from which it follows that $k_{t-1} < k_t$. Similarly, $l_{t-1} < l_t$. Thus λ, μ are well-defined elements of $I_{d,n}$. Further from the definition of λ, μ, it is easily seen that λ (resp. μ) is the unique minimal (resp. maximal) element of $I_{d,n}$ which is greater (resp. less) than both τ and ϕ. □

COROLLARY 7.0.2. *Let $\tau, \phi, \lambda, \mu \in I_{d,n}$ be as in the proof of Proposition 7.0.1. Then $\lambda = \tau \vee \phi$, $\mu = \tau \wedge \phi$.*

COROLLARY 7.0.3. *For $\tau \in I_{d,n}$, let*

$$H_\tau = \{\underline{i} \in I_{d,n} \mid \underline{i} \leqslant \tau\}.$$

Then H_τ is a distributive lattice.

PROPOSITION 7.0.4. *Let $\tau, \phi \in I_{d,n}$ be non-comparable. Let*

$$(*) \qquad p_\tau p_\phi = \sum c_{\alpha,\beta} p_\alpha p_\beta$$

be the expression for $p_\tau p_\phi$ as a sum of standard monomials. Then $p_{\tau\vee\phi} p_{\tau\wedge\phi}$ occurs with coefficient 1 in (*).

PROOF. Let notations be as above. Then, in view of Proposition 3.3.3, Chapter 6, for any (α, β) on the right-hand side of $(*)$, we have

$$\alpha \geqslant \tau \vee \phi, \ \beta \leqslant \tau \wedge \phi.$$

Further, if $\alpha = \tau \vee \phi$, then $\beta = \tau \wedge \phi$ (by weight considerations; note that for $\theta \in I_{d,n}$, say $\theta = (i_1, \cdots, i_d)$, weight of $p_\theta = -(\epsilon_{i_1} + \cdots + \epsilon_{i_d})$). Hence restricting (*) to $X(\tau \vee \phi)$, we obtain

$$p_\tau p_\phi = c p_{\tau\vee\phi} p_{\tau\wedge\phi}, \quad \text{on } X(\tau \vee \phi).$$

Since standard monomials basis is characteristic-free, we conclude that the coefficient c of $p_{\tau\vee\phi} p_{\tau\wedge\phi}$ on the right-hand side of (*) is ± 1.

In order to prove that $c = 1$, we compare the coefficients of the monomial $\underline{m} = x_{i_1 1} \ldots x_{i_d d} x_{j_1 1} \ldots x_{j_d d}$ on both sides (note that a Plücker coordinate $p_{a_1,\ldots,a_d}$ (being the $d \times d$-minor of the generic $n \times n$ matrix (x_{ij}) with row indices $a_1, \ldots, a_d$ and column indices $1, \ldots, d$) is a polynomial in the x_{ij}'s). Let $p_\alpha p_\beta$ appearing on the right-hand side of (*) be such that $\alpha \neq \tau \vee \phi$. This implies $\alpha > \tau \vee \phi$, and $\beta < \tau \wedge \phi$. Let $\alpha = (\alpha_1, \ldots, \alpha_d)$ and $\beta = (\beta_1, \ldots, \beta_d)$. We have $(\alpha_1, \ldots, \alpha_d) > (k_1, \ldots, k_d)$ ($(k_1, \ldots, k_d)$ being as in the proof of Proposition 7.0.1). Let t be the smallest integer $\leqslant d$ such that $\alpha_t > k_t$. This implies, in view of Corollary 7.0.2 (notations being as in that Corollary)

$$\alpha_p = k_p, \quad \beta_p = l_p, \quad p < t,$$
$$k_t \notin \{\alpha_1, \ldots, \alpha_d\}, \quad k_t, l_t \in \{\beta_1, \ldots, \beta_d\}.$$

Hence in the expression for $p_\alpha p_\beta$ as a polynomial in the x_{ij}'s, $x_{k_t t} x_{l_t t}$ will not be a factor in any of the monomials. Hence the monomial $\underline{m}$ does not occur in $p_\alpha p_\beta$.

Now, the term $x_{i_1 1} \ldots x_{i_d d}$ (resp. $x_{j_1 1} \ldots x_{j_d d}$) being the product of the diagonal entries in p_τ (resp. p_ϕ), it occurs with coefficient 1 in p_τ (resp. p_ϕ). Hence $\underline{m}$ occurs with coefficient 1 in $p_\tau p_\phi$. This, together with the fact that $\underline{m}$ does not occur in any $p_\alpha p_\beta$, $\alpha \neq \tau \vee \phi$, implies that $\underline{m}$ should occur with coefficient 1 in $p_{\tau\vee\phi} p_{\tau\wedge\phi}$ (note that in $p_{\tau\vee\phi} p_{\tau\wedge\phi}$ the monomial $\underline{m}$ is realized as the product $\underline{m}_1 \underline{m}_2$, where $\underline{m}_1$ (resp. $\underline{m}_2$) is the product of the diagonal entries in $p_{\tau\vee\phi}$ (resp. $p_{\tau\wedge\phi}$). From this, it follows that $c = 1$. □

Let $\mathfrak{L} = H_\tau$, $V_\tau = \operatorname{Spec} R_{\mathfrak{L}}$. Then by **[69]** (see also Remark 5.0.4) we get

THEOREM 7.0.5. *V_τ is an affine toric variety.*

Combining the above Theorem with Theorem 6.0.2, we obtain a degeneration of $\widehat{X(\tau)}$ to V_τ:

THEOREM 7.0.6. *(cf. **[61]**) There exists a flat family over* $\operatorname{Spec} k[t]$ *with generic fiber (t invertible) equal to $\widehat{X(\tau)}$ and special fiber ($t = 0$) equal to V_τ.*

8. Degeneration of the flag variety to a toric variety

By way of generalizing the results of the previous section to other G/P's, and more generally, to G/Q's, (and their Schubert varieties), we prove the degeneration of $SL(n)/B$ to a normal toric variety (and also the degenerations of a certain class of Schubert varieties in $SL(n)/B$, namely the class of Kempf varieties, cf. **[96]**), as described below.

Let $Q = \bigcap_{i=1}^{r} P_{k_i}$, $H = \bigcup_{i=1}^{r} W^{k_i}$. We follow the notations of §2. As in Section 2, let $R = \bigoplus_{\mathbf{a}} H^0(SL(n)/Q, L^{\mathbf{a}})$. As an algebra R is generated by $\{p_\tau, \tau \in H\}$.

We take the canonical partial order on H, namely, given $\tau, \phi \in H$, say $\tau = (i_1, \dots, i_c)$, $\phi = (j_1, \dots, j_d)$,

$$\tau \geqslant \phi \iff c \leqslant d \text{ and } i_t \geqslant j_t,\ 1 \leqslant t \leqslant c,$$

i.e., the Young Diagram (ϕ, τ) is semistandard.

Note that H is a distributive lattice. From Theorem 2.0.1, we have that the quadratic relations among the p_τ's generate all other relations. Let

$$(*) \qquad p_\tau p_\phi = \sum c_{\alpha\beta} p_\alpha p_\beta$$

be a typical quadratic relation, where we suppose that τ and ϕ are two non-comparable elements of H, and for each (α, β) on the right-hand side, $\alpha \geqslant \beta$. Then, by Proposition 3.2.1, Chapter 10, we have

$$(**) \qquad \alpha > \tau, \phi > \beta.$$

Further, we have (by weight considerations, for example)

$$(***) \qquad \tau \dot{\cup} \phi = \alpha \dot{\cup} \beta$$

(note that if $\tau = (i_1, \dots, i_c)$, $\phi = (j_1, \dots, j_d)$, then weight of $p_\tau p_\phi = -(\epsilon_{i_1} + \cdots + \epsilon_{i_c} + \epsilon_{j_1} + \cdots + \epsilon_{j_d})$). Towards proving the crucial result of this section (Proposition 8.0.4 below), we need the following two lemmas:

LEMMA 8.0.1. *Let $\tau = (i_1, \dots, i_c)$, $\phi = (j_1, \dots, j_d)$, $\lambda = \tau \vee \phi$, $\mu = \tau \wedge \phi$. Then $\lambda = (k_1, \dots, k_c)$, $\mu = (l_1, \dots, l_d)$, where for $t \leqslant c$, $k_t = \max\{i_t, j_t\}$, and $l_t = \min\{i_t, j_t\}$, and for $t > c$, $l_t = j_t$.*

PROOF. We just need to check that $(k_1, \dots, k_c)$ and $(l_1, \dots, l_d)$ belong to H, i.e. the k_i's are distinct, and the l_j's are distinct. The fact that the k_i's are distinct is clear, since for $r < s \leqslant c$, $\max\{i_r, j_r\} < \max\{i_s, j_s\}$. Regarding the l_i's, it is clear that $l_r \neq l_s$ for $r < s \leqslant c$, for similar reasons as above. Also, we have $l_r = j_r$, $r > c$.

Hence it suffices to check that $\{l_1, \ldots, l_c\} \cap \{j_{c+1}, \ldots, j_d\} = \emptyset$. But this is obvious, since $l_c \leqslant j_c < j_{c+1} < \cdots < j_d$. □

LEMMA 8.0.2. *Let* $\alpha = (a_1, \ldots, a_c) \in W^c$, *and* $\beta = (b_1, \ldots, b_d) \in W^d$. *Let* $\alpha \geq \beta$. *Let* $\theta = (\alpha, \beta)^{min}$ *(the unique minimal defining sequence of* (α, β) *(cf. Chapter 7, Corollary 3.1.4)), say* $\theta = (\theta_1, \theta_2)$, $\theta_i \in W^Q$, $i = 1, 2$, *where* $Q = P_c \cap P_d$. *Then*

$$\theta_1^{(c)} = (a_1, \ldots, a_c), \qquad \theta_2^{(c)} = (b_1, \ldots, b_c),$$

$$\theta_1^{(d)} = (a_1, \ldots, a_c, s_1, \ldots, s_e), \qquad \theta_2^{(d)} = (b_1, \ldots, b_d),$$

where $e = d - c$, *and* $\{s_1, \ldots, s_e\}$ *is the subset of* $\{b_1, \ldots, b_d\}$ *with the property that the complement of* $\{s_1, \ldots, s_e\}$ *in* $\{b_1, \ldots, b_d\}$ *is the largest c-tuple which is* $\leqslant \alpha$.

The proof is immediate from the definition of $(\alpha, \beta)^{\min}$.

REMARK 8.0.3. Let notations be as in Lemma 8.0.1. Let $\theta = (\lambda, \phi)^{\min}$, say $\theta = (\theta_1, \theta_2)$. Then we have (by Lemma 8.0.2)

(1) $\theta_1 \geqslant \theta_2$.

(2) $\theta_1^{(c)} = \lambda$.

(3) $\theta_2^{(d)} = \phi$.

In particular we have

$$(\dagger) \qquad p_\tau\big|_{X(\theta_1)} \neq 0, \quad p_\phi\big|_{X(\theta_1)} \neq 0.$$

PROPOSITION 8.0.4. $p_{\tau\vee\phi}p_{\tau\wedge\phi}$ *occurs on the right-hand side of (*) with coefficient* 1.

PROOF. Let notations be as above. Then, in view of (**) in §8, for any (α, β) on the right-hand side of (*) in §8, we have

$$\alpha \geqslant \tau \vee \phi.$$

Further, if $\alpha = \tau \vee \phi$, then in view of (***) in §8, we have $\beta = \tau \wedge \phi$. Hence for any p_α appearing on the right-hand side of (*) in §8, such that $\alpha \neq \tau \vee \phi$, we have, $\alpha > \tau \vee \phi (= \lambda = \theta_1^{(c)})$; for such a p_α, we have,

$$p_\alpha\big|_{X(\theta_1)} = 0.$$

Now we restrict (*) in §8 to $X(\theta_1)$. Then, in view of (†) in Remark 8.0.3, the restriction of the left-hand side of (*) to $X(\theta_1)$ is nonzero, while on the right-hand side of (*), $p_\alpha p_\beta\big|_{X(\theta_1)} = 0$, if $(\alpha, \beta) \neq (\tau \vee \phi, \tau \wedge \phi)$. Hence we obtain

$$p_\tau p_\phi = c p_{\tau\vee\phi} p_{\tau\wedge\phi}, \quad \text{on } X(\theta_1).$$

Since standard monomials basis is characteristic-free, we conclude that the coefficient c of $p_{\tau\vee\phi}p_{\tau\wedge\phi}$ on the right-hand side of (*) is ± 1. In order to prove that $c = 1$, we compare the coefficients of the monomial $\underline{m} = x_{i_1 1} \cdots x_{i_c c} x_{j_1 1} \cdots x_{j_d d}$ on both sides (note that a Plücker coordinate $p_{a_1, \ldots, a_s}$ (being the $s \times s$ minor of the generic $n \times n$ matrix (x_{ij}) with row indices $a_1, \ldots, a_s$ and column indices $1, \ldots, s$) is a polynomial in the x_{ij}'s). Let $p_\alpha p_\beta$ appearing on the right-hand side of (*) be such that $\alpha \neq \tau \vee \phi$.

This implies $\alpha > \tau \vee \phi$, and $\beta < \tau \wedge \phi$. Let $\alpha = (\alpha_1, \ldots, \alpha_c)$ and $\beta = (\beta_1, \ldots, \beta_d)$. We have $(\alpha_1, \ldots, \alpha_c) > (k_1, \ldots, k_c)$. Let t be the smallest integer $\leqslant c$ such that $\alpha_t > k_t$. This implies, in view of (***)

$$\alpha_p = k_p, \quad \beta_p = l_p, \quad p < t,$$
$$k_t \notin \{\alpha_1, \ldots, \alpha_c\}, \quad k_t, l_t \in \{\beta_1, \ldots, \beta_d\}.$$

Hence in the expression for $p_\alpha p_\beta$ as a polynomial in the x_{ij}'s, $x_{k_t t} x_{l_t t}$ will not be a factor in any of the monomials. Hence the monomial $\underline{m}$ does not occur in $p_\alpha p_\beta$. Now the term $x_{i_1 1} \ldots x_{i_c c}$ (resp. $x_{j_1 1} \ldots x_{j_d d}$) being the product of the diagonal entries in p_τ (resp. p_ϕ), it occurs with coefficient 1 in p_τ (resp. p_ϕ). Hence $\underline{m}$ occurs with coefficient 1 in $p_\tau p_\phi$. This, together with the fact that $\underline{m}$ does not occur in any $p_\alpha p_\beta$, $\alpha \neq \tau \vee \phi$, implies that $\underline{m}$ should occur with coefficient 1 in $p_{\tau\vee\phi} p_{\tau\wedge\phi}$ (note that in $p_{\tau\vee\phi} p_{\tau\wedge\phi}$ the monomial $\underline{m}$ is realized as the product $\underline{m}_1 \underline{m}_2$, where $\underline{m}_1$ (resp. $\underline{m}_2$) is the product of the diagonal entries in $p_{\tau\vee\phi}$ (resp. $p_{\tau\wedge\phi}$). Now $\underline{m}$ occurs with coefficient 1 on the left-hand side of (*). Hence $\underline{m}$ should appear with coefficient 1 in $p_{\tau\vee\phi} p_{\tau\wedge\phi}$. From this, it follows that $c = 1$. □

To an element $\tau = (i_1, \ldots, i_d) \in H$, we associate the number

$$n_\tau = \sum_{t=1}^{d} N^{d-t} i_t,$$

where $N \gg 0$. We now carry out the flat deformation in the same manner as described in Section 6, and obtain (using **[69]**)

THEOREM 8.0.5. *The flag variety $SL(n)/Q$ degenerates to a (normal) toric variety Y. Further, the Gröbner basis for $SL(n)/Q$ as constructed in Section 2 descends to a Gröbner basis for Y.*

9. Degenerations of Kempf varieties to toric varieties

We first recall (cf. **[96]**) the definition of Kempf varieties, in $SL(n)/B$. For defining these varieties, we need to consider the projections $\pi_r : G/B \to G/P$, $\pi_l : B\backslash G \to P\backslash G$, where P is a given parabolic subgroup of $G = SL(n)$. For $w \in W$, we shall denote by $X(w)_r$ (resp. $X(w)_l$) the Schubert variety in G/B (resp. $B\backslash G$). Let P_1 be the maximal parabolic subgroup corresponding to α_1, and π_l be the projection $B\backslash G \to P_1\backslash G$.

DEFINITION 9.0.1. (cf. **[96]**) A Schubert variety $X(w)_r$ in G/B is a Kempf variety if

(1) $\pi_l|_{X(w)_l} : X(w)_l \to \mathrm{Im} X(w)_l$ is equidimensional.
(2) $X(w)_l \cap B\backslash P_1$ is irreducible.
(3) Let $X(w)_l \cap B\backslash P_1 = Y(w')_l$. Then $Y(w')_r$ is a Kempf variety in P_1/B.

REMARK 9.0.2. Note that $SL(n)/B$ itself is a Kempf variety.

REMARK 9.0.3. It is possible to define a Kempf variety $X(w)$ purely algebraically, in terms of a reduced expression for w, as well as purely combinatorially, in terms of the permutation corresponding to w (cf. **[96]**).

REMARK 9.0.4. Recall (Chapter 7, §2.1) that for $SL(n)/P$ (P being a maximal parabolic subgroup), we have the nice phenomenon that a monomial $\mathbf{m}$ standard on G/P when restricted to a Schubert variety X, either vanishes on X or remains standard on X. This no longer holds when P is not a maximal parabolic subgroup. For example, in $SL(3)/B$, the monomial p_2p_{13} while being standard on $SL(3)/B$ is not standard on $X(w)$ where w is the permutation (312) (since the minimal defining tableau for $\lambda = \{(2),(13)\}$ is given by $\{(231),(132)\}$, and $X(w) \not\supseteq X(231)$). In fact, it is easy to see that on $X(w)$, $p_2p_{13} = p_3p_{12}$.

It is shown in **[76]** that a Kempf variety has the property that any standard monomial on $SL(n)/B$, when restricted to $X(w)$, either remains standard on $X(w)$, or is identically zero. As a consequence, given a Kempf element $w \in W$, setting $H_w = \cup H_w^i$, where $H_w^i = \{\tau \in W^i, \tau \leqslant w^{(i)}\, 1 \leqslant i \leqslant l\}$ ($w^{(i)} \in W^i$ being the element representing the coset wW_i), we have, H_w is a distributive lattice for the partial order induced from H (note that for an arbitrary w, H_w is not a distributive lattice!). Hence proceeding as in the previous section, we obtain

THEOREM 9.0.5. *Let X be a Kempf variety. Then X degenerates to a (normal) toric variety Y. Further, the Gröbner basis for X as constructed in Section 2 descends to a Gröbner basis for Y.*

CHAPTER 12

Some affine varieties related to Schubert varieties

In this chapter, we consider two classes of affine varieties, namely a certain class of ladder determinantal varieties and a certain class of quiver varieties, for both of which we conclude the geometric properties of normality and Cohen-Macaulayness by identifying them with "opposite cells" in suitable Schubert varieties in suitable $SL(n)/Q$. We also determine the singular loci of these ladder determinantal varieties and certain of these quiver varieties. This in turn gives new information on the irreducible components of the singular loci of the associated Schubert varieties. We have also included a self-contained exposition of the classical determinantal varieties from the view-point of their relationship to Schubert varieties. After reviewing opposite cells in §1, we present results for determinantal varieties in §2; we also present results from classical invariant theory. In §3, §4, we present results for ladder determinantal varieties. In §5, §6, we present results for quiver varieties, varieties of complexes respectively. In §7, we verify the conjecture of Chapter 8, §16 for a class of Schubert varieties associated with certain varieties of complexes. We also verify this conjecture for Schubert varieties related to ladder determinantal varieties in §3, §4.

1. Opposite cells in Schubert varieties

Let $G = SL(n)$, the special linear group of rank $n-1$. Let T be the maximal torus consisting of all the diagonal matrices in G, and B the Borel subgroup consisting of all the upper triangular matrices in G. The Weyl group W gets identified with $\mathcal{S}_n$, the symmetric group on n letters. We preserve the notations of Chapters 6, 7. We recall the following from Chapters 6, 7.

1.1. The partially ordered set $I_{a_1,\dots,a_k}$

Let Q be a parabolic subgroup in $SL(n)$. Let $1 \leqslant a_1 < \cdots < a_k \leqslant n$, such that $S_Q = S \setminus \{\alpha_{a_1}, \dots, \alpha_{a_k}\}$ (we follow **[27]** for indexing the simple roots). Then $Q = P_{a_1} \cap \cdots \cap P_{a_k}$, and $W_Q = \mathcal{S}_{a_1} \times \mathcal{S}_{a_2-a_1} \times \cdots \times \mathcal{S}_{n-a_k}$. Let

$$I_{a_1,\dots,a_k} = \{(\underline{i}_1, \dots, \underline{i}_k) \in I_{a_1,n} \times \cdots \times I_{a_k,n} \mid \underline{i}_t \subset \underline{i}_{t+1} \text{ for all } 1 \leqslant t \leqslant k-1\}.$$

Then $W_Q^{\min}$ may be identified with $I_{a_1,\dots,a_k}$.

The partial order on the set of Schubert varieties in G/Q (given by inclusion) induces a partial order $\geqslant$ on $I_{a_1,\dots,a_k}$, namely, for $\mathbf{i} = (\underline{i}_1, \dots, \underline{i}_k), \mathbf{j} = (\underline{j}_1, \dots, \underline{j}_k) \in$

$I_{a_1,\ldots,a_k}$, $\mathbf{i} \geqslant \mathbf{j} \iff \underline{i}_t \geqslant \underline{j}_t$ for all $1 \leqslant t \leqslant k$ (the partial order on $I_{d,n}$ being as in Chapter 6).

1.2. The minimal and maximal representatives as permutations

Let $w \in W_Q$, and let $\mathbf{i} = (\underline{i}_1, \ldots, \underline{i}_k)$ be the element in $I_{a_1,\ldots,a_k}$ which corresponds to $w_Q^{\min}$. As a permutation, the element $w_Q^{\min}$ is given by $\underline{i}_1$, followed by $\underline{i}_2 \setminus \underline{i}_1$ arranged in ascending order, and so on, ending with $\{1, \ldots, n\} \setminus \underline{i}_k$ arranged in ascending order. Similarly, as a permutation, the element $w_Q^{\max}$ is given by $\underline{i}_1$ arranged in descending order, followed by $\underline{i}_2 \setminus \underline{i}_1$ arranged in descending order, etc..

1.3. The opposite big cell in G/Q

Let $Q = \bigcap_{t=1}^k P_{a_t}$. Let $a = n - a_k$, and Q be the parabolic subgroup consisting of all the elements of G of the form

$$\begin{pmatrix} A_1 & * & * & \cdots & * & * \\ 0 & A_2 & * & \cdots & * & * \\ \vdots & \vdots & \vdots & & \vdots & \vdots \\ 0 & 0 & 0 & \cdots & A_k & * \\ 0 & 0 & 0 & \cdots & 0 & A \end{pmatrix},$$

where A_t is a matrix of size $c_t \times c_t$, $c_t = a_t - a_{t-1}$, $1 \leqslant t \leqslant k$ (here $a_0 = 0$), A is a matrix of size $a \times a$, and $x_{ml} = 0$, $m > a_t$, $l \leqslant a_t$, $1 \leqslant t \leqslant k$. Denote by O^- the subgroup of G generated by $\{U_\alpha \mid \alpha \in R^- \setminus R_Q^-\}$. Then O^- consists of the elements of G of the form

$$\begin{pmatrix} I_1 & 0 & 0 & \cdots & 0 & 0 \\ * & I_2 & 0 & \cdots & 0 & 0 \\ \vdots & \vdots & \vdots & & \vdots & \vdots \\ * & * & * & \cdots & I_k & 0 \\ * & * & * & \cdots & * & I_a \end{pmatrix},$$

where I_t is the $c_t \times c_t$ identity matrix, $1 \leqslant t \leqslant k$, I_a is the $a \times a$ identity matrix, and if $x_{ml} \neq 0$, with $m \neq l$, then $m > a_t$, $l \leqslant a_t$ for some t, $1 \leqslant t \leqslant k$. Further, the restriction of the canonical morphism $f : G \to G/Q$ to O^- is an open immersion, and $f(O^-) \simeq B^- e_{\mathrm{id},Q}$. Thus $B^- e_{\mathrm{id},Q}$ gets identified with O^-.

1.4. Plücker coordinates on the Grassmannian

Let $G_{d,n}$ be the Grassmannian variety, consisting of d-dimensional subspaces of an n-dimensional vector space V. Let us identify V with k^n, and denote the standard basis of k^n by $\{e_i \mid 1 \leqslant i \leqslant n\}$. Consider the Plücker embedding

$$f_d : G_{d,n} \hookrightarrow \mathbb{P}(\wedge^d V),$$

where $\wedge^d V$ is the d-th exterior power of V. For $\underline{i} = (i_1, \ldots, i_d) \in I_{d,n}$, let $e_{\underline{i}} = e_{i_1} \wedge \ldots \wedge e_{i_d}$. Then the set $\{e_{\underline{i}} \mid \underline{i} \in I_{d,n}\}$ is a basis for $\wedge^d V$. Let us denote the basis of $(\wedge^d V)^*$ (the linear dual of $\wedge^d V$) dual to $\{e_{\underline{i}} \mid \underline{i} \in I_{d,n}\}$ by $\{p_{\underline{j}} \mid \underline{j} \in I_{d,n}\}$. Then

$\{p_{\underline{j}} \mid \underline{j} \in I_{d,n}\}$ gives a system of coordinates for $\mathbb{P}(\wedge^d V)$. Recall (cf. Chapter 6) that these are the so-called *Plücker coordinates*.

1.5. Schubert varieties in the Grassmannian

Let $Q = P_d$. We have

$$G_{d,n} \simeq G/P_d.$$

Let $\underline{i} = (i_1, \ldots, i_d) \in I_{d,n}$. Then the T-fixed point $e_{\underline{i},P_d}$ is simply the d-dimensional span of $\{e_{i_1}, \ldots, e_{i_d}\}$. Thus $X_{P_d}(\underline{i})$, the Schubert variety associated to $\underline{i}$, is simply the Zariski closure of $B[e_{i_1} \wedge \ldots \wedge e_{i_d}]$ in $\mathbb{P}(\wedge^d V)$.

In view of the Bruhat decomposition for $X_{P_d}(\underline{i})$, we have

$$p_{\underline{j}}|_{X_{P_d}(\underline{i})} \neq 0 \iff \underline{i} \geqslant \underline{j}.$$

1.6. Evaluation of Plücker coordinates on the opposite big cell in G/P_d

Consider the morphism $\phi_d : G \to \mathbb{P}(\wedge^d V)$, where $\phi_d = f_d \circ \theta_d$, θ_d being the natural projection $G \to G/P_d$. Then $p_{\underline{j}}(\phi_d(g))$ is simply the minor of g consisting of the first d columns and the rows with indices $j_1, \ldots, j_d$. Now, denote by Z_d the unipotent subgroup of G generated by $\{U_\alpha \mid \alpha \in R^- \setminus R^-_{P_d}\}$. We have,

$$Z_d = \left\{ \begin{pmatrix} I_d & 0_{d\times(n-d)} \\ A_{(n-d)\times d} & I_{n-d} \end{pmatrix} \in G \right\}$$

As in Chapter 8, §9, we identify Z_d with the opposite big cell in G/P_d. Then, given $z \in Z_d$, the Plücker coordinate $p_{\underline{j}}$ evaluated at z is simply a certain minor of A, which may be explicitly described as follows. Let $\underline{j} = (j_1, \ldots, j_d)$, and let j_r be the largest entry $\leqslant d$. Let $\{k_1, \ldots, k_{d-r}\}$ be the complement of $\{j_1, \ldots, j_r\}$ in $\{1, \ldots, d\}$. Then this minor of A is given by column indices $k_1, \ldots k_{d-r}$, and row indices $j_{r+1}, \ldots, j_d$ (here the rows of A are indexed as $d+1, \ldots, n$). Conversely, given a minor of A, say, with column indices $b_1, \ldots, b_s$, and row indices $i_{d-s+1}, \ldots, i_d$, it is the evaluation of the Plücker coordinate $p_{\underline{i}}$ at z, where $\underline{i} = (i_1, \ldots, i_d)$ may be described as follows: $\{i_1, \ldots, i_{d-s}\}$ is the complement of $\{b_1, \ldots, b_s\}$ in $\{1, \ldots, d\}$, and $i_{d-s+1}, \ldots, i_d$ are simply the row indices (again, the rows of A are indexed as $d+1, \ldots, n$).
Convention. If $\underline{j} = (1, \ldots, d)$, then $p_{\underline{j}}$ evaluated at z is 1. We shall consider the element 1 (in $k[Z_d]$) as the minor of A with row indices (and column indices) given by the empty set.

1.7. Evaluation of the Plücker coordinates on the opposite big cell in G/Q

Consider

$$f : G \longrightarrow G/Q \hookrightarrow G/P_{a_1} \times \cdots \times G/P_{a_k} \hookrightarrow \mathbf{P}_1 \times \cdots \times \mathbf{P}_k,$$

where $\mathbf{P}_t = \mathbb{P}(\wedge^{a_t} V)$. Denoting the restriction of f to O^- also by just f (O^- being as in §1.3), we obtain an embedding $f : O^- \hookrightarrow \mathbf{P}_1 \times \cdots \times \mathbf{P}_k$, O^- having been identified with the opposite big cell in G/Q. For $z \in O^-$, the multi-Plücker coordinates of $f(z)$ are simply all the $a_t \times a_t$ minors of z with column indices $\{1, \ldots, a_t\}$, $1 \leqslant t \leqslant k$.

1.8. Equations defining the cones over Schubert varieties in $G_{d,n}$

Let $Q = P_d$. Given a d-tuple $\underline{i} = (i_1, \dots, i_d) \in I_{d,n}$, let us denote the associated element of $W_{P_d}^{\min}$ by $\theta_{\underline{i}}$. For simplicity of notation, let us denote P_d by just P, and $\theta_{\underline{i}}$ by just θ. Let A (resp. A_θ) be the homogeneous coordinate ring of $G_{d,n}$ (resp. $X(\theta)$) for the Plücker embedding. Recall (cf. Chapter 6) that the restriction map $A \to A_\theta$ is surjective, and the kernel is generated as an ideal by $\{p_{\underline{j}} \mid \underline{i} \not\geqslant \underline{j}\}$.

1.9. Equations defining multicones over Schubert varieties in G/Q

Let Q be a parabolic subgroup, say $Q = P_{a_1} \cap \dots \cap P_{a_h}$, P_{a_i}, $1 \leqslant i \leqslant h$ being the maximal parabolic subgroups containing Q. Let $X_Q(w) \subset G/Q$. For $\mathbf{b} = (b_1, \dots, b_h) \in \mathbb{Z}_+^r$, let $(R_w)_{\mathbf{b}} = H^0(X(w), L^{\mathbf{b}})$, where $L^{\mathbf{b}} = L_{a_1}^{b_1} \otimes \dots \otimes L_{a_h}^{b_h}$, L_{a_i} being the ample generator of PicG/P_{a_i}. Now define

$$R_w = \bigoplus_{\mathbf{b} \in \mathbb{Z}_+^r} H^0(X(w), L^{\mathbf{b}})$$

For $X(w) = G/Q$, we shall denote R_w (resp. $(R_w)_{\mathbf{b}}$) by just R (resp. $R_{\mathbf{b}}$). The kernel of the restriction map $R \to R_w$ is generated by the kernel of $R_1 \to (R(w))_1$ (cf. Chapter 7, §5.1); but now, in view of §1.8, this kernel is the span of

$$\{p_{\underline{i}} \mid \underline{i} \in I_{d,n}, d \in \{a_1, \dots, a_h\},\ w^{(d)} \not\geqslant \underline{i}\},$$

where $w^{(d)}$ is the d-tuple corresponding to the Schubert variety which is the image of $X_Q(w)$ under the projection $G/Q \to G/P_{a_t}$, $1 \leqslant t \leqslant h$.

1.10. Ideal of the opposite cell in $X_Q(w)$

Let us denote $B^- e_{\mathrm{id},Q} \cap X_Q(w)$ by Y_w. We consider Y_w as a closed subvariety of O^-. In view of §1.9, we obtain that the ideal defining A_w in O^- is generated by

$$\{p_{\underline{i}}|_{O^-} \mid \underline{i} \in I_{d,n}, d \in \{a_1, \dots, a_h\},\ w^{(d)} \not\geqslant \underline{i}\}.$$

1.11. Two lemmas related to Plücker coordinates on the opposite cell

Let $G = SL(n)$, $1 \leqslant a_1 < \dots < a_h \leqslant n$, $Q = P_{a_1} \cap \dots \cap P_{a_h}$. Let O^- be the opposite big cell in G/Q. Let $X = (x_{ba})$, $1 \leqslant b, a \leqslant n$ be a generic $n \times n$ matrix, and $H = \{(i,j) \mid i > j\}$. Let $X^- = (x_{ba}^-)$, $1 \leqslant b, a \leqslant n$, where

$$x_{ba}^- = \begin{cases} x_{ba}, & \text{if } (b,a) \in H \\ 1, & \text{if } b = a \\ 0, & \text{otherwise.} \end{cases}$$

Note that, given $\tau \in W^{a_i}$, for some i, $1 \leqslant i \leqslant h$, the function $p_\tau|_{O^-}$ represents the determinant of the $a_i \times a_i$ submatrix T of X^- whose row indices are $\{\tau(1), \dots, \tau(a_i)\}$, and column indices are $\{1, \dots, a_i\}$.

Let $H_i = \{x_{ba} \mid a_i + 1 \leqslant b \leqslant n, 1 \leqslant a \leqslant a_i\}$, $1 \leqslant i \leqslant h$.

LEMMA 1.11.1. *Let M be a $t \times t$ matrix contained in H_i, for some i, $1 \leqslant i \leqslant h$, with row indices $r_1 < \cdots < r_t$. Then $\det M$ belongs to the ideal of $k[H](= k[x_{ij}, i > j])$ generated by $p_\phi|_{O^-}$, with $\phi \in W^{a_i}$ such that*

$$\{\phi(1), \ldots, \phi(a_i)\} \cap \{a_i + 1, \ldots, n\} = \{r_1, \ldots, r_t\}.$$

PROOF. Denote by $c_1 < \cdots < c_t$ the column indices of M.

Let $\tau = (\{1, \ldots, a_i\} \setminus \{c_1, \ldots, c_t\}) \cup \{r_1, \ldots, r_t\}$. Then $\tau \in W^{a_i}$, and

$$p_\tau|_{O^-} = \det T,$$

where T is the $a_i \times a_i$ submatrix of X^- with row indices $\{\tau(1), \ldots, \tau(a_i)\}$ and column indices $\{1, \ldots, a_i\}$. Using Laplace expansion with respect to the last t rows of T, we obtain

$$(*) \qquad \det T = \sum \pm \det N_{c'_1,\ldots,c'_t} \det M_{c'_1,\ldots,c'_t},$$

the sum being taken over all subsets with t elements $\{c'_1, \ldots, c'_t\}$ of $\{1, \ldots, a_i\}$, where $N_{c'_1,\ldots,c'_t}$ is the $(a_i - t) \times (a_i - t)$ submatrix of X^- with row indices $\{1, \ldots, a_i\} \setminus \{c_1, \ldots, c_t\}$ and column indices $\{1, \ldots, a_i\} \setminus \{c'_1, \ldots, c'_t\}$, and $M_{c'_1,\ldots,c'_t}$ is the $t \times t$ submatrix of X^- with row indices $\{r_1, \ldots, r_t\}$ and column indices $\{c'_1, \ldots, c'_t\}$. Note that $M_{c_1,\ldots,c_t} = M$, and $N_{c_1,\ldots,c_t}$ is a lower triangular matrix, with all diagonal entries equal to 1, and hence $\det M$ appears in $(*)$, and its coefficient is ± 1. Also note that $N_{c'_1,\ldots,c'_t}$ is obtained from $N_{c_1,\ldots,c_t}$ by replacing the columns with indices $c'_1, \ldots, c'_t$ by the columns with indices $c_1, \ldots, c_t$.

Let $\geqslant$ denote the partial order on I_{t,a_i}, namely $(d_1, \ldots, d_t) \geqslant (c_1, \ldots, c_t)$ if $d_j \geqslant c_j$ for all $1 \leqslant j \leqslant t$. We prove the lemma by decreasing induction with respect to the order $\geqslant$ on the t-tuple $(c_1, \ldots, c_t)$ consisting of the column indices of M.

If $c_j > a_{i-1}$ for all $1 \leqslant j \leqslant t$, then for $\{c'_1, \ldots, c'_t\} \neq \{c_1, \ldots, c_t\}$ we have $\det N_{c'_1,\ldots,c'_t} = 0$, since at least one of $c_1, \ldots, c_t$ is an index for a column in $N_{c'_1,\ldots,c'_t}$, and all entries of this column are 0. Thus, in this case $(*)$ reduces to $\det T = \pm \det M$, i.e. $\det M = \pm p_\tau|_{O^-}$, with $\tau \in W^{a_i}$ such that

$$\{\tau(1), \ldots, \tau(a_i)\} \cap \{a_i + 1, \ldots, n\} = \{r_1, \ldots, r_t\}.$$

Assume now that the assertion is true for all matrices with row indices $r_1 < \cdots < r_t$ and column indices $d_1 < \cdots < d_t$ such that $(d_1, \ldots, d_t) > (c_1, \ldots, c_t)$ (i.e. such that $d_j \geqslant c_j$ for all $1 \leqslant j \leqslant t$ and $(d_1, \ldots, d_t) \neq (c_1, \ldots, c_t)$). We shall now prove it for the matrix M with row indices $r_1 < \cdots < r_t$ and column indices $c_1 < \cdots < c_t$. Consider a typical $N_{c'_1,\ldots,c'_t}$ in $(*)$. If there exists a j such that $c'_j < c_j$, then the column with index c_j is replacing the column with index c'_j while obtaining $N_{c'_1,\ldots,c'_t}$ from $N_{c_1,\ldots,c_t}$; hence $N_{c'_1,\ldots,c'_t}$ is still lower triangular, but the diagonal entry in the column with index c_j is 0, which implies that $\det N_{c'_1,\ldots,c'_t} = 0$. Consequently we obtain

$$\det T = \pm \det M + \sum \pm \det N_{c'_1,\ldots,c'_t} \det M_{c'_1,\ldots,c'_t},$$

and hence

$$\det M = \pm p_\tau|_{O^-} + \sum \pm \det N_{c'_1,\ldots,c'_t} \det M_{c'_1,\ldots,c'_t},$$

the sum being taken over all $(c_1, \ldots c_t') \in I_{t,a_i}$ such that $(c_1', \ldots, c_t') > (c_1, \ldots, c_t)$. The required result now follows by induction hypothesis. □

LEMMA 1.11.2. *Let* $1 \leqslant t \leqslant a \leqslant a_i$, $1 \leqslant s \leqslant n$ *and* $\tau \in W^{a_i}$ *such that* $\tau(a-t+1) \geqslant s$. *Then* $p_\tau|_{O^-}$ *belongs to the ideal of* $k[H]$ *generated by t-minors in* X^- *with row indices* $\geqslant s$ *and column indices* $\leqslant a$.

PROOF. Let T be the $a_i \times a_i$ submatrix of X^- with row indices $\{\tau(1), \ldots, \tau(a_i)\}$ and column indices $\{1, \ldots, a_i\}$. Then $p_\tau|_{O^-} = \det T$. Using Laplace expansion with respect to the first a columns, we have $\det T = \sum_p \det A_p \det B_p$, where A_p (resp. B_p) is an $a \times a$ (resp. $(a_i - a) \times (a_i - a)$) matrix. Clearly, all the column indices of a typical A_p are $\leqslant a$, and since $\tau(a-t+1) \geqslant s$, at least t of the row indices of A_p are $\geqslant s$. Using Laplace expansion for A_p with respect to t rows with indices $\geqslant s$, we obtain $\det A_p = \sum_q \det C_q \det D_q$, where C_q (resp. D_q) is a $t \times t$ (resp. $(a-t) \times (a-t)$) matrix, the row indices of C_q are $\geqslant s$, and column indices of C_q are $\leqslant a$. The required result follows from this. □

2. Determinantal varieties

Even though determinantal varieties are classically well-known, their relationship to Schubert varieties has been developed only recently (cf. **[104]**, **[129]**). In this section, we present this aspect of determinantal varieties.

Let $Z = M_{m,n}$, the space of all $m \times n$ matrices with entries in k. We shall identify Z with $\mathbb{A}^{mn}$. We have $k[Z] = K[x_{ij},\ 1 \leqslant i \leqslant m,\ 1 \leqslant j \leqslant n]$.

2.1. The variety D_t

Let $X = (x_{ij})$, $1 \leqslant i \leqslant m$, $1 \leqslant j \leqslant n$ be a $m \times n$ matrix of indeterminates. Let A, B, $A \subset \{1, \cdots, m\}$, $B \subset \{1, \cdots, n\}$, $\#A = \#B = s$, where $s \leqslant \min\{m, n\}$. We shall denote by $[A|B]$ the s-minor of X with row indices given by A, and column indices given by B. For t, $1 \leqslant \min\{m,n\}$, let $I_t(X) = ([A|B],\ A \subset \{1, \cdots, m\},\ B \subset \{1, \cdots, n\},\ \#A = \#B = t)$. Let D_t be the determinantal variety, a closed subvariety of Z, with $I_t(X)$ as the defining ideal.

2.2. The bijection θ

Let $n = r + d$. Given $\underline{i} \in I_{d,n}$, let m be such that $i_m \leqslant d$, $i_{m+1} > d$. Set

$$A_{\underline{i}} = \{n+1-i_d, n+1-i_{d-1}, \cdots, n+1-i_{m+1}\},$$
$$B_{\underline{i}} = \text{ the complement of } \{i_1, i_2 \cdots, i_m\} \text{ in } \{1, 2, \cdots, d\}.$$

Define $\theta : I_{d,n} \setminus \{(12 \cdots d)\} \to \{\text{all minors of } X\}$ by setting $\theta(\underline{i}) = [A_{\underline{i}}|B_{\underline{i}}]$. Clearly, θ is a bijection.

2.3. The set Z_t

Let $G = SL(n)$. Let r, d be such that $r + d = n$. Let X be a $r \times d$ matrix of indeterminates.

Let us identify $\mathcal{O}_d$, the opposite cell in G/P_d as

$$\mathcal{O}_d = \left\{ \begin{pmatrix} I_d \\ X \end{pmatrix} \right\}.$$

As in §2.2, we have a bijection between {Plücker coordinates $p_{\underline{i}}$, $\underline{i} \neq \{1, 2, \cdots, d\}$} and {minors of X} (note that if $\underline{i} = \{1, 2, \cdots, d\}$, then $p_{\underline{i}}$ = the constant function 1). For example, take $r = 3 = d$. We have,

$$\mathcal{O}_3 = \left\{ \begin{pmatrix} I_3 \\ X_{3\times 3} \end{pmatrix} \right\}.$$

We have, $p_{(1,2,4)} = [\{1\}|\{3\}]$, $p_{(2,4,6)} = [\{1,3\}|\{1,3\}]$.
In the sequel, given a Plücker coordinate $p_{\underline{i}}$ (on $\mathcal{O}_d$), we shall denote the associated minor (of $X_{r,d}$) by $\Delta_{\underline{i}}$.
Let $Z_t = \{\underline{i} \in I_{d,n} \mid \Delta_{\underline{i}}$ is a t minor of $X\}$. The partial order on the Plücker coordinates induces a partial order on Z_t.

LEMMA 2.3.1. *Let τ be the d-tuple, $\tau = (1, 2, \cdots, d-t, d+1\, d+2, \cdots, d+t)$. Then $\tau \in Z_t$. Further, τ is the unique smallest element in Z_t.*

PROOF. Clearly $\tau \in Z_t$. Let Δ be a t-minor of X, and $p_{\underline{i}}$ the associated Plücker coordinate. Let $\underline{i} = (i_1, \cdots, i_d)$. We have, for $1 \leqslant k \leqslant d-t$, $i_k \leqslant d$, and for $d-t+1 \leqslant k \leqslant d$, $i_k > d$. Clearly τ is the smallest such d-tuple. □

REMARK 2.3.2. With τ as in Lemma 2.3.1, note that the associated minor Δ_τ of X is $[\{d+1, d+2, \cdots, d+t\} \mid \{d+1-t, d+2-t, \cdots, d\}]$ i.e., the right most top corner t minor of X, the rows of X being indexed as $d+1, \ldots, n$.

REMARK 2.3.3. With τ as in Lemma 2.3.1, note that τ is the smallest d-tuple $(j_1, \cdots, j_d)$ such that $j_{d-t+1} \geqslant d+1$.

LEMMA 2.3.4. *Let τ be as in Lemma 2.3.1. Let*

$$N_t = \{\underline{i} \in I_{d,n} \mid p_\tau |_{X_{P_d}(\underline{i})} = 0\}.$$

Let ϕ be the d-tuple, $\phi = (t, t+1, \cdots, d, n+2-t, n+3-t, \cdots, n)$ (note that ϕ consists of the two blocks $[t, d]$, $[n+2-t, n]$ of consecutive integers - here, for $i < j$, $[i, j]$ denotes the set $\{i, i+1, \cdots, j\}$). Then ϕ is the unique largest element in N_t.

PROOF. Let $\underline{i} \in N_t$, say, $\underline{i} = (i_1, \cdots, i_d)$. We have, $p_\tau |_{X_{P_d}(\underline{i})} = 0$ if and only if $\underline{i} \not\geqslant \tau$, i.e., if and only if $i_{d-t+1} \not\geqslant d+1$ (cf. Remark 2.3.3), i.e., $i_{d-t+1} \leqslant d$. Now it is easily checked that ϕ is the largest d-tuple $(j_1, \cdots, j_d)$ such that $j_{d-t+1} \leqslant d$. □

REMARK 2.3.5. Let ϕ be as in Lemma 2.3.4. As observed in the proof of Lemma 2.3.4, we have ϕ is the largest d-tuple $(j_1, \cdots, j_d)$ such that $j_{d-t+1} \leqslant d$.

COROLLARY 2.3.6. *Let $\underline{i} \in Z_t$. Then $p_{\underline{i}} |_{X_{P_d}(\phi)} = 0$, ϕ being as in Lemma 2.3.4.*

PROOF. We have (cf. Lemma 2.3.4), $p_\tau\,|_{X_{P_d}(\phi)} = 0, \tau$ being as in Lemma 2.3.1 and hence $\phi \not\geqslant \tau$ which in turn implies (in view of Lemma 2.3.1), $\phi \not\geqslant \underline{i}$. Hence we obtain $p_{\underline{i}}\,|_{X_{P_d}(\phi)} = 0$. □

2.4. The identification of D_t with Y_ϕ

THEOREM 2.4.1. *Let ϕ be as in Lemma 2.3.4, and let $Y_\phi = \mathcal{O}_d \cap X_{P_d}(\phi)$ (the opposite cell in $X_{P_d}(\phi)$). Then $D_t \cong Y_\phi$.*

PROOF. Let I_ϕ be the ideal defining Y_ϕ in $\mathcal{O}_d$. We have (cf. §1.10), I_ϕ is generated by $M_t := \{p_{\underline{i}},\ \underline{i} \not\leqslant \phi\}$. Also, $I_t(X)$ (the ideal defining D_t in $\mathcal{O}_d$) is generated by $\{p_{\underline{i}},\ \underline{i} \in Z_t\}$.
Let $\underline{i} \in Z_t$, say, $\underline{i} = (i_1, \cdots, i_d)$. We have (cf. Lemma 2.3.4) $\underline{i} \not\leqslant \phi$, and hence $p_{\underline{i}} \in M_t$.
Let now $p_{\underline{i}} \in M_t$, say, $\underline{i} = (i_1, \cdots, i_d)$. This implies that $i_{d-t+1} \geqslant d+1$ (cf. Remark 2.3.5), and hence it corresponds to a s-minor in X, where $s \geqslant t$. From this it follows that $p_{\underline{i}} \in$ the ideal generated by $\{p_{\underline{i}},\ \underline{i} \in Z_t\}$, i.e., $p_{\underline{i}}$ belongs to $I_t(X)$.
Thus we have shown $I_\phi = I_t(X)$ and the result follows from this. □

COROLLARY 2.4.2. *$dimD_t = (t-1)(n-(t-1))$, where note that $n = r + d$.*

PROOF. We have, $\dim D_t = \dim X(\phi) = (t-1)(n-(t-1))$ (cf. Chapter 6, Proposition 2.2.1). □

COROLLARY 2.4.3. *D_t is normal, Cohen-Macaulay, and has rational singularities.*

PROOF. This follows from the fact that Schubert varieties are normal, Cohen-Macaulay, and have rational singularities (cf. Chapter 6). □

2.4.4. Cogeneration of an ideal. As above, let X be a matrix of variables of size $r \times d$, and let $\geqslant$ be a partial order on the set of all minors of X. Given a minor M of X, the ideal I_M of $k[x_{ij},\ 1 \leqslant i \leqslant r,\ 1 \leqslant j \leqslant d]$ generated by $\{M' \mid M'$ a minor of X such that $M' \not\geqslant M\}$ is called the ideal *cogenerated* by M (for the given partial order).

2.4.5. The partial order $\succeq$. We introduce a partial order on the set of all minors of X as follows: let $[A|B]$, $[A'|B']$ be two minors of size s, s' respectively. Let $A = \{a_1, \cdots, a_s\}$, $B = \{b_1, \cdots, b_s\}$, $A' = \{a'_1, \cdots, a'_s\}$, $B = \{b'_1, \cdots, b'_s\}$. We define $[A|B] \succeq [A'|B']$ if $s \leqslant s'$, $a_j \geqslant a'_j$, $b_j \geqslant b'_j$, $1 \leqslant j \leqslant s$. Note that the bijection θ defined in §2.2 reverses the respective partial orders, i.e., given $\underline{i}, \underline{i}' \in I_{d,n}$, we have, $\underline{i} \leqslant \underline{i}' \iff \theta(\underline{i}) \succeq \theta(\underline{i}')$.

PROPOSITION 2.4.6. *Given a minor M of X, let τ_M be the element of $I_{d,n}$ such that $\theta(\tau_M) = M$. Then $I_M = I(Y_{\tau_M})$, where $Y_{\tau_M} = \mathcal{O}_d^- \cap X_{P_d}(\tau_M)$ and $I(Y_{\tau_M})$ is the ideal defining Y_{τ_M} as a closed subvariety of $\mathcal{O}_d^-$.*

PROOF. We have, $I(Y_{\tau_M})$ is generated by $\{p_{\underline{i}} \mid \underline{i} \in I_{d,n}, \mid \tau_M \not\geqslant \underline{i}\}$. The required result follows from this in view of the bijection θ. □

COROLLARY 2.4.7. *$\{I_M,\ M$ a minor of $X\}$ is precisely the set of ideals of the opposite cells in Schubert varieties in $G_{d,n}$.*

PROOF. The result follows in view of Proposition 2.4.6 and the bijection θ (cf. §2.2). □

COROLLARY 2.4.8. *Let M be the $(t-1)$-minor of X consisting of the first $(t-1)$-rows of X and the first $(t-1)$-columns of X. Then $I_t(X)$ $(=$ the ideal of $D_t)$ equals I_M.*

PROOF. We have (cf. Theorem 2.4.1), $I_t(X) = I(Y_\phi)$, ϕ being as in Theorem 2.4.1. Now, under the bijection θ, we have

$$\theta(\phi) = \big[\{1, 2, \cdots, t-1\} \mid \{1, 2, \cdots, t-1\}\big].$$

The result now follows from Proposition 2.4.6. □

In the above proof the cogeneration of $I_t(X)$ is obtained by relating D_t to the associated Schubert variety (as in Theorem 2.4.1). In the next subsection, we give an alternate proof of the cogeneration of $I_t(X)$ which does not use the theory of Schubert varieties; to achieve this, we need to introduce a partial order among minors of X different from the one considered in §2.4.5.

2.5. The partial order among minors

As above, let $X = (x_{ba})$, $1 \leqslant b \leqslant m$, $1 \leqslant a \leqslant n$ be a $m \times n$ matrix of indeterminates. We introduce a partial order on the set of all minors of X as follows: $[i_1, \dots, i_r | j_1, \dots, j_r] \leqslant [i'_1, \dots, i'_s | j'_1, \dots, j'_s]$ if

$$r \geqslant s \quad \text{and} \quad \begin{array}{l} i_r \geqslant i'_s,\ i_{r-1} \geqslant i'_{s-1}, \dots, i_{r-s+1} \geqslant i'_1, \\ j_1 \leqslant j'_1,\ j_2 \leqslant j'_2, \dots, j_s \leqslant j'_s. \end{array}$$

We say that an ideal I of $k[X]$ is *cogenerated* by a given minor M if I is generated by $\{M' \mid M'$ a minor of X such that $M' \not\geqslant M\}$.

LEMMA 2.5.1. *Let t, $1 \leqslant t \leqslant \min\{m, n\}$. Let M be the $(t-1)$-minor $[A|B]$ of X, where $A = \{m+2-t, m+3-t, \cdots, m\}$, $B = \{1, 2, \cdots, t-1\}$ (note that M is the $(t-1)$-minor of X consisting of the last $(t-1)$ rows of X, and the first $(t-1)$ columns of X). Then $I_t(X)$ is cogenerated by M.*

PROOF. Let us write $A = \{i_1, \cdots, i_{t-1}\}$, $B = \{j_1, \cdots, j_{t-1}\}$. Let I be the ideal cogenerated by M. Then I is generated by $\mathcal{G}_t$, where

$$\mathcal{G}_t = \{ \text{ all minors } M' \mid M' \not\geqslant M\}.$$

Let $N = [i'_1, \dots, i'_s | j'_1, \dots, j'_s]$ be a minor of X of size s, $N \neq M$. If $s \geqslant t$, then clearly $N \not\geqslant M$, and hence $N \in \mathcal{G}_t$. Let now $s \leqslant t-1$. We have clearly $N \geqslant M$, since $i'_s \leqslant m\,(= i_{t-1}), i'_{s-1} \leqslant m-1\,(= i_{t-2}), \cdots, i'_1 \leqslant m+1-s\,(=$

i_{t-s}), and $j'_l \geqslant j_l(=l)$, $1 \leqslant l \leqslant s$; hence $N \notin \mathcal{G}_t$. From this it follows that $\mathcal{G}_t = \{M' \mid M' \text{ is a minor of } X \text{ of size } \geqslant t\}$. But then the ideal generated by $\mathcal{G}_t$ is precisely $I_t(X)$. Thus we obtain $I = I_t(X)$ and the result follows. □

2.6. A Gröbner basis for D_t

We introduce a total order on the variables as follows:

$$x_{m1} > x_{m2} > \cdots > x_{mn} > x_{m-1\,1} > x_{m-1\,2} > \cdots > x_{m-1\,n} > \cdots$$
$$\cdots > x_{11} > x_{12} > \cdots > x_{1n}.$$

This induces a total order, namely the lexicographic order, on the set of monomials in $k[X] = k[x_{11}, \ldots, x_{mn}]$, denoted by $\prec$. The largest monomial (with respect to $\prec$) present in a polynomial $f \in k[X]$ is called the *initial term* of f, and is denoted by $in(f)$. Note that the initial term (with respect to $\prec$) of a minor of X is equal to the product of its elements on the skew diagonal.

Given an ideal $I \subset k[X]$, recall that we denote by $in(I)$ the ideal generated by the initial terms of the elements in I. A set $G \subset I$ is called a *Gröbner basis* of I (with respect to the monomial order $\prec$), if $in(I)$ is generated by the initial terms of the elements in G. Note that a Gröbner basis of I generates I as an ideal.

We recall the following (see **[68]**)

THEOREM 2.6.1. *Let $M = [i_1, \ldots, i_r | j_1, \ldots, j_r]$ be a minor of X, and I the ideal of $K[X]$ cogenerated by M. For $1 \leqslant s \leqslant r+1$, let G_s be the set of all s-minors $[i'_1, \ldots, i'_s | j'_1, \ldots, j'_s]$ satisfying the conditions*

(1) $$i'_s \leqslant i_r,\ i'_{s-1} \leqslant i_{r-1}, \ldots,\ i'_2 \leqslant i_{r-s+2},$$
$$j'_{s-1} \geqslant j_{s-1}, \ldots,\ j'_2 \geqslant j_2,\ j'_1 \geqslant j_1$$

(2) $$\text{if } s \leqslant r, \text{ then } i'_1 > i_{r-s+1} \text{ or } j'_s < j_s.$$

Then the set $G = \bigcup_{i=1}^{r+1} G_i$ is a Gröbner basis for the ideal I with respect to the monomial order $\prec$.

THEOREM 2.6.2. *Let t, $1 \leqslant t \leqslant \min\{m, n\}$. Let*

$$\mathcal{F}_t = \{M' \mid M' \text{ is a minor of } X \text{ of size } t\}.$$

Then $\mathcal{F}_t$ is a Gröbner basis for $I_t(X)$.

PROOF. We have D_t is generated by $\mathcal{F}_t$. Let M be as in Lemma 2.5.1, namely, the $(t-1)$-minor $[A|B]$ of X, where

$$A = \{m+2-t, m+3-t, \cdots, m\},$$
$$B = \{1, 2, \cdots, t-1\}.$$

Let us write $A = \{i_1, \cdots, i_r\}$, $B = \{j_1, \cdots, j_r\}$. We have $r = t-1$, and for $1 \leqslant l \leqslant r$ $i_l = m + l - r$, $j_l = l$.

Let $M' \in G$, G being as in Theorem 2.6.1, say, $M' = [i'_1, \ldots, i'_s | j'_1, \ldots, j'_s]$ be a minor of X of size s, $s \leqslant r+1 (= t)$.The inequalities regarding i's and i''s in condition (1) of Theorem 2.6.1 are redundant, since $i_l = m + l - r$, $1 \leqslant l \leqslant r$. Similarly, the inequalities regarding j's and j''s in condition (1) of Theorem 2.6.1 are again redundant (since $j_l = l$, $1 \leqslant l \leqslant r$); also, condition (2) reduces to the condition that if $s \leqslant r (= t-1)$, then $i'_1 > i_{r-s+1}$ (since $j_s = s$). Therefore, for the above choice of M, the conditions (1) and (2) of Theorem 2.6.1 are equivalent to

$$\text{if } s \leqslant r\ (= t-1), \text{ then } i'_1 > i_{r-s+1}\ (= i_{t-s}).$$

Let $s \leqslant t-1$. The condition that $i'_1 > i_{t-s} (= m+1-s)$ implies $i'_s > m+1$, which is not possible. Hence $G_s = \emptyset$, if $s \leqslant t-1$; thus we obtain $G = G_t$. Now for $s = t$, as discussed above, condition (1) Theorem 2.6.1 is satisfied by any t-minor M' while condition (2) is vacuous. Hence we obtain G_t is the set of all t-minors, i.e., $G = \mathfrak{F}_t$. The result now follows from Lemma 2.5.1 and Theorem 2.6.1 . □

2.7. The singular locus of D_t

As in Chapter 8, §12, in this section, we represent the Schubert varieties by Young diagrams. To each $(a_1, \ldots, a_d) \in W^{P_d}$, we associate a partition $\mathbf{a} = (\mathbf{a}_1, \ldots, \mathbf{a}_d)$, where $\mathbf{a}_i = a_{d-i+1} - (d-i+1)$. For any partition $\mathbf{a} = (\mathbf{a}_1, \ldots, \mathbf{a}_d)$, we will denote by $X_{\mathbf{a}}$ the Schubert variety corresponding to $(a_1, \ldots, a_d)$.

Let $\mathbf{a}$ consist of r rectangles, say,

$$\mathbf{a} = (\mathbf{p}_1^{q_1}, \ldots, \mathbf{p}_r^{q_r}) = (\underbrace{\mathbf{p}_1, \ldots, \mathbf{p}_1}_{q_1}, \underbrace{\mathbf{p}_2, \ldots, \mathbf{p}_2}_{q_2}, \underbrace{\mathbf{p}_r, \ldots, \mathbf{p}_r}_{q_r}).$$

THEOREM 2.7.1. *(cf.* **[108]**, *see also Chapter 8, Theorem 12.0.3).* Sing $X_{\mathbf{a}}$ *has* $r-1$ *components* $X_{\mathbf{a}'_1}, \cdots, X_{\mathbf{a}'_{r-1}}$, *where* $1 \leqslant i \leqslant r-1$ *and*

$$\mathbf{a}'_i = (p_1^{q_1}, \cdots, p_{i-1}^{q_{i-1}}, p_i^{q_i - 1}, (p_{i+1} - 1)^{q_{i+1}+1}, p_{i+2}^{q_{i+2}}, \cdots, p_r^{q_r}),$$

(Note that $\mathbf{a}/\mathbf{a}'_i$, $1 \leqslant i \leqslant r-1$ are simply the hooks in the Young diagram $\mathbf{a}$. For example, if $\mathbf{a} = (7, 5^4, 3^2, 1^3)$, then $\mathbf{a}'_2 = (7, 5^3, 2^3, 1^3)$)

COROLLARY 2.7.2. $X_{\mathbf{a}}$ *is smooth if and only if* $\mathbf{a}$ *consists of one rectangle.*

THEOREM 2.7.3. Sing $D_t = D_{t-1}$.

PROOF. We have (cf. Theorem 2.4.1) $D_t \cong Y_\phi$ where $\phi = (t, t+1, \cdots, d, n+2-t, n+3-t, \cdots, n)$, and Y_ϕ is the opposite cell in $X(\phi)$ $(\subset G_{d,n})$, notations being as in §2 (note that ϕ consists of the two blocks $[t, d]$, $[n+2-t, n]$ of consecutive integers; here, for $a, b \in \mathbb{N}$, $a < b$, $[\![a, b]\!]$ denotes the set $\{a, a+1, \cdots, b\}$). The partition determined by ϕ is given by $\mathbf{a} = (\underbrace{n-d, \cdots, n-d}_{t-1 \text{ times}}, \underbrace{t-1, \cdots, t-1}_{d+1-t \text{ times}})$. Now by Theorem 2.7.1, Sing $X_{\mathbf{a}} = X_{\mathbf{b}}$, where $\mathbf{b}$ is the partition

$$\mathbf{b} = (\underbrace{n-d, \cdots, n-d}_{t-2 \text{ times}}, \underbrace{t-2, \cdots, t-2}_{d+2-t \text{ times}}).$$

Now the d-tuple τ associated to $\mathbf{b}$ is given by $\tau = (t-1, t, \cdots, d, n+3-t, n+4-t, \cdots, n) = ([t-1, d], [n+3-t, n])$. Hence $\operatorname{Sing} Y_\phi = Y_\tau$. Hence we obtain $\operatorname{Sing} D_t = D_{t-1}$ (in view of Theorem 2.4.1). □

Alternate proof. We can give an independent proof of the above result (not using Schubert varieties) as follows. Let $\mathfrak{J}$ be the Jacobian matrix of D_t (considered as a subvariety of $M_{r,d} \cong \mathbb{A}^{rd}$). Let $\mathcal{M}_t$ denote the set of all t-minors of the generic $r \times d$ matrix X. We shall index the rows of $\mathfrak{J}$ by $\mathcal{M}_t$ and the columns by the set of all variables in X. Given $M \in \mathcal{M}_t$, and τ, a variable in X, the (M, τ)-th entry in $\mathfrak{J}$ is non-zero if and only if τ is an entry in M, in which case it is equal to $\pm \det M'$, where M' is the matrix obtained from M by deleting the row and column containing τ.

Let $z \in Z$ and let $\mathfrak{J}_z$ be the Jacobian matrix $\mathfrak{J}$ evaluated at z.

(1) Let z be such that all $(t-1)$-minors of X vanish at z. Then clearly $\mathfrak{J}_z = 0$, and z is a singular point of D_t.

(2) Let z be such that some $(t-1)$-minor M of X is non-zero at z. We shall now show that z is a smooth point of D_t. Let $\mathcal{B}$ denote the set of all variables in X not appearing in any row or column of X given by the rows and columns of M. Let $\tau \in \mathcal{B}$, and let M_τ be the t-minor of X obtained from M by adding the row and column of X through τ to M. Then the (M_τ, τ)-th entry in $\mathfrak{J}_z$ is equal to $\pm(\det M)(z)$ and hence is non-zero; also for $\sigma \in \mathcal{B}$, $\sigma \neq \tau$, the (M_τ, σ)-th entry in $\mathfrak{J}_z$ is 0. From this, it follows that the minor in $\mathfrak{J}_z$ with row indices given by $\{M_\tau,\ \tau \in \mathcal{B}\}$ and column indices given by $\mathcal{B}$ is non-zero. Hence we obtain rank $\mathfrak{J}_z \geqslant (r-(t-1))(d-(t-1)) (= \#\mathcal{B}) = rd - (t-1)(r+d-(t-1)) = \operatorname{codim}_Z D_t$. From this it follows that rank $\mathfrak{J}_z = \operatorname{codim}_Z D_t$, and hence z is a smooth point of D_t (by Jacobian criterion (cf. Chapter 8, §4)).

2.8. Degenerations of determinantal varieties to toric varieties

Let D_{n+1} be the determinantal variety, the subvariety in the space $M_{m,s}$ of $m \times s$ matrices with entries in k, for some $m, s > n$, defined by the vanishing of all $(n+1) \times (n+1)$ minors. Then D_{n+1} gets identified with the opposite big cell, $B^- e_{\mathrm{id}} \cap X(w)$ of $X(w)$, where $X(w)$ is the Schubert variety $X(w)$ in $G_{s,m+s} (= SL_{m+s}/P_s)$, given by

$$w = (n+1, n+2, \ldots, s, m+s+1-n, m+s+2-n, \ldots, m+s)$$

(here B^- is the subgroup in $SL(n)$ consisting of all the lower triangular matrices).

Let $R = k[D_{n+1}]$ and $\check{H}_w = H_w \setminus \{(1, \ldots, s)\}$ where

$$H_w = \{\tau \in I_{s,m+s} \mid \tau \leqslant w\}.$$

Then R is a quotient of the polynomial ring $k[x_\alpha, \alpha \in \check{H}_w]$. The relations are simply induced from those on $X(w)$. To be very precise

$$R = k[x_\alpha, \alpha \in \check{H}_w]/I$$

where I is generated by the straightening relations of the following form. Let τ, ϕ be two noncomparable elements in $\check{H}_w$. Let

$$p_\tau p_\phi = \sum c_{\alpha\beta} p_\alpha p_\beta$$

be the straightening relation for $p_\tau p_\phi$ on $X(w)$. Restricting this relation to D_n, we obtain a relation

$$X_\tau X_\phi = X_{\tau\vee\phi} X_{\tau\wedge\phi} + \sum c_{\alpha\beta} X_\alpha X_\beta,$$

where X_β is understood as 1 if $\beta = (1, \ldots, s)$ (here, for $\alpha \in \check{H}_w$, $X_\alpha = x_\alpha (mod\, I)$). We carry out the flat deformation as in Chapter 11, §6; to each $\tau \in \check{H}_w$, we associate the number

$$m_\tau = \sum_{t=1}^{s} N^{m-t}(i_t - t),$$

where N is a sufficiently large integer.

We first observe that given τ, ϕ two non-comparable elements in $\check{H}_w$,

$$\tau \wedge \phi > \text{id} = (1, \ldots, s)$$

(since in $G_{s,m+s}$, there is a unique one-dimensional Schubert variety, namely $X(\theta)$, $\theta = (1, \ldots, s-1, s+1)$. Also, for the same reasons, we have, $\check{H}_w$ is a distributive lattice (the element 0 being $(1, \ldots, s-1, s+1)$). The proof of Theorem 6.0.3, Chapter 11 goes through in this case . (Note that while proving the generation of $\mathcal{R}$ as a $k[t]$-module by $\mathcal{S}$ (notations being as in the proof of Theorem 6.0.2, Chapter 11), if $F = X_{\tau_1} \ldots X_{\tau_r}$ is such that $\tau_i \not\geq \tau_{i+1}$ for some i, then in the relation

$$X_\tau X_\phi = \sum c_{\alpha\beta} X_\alpha X_\beta t^{m_\alpha + m_\beta - m_\tau - m_\phi},$$

where $\tau = \tau_i$, $\phi = \tau_{i+1}$, we can have a term of the form $X_\alpha t^{m_\alpha - m_\tau - m_\phi}$ (corresponding to $\beta = (1, \ldots, s)$). Nevertheless, we have,

$$X_{\tau_1} \ldots X_{\tau_{i-1}} X_\alpha X_{\tau_{i+2}} \ldots X_{\tau_r} \preceq_{lex} F.$$

Thus $F = \sum a_i F_i$, where each $F_i \preceq_{lex} F$. The rest of the argument is as in Theorem 6.0.2, Chapter 11. Thus we obtain (as in Chapter 11, §6)

THEOREM 2.8.1. *The determinantal variety D_{n+1} in $\mathcal{M}_{m,s}$ degenerates to a (normal) toric variety Y.*

2.9. Determinanatal varieties and Invariant theory

In this section, we present two results of classical invariant theory - the first and second fundamental Theorems (cf. [**162**]) - in the form given to them by De Concini-Procesi (cf. [**40**]). Our presentation below uses the identification of a determinantal variety with the "opposite cell" in a certain Schubert variety in a suitable Grassmannian (cf. §2.4)

We recall that the Grassmannian $G_{d,n}$ is covered by the affine open sets U_τ, $\tau \in I_{d,n}$, with each U_τ being identified with the set of $n \times d$ matrices of rank d whose

τ-th minor is the identity matrix; moreover, in the Plücker embedding of $G_{d,n}$, each U_τ is identified with the open subset defined by $p_\tau \neq 0$. We have, $U_{\tau^{\max}}$, where $\tau^{\max} = (n-d+1, n-d+2, \ldots, n)$, is the only open set among the U_τ's which is a Schubert cell, (it is in fact the big cell in $G_{d,n}$). On the other hand, the affine open set $U_{\tau^{\min}}$, where $\tau^{\min} = (1, 2, \cdots, d)$, is the opposite big cell $\mathcal{O}_d^-$ in $G_{d,n}$.

Recall that $\mathcal{O}_d^-$ can be identified with the set of $n \times d$ matrices of the form

$$\begin{pmatrix} I_d \\ A \end{pmatrix},$$

where I_d is the $d \times d$ identity matrix and A is an $(n-d) \times d$ matrix. Set $r = n - d$. Then, this gives an identification of $\mathcal{O}_d^-$ with $M(r, d)$, the space of $r \times d$ matrices. Thus, the opposite big cell is an affine $(n-d)d$-space whose origin is the point cell consisting of the point $\begin{pmatrix} I_d \\ 0 \end{pmatrix}$. Since the point cell is contained in every Schubert variety, the opposite big cell has nonempty intersection with every Schubert variety, and the intersection is a closed subvariety of the affine $d(n-d)$-space.

Further, since the opposite big cell is defined by $p_{\tau^{\min}} \neq 0$ in $G_{d,n}$, its coordinate ring $k[U_{\tau^{\min}}]$ is the homogeneous localization

$$k[p_\tau, \tau \in I_{d,n}]_{(p_{\tau^{\min}})} = k\Big[\frac{p_\tau}{p_{\tau^{\min}}}, \tau \in I_{d,n}\Big].$$

As a k-algebra, $k[U_{\tau^{\min}}]$ is generated by the regular functions $p_\tau / p_{\tau^{\min}}$, which are nothing but the restrictions of the Plücker coordinates to $U_{\tau^{\min}}$. We shall denote the restrictions of Plücker coordinates p_θ, $\theta \in I_{d,n}$ to $U_{\tau^{\min}}$ also by just p_θ and call them the Plücker coordinates on $U_{\tau^{\min}}$. Recall from Chapter 6:

PROPOSITION 2.9.1. (1) *The coordinate ring $k[U_{\tau^{\min}}]$ of $U_{\tau^{\min}}$ has a basis consisting of standard monomials in the Plücker coordinates p_τ on $U_{\tau^{\min}}$.*

(2) *For a Schubert variety $X(\phi)$, let $Y_\phi = X(\phi) \cap U_{\tau^{\min}}$. The ideal of Y_ϕ in $k[U_{\tau^{\min}}]$ is generated by the Plücker coordinates p_τ on $U_{\tau^{\min}}$ with $\tau \not\leqslant \phi$.*

We first fix some notations: We fix positive integers r, d. We shall denote $r + d$ by n. For a p-tuple $\{a_1, \cdots, a_p\}$, $a_p \in \mathbb{Z}$, we shall denote by $\{a_1, \cdots, a_p\} \uparrow$ the p-tuple obtained from $\{a_1, \cdots, a_p\}$ by arranging the entries in ascending order. For two positive integers $p \leqslant q$, $I_{p,q}$ will denote $\{(a_1, \cdots, a_p) \mid 1 \leqslant a_1 < \cdots < a_p \leqslant q\}$.

(1) Given $\tau = (\tau_1, \ldots, \tau_d) \in I_{d,n}$, and a constant sequence $(c) := (c, c, \ldots, c)$, we shall write $\tau \pm (c)$ for the sequence $(\tau_1 \pm c, \ldots, \tau_d \pm c)$.

(2) Let A be an $r \times d$ matrix and s a positive integer with $s \leqslant \min\{r, d\}$. For $\tau = (\tau_1, \ldots, \tau_s)$, $1 \leqslant \tau_i \leqslant r$ and $\beta = (\beta_1, \ldots, \beta_s)$, $1 \leqslant \beta_j \leqslant s$, we shall denote by $p_{\tau,\beta}(A)$ the s- minor of A with row indices τ_i, $1 \leqslant i \leqslant s$ and column indices β_j, $1 \leqslant j \leqslant s$. Thus for each pair (τ, ϕ) in $I_{s,r} \times I_{s,d}$, $p_{\tau,\phi}$ is a polynomial function on $M(r, d)$, the affine space of $r \times d$ matrices.

(3) Let $\overline{A} = \begin{pmatrix} I_d \\ A \end{pmatrix}$ (where $A \in M(r, d)$) be an element of the opposite big cell in $G_{d,n}$ (note that $n = r + d$). Let $\tau \in I_{d,n}$ with $\tau \neq \tau^{\min}\big(= (1, 2, \cdots, d)\big)$ and let s

be the integer such that $\tau_s \leqslant d$ and $\tau_{s+1} > d$. If

$$\lambda = (\tau_{s+1}, \tau_{s+2}, \ldots, \tau_d),$$
$$\mu = \lambda - (d) = (\tau_{s+1} - d, \ldots, \tau_d - d),$$
$$\nu = (\{1, \ldots, d\} \setminus \{\tau_1, \ldots, \tau_s\}) \uparrow .$$

Then

$$p_\tau(\overline{A}) = \pm p_{\mu,\nu}(A).$$

2.9.2. A standard monomial basis for D_t. Fix $t \leqslant \min\{r, d\}$, and consider D_t, the determinantal variety in $M(r, d)$ defined by the vanishing of all t-minors. Let $\phi \in I_{d,n}$ $(n = r + d)$ be as in Theorem 2.4.1. Then by Theorem 2.4.1, we have, $D_t \cong Y_\phi$. Hence the standard monomial basis for $R(\phi)$ (the homogeneous coordinate ring of $X(\phi)$ for the Plücker embedding) induces a basis for $k[D_t]$. In order to have compatibility of the partial order on $[id, \phi]$ (given by Bruhat order) and the partial order among the minors of a generic $r \times d$ matrix, we need to take a slightly different identification of $M(r, d)$ with $\mathcal{O}_d^-$ (the opposite cell in $G_{d,n}$, $n = r + d$) than that considered above, which we describe now.

LEMMA 2.9.3. *Let J denote the $r \times r$ matrix*

$$\begin{pmatrix} 0 & 0 & \ldots & 0 & 1 \\ 0 & 0 & \ldots & 1 & 0 \\ \vdots & \vdots & & \vdots & \vdots \\ 0 & 1 & \ldots & 0 & 0 \\ 1 & 0 & \ldots & 0 & 0 \end{pmatrix}.$$

Let $s \leqslant min\{r, d\}$, $\tau, \beta \in I_{s,r}$. Let $\gamma = (r + 1 - \tau_s, r + 1 - \tau_{s-1}, \ldots, r + 1 - \tau_1)$. If A is any $r \times d$ matrix and $B = JA$, then

$$p_{\gamma,\beta}(B) = \pm p_{\tau,\beta}(A).$$

PROOF. It is easy to see that multiplication by J on the left reverses the order of rows of A; in other words, the i-th row of A is the $(r + 1 - i)$-th row of B. From this it is clear that

$$p_{\gamma,\beta}(B) = \pm p_{\tau,\beta}(A).$$

□

2.9.4. The partial order. Let s, t, r, d be positive integers with $s \leqslant r$ and $t \leqslant d$. For $\tau \in I_{s,r}$ and $\beta \in I_{t,d}$, we define

$$\tau \leqslant \beta \quad \text{if} \quad s \geqslant t, \text{ and } \tau_i \leqslant \beta_i \text{ for all } i,\ 1 \leqslant i \leqslant t.$$

Let s_i, t_i, r_i, d_i, $i = 1, 2$, be positive integers with $s_i \leqslant r_i$ and $t_i \leqslant d_i$. For $(\tau^1, \tau^2) \in I_{s_1,r_1} \times I_{s_2,r_2}$ and $(\beta^1, \beta^2) \in I_{t_1,d_1} \times I_{t_2,d_2}$, we define

$$(\tau^1, \tau^2) \leqslant (\beta^1, \beta^2), \quad \text{if} \quad \tau^1 \leqslant \beta^1 \text{ and } \tau^2 \leqslant \beta^2.$$

Given positive integers r and d, let

$$P(r,d) = \bigcup_{t=1}^{\min\{r,d\}} I_{t,r} \times I_{t,d}.$$

We have a partial order $\leqslant$ on $P(r,d)$ as defined above, and for each $(\tau,\beta) \in P(r,d)$, we have a regular function $p_{\tau,\beta}$ on $M(r,d)$.

DEFINITION 2.9.5. A monomial $p_{\tau^1,\beta^1} p_{\tau^2,\beta^2} \dots p_{\tau^m,\beta^m}$, $(\tau^i,\beta^i) \in P(r,d)$ is *standard* (or *standard on* $M(r,d)$) if $(\tau^1,\beta^1) \leqslant (\tau^2,\beta^2) \leqslant \dots \leqslant (\tau^m,\beta^m)$. The sequence

$$(\tau^1,\beta^1),(\tau^2,\beta^2),\dots,(\tau^m,\beta^m)$$

satisfying the above inequalities is called a *double standard tableau* in the sense of Doubilet-Rota-Stein (cf. [**46**]).

Note that a standard monomial on $M(r,d)$ gives an element of the coordinate ring $k[M(r,d)]$.

For $\tau \in I_{d,n}$ with $\tau \neq \tau^{\min}$ $(= (1,2,\cdots,d))$, let s be the largest integer such that $\tau_s \leqslant d$, and let

$$\tau' = (n+1-\tau_d, n+1-\tau_{d-1},\dots,n+1-\tau_{s+1}),$$
$$\tau'' = (\{1,\dots,d\} \setminus \{\tau_1,\dots,\tau_s\}) \uparrow .$$

Note that $\tau' \in I_{d-s,r}$ (since for any $j > s$, we have, $n+1-\tau_j < n+1-d = r+1$), and $\tau'' \in I_{d-s,d}$. The pair (τ',τ'') is called *the dual pair associated to* τ (cf. [**104**]). Thus we have a map $h : I_{d,n} \setminus \{(1,2,\cdots,d)\} \to P(r,d)$ defined by $h(\tau) = (\tau',\tau'')$.

LEMMA 2.9.6. *The above defined map is an order reversing bijection, i.e.*

$$\tau \leqslant \beta \iff (\beta',\beta'') \leqslant (\tau',\tau'').$$

Moreover, if $(\lambda,\mu) \in P(r,d)$, *say* $(\lambda,\mu) \in I_{t,r} \times I_{t,d}$ *for some* t, $1 \leqslant t \leqslant \min\{r,d\}$, *then the element* $\tau \in I_{d,n}$ *such that* $h(\tau) = (\lambda,\mu)$ *is given as follows:*

$$(\tau_1,\dots,\tau_{d-t}) = (\{1,\dots,d\} \setminus \mu) \uparrow,$$
$$(\tau_{d-t+1},\dots,\tau_d) = (n+1-\lambda_t,\dots,n+1-\lambda_1),$$

where recall that $n = r + d$.
The proof is easy.

THEOREM 2.9.7. *There is an isomorphism between the affine space* $M(r,d)$ *and the opposite big cell* $\mathcal{O}_d^-$ *(in* $G_{d,n}$*) such that the induced isomorphism between their coordinate rings identifies a Plücker coordinate* p_τ *on* $\mathcal{O}_d^-$ *with the regular function* $\pm p_{\tau',\tau''}$ *on* $M(r,d)$.

PROOF. Let $f : M(r,d) \to \mathcal{O}_d^-$ be the map defined by

$$f(A) = \begin{pmatrix} I_d \\ JA \end{pmatrix},$$

where I_d is the $d \times d$ identity matrix and J is the $r \times r$ matrix with 1 along the anti-diagonal and zeros elsewhere. Since J is an invertible matrix, f is an isomorphism.

Let $f^* : k[\mathcal{O}_d^-] \to k[M(r,d)]$ be the induced isomorphism of the coordinate rings. If p_τ is a Plücker coordinate on $\mathcal{O}_d^-$ and $A \in M(r,d)$, then in view of Lemmas 2.9.3, 2.9.6 it is easy to see that

$$f^*(p_\tau)(A) = p_\tau(\phi(A)) = \pm p_{\tau',\tau''}(A).$$

Hence $f^*(p_\tau) = \pm p_{\tau',\tau''}$. □

THEOREM 2.9.8. (1) *For positive integers r and d, the coordinate ring $k[M(r,d)]$ of $M(r,d)$ has a basis consisting of standard monomials in the regular functions $p_{\tau,\beta}$ on $M(r,d)$.*

(2) *The coordinate ring of D_t ($\subset M(r,d)$) has a basis consisting of standard monomials in the regular functions $p_{\lambda,\mu}$ on D_t with $\#\lambda \leqslant t-1$.*

PROOF. (1) By Theorem 2.9.7, we have an isomorphism $f^* : k[\mathcal{O}_d^-] \cong k[M(r,d)]$ such that $f^*(p_\tau) = \pm p_{\tau',\tau''}$. Moreover, since $\tau \mapsto (\tau',\tau'')$ is an order reversing bijection, f^* establishes a bijection between the set of standard monomials in the Plücker coordinates p_τ on $\mathcal{O}_d^-$ and the set of standard monomials in the regular functions $p_{\tau,\beta}$ on $M(r,d)$. Now the theorem immediately follows from the fact that $k[\mathcal{O}_d^-]$ has a basis consisting of standard monomials in p_τ's (cf. Proposition 2.9.1).

(2) Under the isomorphism $f : M(r,d) \cong \mathcal{O}_d^-$, we have (cf. Theorem 2.4.1) f maps D_t onto the affine variety Y_ϕ ($= X(\phi) \cap \mathcal{O}_d^-$) ($\phi$ being as in Theorem 2.4.1). Since the ideal of Y_ϕ in $k[\mathcal{O}_d^-]$ is generated by the Plücker coordinates p_τ with $\tau \not\leqslant \phi$ (cf. Proposition 2.9.1), it follows from Lemma 2.9.6 that the ideal of D_t is generated by the polynomials $p_{\tau',\tau''}$ in $k[M(r,d)]$ with $(\tau',\tau'') \not\geqslant (\phi',\phi'')$, where (ϕ',ϕ'') is the dual pair associated to ϕ. We have (cf. Lemma 2.3.4),

$$\phi = (t, t+1, \cdots, d, n+2-t, n+3-t, \cdots, n),$$

and hence $\phi' = \phi'' = (1, 2, \cdots t-1)$. Now the assertion follows from the definition of partial order (cf. §2.9.4). □

REMARK 2.9.9. Part (1) in Theorem 2.9.8 is due to Doubilet-Rota-Stein (cf. [**46**]), while part (2) in Theorem 2.9.8 is due to De Concini-Procesi (cf. [**40**]).

2.10. Invariant Theory

For the material discussed in this subsection, we need some basic definitions and results from Geometric Invariant Theory (cf. [**127, 132**]) which we briefly recall below. Let G be an algebraic group acting morphically on an algebraic variety X, i.e., we have an action of G on X and the map $G \times X \to X$, $(g,x) \mapsto gx$, $g \in G$, $x \in X$, is a morphism of varieties.

DEFINITION 2.10.1. A *categorical quotient* of X by G is a pair (Y,π), where Y is a variety and $\pi : X \to Y$ is a morphism such that

(i) π is G-invariant, $\pi(g \cdot x) = \pi(x)$, $g \in G$, $x \in X$,

(ii) for any variety Z and any G-invariant morphism $\phi : X \to Z$, there exists a unique morphism $\chi : Y \to Z$ such that $\chi \circ \pi = \phi$.

DEFINITION 2.10.2. A *good quotient* of X by G is a pair (Y, ϕ) such that $\phi : X \to Y$ is a surjective, G-invariant, affine (i.e., for an open affine subset V of Y, $\phi^{-1}(V)$ is affine) morphism such that

(i) if V is open in Y, then $\phi^* : k[V] \to k[\phi^{-1}(V)]$ maps $k[V]$ isomorphically onto $k[\phi^{-1}(V)]^G$,

(ii) if U is a closed G-stable subset of X, then $\phi(U)$ is closed,

(iii) if U_1, U_2 are two G-stable disjoint closed subsets of X, then $\phi(U_1)$ and $\phi(U_2)$ are disjoint.

DEFINITION 2.10.3. A *geometric quotient* of X by G is a good quotient (Y, ϕ) which is also an orbit space , i.e., fibers of ϕ are the G-orbits in X.
We have the following important
Fact. (cf. [**132**], Proposition 3.11) *A good quotient of X by G is in particular a categorical quotient.*

Affine quotients

Let G be a reductive group acting on an affine variety $X = \operatorname{Spec} R$. Then G acts in a natural way on R, namely,

$$(g \cdot f)(x) = f(g^{-1}x), \; g \in G, x \in X.$$

We shall denote by R^G the subring of R consisting of the G-invariant functions. One knows (cf. [**127, 132**]) that R^G is a finitely generated k-algebra. Hence $Y := \operatorname{Spec} R^G$ is an affine variety. The inclusion $R^G \hookrightarrow R$ induces a surjective morphism $\pi : X \to Y$. We have

THEOREM 2.10.4. *(cf. [**132**]) G, X, Y being as above, (Y, π) is a categorical quotient of X by G.*

Projective quotients

Let $X \hookrightarrow \mathbb{P}_k^n$ be a projective variety. Let G be a reductive group acting linearly on X, i.e., there is given a linear action $\rho : G \to \operatorname{Aut}(k^{n+1})$ and the action of G on X is induced by ρ (via the action of G on $\mathbb{P}_k^n$). For a point $x \in X$, we shall denote the orbit through x by $O(x)$.

DEFINITION 2.10.5. A point $x \in X$ is called *semistable* if there exists a G-invariant homogeneous polynomial of degree $\geqslant 1$ such that $f(x) \neq 0$. Equivalently, if $\widehat{x} \in k^{n+1}$ lies over x, then $0 \notin \overline{O(\widehat{x})}$.
We shall denote the set of semistable points by X^{ss}.

DEFINITION 2.10.6. A point $x \in X$ is called *stable* if there exists a G-invariant homogeneous polynomial of degree $\geqslant 1$ such that $f(x) \neq 0$, $O(x)$ is closed in $X_f(=$

$\{y \in X \mid f(y) \neq 0\}$), and $\dim O(x) = \dim G$. Equivalently, if $\widehat{x} \in k^{n+1}$ lies over x, then $O(\widehat{x})$ is closed and $\dim O(x) = \dim G$.
We shall denote the set of stable points by X^{s}.

THEOREM 2.10.7. *(cf.* **[132, 127]***) Let $X \hookrightarrow \mathbb{P}^n$ be a projective variety on which a reductive group G acts linearly. Then*

(1) *there exists a good quotient (Y, ϕ) of X^{ss} by G, and Y is a projective variety.*

(2) *there exists an open subset Y^{s} of Y such that $\phi^{-1}(Y^{\mathrm{s}}) = X^{\mathrm{s}}$, and (Y^{s}, ϕ) is a geometric quotient of X^{s} by G. (In particular note that the quotient of X^{s} by G is quasiprojective.)*

Now we proceed to prove two results of classical invariant theory - the first and second fundamental theorems (cf. **[162]**) - in the form given to them by De Concini and Procesi (cf. **[40]**).

THEOREM 2.10.8. *Let $X = M(m,n)$, where $m > n$, and let $A = k[x_{ij}, 1 \leqslant i \leqslant m, 1 \leqslant j \leqslant n]$ be its coordinate ring. Consider the canonical action of $G = SL_m(k)$ on A induced by the action of G on X by left multiplication. If $\widehat{G}_{n,m}$ denotes the cone over the Grassmannian $G_{n,m}$, then* Spec A^G *is isomorphic to $\widehat{G}_{n,m}$; in other words, the ring of invariants A^G is generated as a k-algebra by the (determinant) functions $\pm p_\tau$, $\tau \in I_{n,m}$.*

PROOF. Let R be the homogeneous coordinate ring of $G_{n,m}$. Clearly, each p_τ, $\tau \in I_{n,m}$ is a G-invariant. Hence we have the following inclusions $R \subset A^G \subset A$, which give the following commutative diagram:

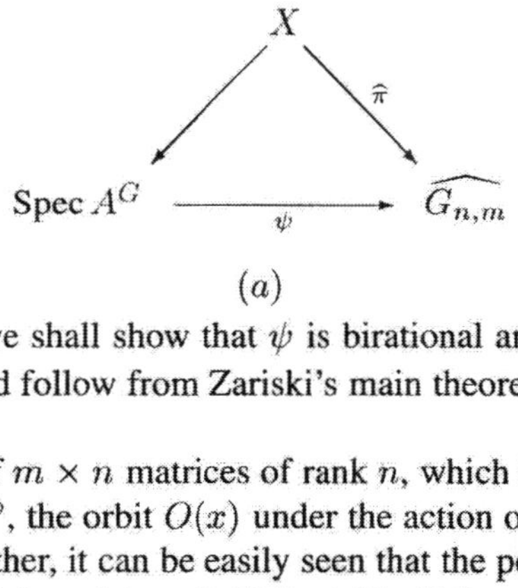

(a)

To prove the theorem, we shall show that ψ is birational and bijective. Then, since $\widehat{G}_{n,m}$ is normal, it would follow from Zariski's main theorem (cf. **[67]**) that ψ is an isomorphism.

Let X° be the set of $m \times n$ matrices of rank n, which is clearly an open subset of X. For every $x \in X^\circ$, the orbit $O(x)$ under the action of G is a closed subset of X isomorphic to G. Further, it can be easily seen that the points of X° are precisely the "stable points" of X (a point $x \in X$ is stable if the orbit $O(x)$ is closed and has dimension equal to $\dim G$). Therefore, by Theorem 2.10.7, the quotient variety $U = G\backslash X^\circ$ exists and is an open subset of Spec A^G. Thus we have the following commutative diagrams:

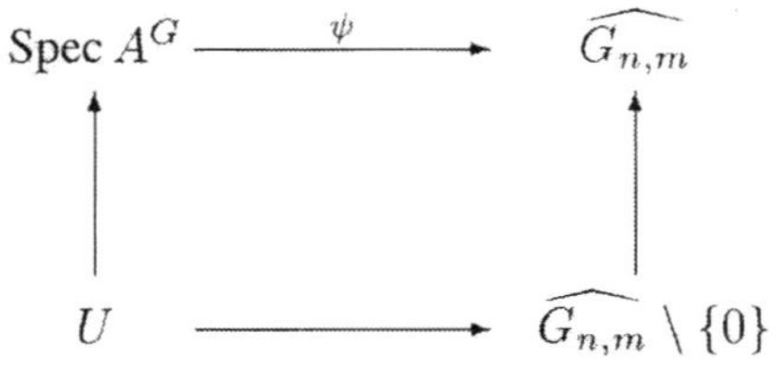

(b)

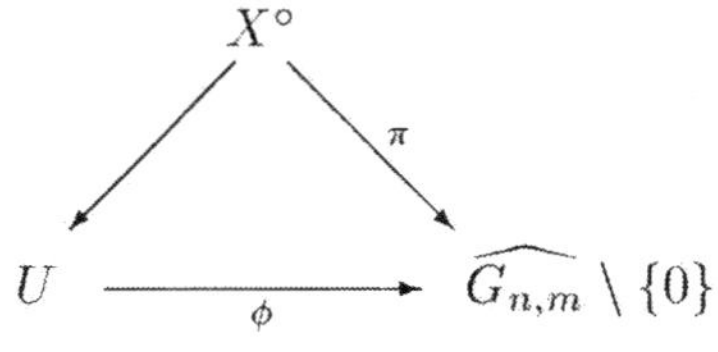

(c)

where 0 denotes the vertex of the cone $\widehat{G}_{n,m}$. For $B \in X$, we have, $\widehat{\pi}(B) = (p_\tau(B), \tau \in I_{n,m})$. Therefore, $\widehat{\pi} : X^\circ \to \widehat{G}_{n,m} \setminus \{0\}$ defines a principal G-bundle on $\widehat{G}_{n,m} \setminus \{0\}$. Now in view of the above commutative diagram, we have, $\phi : U \to \widehat{G}_{n,m} \setminus \{0\}$ is an isomorphism, which shows that ψ : Spec $A^G \to \widehat{G}_{n,m}$ is birational.

Since $X \setminus X^\circ$ consists of unstable points, its image under the map $X \to$ Spec A^G is a point (see [**132**]). However, since $X \to$ Spec A^G is surjective, Spec $A^G \setminus U$ consists of one point and it is clear from the diagram (a) that this point is mapped to 0 by ψ. This proves that ψ is bijective, and the required result follows. □

Let V be a vector space of dimension n and let V^* denote the dual of V. We have canonical actions of $G = GL(V)$ on V and V^*. Recall that if we fix a basis $\{e_1, \ldots, e_n\}$ of V and if the action of $g \in G$ on V is given by a matrix M_g with respect to this basis, then the action of g on V^* is given by the matrix $({}^t M_g)^{-1}$ with respect to the dual basis $\{e_1{}^*, \ldots, e_n{}^*\}$ of V^*. Let

$$X = \underbrace{V \times \cdots \times V}_{l \text{ copies}} \times \underbrace{V^* \times \cdots \times V^*}_{m \text{ copies}} \text{ with } n < \min(l, m),$$

and let $R = k[X]$ be the coordinate ring of X. Denoting by $(\underline{x}, \underline{\xi})$, an element $(x_1, \ldots, x_l, \xi_1, \ldots, \xi_m)$ of X we define the map

$$F : X \longrightarrow M(l, m)$$
$$(\underline{x}, \underline{\xi}) \longmapsto (\langle x_i, \xi_j \rangle),$$

where $\langle x_i, \xi_j \rangle = \xi_j(x_i)$. Consider the diagonal action of G on X, namely

$$g(x_1, \ldots, x_l, \xi_1, \ldots, \xi_m) = (gx_1, \ldots, gx_l, g\xi_1, \ldots, g\xi_m).$$

Since $(\underline{x}, \underline{\xi}) \mapsto \langle x_i, \xi_j \rangle$ is a G-invariant function on X, F is a G-invariant morphism.

LEMMA 2.10.9. *Let X be an affine space on which a reductive algebraic group G acts linearly. Let $\mathbb{A}^N$ be an N-dimensional affine space and let $\psi : X \to \mathbb{A}^N$ be a G-invariant graded morphism (i.e. $\psi(tx) = t^d \psi(x)$ for $x \in X$ and $t \in k$). Let D be a closed subvariety of $\mathbb{A}^N$ such that $\psi(X) \subset D$. Then D is the categorical quotient of X by G, i.e., $D =$ Spec R^G, and $\psi : X \to D$ is the canonical quotient map, provided the following conditions are satisfied:*

(i) For $x \in X^{\mathrm{ss}}$, $\psi(x) \neq (0)$ (here X^{ss} denotes the set of semi-stable points in X; a point $x \in X$ is semi-stable if the origin does not belong to $\overline{O(x)}$).

(ii) There is a G-stable open subset U of X^{ss} such that G operates freely on U and ψ induces an immersion $U/G \to \mathbb{A}^N$ (i.e. an injective morphism with the induced maps between tangent spaces injective) (here U/G denotes the quotient of U by G).

(iii) $\dim D = \dim U/G$.

(iv) D is normal.

PROOF. Let $X = \mathbb{A}^{r+1}$. Let R be the coordinate ring of X and let $Y = \operatorname{Spec} R^G$. Since the canonical morphism $\pi : X \to Y$ is a categorical quotient and $\psi : X \to \mathbb{A}^N$ is G-invariant, there is a canonical morphism $\rho : Y \to \mathbb{A}^N$ such that the following diagram is commutative:

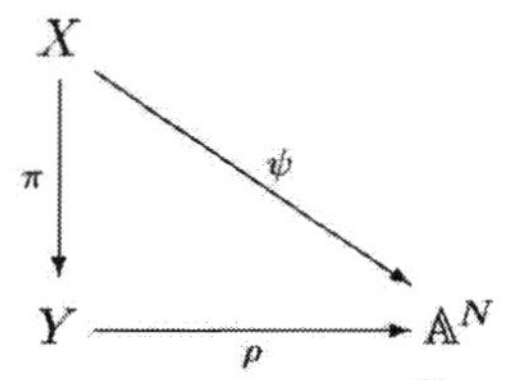

Let $X_1 = \operatorname{Proj} R (= \mathbb{P}^r)$ and $Y_1 = \operatorname{Proj} R^G$. Since ψ is graded, ψ defines a rational map $\psi_1 : X_1 \to \mathbb{P}^{N-1}$. Let X_1^{ss} be the set of semi-stable points in X_1. The hypotheses (i) implies that the restriction of ψ_1 to X_1^{ss} is a morphism. Further, since the morphism $\pi_1 : X_1^{\mathrm{ss}} \to Y_1$ (the restriction of the canonical map $X_1 \to Y_1$) is the categorical quotient (see [**132**], Theorem 3.14) and ψ_1 is G-invariant, we again get a morphism $\rho_1 : Y_1 \to \mathbb{P}^{N-1}$ such that the following diagram is commutative:

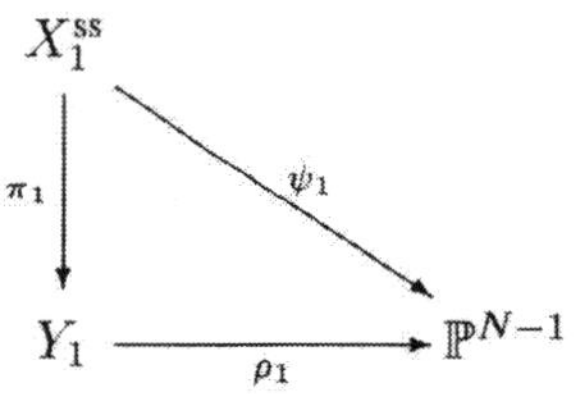

We claim that ρ_1 is an immersion. In order to prove this, we first show that the line bundle $\mathcal{L} = \rho_1{}^*(\mathcal{O}_{\mathbb{P}^{N-1}}(1))$ on Y_1 is ample. In fact, if $\mathcal{M} = \psi_1{}^*(\mathcal{O}_{\mathbb{P}^{N-1}}(1))$, then for a given $x \in X_1^{\mathrm{ss}}$ we can find an invariant section f of $\mathcal{M}^r$ for some $r \geqslant 1$, such that $f(x) \neq 0$ and $(X_1)_f = \{z \in X_1 \mid f(z) \neq 0\}$ is affine. This implies that given $y = \pi_1(x) \in Y_1$, there is a section h of $\mathcal{L}^r$ for some $r \geqslant 1$, such that $h(y) \neq 0$ and $(Y_1)_h = \{x \in Y_1 \mid h(z) \neq 0\}$ is affine. But this means that $\mathcal{L}$ is ample (see for instance [**132**], Lemma 3.20).

Let $\mathbb{P}^{N-1} = \mathbb{P}(V)$, where V is given by $V^* = H^0(\mathbb{P}^{N-1}, L)$, $L = \mathcal{O}_{\mathbb{P}^{N-1}}(1)$. Now, the fact that $\mathcal{L}$ is ample implies that $\mathcal{L}^p$ is very ample for some positive integer $p \geqslant 1$, i.e., there exists an immersion $i : Y_1 \hookrightarrow \mathbb{P}(U)$, where U is given by $U^* = S^pV^*$; further, $i^*\mathcal{O}_{\mathbb{P}}(1) = \mathcal{L}^p$. But we also have an immersion $j : \mathbb{P}(V) \hookrightarrow \mathbb{P}(U)$, the Veronese embedding. Further, $i = j \circ \rho_1$. From this, it follows that ρ_1 is an immersion.

Now, if P is the coordinate ring of $\mathbb{A}^N$, then $\rho : Y \to \mathbb{A}^N$ and $\rho_1 : Y_1 \to \mathbb{P}^{N-1}$ are induced by a graded homomorphism $P \to R^G$ (note that $R^G = \bigoplus_{m\geqslant 0} R^G_m$). Now ρ_1 being an immersion, is a finite morphism, and hence there exists a positive integer m_0 such that $\bigoplus_{m\geqslant m_0} R^G_m$ is a finitely generated P-module. However, since $\bigoplus_{m<m_0} R^G_m$ is finite dimensional, R^G itself is a finitely generated P-module, which shows that $\rho : Y \to \mathbb{A}^N$ is a finite morphism. Now the hypotheses (ii) and (iii) imply that ρ induces a birational morphism $\rho : Y \to D$. Since Y and D are normal (cf. (iv)), it follows that ρ is an isomorphism. □

THEOREM 2.10.10. *The morphism $F : X \to M(l,m)$ maps X into the determinantal variety $D_{n+1}(l,m)$, and the induced homomorphism*

$$F^* : k\big[D_{n+1}(l,m)\big] \longrightarrow k[X]$$

between the coordinate rings induces an isomorphism

$$F^* : k\big[D_{n+1}(l,m)\big] \longrightarrow k[X]^G,$$

i.e. the determinantal variety $D_{n+1}(l,m)$ is the categorical quotient of X by G.

PROOF. We can suppose without loss of generality that l and m are sufficiently large, and hence that $l = m$. For, Let

$$X' = \underbrace{V \times \cdots \times V}_{l' \text{ copies}} \times \underbrace{V^* \times \cdots \times V^*}_{m' \text{ copies}},$$

where $l' > l$ and $m' > m$. Then the projection map $p : X' \to X$ induces the inclusion $k[X] \hookrightarrow k[X']$ and we have $k[X]^G = k[X']^G \cap k[X]$. Thus we may assume that $l = m$.

Clearly, F maps X into the determinantal variety $D_{n+1}(m)$. The theorem would immediately follow from the above lemma, if we show that conditions (i)-(iv) of the lemma are satisfied. We check these conditions one by one.

(i) Let $(\underline{x}, \underline{\xi}) = (x_1, \ldots, x_m, \xi_1, \ldots, \xi_m) \in X^{ss}$. Let $W_{\underline{x}}$ be the subspace of V spanned by x_i's and $W_{\underline{\xi}}$ the subspace of V^* spanned by ξ_j's.

Suppose that $\psi((\underline{x}, \underline{\xi})) = 0$, i.e. $\langle x_i, \xi_j\rangle = 0$ for all i, j.

Case (a): $W_{\underline{\xi}} = 0$, i.e. $\xi_j = 0$ for all j.

Consider the one parameter subgroup $\Gamma = \{g_t, t \neq 0\}$ of $GL(V)$, where $g_t = tI_m$, I_m being the $m \times m$ identity matrix. Then $g_t \cdot (\underline{x}, \underline{0}) = (t\underline{x}, \underline{0})$, so that

$$\lim_{t\to 0} g_t \cdot (\underline{x}, \underline{0}) = (0).$$

Thus the origin 0 is in the closure of $O((\underline{x}, \underline{\xi}))$, and consequently $(\underline{x}, \underline{\xi})$ is not semi-stable, which is a contradiction.

Case (b): $W_{\underline{\xi}} \neq 0$.

By Case (a), we may suppose that $W_{\underline{x}} \neq 0$. Since $\langle x_i, \xi_j\rangle = 0$ for all i, j, we can choose a basis $\{e_1, \ldots, e_n\}$ of V such that $W_{\underline{x}} = \langle e_1, \ldots, e_r\rangle$, $r < n$, and

$W_{\underline{\xi}} = \langle e_{r+1}{}^*, \ldots, e_n{}^* \rangle$, where $\{e_1{}^*, \ldots, e_n{}^*\}$ is the dual basis in V^*. Consider the one parameter subgroup $\Gamma = \{g_t, t \neq 0\}$ of $GL(V)$, where

$$g_t = \begin{pmatrix} tI_r & 0 \\ 0 & t^{-1} I_{m-r} \end{pmatrix}.$$

We have $g_t \cdot (\underline{x}, \underline{\xi}) = (t\underline{x}, t\underline{\xi}) \to 0$ as $t \to 0$. Thus, by the same reasoning as in Case (a), the point $(\underline{x}, \underline{\xi})$ is not semi-stable, which leads to a contradiction.

From Cases (a) and (b), we obtain $\psi((\underline{x}, \underline{\xi})) \neq 0$.

(ii) Let $U = \{(\underline{x}, \underline{\xi}) \in X \mid x_1, \ldots, x_n \text{ are linearly independent in } V\}$. Clearly, U is a G-stable open subset of X and G operates freely on it. Let $\{e_1, \ldots, e_n\}$ be a basis of V and let

$$U' = \{(\underline{x}, \underline{\xi}) \mid x_i = e_i, 1 \leqslant i \leqslant n\}.$$

Then U' can be identified with U/G and we have a commutative diagram

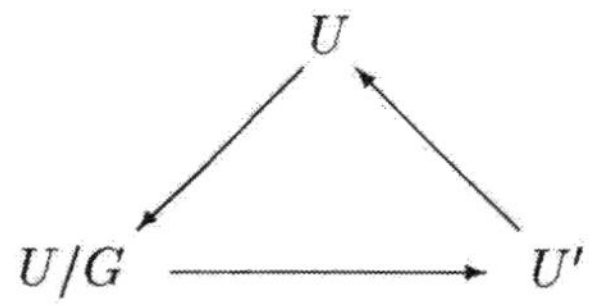

Thus, in order to verify (ii), we only have to show that the map $F : U' \to M(m, m)$ and its differential dF are both injective. Indeed, if $(\underline{x}, \underline{\xi})$, $(\underline{x}', \underline{\xi}')$ are in U' such that $F((\underline{x}, \underline{\xi}')) = F((\underline{x}', \underline{\xi}'))$, then

$$\langle e_i, \xi_j \rangle = \langle e_i, \xi_j' \rangle \text{ for all } i, j, \text{ implies } \xi_j = \xi_j', \text{ for all j.}$$

Since $\{e_1, \ldots, e_n\}$ is a basis of V, the equality $\langle x_i, \xi_j \rangle = \langle x_i', \xi_j \rangle$ for all i, j implies that $x_i = x_i'$ for all i. This shows that $F : U' \to M(m, m)$ is injective. To prove that its differential is injective, we merely note that the above argument remains valid for the points over $k[\epsilon]$, the algebra of dual numbers ($= k \oplus k\epsilon$, the k-algebra with one generator ϵ, and one relation $\epsilon^2 = 0$), i.e. it remains valid if we replace k by $k[\epsilon]$, or in fact by any k-algebra.

(iii) We have

$$\dim U/G = \dim U - \dim G = 2mn - n^2 = \dim D_{n+1}(m).$$

(iv) This follows from Corollary 2.4.3. □

Combining the above Theorem with Theorem 2.9.8, (2), we obtain the following

COROLLARY 2.10.11. *Let X and G be as above. Let f_{ij} denote the regular function $(\underline{x}, \underline{\xi}) \mapsto \langle x_i, \xi_j \rangle$ on X, $1 \leqslant i \leqslant l$, $1 \leqslant j \leqslant m$, and let f denote the $l \times m$ matrix (f_{ij}). The ring of invariants $k[X]^G$ has a basis consisting of standard monomials in the regular functions $p_{\lambda,\mu}(f)$ with $\#\lambda \leqslant n$, where $\#\lambda = t$ is the number of elements in the sequence $\lambda = (\lambda_1, \ldots, \lambda_t)$ and $p_{\lambda,\mu}(f)$ is the t-minor with row indices $\lambda_1, \ldots, \lambda_t$ and column indices $\mu_1, \ldots, \mu_t$.*

As a consequence of the above Theorem, we obtain the first and second fundamental Theorems of classical invariant theory (cf. [**162**])

THEOREM 2.10.12. *Let* $V = k^n$, $X = \underbrace{V \times \cdots \times V}_{m \text{ copies}} \times \underbrace{V^* \times \cdots \times V^*}_{m \text{ copies}}$, *where* $m > n$. *Let* f_{ij} *denote the regular function* $(\underline{x}, \underline{\xi}) \mapsto \langle x_i, \xi_j \rangle$ *on* X, $1 \leqslant i, j \leqslant m$. *Consider the diagonal action of* $GL(V)$ *on* X.

(1) **First fundamental theorem** *The ring of invariants* $k[X]^G$ *is generated by* f_{ij}, $1 \leqslant i, j \leqslant m$.
(2) **Second fundamental theorem** *The ideal of relations among the generators in (1) is generated by the* $(n+1)$*-minors of the* $m \times m$*-matrix* (f_{ij}).
(3) **A standard monomial basis for the ring of invariants** *The ring of invariants* $k[X]^G$ *has a basis consisting of standard monomials in the regular functions* $p_{\lambda,\mu}(f)$ *with* $\#\lambda, \#\mu \leqslant n$, *where* $p_{\lambda,\mu}(f)$ *is the t-minor of* (f_{ij}) *with row indices* $\lambda_1, \ldots, \lambda_t$ *and column indices* $\mu_1, \ldots, \mu_t$ *(here,* $t = \#\lambda = \#\mu$*).*

3. Ladder determinantal varieties

Let $X = (x_{ba})$, $1 \leqslant b \leqslant m$, $1 \leqslant a \leqslant n$, be a $m \times n$ matrix of indeterminates.

Given $1 \leqslant b_1 < \cdots < b_h < m$, $1 < a_1 < \cdots < a_h \leqslant n$, we consider the subset of X, defined by

$$L = \{x_{ba} \mid \text{there exists } 1 \leqslant i \leqslant h \text{ such that } b_i \leqslant b \leqslant m, 1 \leqslant a \leqslant a_i\}.$$

We call L a *one-sided ladder* in X, defined by the *outside corners* $\omega_i = x_{b_i a_i}$, $1 \leqslant i \leqslant h$. For example, the following is a picture of a ladder (or Ferrers shape) with 4 outside corners.

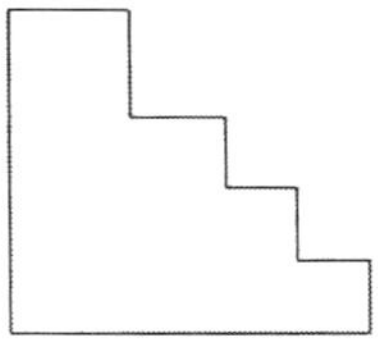

For simplicity of notation, we identify the variable x_{ba} with (b, a).

Let $2 \leqslant t \leqslant \min\{m - b_i + 1, a_i\}$, for all $1 \leqslant i \leqslant h$. For $1 \leqslant i \leqslant h$, let

$$L_i = \{x_{ba} \in L \mid b_i \leqslant b \leqslant m\}.$$

Let $k[L]$ denote the polynomial ring $k[x_{ba} \mid x_{ba} \in L]$, and let $\mathbb{A}(L) = \mathbb{A}^{|L|}$ be the associated affine space. Let $I_t(L)$ be the ideal in $k[L]$ generated by all the t-minors contained in L, and $D_t(L) \subset \mathbb{A}(L)$ the variety defined by the ideal $I_t(L)$. We call $D_t(L)$ a *ladder determinantal variety* (associated to a one-sided ladder).

For $1 \leqslant i \leqslant h$, let

$$L(i) = \{x_{ba} \in L_i \mid 1 \leqslant a \leqslant a_i\}.$$

REMARK 3.0.1. Given a t-minor Δ, say with row indices $\{r_1, \cdots, r_t\}$, and column indices $\{c_1, \cdots, c_t\}$, let i, $1 \leqslant i \leqslant h$ be such that $x_{r_1 c_t}$ belongs to $L(i)$ (note that $L = \cup_i L(i)$). Then Δ is contained in $L(i)$. Thus the ideal $I_t(L)$ is generated by the t-minors of X contained in $L(i)$, $1 \leqslant i \leqslant h$.

From now on we shall suppose that $m = n$, with n large enough such that L is situated below the main diagonal, i.e. $b_i \geqslant a_i + 1$, $1 \leqslant i \leqslant h$. Let $G = SL(n)$, $Q = P_{a_1} \cap \cdots \cap P_{a_h}$. Let O^- be the opposite big cell in G/Q. Let H be the one-sided ladder defined by the outside corners $(a_i + 1, a_i)$, $1 \leqslant i \leqslant h$.

Let Z be the variety in $\mathbb{A}(H) \cong O^-$ defined by the vanishing of the t-minors in $L(i)$, $1 \leqslant i \leqslant h$. Note that $Z \cong D_t(L) \times \mathbb{A}(H \setminus L) \cong D_t(L) \times \mathbb{A}^r$, where $r = \dim SL(n)/Q - |L|$.

We shall now define an element $w \in W_Q^{\min}$, such that the variety Z identifies with the opposite cell in the Schubert variety $X(w)$ in G/Q. We define $w \in W_Q^{\min}$ by specifying $w^{(a_i)} \in W^{a_i}$ $1 \leqslant i \leqslant h$, where $\pi_i(X(w)) = X(w^{(a_i)})$ under the projection $\pi_i : G/Q \to G/P_{a_i}$.

Define $w^{(a_i)}$, $1 \leqslant i \leqslant h$, inductively, as the (unique) maximal element in W^{a_i} such that

(1) $w^{(a_i)}(a_i - t + 1) = b_i - 1$.

(2) if $i > 1$, then $w^{(a_{i-1})} \subset w^{(a_i)}$.

Note that $w^{(a_i)}$, $1 \leqslant i \leqslant h$, is well defined in W^i, and w is well defined as an element in $W_Q^{\min}$.

REMARK 3.0.2. Note that $w^{(a_i)}$, $1 \leqslant i \leqslant h$, consists of several blocks of consecutive integers ending with $b_j - 1$ at the $(a_j - t + 1)$-th place for some j's belonging to $\{1, \cdots, i\}$, and a last block (of length $(t-1)$) ending with n at the a_j-th place.

Example. Let $n = 13$, $h = 3$, $t = 3$, $(b_1, a_1) = (6, 4)$, $(b_2, a_2) = (8, 5)$, $(b_3, a_3) = (11, 9)$. Then

$$w^{(4)} = (4, 5,\ 12, 13),\ w^{(5)} = (4, 5,\ 7,\ 12, 13),\ w^{(9)} = (4, 5, 6, 7, 8, 9, 10,\ 12, 13)$$

THEOREM 3.0.3. *The variety Z $(= D_t(L) \times \mathbb{A}^r)$ identifies with the opposite cell in $X(w)$, i.e. $Z = X(w) \cap O^-$ (scheme theoretically).*

PROOF. Let $f = \det M$, where M is a $t \times t$ matrix contained in $L(i)$ for some $1 \leqslant i \leqslant h$, be a generator of $I(Z)$ (cf. Remark 3.0.1). By Lemma 1.11.1, f can be written in the form $f = \sum g_\phi p_\phi|_{O^-}$, with $\phi \in W^{a_i}$ such that

$$\{\phi(1), \ldots, \phi(a_i)\} \cap \{a_i + 1, \ldots, n\} = \{r_1, \ldots, r_t\},$$

and $g_\phi \in k[H]$ (here $r_1, \ldots, r_t$ are the row indices of M). In particular, we have $\phi(a_i - t + 1) = r_1$. Since M is contained in $L(i)$, we have $r_1 \geqslant b_i$, and hence $\phi(a_i - t + 1) \geqslant b_i$. We have $w^{(a_i)}(a_i - t + 1) = b_i - 1$, and hence $\phi(a_i - t + 1) > w^{(a_i)}(a_i - t + 1)$. This shows that $\phi \not\leqslant w^{(a_i)}$, and therefore $p_\phi \in I(X(w) \cap O^-)$. Thus $f \in I(X(w) \cap O^-)$.

Let now g be a generator of the ideal $I(X(w) \cap O^-)$, i.e. $g = p_\tau|_{O^-}$, with $\tau \in W^{a_i}$ for some i, $1 \leqslant i \leqslant h$, such that $\tau \not\leqslant w^{(a_i)}$. Since $w^{(a_i)}$ consists of several blocks of consecutive integers ending with $b_j - 1$ at the a_j-th place for some $j \in \{1, \cdots, i\}$, and a last block ending with n at the a_j-th place (cf. Remark 3.0.2), it follows that $\tau(a_j - t + 1) \geqslant b_j$ for some $j \in \{1, \cdots, i\}$. Using Lemma 1.11.2, we deduce that $p_\tau|_{O^-}$ belongs to the ideal of $k[H]$ generated by t-minors in L with row indices $\geqslant b_j$, and column indices $\leqslant a_j$. Thus $p_\tau|_{O^-}$ belongs to the ideal generated by t-minors contained in $L(j)$, which shows that $g \in I(Z)$. □

Since the Schubert varieties are irreducible, normal, Cohen-Macaulay, and have rational singularities (cf. **[85]**, **[135]**, **[137]**, **[136]**), as a consequence of Theorem 4.0.5 we obtain

THEOREM 3.0.4. *The variety $D_t(L)$ is irreducible, normal, Cohen-Macaulay, and has rational singularities.*

3.1. A Gröbner basis for $D_t(L)$

We recall the following well-known

LEMMA 3.1.1. *Let $X = (x_{ij})$ be a generic $m \times n$ matrix. Let $k[X]$ be the polynomial ring in the set of indeterminates x_{ij}, I an ideal of $k[X]$, and G a Gröbner basis of I with respect to a certain monomial order. Let $L \subset X$ such that*

$$\text{if } f \in G \text{ and } in(f) \in k[L], \text{ then } f \in k[L].$$

Then the set $G \cap k[L]$ is a Gröbner basis of the ideal $I \cap k[L]$.

PROOF. Let $g \in I \cap k[L]$. Since G is a Gröbner basis of I, there exists $f \in G$ such that $\text{in}(g) = \langle \text{in}(f) \rangle$. Since $g \in k[L]$, we have $\text{in}(g) \in k[L]$, and hence $\text{in}(f) \in k[L]$. By hypothesis, $f \in k[L]$, and hence $f \in G \cap k[L]$. Therefore, the initial terms of the elements of $G \cap k[L]$ generate the ideal $\text{in}(I \cap k[L])$. □

As a direct consequence, we obtain the following

PROPOSITION 3.1.2. *Let $L \subset X$ be a one-sided ladder and $\mathcal{G}_t(L)$ be the set of all t-minors in L. Then $\mathcal{G}_t(L)$ is a Gröbner basis of $I_t(L)$ with respect to the monomial order $\prec$.*

PROOF. By Theorem 2.6.2, $\mathcal{G}_t$ $(= \{$ all t-minors in $X\})$ is a Gröbner basis of $I_t(X)$. Let $\Delta \in \mathcal{G}_t$ be such that $in(\Delta) \in k[L]$. Now $in(\Delta)$ (with respect to the monomial order $\prec$) is simply the product of the entries on the anti-diagonal of Δ. Hence the fact that $in(\Delta) \in k[L]$ implies that $(\Delta) \in k[L]$. From this it follows (by Lemma 3.1.1), $\mathcal{G}_t(L) = (\mathcal{G}_t \cap k[L])$ is a Gröbner basis of the ideal $I_t(X) \cap k[L]$. On the other hand we have that $\mathcal{G}_t(L)$ generates $I_t(L)$, and the result follows. □

3.2. The dimension of $D_t(L)$

Let $\mathcal{C}_t(X)$ be the set of variables in the submatrix obtained from X by deleting the first $t - 1$ columns and the last $t - 1$ rows. Set $\mathcal{C}_t(L) = \mathcal{C}_t(X) \cap L$. For $i \leqslant h$, let $X_i = \{x_{ba} \in X \mid b_i \leqslant b \leqslant m\}$, $\tilde{X}_i = X_i \setminus X_{i+1}$ (here, if $i = h$, then X_{i+1} is supposed to be the empty set). Let $\mathcal{C}_i(X) = \mathcal{C}_t(X) \cap \tilde{X}_i$ (note that $\mathcal{C}_t(X) = \dot{\cup}\mathcal{C}_i(X)$).

For a one-sided ladder $L \subset X$, let $\mathcal{C}_i(L) = \mathcal{C}_i(X) \cap L$.

Note that in a solid minor in L (i.e. a minor with consecutive row indices and consecutive column indices), the smallest (for the order in §2.6) element belongs to $\mathcal{C}_t(L)$, and conversely, an element $\alpha \in \mathcal{C}_t(L)$ determines uniquely a solid minor in L having α as the smallest element. Hence the number of elements in $\mathcal{C}_t(L)$ is equal to the number of solid minors in the set L.

The following is a generalization of Proposition 8 in [**59**].

PROPOSITION 3.2.1. *Let $L \subset X$ an one-sided ladder. Then*

$$\operatorname{codim}_{\mathbb{A}(L)} D_t(L) = |\mathcal{C}_t(L)|.$$

PROOF. By Proposition 3.1.2, the ideal $I_t(L)$ and the ideal $J_t(L)$ of its initial terms determine graded quotient rings of $k[L]$ having the same Hilbert series, and hence the codimension of the variety $D_t(L)$ is equal to the height of the monomial ideal $J_t(L)$. In general, the height of a monomial ideal J in a polynomial ring $k[x_1, \dots, x_N]$ is equal to the minimal cardinality of a set $\mathcal{C} \subset \{x_1, \dots, x_N\}$ of variables such that

(∗) each monomial in a set of monomial generators for J contains a variable from $\mathcal{C}$.

Let $J = J_t(L)$ and $\mathcal{C} = \mathcal{C}_t(L)$. Then it is easy to see that $\mathcal{C}$ satisfies (∗), the set of monomial generators being the set of the initial terms of all the t-minors in L (note that for a t-minor $\Delta \in L$, $\operatorname{in}(\Delta)$ (with respect to the monomial order $\prec$) is simply the product of the entries on the anti-diagonal of Δ). Let us denote

$$\Delta_k = \{x_{ba} \in L \mid b + a = k + 1\}, \quad k \geqslant 1.$$

Then $L = \dot{\bigcup}_{k\geqslant 1}\Delta_k$, and $\mathcal{C} = \dot{\bigcup}_{k\geqslant 1}(\mathcal{C} \cap \Delta_k)$.

Let now $\mathcal{C}' \subset \{x_{ba} \mid x_{ba} \in L\}$ be a set such that $|\mathcal{C}'| < |\mathcal{C}|$. Then there exists a k such that $|\mathcal{C}' \cap \Delta_k| < |\mathcal{C} \cap \Delta_k|$ (in particular $\mathcal{C} \cap \Delta_k \neq \emptyset$). Let $i \in \{1, \dots, l\}$ be the largest such that $\Delta_k \cap \mathcal{C} \subset L_i$. Then

$$|\mathcal{C}' \cap \Delta_k| < |\mathcal{C} \cap \Delta_k| = |\Delta_k| - (t - 1).$$

Therefore there exist t distinct variables in $\Delta_k \setminus \mathcal{C}'$. Thus the initial term of the t-minor in L_i having these elements on the skew diagonal does not contain any variable in $\mathcal{C}'$, and hence $\mathcal{C}'$ does not satisfy (∗).

Therefore $\mathcal{C}$ is a set of minimal cardinality among the sets satisfying (∗), and the required result follows. □

3.3. The singular locus of $D_t(L)$

Let $X = (x_{ba})$, $1 \leqslant b < m$, $1 < a \leqslant n$ be a $m \times n$ matrix of indeterminates. Let $L \subset X$ be a one-sided ladder defined by the outside corners $\omega_i = x_{b_i a_i}$, $1 \leqslant i \leqslant h$, $1 \leqslant b_1 < \cdots < b_h \leqslant m$, $1 \leqslant a_1 < \cdots < a_h \leqslant n$. Let $V = D_t(L)$, $\mathcal{C} = \mathcal{C}_t(L)$.

For $1 \leqslant i \leqslant h$, let $V_i \subset \mathbb{A}(L)$ be the variety defined by the vanishing of the t-minors in $L(j)$, with $j \in \{1, \ldots, l\} \setminus \{i\}$, and the $(t-1)$-minors in $L(i)$.

THEOREM 3.3.1. *With notations as above, we have*

$$\operatorname{Sing} V = \bigcup_{i=1}^{h} V_i.$$

PROOF. For simplicity of notation, we identify the variable x_{ba} with the element (b, a).

First, we prove that $V_i \subset \operatorname{Sing} V$, for all $1 \leqslant i \leqslant h$. Let $x \in V_i$ for some $1 \leqslant i \leqslant h$. Let $\mathcal{J}$ be the Jacobian matrix associated to the variety $V \subset \mathbb{A}(L)$, evaluated at x. Then the rows of $\mathcal{J}$ are indexed by t-minors in $L(j)$, $1 \leqslant j \leqslant h$, and the columns are indexed by the elements $\alpha \in L$. The (M, α)-th entry in $\mathcal{J}$ is equal to $\pm(\det M')(x)$, where M' is the matrix obtained from M by deleting the row and the column containing α, if α appears in M, and 0 otherwise.

Fix i, $1 \leqslant i \leqslant h$. It is easily seen that $\omega_i \in \mathcal{C}_t(L)$. Now consider the one-sided ladder L' obtained from L by deleting the element ω_i, i.e. the one-sided ladder defined by the outside corners

$$\begin{aligned} &\omega_1 = (b_1, a_1), \ldots, \omega_{i-1} = (b_{i-1}, a_{i-1}), \omega_{i^-} = (b_i, a_i - 1), \\ &\omega_{i^+} = (b_i + 1, a_i), \omega_{i+1} = (b_{i+1}, a_{i+1}), \ldots, \omega_h = (b_h, a_h), \end{aligned}$$

where ω_{i^-} is present only if $a_i - 1 > a_{i-1}$, and ω_{i^+} is present only if $b_i + 1 < b_{i+1}$.

Since $x \in V_i$, a row of $\mathcal{J}$ indexed by a t-minor involving $\omega_i = x_{b_i a_i}$ is 0 (note that such a t-minor is completely contained in $L(i)$). Also, the column of $\mathcal{J}$ indexed by ω_i is 0 (note again that a t-minor involving ω_i is completely contained in $L(i)$). Let $\mathcal{J}'$ be the matrix obtained from $\mathcal{J}$ by deleting the column indexed by ω_i and the rows indexed by t-minors containing ω_i. Then

$$\operatorname{rank} \mathcal{J} = \operatorname{rank} \mathcal{J}',$$

since $\mathcal{J}'$ is obtained from $\mathcal{J}$ by deleting rows and columns consisting of zero entries. Let $x = (x_\alpha)_{\alpha \in L}$, $x' = (x_\alpha)_{\alpha \in L'}$. Then $x' \in D_t(L')$, and $\mathcal{J}'$ is the Jacobian matrix associated to the variety $D_t(L') \subset \mathbb{A}(L')$, evaluated at x'. Thus

$$\operatorname{rank} \mathcal{J}' \leqslant \operatorname{codim}_{\mathbb{A}(L')} D_t(L').$$

Now, using Proposition 3.2.1 we obtain

$$\operatorname{codim}_{\mathbb{A}(L')} D_t(L') = |\mathcal{C}_t(L')| = |\mathcal{C}_t(L) \setminus \{\omega_i\}| < |\mathcal{C}_t(L)| = \operatorname{codim}_{\mathbb{A}(L)} D_t(L).$$

Hence $\operatorname{rank} \mathcal{J}' < \operatorname{codim}_{\mathbb{A}(L)} V$, which implies $\operatorname{rank} \mathcal{J} < \operatorname{codim}_{\mathbb{A}(L)} V$, i.e. $x \in \operatorname{Sing} V$.

Now we prove that $\operatorname{Sing} V \subset \bigcup_{i=1}^{l} V_i$. Let $\mathcal{C} = \mathcal{C}_t(L)$ (note that $\mathcal{C} = \bigcup_{i=1}^{h} \mathcal{C}_i$, as defined in §3.2).

We introduce a total order on the set of minors of L of size r, with $r \geqslant 1$ fixed, as follows: $[i_1, \ldots, i_r | j_1, \ldots, j_r] < [i'_1, \ldots, i'_r | j'_1, \ldots, j'_r]$ if there exists $1 \leqslant k \leqslant r$ such that

$$\text{either } i_1 = i'_1, \ldots, i_{k-1} = i'_{k-1}, i_k < i'_k,$$

$$\text{or } i_1 = i'_1, \ldots, i_r = i'_r, j_1 = j'_1, \ldots, j_{k-1} = j'_{k-1}, j_k < j'_k$$

(this is simply the lexicographic order on $\{i_1, \ldots, i_r, j_1, \ldots, j_r\}$).

Let $x \in V \setminus \bigcup_{i=1}^{h} V_i$. For each $1 \leqslant i \leqslant h$, let M_i be the largest $(t-1)$-minor in $L(i)$ such that $(\det M_i)(x) \neq 0$. Let $\mathcal{T}_h$ be the set of elements in $L(h)$ not in the rows or the columns given by the rows and the columns of M_h. Clearly, $|\mathcal{T}_h| = |\mathcal{C}_h|$. By (decreasing) induction on i, suppose that, for some i, $1 < i \leqslant h$, the sets $\mathcal{T}_i, \ldots, \mathcal{T}_h$ have been constructed, such that

$(1)_i$ $\mathcal{T}_j \subset L(j)$, $i \leqslant j \leqslant h$,

$(2)_i$ the sets $\mathcal{T}_i, \ldots, \mathcal{T}_l$ are pairwise disjoint,

$(3)_i$ $|\mathcal{T}_j| = |\mathcal{C}_j|$, $i \leqslant j \leqslant h$,

$(4)_i$ $\mathcal{T}_j$ contains no elements appearing in the rows or in the columns of L given by the rows and the columns of M_j, $i \leqslant j \leqslant h$,

$(5)_i$ there exist $t-1$ rows in $L(i)$ not containing any element from $\mathcal{T}_i \cup \cdots \cup \mathcal{T}_h$.

We define the set $\mathcal{T}_{i-1}$ as follows. Let r be the number of the rows of M_{i-1} contained in $\tilde{L}(i-1) := L(i-1) \setminus L(i)$ (note that $r \leqslant t-1$). Then $\mathcal{T}_{i-1}$ is obtained from $\tilde{L}(i-1)$ by deleting the r rows given by the rows of M_{i-1}, then adding r rows from the $t-1$ rows of $L(i)$ in $(5)_i$ which are not rows of M_{i-1}, intersected with $L(i-1)$ (this is possible, since there are $t-1-r$ rows of M_{i-1} in $L(i)$, and hence at least $(t-1)-(t-1-r) = r$ rows from the $t-1$ rows of $L(i)$ in $(5)_i$ are not rows of M_{i-1}), followed by the deletion of the $t-1$ columns in $\tilde{L}(i-1)$ obtained by intersecting the columns of M_{i-1} with $\tilde{L}(i-1)$.

Again, the properties $(1)_{i-1}$-$(4)_{i-1}$ are obvious; the r rows of M_{i-1} which were deleted from $\tilde{L}(i-1)$, and the $(t-1)-r$ rows from the $t-1$ rows in $(5)_i$ which were not used while defining $\mathcal{T}_{i-1}$, intersected with $L(i-1)$, give $t-1$ rows of $L(i-1)$ not containing any elements in $\mathcal{T}_{i-1} \cup \mathcal{T}_i \cup \cdots \cup \mathcal{T}_h$, so that we have $(5)_{i-1}$.

Thus, using induction, we obtain the disjoint sets $\mathcal{T}_j \subset L(j)$, $1 \leqslant j \leqslant h$, such that $|\mathcal{T}_j| = |\mathcal{C}_j|$, and $\mathcal{T}_j$ contains no elements in the rows or columns of L given by the rows and columns of M_j.

For $\tau \in \mathcal{T}_i \subset \mathcal{T}$, $1 \leqslant i \leqslant h$, let M^τ be the t-minor obtained from M_i by adding the row and the column containing τ. Obviously, $M^\tau \neq M^{\tau'}$ for $\tau, \tau' \in \mathcal{T}$, with $\tau \neq \tau'$.

We now take a total order on $\mathcal{T}$, namely $(b, a) > (b', a')$ if either $b > b'$, or $b = b'$ and $a > a'$.

Let us fix $\tau \in \mathcal{T}$, say $\tau \in \mathcal{T}_i$ for some i, $1 \leqslant i \leqslant l$. Then the (M^τ, τ)-th entry in $\mathcal{J}$ is equal to $\pm(\det M_i)(x)$, so it is nonzero. Let now $\sigma \in \mathcal{T}$, $\sigma < \tau$. If σ is not an entry of M^τ, then the (M^τ, σ)-th entry of $\mathcal{J}$ is equal to 0. Assume now that σ is the (r,s)-th entry of M^τ. Then the (M^τ,σ)-th entry of $\mathcal{J}$ is equal to $\pm(\det M')(x)$, where M' is the $(t-1)\times(t-1)$ matrix obtained from M^τ by deleting the r-th row and the s-th column.

Let $\tau=(b,a)$, $\sigma=(b',a')$. If $b'<b$, then the indices of the first $r-1$ rows of M' and M_i are the same, while the index of the r-th row of M' is $>b'$, which is the index of the r-th row of M_i. Thus, $M'>M_i$, and by the maximality of M_i, we obtain $(\det M')(x)=0$. If $b'=b$, then $a'<a$. The indices of all the rows and those of the first $s-1$ columns are the same, while the index of the s-th column in M' is $>a'$, which is the index of the s-th column of M_i. Thus $M'>M_i$, and the maximality of M_i implies that $(\det M')(x)=0$. Thus, for $\sigma<\tau$, the (M^τ,σ)-th entry in $\mathcal{J}$ is 0.

Let $\mathcal{J}'$ be the submatrix of $\mathcal{J}$given by the rows indexed by M^τ's and the columns indexed by τ's, with $\tau\in\mathcal{T}$. We suppose that both rows and columns of $\mathcal{J}'$ are indexed by the elements in $\mathcal{T}$, and we arrange them increasingly, with respect to the total order on $\mathcal{T}$ defined above. Then $\mathcal{J}'$ is upper triangular, and all the diagonal entries are nonzero. Thus $\det\mathcal{J}'\neq 0$, and this implies that

$$\operatorname{rank}\mathcal{J}'=|\mathcal{T}|=|\mathcal{C}|=\operatorname{codim}_{\mathbb{A}(L)}D_t(L).$$

Consequently $\operatorname{rank}\mathcal{J}=\operatorname{codim}_{\mathbb{A}(L)}V$, i.e. $x\notin\operatorname{Sing}V$. □

3.4. The irreducible components of $\operatorname{Sing}D_t(L)$

We preserve the notations of §3.3.

Let us fix $j\in\{1,\ldots,l\}$, and let $Z_j=V_j\times\mathbb{A}(H\setminus L)$. We shall now define $\theta_j\in W_Q^{\min}$ such that the variety Z_j identifies with the opposite cell in the Schubert variety $X(\theta_j)$ in G/Q.

Note that $w^{(a_j)}(a_j-t+1)=b_j-1$, and b_j-1 is the end of a block of consecutive integers in $w^{(a_j)}$. Also, the beginning of this block is $\geqslant 2$ (if the block started with 1, we would have $a_j-t+1=b_j-1$, which would then imply $2-t=b_j-a_j>0$, i.e., $t<2$ which is not possible, since $t\geqslant 2$). Let u_j+1 be the beginning of this block, where $u_j\geqslant 1$. The fact that $u_j\notin w^{(a_j)}$ implies that $u_j\notin w^{(a_i)}$, $i\leqslant j$. For $i>j$, if b_j-1 is the end of a block in $w^{(a_i)}$, then the beginning of the block is u_j+1 (since $w^{(a_i)}(a_j-t+1)=b_j-1$ $(=w^{(a_j)}(a_j-t+1))$ (cf. Remark 3.0.2) and $w^{(a_i)}\supset w^{(a_j)}$). For each i, $1\leqslant i\leqslant h$, such that $u_j\notin w^{(a_i)}$, let v_i be the smallest entry in $w^{(a_i)}$ which is bigger than b_j-1.

REMARK 3.4.1. Note that for $i\leqslant j$, $v_i=n-t+2$, and for $i>j$ such that $u_j\notin w^{(a_i)}$, $v_i=w^{(a_i)}(a_j-t+2)$.

Define ${\theta_j}^{(a_i)}$, $1\leqslant i\leqslant h$, as follows.

If $b_j-1\notin w^{(a_i)}$ (equivalently $i<j$), let $\theta_j^{(a_i)}=w^{(a_i)}\setminus\{v_i\}\cup\{b_j-1\}$.

If $b_j-1\in w^{(a_i)}$ (equivalently $i\geqslant j$) and $u_j\notin w^{(a_i)}$, then

$$\theta_j^{(a_i)}=w^{(a_i)}\setminus\{v_i\}\cup\{u_j\}.$$

If $b_j-1\in w^{(a_i)}$ and $u_j\in w^{(a_i)}$, then $\theta_j^{(a_i)}=w^{(a_i)}$.

Note that θ_j is well defined as an element in $W_Q^{\min}$, and $\theta_j \leqslant w$.

REMARK 3.4.2. Note that for $i < j$, $\theta_j^{(a_i)}$ differs from $w^{(a_i)}$ only at the $a_i - t + 1$-th place ($w^{(a_i)}(a_i - t + 1) = n - t + 2$, $\theta_j^{(a_i)}(a_i - t + 1) = b_j - 1$). Also, for $i \geqslant j$ such that $u_j \notin w^{(a_i)}$, $\theta_j^{(a_i)}(a_j - t + 2) = b_j - 1$ while $w^{(a_i)}(a_j - t + 2) = v_i$; further, there is a block in $\theta_j^{(a_i)}$ ending with $b_j - 1$ at the $(a_j - t + 2)$-th place. In particular, we have $\theta_j^{(a_j)}(a_j - t + 2) = b_j - 1$

Example. Let $n = 13$, $h = 3$, $t = 3$, $(b_1, a_1) = (6, 4)$, $(b_2, a_2) = (8, 5)$, $(b_3, a_3) = (11, 9)$. Then we have,

$$w^{(4)} = (4, 5, 12, 13),\ w^{(5)} = (4, 5, 7, 12, 13),\ w^{(9)} = (4, 5, 6, 7, 8, 9, 10, 12, 13).$$

We have

$$\theta_1^{(4)} = (3, 4, 5, 13), \quad \theta_1^{(5)} = (3, 4, 5, 12, 13), \quad \theta_1^{(9)} = (3, 4, 5, 7, 8, 9, 10\ 12, 13),$$
$$\theta_2^{(4)} = (4, 5, 7, 13), \quad \theta_2^{(5)} = (4, 5, 6, 7, 13), \quad \theta_2^{(9)} = (4, 5, 6, 7, 8, 9, 10, 12, 13),$$
$$\theta_3^{(4)} = (4, 5, 10, 13), \quad \theta_3^{(5)} = (4, 5, 7, 10, 13), \quad \theta_3^{(9)} = (3, 4, 5, 6, 7, 8, 9, 10, 13).$$

THEOREM 3.4.3. *The subvariety $Z_j \subset Z$ identifies with the opposite cell in $X(\theta_j)$, i.e. $Z_j = X(\theta_j) \cap O^-$ (scheme theoretically).*

PROOF. Let $f = \det M$, M being either a t-minor contained in $L(i)$, $i \in \{1, \ldots, h\} \setminus \{j\}$, or a $(t-1)$-minor contained in $L(j)$ be a generator of $I(Z_j)$. In the former case we have $f \in I(Z)$, and Theorem 4.0.5 implies that

$$f \in I\big(X(w) \cap O^-\big) \subset I\big(X(\theta_j) \cap O^-\big).$$

In the latter case, M is contained in H_j (of the one-sided ladder H with outside corners (a_{i+1}, a_i), $1 \leqslant i \leqslant h$). By Lemma 1.11.1, f can be written in the form $f = \sum g_\phi p_\phi|_{O^-}$, with $\phi \in W^{a_j}$ such that $\{\phi(1), \ldots, \phi(a_j)\} \cap \{a_j + 1, \ldots, n\} = \{r_1, \ldots, r_{t-1}\}$, and $g_\phi \in k[H]$ (here $r_1, \ldots, r_{t-1}$ are the row indices of M). In particular we have $\phi(a_j - t + 2) = r_1$. Since M is contained in $L(j)$, we deduce that $r_1 \geqslant b_j$, and hence $\phi(a_j - t + 2) \geqslant b_j$. We have $\theta_j{}^{(a_j)}(a_j - t + 2) = b_j - 1$, and hence $\phi(a_j - t + 2) > \theta_j{}^{(a_j)}(a_j - t + 2)$. This shows that $\phi \not\leqslant \theta_j{}^{(a_j)}$, and therefore $p_\phi \in I\big(X(\theta_j) \cap O^-\big)$. Thus $f \in I\big(X(\theta_j) \cap O^-\big)$.

Let now $g = p_\tau|_{O^-}$, with $\tau \in W^{a_i}$ for some i, $1 \leqslant i \leqslant h$, such that $\tau \not\leqslant \theta_j^{(a_i)}$, be a generator of the ideal $I(X(\theta_j) \cap O^-)$. Since $I(X(w) \cap O^-) = I(Z)$, $I(Z) \subset I(Z_j)$, we may suppose $p_\tau \notin I(X(w) \cap O^-)$, and thus may suppose $\tau \leqslant w^{(a_i)}$, $\tau \not\leqslant \theta_j^{(a_i)}$.

Let $i \leqslant j$. The facts that $\tau \leqslant w^{(a_i)}$, $\tau \not\leqslant \theta_j^{(a_i)}$, and $i \leqslant j$ imply that $\tau(a_i - t + 2) \geqslant b_j$. Hence using Lemma 1.11.2, we deduce that $p_\tau|_{O^-}$ belongs to the ideal of $k[H]$ generated by $(t-1)$-minors with row indices $\geqslant b_j$, and column indices $\leqslant a_i$ (note that such minors are contained in $L(j)$). Thus $p_\tau|_{O^-}$ belongs to the ideal generated by $(t-1)$-minors contained in $L(j)$, which implies that $g \in I(Z_j)$.

Let $i > j$. We have $w^{(a_i)} \neq \theta_j^{(a_i)}$ (since $\tau \leqslant w^{(a_i)}$, $\tau \not\leqslant \theta_j^{(a_i)}$), and hence $u_j \notin w^{(a_i)}$ (by definition of θ_j). Now $\theta_j^{(a_i)}$ consists of several blocks of consecutive integers ending with $b_m - 1$ at the $(a_m - t+1)$-th place, for some $m \in \{1, \ldots, i\} \setminus \{j\}$, a block ending with $b_j - 1$ at the $(a_j - t + 2)$-th place (cf. Remark 3.4.2), and a last block ending with n at the a_i-th place. As above, the facts that $\tau \leqslant w^{(a_i)}$, $\tau \not\leqslant \theta_j^{(a_i)}$ imply that $\tau(a_j - t + 2) \geqslant b_j$. Using Lemma 1.11.2, we deduce that $p_\tau|_{O^-}$ belongs to the ideal of $k[H]$ generated by $(t-1)$-minors with row indices $\geqslant b_j$, and column indices $\leqslant a_j$. Thus $p_\tau|_{O^-}$ belongs to the ideal generated by $(t-1)$-minors contained in $L(j)$, which implies that $g \in I(Z_j)$. □

THEOREM 3.4.4. *The irreducible components of* Sing $D_t(L)$ *are precisely the* V_j*'s*, $1 \leqslant j \leqslant h$.

PROOF. In view of Theorem 3.4.3, we obtain that V_j, $1 \leqslant j \leqslant h$, is irreducible, and the required result follows from Theorem 3.3.1. □

Let $X(w^{\max})$ (resp. $X(\theta_j^{\max})$, $1 \leqslant j \leqslant h$) be the pull-back in $SL(n)/B$ of $X(w)$ (resp. $X(\theta_j)$, $1 \leqslant j \leqslant h$) under the canonical projection $\pi : SL(n)/B \to SL(n)/Q$. Then using Theorems 3.3.1, 4.0.5 and 3.4.3, we obtain

THEOREM 3.4.5. *The irreducible components of* Sing $X(w^{max})$ *are precisely* $X(\theta_j^{max})$, $1 \leqslant j \leqslant h$.

3.5. Validity of the conjecture of Chapter 8, §16.1 for Sing $X(w^{\max})$

Let Q be as above. Let $\pi : G/B \to G/Q$ be the canonical projection. Let $\tau \in W_Q^{\min}$. We have $\pi^{-1}(X_Q(\tau)) = X_B(\tau^{\max})$, where $\tau^{\max}$, as a permutation, is given by $\tau^{(a_1)}$ arranged in descending order, followed by $\tau^{(a_2)} \setminus \tau^{(a_1)}$ arranged in descending order, etc.. We shall refer to the set $\tau^{(a_i)} \setminus \tau^{(a_{i-1})}$, $1 \leqslant i \leqslant h+1$, arranged in descending order, as the i-th block in $\tau^{\max}$ (here, $\tau^{(a_0)} = \emptyset$, and $\tau^{(a_{h+1})}$ is the set $\{1, \ldots, n\} \setminus \tau^{(a_h)}$ arranged in descending order).

Example. Let $n = 13$, $h = 3$, $t = 3$, $(b_1, a_1) = (6,4)$, $(b_2, a_2) = (8,5)$, $(b_3, a_3) = (11,9)$. Then

$$\begin{aligned} w^{\max} &= (13, 12, 5, 4,\ 7,\ 10, 9, 8, 6,\ 11, 3, 2, 1), \\ \theta_1^{\max} &= (13, 5, 4, 3,\ 12,\ 10, 9, 8, 7,\ 11, 6, 2, 1), \\ \theta_2^{\max} &= (13, 7, 5, 4,\ 6,\ 12, 10, 9, 8,\ 11, 3, 2, 1), \\ \theta_3^{\max} &= (13, 10, 5, 4,\ 7,\ 9, 8, 6, 3,\ 12, 11, 2, 1). \end{aligned}$$

For the rest of this section, w and Q will be as in Section 3.

REMARK 3.5.1. Set $b_{h+1} - 1 = n - t + 1$. All of the entries in the i-th block in $w^{\max}$ are $\leqslant b_i - 1$, $2 \leqslant i \leqslant h+1$. In particular, for $1 \leqslant j \leqslant h$, b_j occurs after $b_j - 1$ in $w^{\max}$.

For $\tau \in W$, let P_τ (resp. Q_τ) be the maximal element of the set of parabolic subgroups which leave $\overline{B\tau B}$ (in G) stable under multiplication on the left (resp. right).

LEMMA 3.5.2. *We have*
(1) $Q_{w^{max}} = Q$.
(2) *Let* $I_{w^{max}} = \{\epsilon_i - \epsilon_{i+1} \mid i = b_j - 1, 1 \leqslant j \leqslant h\}$. *Then* $S_{P_{w^{max}}} = S \setminus I_{w^{max}}$.
The assertions are clear from the description of $w^{\max}$ in view of Remark 3.5.1.

LEMMA 3.5.3. *Let* $P = P_{w^{max}}$, $Q = Q_{w^{max}}$. *Then* $\overline{B\theta_j^{max}B}$ *is P-Q stable.*

PROOF. The Q-stability of $\overline{B\theta_j^{\max}B}$ on the right is obvious.
Regarding the P-stability of $\overline{B\theta_j^{\max}B}$ on the left, we shall now show that if $x \neq b_l$, for any l, $1 \leqslant l \leqslant h$, then $x - 1$ occurs after x in $\theta_j^{\max}$ (note that

$$S_P = \{\epsilon_i - \epsilon_{i+1} \mid i \neq b_j - 1, 1 \leqslant j \leqslant h\}$$

(cf. Lemma 3.5.2)).
Let u_j appear in the k-th block in $w^{\max}$, for some $k > j$.
Fix $x \in \{1, \ldots, n\}$, $x \neq b_l$, for any l, $1 \leqslant l \leqslant h$.
If $x \neq v_i$, $j \leqslant i < k$ (v_i being as in the definition of θ_j), then clearly $x - 1$ occurs after x in $\theta_j^{\max}$ also.
Let then $x = v_i$, for some i, $j \leqslant i < k$. The fact that $v_i \neq b_l$, for any l, $1 \leqslant l \leqslant h$ implies that $v_i - 1 \neq b_l - 1$, for any l, $1 \leqslant l \leqslant h$; in particular, $v_i - 1 \neq b_j - 1$. Hence we conclude (by the definition of v_i) that $v_i - 1$ occurs in the m-th block in $w^{\max}$, for some $i < m \leqslant k$. This implies that for $i \leqslant l < m$, $v_l = v_i$, and if $m < k$, then $v_m \leqslant v_{i-1}$. From this it follows that in $\theta_j^{\max}$, v_i occurs in the m-th block while v_{i-1} occurs in the q-th block where $q \geqslant m$. From this it follows that $v_i - 1$ appears after v_i in $\theta_j^{\max}$ also. □

An occurrence of the form (d, b, c, a) (resp. (c, d, a, b)) will be referred to as a bad pattern of Type I (resp. Type II).

LEMMA 3.5.4. *Any bad pattern in* w^{max} *is of Type I.*

PROOF. Let $w^{\max} = (a_1 \ldots a_n)$. Assume that (c, d, a, b) is a bad pattern of Type II in $w^{\max}$, where $a < b < c < d$. Clearly, c and d (resp. a and b) cannot both appear in the same block, in view of the description of $w^{\max}$. Let then c, d, a, b appear in the r-th, i-th, j-th, k-th blocks respectively, where $r < i \leqslant j < k$. This implies that $a < b < c < d \leqslant b_i - 1$ (cf. Remark 3.5.1). But now, a and b are both $< b_i - 1$, and they both appear after $b_i - 1$; further, a appears before b in $w^{\max}$, which is not possible by the construction of $w^{\max}$ (note that $a < b$). The required result follows from this. □

REMARK 3.5.5. Of course, there are several bad patterns in $w^{\max}$ of Type I. For example, fix some j, $1 \leqslant j \leqslant h$. Observe that b_j appears after $b_j - 1$ (cf. Remark 3.5.1), and u_j appears after b_j in $w^{\max}$ (notations being as in Section 3.4). Take d to be any entry in $\{n - t + 2, \ldots, n\}$, $b = b_j - 1$, $c = b_j$, $a = u_j$. Then d, b, c, a occur in the 1-st, j-th, k-th, m-th blocks respectively, where $m \geqslant k > j$. This provides an example of a Type I bad pattern in $w^{\max}$.

LEMMA 3.5.6. *Let d, b, c, a be a Type I bad pattern in w^{max}, where $a < b < c < d$. Assume that b belongs to the i-th block, for some i (note that $i \leqslant h$, since $b < c$). Then*

(1) $c < n - t + 2$

(2) $b \leqslant b_i - 1$

(3) $d \geqslant n - t + 2$

PROOF. Let d, b, c, a occur in the r-th, i-th, j-th, k-th blocks respectively in $w^{\max}$, where $r \leqslant i < j \leqslant k$. The hypothesis that $b < c$ implies that $j > 1$. Hence we obtain $c \leqslant b_j - 1$ (cf. Remark 3.5.1), and (1) follows. Now, if $i \geqslant 2$, then the assertion (2) follows from Remark 3.5.1. If $i = 1$, then the assertion (2) follows from the fact that $b < c < n - t + 2$.

Claim. $d > b_i - 1$.

PROOF. Assume that $d \leqslant b_i - 1$. Then assumption implies $c < b_i - 1$ (since $c < d$). Now both c and b are $< b_i - 1$, and b belongs to the i-th block in $w^{\max}$. This implies that c should occur before b, which is not possible. Hence our assumption is wrong, and the claim follows. □

Note that the Claim and Remark 3.5.1 imply that $d \geqslant n - t + 2$, and d appears in the first block. □

LEMMA 3.5.7. *Fix j, $1 \leqslant j \leqslant h$. Then θ_j^{max} is the unique maximal element of the set $\{\tau \in W \mid \tau \leqslant w^{max}, \tau^{(a_j)}(a_j - t + 2) \leqslant b_j - 1\}$.*

The proof is clear from the definition of $\theta_j^{\max}$.

PROPOSITION 3.5.8. *The maximal elements in $F_{w^{max}}$ are precisely the θ_j^{max}, for $1 \leqslant j \leqslant h$ (here $F_{w^{max}}$ is as in Ch8, §16.1).*

PROOF. We first observe that $\theta_j^{\max} \in F_{w^{\max}}$; for, corresponding to the bad pattern $d = n - t + 2, b = b_j - 1, c = b_j, a = u_j$ (cf. Remark 3.5.5), we have the bad pattern (b, a, d, c) (note that b, a, d, c occur in that order in $\theta_j^{\max}$). Let us denote $\theta_j^{\max}$ by τ'. Let w' (resp. τ) be the element of $\mathfrak{S}_n$ obtained from τ' (resp. w) by replacing b, a, d, c (resp. d, b, c, a) respectively by d, b, c, a (resp. b, a, d, c). Then clearly $\tau \leqslant \tau'$, and $w' \leqslant w$. Further, $\overline{B\theta_j^{\max}B}$ is P-Q stable (cf. Lemma 3.5.3). Thus $\theta_j^{\max} \in F_{w^{\max}}$.

Let now $\tau' \in F_{w^{\max}}$. In particular, we have $\tau' \in W_Q^{\max}$.

We have a bad occurrence in τ' which has to be of the form (b, a, d, c), $a < b < c < d$, corresponding to the occurrence (d, b, c, a) in $w^{\max}$ (cf. Lemma 3.5.4). Let b, a, d, c occur in the p-th, q-th, r-th, s-th blocks respectively in τ', where $p \leqslant q < r \leqslant s$ (note that $\tau' \in W_Q^{\max}$).

We have

$$w'^{(a_q)}(a_q - t + 1) \leqslant w^{(a_q)}(a_q - t + 1) = b_q - 1$$

(here w' is as in Chapter 8, §16.1). Further, $\tau'^{(a_q)}$ is obtained from $w'^{(a_q)}$ by replacing d by a, where $a(< b) < n - t + 2 \leqslant d$ (cf. Lemma 3.5.6). Hence we obtain $a \leqslant b_q - 1$

(since $\tau'^{(a_q)} \leqslant w^{(a_q)}$), and

$$\tau'^{(a_q)}(a_q - t + 2) \leqslant w'^{(a_q)}(a_q - t + 1) \leqslant b_q - 1.$$

This implies $\tau' \leqslant \theta_q^{\max}$ (cf. Lemma 3.5.7) □

THEOREM 3.5.9. *The conjecture of Chapter 8, §16.1 holds for $X(w^{max})$.*

PROOF. In view of Theorem 3.4.5, $X(\theta_j^{\max})$, $1 \leqslant j \leqslant h$ are precisely the irreducible components of $X(w^{\max})$. On the other hand, we have (cf. Proposition 3.5.8) that the maximal elements in $F_{w^{\max}}$ are precisely $\theta_j^{\max}$, $1 \leqslant j \leqslant h$. Hence the irreducible components of Sing $X(w^{\max})$ are precisely

$$\{X(\theta) \mid \theta \text{ a maximal element of } F_{w^{\max}}\}.$$

Thus the conjecture holds for $X(w^{\max})$. □

4. More general ladder determinantal varieties

Let $X = (x_{ba})$, $1 \leqslant b \leqslant m$, $1 \leqslant a \leqslant n$ be a $m \times n$ matrix of indeterminates.

Given $1 \leqslant b_1 < \cdots < b_h < m$, $1 < a_1 < \cdots < a_h \leqslant n$, we consider the subset of X, defined by

$$L = \{x_{ba} \mid \text{ there exists } 1 \leqslant i \leqslant h \text{ such that } b_i \leqslant b \leqslant m, 1 \leqslant a \leqslant a_i\}.$$

We call L an *one-sided ladder* in X, defined by the *outside corners* $\omega_i = x_{b_i a_i}$, $1 \leqslant i \leqslant h$. For simplicity of notation, we identify the variable x_{ba} with (b, a).

Let $\mathbf{s} = (s_1, s_2 \ldots, s_l) \in \mathbb{Z}_+^l$, $\mathbf{t} = (t_1, t_2 \ldots, t_l) \in \mathbb{Z}_+^l$ such that

$$b_1 = s_1 < s_2 < \cdots < s_l \leqslant m,$$

(L1) $$t_1 \geqslant t_2 \geqslant \ldots \geqslant t_l,\ 1 \leqslant t_i \leqslant \min\{m - s_i + 1, a_{i^*}\} \text{ for } 1 \leqslant i \leqslant l, \text{ and}$$

$$s_i - s_{i-1} > t_{i-1} - t_i \text{ for } 1 < i \leqslant l,$$

where i^* denotes the largest integer such that $b_{i^*} \leqslant s_i$.

For $1 \leqslant i \leqslant l$, let

$$L_i = \{x_{ba} \in L \mid s_i \leqslant b \leqslant m\}.$$

Let $k[L]$ denote the polynomial ring $k[x_{ba} \mid x_{ba} \in L]$, and let $\mathbb{A}(L) = \mathbb{A}^{|L|}$ be the associated affine space. Let $I_{\mathbf{s},\mathbf{t}}(L)$ be the ideal in $k[L]$ generated by all the t_i-minors contained in L_i, $1 \leqslant i \leqslant l$, and $D_{\mathbf{s},\mathbf{t}}(L) \subset \mathbb{A}(L)$ the variety defined by the ideal $I_{\mathbf{s},\mathbf{t}}(L)$. We call $D_{\mathbf{s},\mathbf{t}}(L)$ a *ladder determinantal variety* (associated to a one-sided ladder).

Let $\Omega = \{\omega_1, \ldots, \omega_h\}$ and for each $1 < j \leqslant l$, let

$$\Omega_j = \{\omega_i \mid 1 \leqslant i \leqslant h \text{ such that } s_{j-1} < b_i < s_j \text{ and } s_j - b_i \leqslant t_{j-1} - t_j\}.$$

Let

$$\Omega' = (\Omega \setminus \bigcup_{j=2}^{l} \Omega_j) \bigcup_{\Omega_j \neq \emptyset} \{(s_j, a_{j^*})\}.$$

Let L' be the one-sided ladder in X defined by the set of outside corners Ω'. Then it is easily seen that $D_{\mathbf{s},\mathbf{t}}(L) \simeq D_{\mathbf{s},\mathbf{t}}(L') \times \mathbb{A}^d$, where $d = |L| - |L'|$.

Let $\omega'_k = (b'_k, a'_k) \in \Omega'$, for some k, $1 \leqslant k \leqslant h'$, where $h' = |\Omega'|$. If $b'_k \notin \{s_1, \ldots, s_l\}$, then $b'_k = b_i$ for some i, $1 \leqslant i \leqslant h$, and we define $s_{j-} = b_i$, $t_{j-} = t_{j-1}$, $s_{j+} = s_j$, $t_{j+} = t_j$, where j is the unique integer such that $s_j < b_i < s_{j+1}$. Let $\mathbf{s}'$ (resp. $\mathbf{t}'$) be the sequence obtained from $\mathbf{s}$ (resp. $\mathbf{t}$) by replacing s_j (resp. t_j) with s_{j-} and s_{j+} (resp. t_{j-} and t_{j+}) for all k such that $b'_k \notin \{s_1, \ldots, s_l\}$, j being the unique integer such that $s_{j-1} < b_i < s_j$, and i being given by $b'_k = b_i$. Let $l' = |\mathbf{s}'|$. Then $\mathbf{s}'$ and $\mathbf{t}'$ satisfy (L1), and in addition we have $\{b'_1, \ldots, b'_{h'}\} \subset \{s'_1, \ldots, s'_{l'}\}$. It is easily seen that $D_{\mathbf{s},\mathbf{t}}(L') = D_{\mathbf{s}',\mathbf{t}'}(L')$, and hence

$$D_{\mathbf{s},\mathbf{t}}(L) \simeq D_{\mathbf{s}',\mathbf{t}'}(L') \times \mathbb{A}^d.$$

Therefore it is enough to study $D_{\mathbf{s},\mathbf{t}}(L)$ with $\mathbf{s}, \mathbf{t} \in \mathbb{Z}^l_+$ such that

(L2) $$\{s_1, \ldots, s_l\} \supset \{b_1, \ldots, b_h\}.$$

Without loss of generality, we can also assume that

(L3) $$t_l \geqslant 2, \text{ and } t_{i-1} > t_i \text{ if } s_i \notin \{b_1, \ldots, b_h\}, 1 < i \leqslant l.$$

For $1 \leqslant i \leqslant l$, let

$$L(i) = \{x_{ba} \mid s_i \leqslant b \leqslant m, 1 \leqslant a \leqslant a_{i^*}\}.$$

Note that the ideal $I_{\mathbf{s},\mathbf{t}}(L)$ is generated by the t_i-minors of X contained in $L(i)$, $1 \leqslant i \leqslant l$.

The ladder determinantal varieties (associated to one-sided ladders) get related to Schubert varieties(cf. [**62**]). We describe below the main results of [**62**].

Let $G = SL(n)$, $Q = P_{a_1} \cap \cdots \cap P_{a_h}$. Let O^- be the opposite big cell in G/Q. Let H be the one-sided ladder defined by the outside corners $(a_i + 1, a_i)$, $1 \leqslant i \leqslant h$. Let $\mathbf{s}, \mathbf{t} \in \mathbb{Z}^l_+$ satisfy (L1), (L2), (L3) above. Let Z be the variety in $\mathbb{A}(H) \simeq O^-$ defined by the vanishing of the t_i-minors in $L(i)$, $1 \leqslant i \leqslant l$. Note that $Z \simeq D_{\mathbf{s},\mathbf{t}}(L) \times \mathbb{A}(H \setminus L) \simeq D_{\mathbf{s},\mathbf{t}}(L) \times \mathbb{A}^r$, where $r = \dim SL(n)/Q - |L|$.

We shall now define an element $w \in W_Q^{\min}$, such that the variety Z identifies with the opposite cell in the Schubert variety $X(w)$ in G/Q. We define $w \in W_Q^{\min}$ by specifying $w^{(a_i)} \in W^{a_i}$ $1 \leqslant i \leqslant h$, where $\pi_i(X(w)) = X(w^{(a_i)})$ under the projection $\pi_i : G/Q \to G/P_{a_i}$.

Define $w^{(a_i)}$, $1 \leqslant i \leqslant h$, inductively, as the (unique) maximal element in W^{a_i} such that

(1) $w^{(a_i)}(a_i - t_j + 1) = s_j - 1$ for all $j \in \{1, \ldots, l\}$ such that $s_j \geqslant b_i$, and $t_j \neq t_{j-1}$ if $j > 1$.

(2) if $i > 1$, then $w^{(a_{i-1})} \subset w^{(a_i)}$.

Note that $w^{(a_i)}$, $1 \leqslant i \leqslant h$, is well defined in $W^{(}a_i)$, and w is well defined as an element in $W_Q^{\min}$.

Let us denote the distinct elements in $\{t_1, \ldots, t_l\}$ by $t_1 = t_{i_1} > t_{i_2} > \cdots > t_{i_m} = t_l$, where $t_{i_k - 1} > t_{i_k}$ for $2 \leqslant k \leqslant m$. For $2 \leqslant k \leqslant m$, let $I_k = [\![e_{i_k}, s_{i_k} - 1]\!]$, where $e_{i_k} = s_{i_k} - (t_{i_{k-1}} - t_{i_k})$. Let

$$I_1 = [\![b_1 - (a_1 - t_1 + 1), b_1 - 1]\!],$$
$$I_{m+1} = [\![n - t_l + 2, n]\!]$$

REMARK 4.0.1. Fix j, $1 \leqslant j \leqslant h$. Let $b_j = s_c$, for some c, $1 \leqslant c \leqslant l$. Let i_k be the smallest such that $s_{i_k} > b_j$. Then in $w^{(a_j)}$, $b_j - 1$ appears at the $(a_i - t_c + 1)$-th place, and is followed by the blocks $I_k, I_{k+1}, \dots, I_{m+1}$.

LEMMA 4.0.2. *We have*
(1) $w^{(a_1)} = I_1 \cup I_2 \cup \cdots \cup I_{m+1}$.
(2) $I_j \subset w^{(a_i)}$, $1 \leqslant j \leqslant m+1$, $1 \leqslant i \leqslant h$.
(3) *The entries in* $w^{(a_i)} \setminus w^{(a_{i-1})}$ *are* $\leqslant b_i - 1$, $1 \leqslant i \leqslant h$.
All the assertions are clear from the definition of w.

LEMMA 4.0.3. *Fix* j, $1 \leqslant j \leqslant l$.
(1) *We have* $s_j \notin I_r$, $1 \leqslant r \leqslant m+1$.
(2) *Let* $t_j = t_{i_{k-1}}$, *for some* k, $2 \leqslant k \leqslant m+1$. *Then* $e_{i_k} > s_j$ *(here,* $e_{i_{m+1}} = n - t_l + 2$*).*

PROOF. If $k = m+1$, then $t_j = t_l$, $e_{i_{m+1}} = n - t_l + 2 > s_j$ (since $t_j < n - s_j + 1$). Further, $s_j \geqslant s_{i_m}$, and hence $s_j \notin I_r$ for any $1 \leqslant r \leqslant m+1$. Let then $k \leqslant m$. We have $s_{i_k} - s_j > t_j - t_{i_k} = t_{i_{k-1}} - t_{i_k}$. This implies $e_{i_k} > s_j$. Hence $s_j \notin I_r$, $r \geqslant k$. Also the fact that $s_j \geqslant s_{i_{k-1}}$ implies that $s_j \notin I_r$, $r \leqslant k-1$. □

REMARK 4.0.4. Consider a block of consecutive integers in $w^{(a_i)}$, $1 \leqslant i \leqslant h$, ending with $s_j - 1$ at the $(a_k - t_j + 1)$-th place, for some $k \leqslant i$. Then either $k = i$, or $k = j^*$; in other words, k is the largest integer in $\{1, \dots, i\}$ such that $b_k \leqslant s_i$. In particular, if $j^* \leqslant i$, then $k = j^*$.

THEOREM 4.0.5. *The variety* Z $(= D_{\mathbf{s},\mathbf{t}}(L) \times \mathbb{A}^r)$ *identifies with the opposite cell in* $X(w)$, *i.e.* $Z = X(w) \cap O^-$ *(scheme theoretically).*

PROOF. Let $f = \det M$, where M is a $t_i \times t_i$ matrix contained in $L(i)$ for some $1 \leqslant i \leqslant l$, be a generator of $I(Z)$. Let $k = i^*$, i.e. k is the largest integer such that $b_k \leqslant s_i$. Then M is contained in H_k. By Lemma 1.11.1, f can be written in the form $f = \sum g_\phi p_\phi|_{O^-}$, with $\phi \in W^{a_k}$ such that $\{\phi(1), \dots, \phi(a_k)\} \cap \{a_k + 1, \dots, n\} = \{r_1, \dots, r_{t_i}\}$, and $g_\phi \in k[H]$ (here $r_1, \dots, r_{t_i}$ are the row indices of M). In particular, we have $\phi(a_k - t_i + 1) = r_1$. Since M is contained in $L(i)$, we have $r_1 \geqslant s_i$, and hence $\phi(a_k - t_i + 1) \geqslant s_i$. We have $w^{(a_k)}(a_k - t_i + 1) = s_i - 1$, and hence $\phi(a_k - t_i + 1) > w^{(a_k)}(a_k - t_i + 1)$. This shows that $\phi \not\leqslant w^{(a_k)}$, and therefore $p_\phi \in I(X(w) \cap O^-)$. Thus $f \in I(X(w) \cap O^-)$.

Let now g be a generator of the ideal $I(X(w) \cap O^-)$, i.e. $g = p_\tau|_{O^-}$, with $\tau \in W^{a_i}$ for some i, $1 \leqslant i \leqslant h$, such that $\tau \not\leqslant w^{(a_i)}$. Since $w^{(a_i)}$ consists of several blocks of consecutive integers ending with $s_m - 1$ at the $(a_k - t_m + 1)$-th place, for some $m \in \{1, \dots, l\}$, where $k \in \{1, \dots, i\}$ is the largest such that $b_k \leqslant s_m$, and a last block ending with n at the a_i-th place, it follows that $\tau(a_k - t_m + 1) \geqslant s_m$

for some m, where $k \in \{1, \ldots, i\}$ is the largest such that $s_m \geqslant b_k$. Using Lemma 1.11.2, we deduce that $p_\tau|_{O^-}$ belongs to the ideal of $k[H]$ generated by t_m-minors in L with row indices $\geqslant s_m$, and column indices $\leqslant a_k$. Thus $p_\tau|_{O^-}$ belongs to the ideal generated by t_m-minors contained in $L(m)$, which shows that $g \in I(Z)$. □

Since the Schubert varieties are irreducible, normal, Cohen-Macaulay, and have rational singularities (cf. **[85]**, **[135]**, **[137]**, **[136]**), as a consequence of Theorem 4.0.5 we obtain

THEOREM 4.0.6. *The variety $D_{\mathbf{s},\mathbf{t}}(L)$ is irreducible, normal, Cohen-Macaulay, and has rational singularities.*

4.1. A Gröbner basis for $D_{\mathbf{s},\mathbf{t}}(L)$

Let $X = (x_{ba})$, $1 \leqslant b \leqslant m$, $1 \leqslant a \leqslant n$ be a $m \times n$ matrix of indeterminates.

We take the total order on the variables $\{x_{ba},\ 1 \leqslant b \leqslant m,\ 1 \leqslant a \leqslant n\}$ as described in §2.6 and as in §2.6, we shall denote by $\prec$ the induced total order, namely the lexicographic order, on the set of monomials in $k[X]$ $(= k[x_{ba},\ 1 \leqslant b \leqslant m,\ 1 \leqslant a \leqslant n])$. Also, we take the partial order on the set of minors of X as defined in §2.5. Note that the initial term (with respect to $\prec$) of a minor of X is equal to the product of elements in the skew diagonal. The matrix X can be viewed as a one-side ladder with a unique outside corner, namely $(1, n)$. Let $b_1 = 1$, and let $\mathbf{s}, \mathbf{t} \in \mathbb{Z}_+^l$ satisfy (L1) (defined in the beginning of this Section). Let $I_{\mathbf{s},\mathbf{t}}(X)$ be defined as in the beginning of this Section for $L = X$. In other words, $I_{\mathbf{s},\mathbf{t}}(X)$ is the ideal of $k[X]$ generated by the t_i-minors in $X_i = \{x_{ba} \mid s_i \leqslant b \leqslant m\}$, $1 \leqslant i \leqslant l$. For $1 \leqslant i < l$, let $\mathcal{G}_i$ be the set consisting of the t_i minors in X_i such that the number of rows contained in X_j is less than t_j, for all j, $i < j \leqslant l$, and $\mathcal{G}_l$ the set consisting of the t_l minors in X_l. Let $\mathcal{G} = \cup_{i=1}^{l} \mathcal{G}_i$. Clearly, $I_{\mathbf{s},\mathbf{t}}(X)$ is generated by $\mathcal{G}$.

PROPOSITION 4.1.1. *Let $\mathbf{s}, \mathbf{t} \in \mathbb{Z}_+^l$ satisfy (L1), and let $\mathcal{G}$ be as above. Then $\mathcal{G}$ is a Gröbner basis of $I_{\mathbf{s},\mathbf{t}}(X)$, with respect to the monomial order $\prec$.*

PROOF. Let $M_{\mathbf{s},\mathbf{t}}$ be the minor of X of size $t_1 - 1$ given by the last $t_i - t_{i+1}$ rows of $X_i \setminus X_{i+1}$, $1 \leqslant i < l$ and the last $t_l - 1$ rows of X_l, and the first $t_1 - 1$ columns of X. First we show that the ideal $I_{\mathbf{s},\mathbf{t}}(X)$ is cogenerated by $M_{\mathbf{s},\mathbf{t}}$. Let

$$M_{\mathbf{s},\mathbf{t}} = [i_1, \ldots, i_{t_1-1} | j_1, \ldots, j_{t_1-1}],$$

and $\mathcal{F} = \{M' \mid M' \not\geqslant M_{\mathbf{s},\mathbf{t}}\}$. Note that $M' \geqslant M_{\mathbf{s},\mathbf{t}}$ if and only if M' contains at most $t_i - 1$ rows in X_i, $1 \leqslant i \leqslant l$. Thus $\mathcal{F} = \bigcup_{i=1}^{l} \mathcal{F}_i$, where $\mathcal{F}_i = \{M' \mid M'$ contains at least t_i rows in $X_i\}$. Now $\mathcal{F}_i \subset I_{\mathbf{s},\mathbf{t}}(X)$, $1 \leqslant i \leqslant l$, and hence $\langle \mathcal{F} \rangle \subset I_{\mathbf{s},\mathbf{t}}(X)$. On the other hand, $\mathcal{G}_i \subset \mathcal{F}_i$, $1 \leqslant i \leqslant l$, and $\langle \mathcal{G} \rangle = I_{\mathbf{s},\mathbf{t}}(X)$. Therefore $I_{\mathbf{s},\mathbf{t}}(X) = \langle \mathcal{F} \rangle$, i.e. $I_{\mathbf{s},\mathbf{t}}(X)$ is cogenerated by $M_{\mathbf{s},\mathbf{t}}$.

The inequalities regarding j's in condition (1) of Theorem 2.6.1 are redundant in our case (since $j_t = t$, $1 \leqslant t \leqslant t_1 - 1$); also, condition (2) reduces to the condition that if $t \leqslant r$, then $i'_1 > i_{r-t+1}$ (since $j_t = t$, and hence $j'_t \geqslant j_t$ for all t, $1 \leqslant t \leqslant t_1 - 1$).

Therefore, in our case the conditions (1) and (2) are equivalent to

$$i'_t \leqslant i_{t_1-1}, i'_{t-1} \leqslant i_{t_1-2}, \ldots, i'_2 \leqslant i_{t_1-t+1}, \text{ and if } t \leqslant t_1 - 1, \text{ then } i'_1 > i_{t_1-t}.$$

Note that the above inequalities imply $i_{t_1-t+1} \geqslant i'_2 > i'_1 > i_{t_1-t}$; now, if $t \notin \{t_1, \ldots, t_l\}$, then this is not possible, since $i_{t_1-t+1} = i_{t_1-t} + 1$. Hence $G_t = \emptyset$ for $t \in \{1, \ldots, t_1\} \setminus \{t_1, \ldots, t_l\}$ (here G_t is as in Theorem 2.6.1). It is easily seen that $G_{t_i} = \mathcal{G}_i$, for $1 \leqslant i \leqslant l$. Therefore Theorem 2.6.1 implies that $\mathcal{G}$ is a Gröbner basis for $I_{\mathbf{s},\mathbf{t}}(X)$ with respect to the monomial order $\prec$. □

PROPOSITION 4.1.2. *Let $L \subset X$ be a one-sided ladder and $\mathbf{s}, \mathbf{t} \in \mathbb{Z}_+^l$ satisfying (L1). Then $I_{\mathbf{s},\mathbf{t}}(L) = I_{\mathbf{s},\mathbf{t}}(X) \cap k[L]$, and $\mathcal{G}_L = \mathcal{G} \cap k[L]$ is a Gröbner basis of $I_{\mathbf{s},\mathbf{t}}(L)$ with respect to the monomial order $\prec$.*

PROOF. By Proposition 4.1.1, $\mathcal{G}$ is a Gröbner basis of $I_{\mathbf{s},\mathbf{t}}(X)$. By Lemma 3.1.1, $\mathcal{G}_L$ is a Gröbner basis of the ideal $I_{\mathbf{s},\mathbf{t}}(X) \cap k[L]$. On the other hand it is easily seen that $\mathcal{G}_L$ generates $I_{\mathbf{s},\mathbf{t}}(L)$, and the result follows. □

4.2. The dimension of $D_{\mathbf{s},\mathbf{t}}(L)$

We construct a set $\mathcal{C}_{\mathbf{s},\mathbf{t}}(X) \subset X$ as follows. Let $\mathcal{C}_l(X)$ be the submatrix obtained from X_l by deleting the first $t_l - 1$ columns and the last $t_l - 1$ rows. For $i < l$, let $\mathcal{C}_i(X)$ be the matrix obtained from $\tilde{X}_i = X_i \setminus X_{i+1}$ by deleting the first $t_i - 1$ columns and the last $t_i - t_{i+1}$ rows. Now let $\mathcal{C}_{\mathbf{s},\mathbf{t}}(X) = \cup_{i=1}^l \mathcal{C}_i(X)$.

For a one-sided ladder $L \subset X$, and $\mathbf{s}, \mathbf{t} \in \mathbb{Z}_+^l$ satisfying (L1), we define $\mathcal{C}_i(L) = \mathcal{C}_i(X) \cap L$, $\mathcal{C}_{\mathbf{s},\mathbf{t}}(L) = \mathcal{C}_{\mathbf{s},\mathbf{t}}(X) \cap L$.

Note that in a solid minor in $\mathcal{G}_L$ (i.e. a minor with consecutive row indices and consecutive column indices), the smallest (for the order in §2.6) element belongs to $\mathcal{C}_{\mathbf{s},\mathbf{t}}(L)$, and conversely, an element $\alpha \in \mathcal{C}_{\mathbf{s},\mathbf{t}}(L)$ determines uniquely a solid minor in $\mathcal{G}_L$ having α as the smallest element. Hence the number of elements in $\mathcal{C}_{\mathbf{s},\mathbf{t}}(L)$ is equal to the number of solid minors in the set $\mathcal{G}_L$.

PROPOSITION 4.2.1. *Let $L \subset X$ an one-sided ladder, and $\mathbf{s}, \mathbf{t} \in \mathbb{Z}_+^l$ satisfying (L1). Then*

$$\operatorname{codim}_{\mathbb{A}(L)} D_{\mathbf{s},\mathbf{t}}(L) = |\mathcal{C}_{\mathbf{s},\mathbf{t}}(L)|.$$

PROOF. By Proposition 4.1.2, the ideal $I_{\mathbf{s},\mathbf{t}}(L)$ and the ideal $J_{\mathbf{s},\mathbf{t}}(L)$ of its initial terms determine graded quotient rings of $k[L]$ having the same Hilbert series, and hence the codimension of the variety $D_{\mathbf{s},\mathbf{t}}(L)$ is equal to the height of the monomial ideal $J_{\mathbf{s},\mathbf{t}}(L)$. In general, the height of a monomial ideal J in a polynomial ring $k[x_1, \ldots, x_N]$ is equal to the minimal cardinality of a set $\mathcal{C} \subset \{x_1, \ldots, x_N\}$ of variables such that

(*) each monomial in a set of monomial generators for J contains a variable from $\mathcal{C}$.

Let $J = J_{\mathbf{s},\mathbf{t}}(L)$ and $\mathcal{C} = \mathcal{C}_{\mathbf{s},\mathbf{t}}(L)$. Then it is easy to see that $\mathcal{C}$ satisfies $(*)$, the set of monomial generators being the set of the initial terms of all the t_i-minors in L_i, $1 \leqslant i \leqslant l$. Let us denote $\Delta_k = \{x_{ba} \in L \mid b + a = k + 1\}$, $k \geqslant 1$. Then $L = \dot{\bigcup}_{k \geqslant 1} \Delta_k$, and $\mathcal{C} = \dot{\bigcup}_{k \geqslant 1} (\mathcal{C} \cap \Delta_k)$.

Let now $\mathcal{C}' \subset \{x_{ba} \mid x_{ba} \in L\}$ be a set such that $|\mathcal{C}'| < |\mathcal{C}|$. Then there exists a k such that $|\mathcal{C}' \cap \Delta_k| < |\mathcal{C} \cap \Delta_k|$ (in particular $\mathcal{C} \cap \Delta_k \neq \emptyset$). Let $i \in \{1, \ldots, l\}$ be the largest such that $\Delta_k \cap \mathcal{C} \subset L_i$. Then

$$|\mathcal{C}' \cap (\Delta_k \cap L_i)| \leqslant |\mathcal{C}' \cap \Delta_k| < |\mathcal{C} \cap \Delta_k| = |\Delta_k \cap L_i| - (t_i - 1).$$

Therefore there exist t_i distinct variables in $(\Delta_k \cap L_i) \setminus \mathcal{C}'$. Thus the initial term of the t_i-minor in L_i having these elements on the skew diagonal does not contain any variable in $\mathcal{C}'$, and hence $\mathcal{C}'$ does not satisfy $(*)$.

Therefore $\mathcal{C}$ is a set of minimal cardinality among the sets satisfying $(*)$, and the required result follows. □

4.3. The singular locus of $D_{\mathbf{s},\mathbf{t}}(L)$

Let $X = (x_{ba})$, $1 \leqslant b < m$, $1 < a \leqslant n$ be a $m \times n$ matrix of indeterminates. Let $L \subset X$ be a one-sided ladder defined by the outside corners $\omega_i = x_{b_i a_i}$, $1 \leqslant i \leqslant h$, $1 \leqslant b_1 < \cdots < b_h \leqslant m$, $1 \leqslant a_1 < \cdots < a_h \leqslant n$. Let $\mathbf{s}, \mathbf{t} \in \mathbb{Z}_+^l$ satisfy (L1), (L2) and (L3) of Section 4. We preserve the notations of Section 4. Let $V = D_{\mathbf{s},\mathbf{t}}(L)$, $\mathcal{C} = \mathcal{C}_{\mathbf{s},\mathbf{t}}(L)$.

For $1 \leqslant i \leqslant l$, let $V_i \subset \mathbb{A}(L)$ be the variety defined by the vanishing of the t_j-minors in $L(j)$, with $j \in \{1, \ldots, l\} \setminus \{i\}$, and the $(t_i - 1)$-minors in $L(i)$.

THEOREM 4.3.1. *With notations as above, we have*

$$\operatorname{Sing} V = \bigcup_{i=1}^{l} V_i.$$

PROOF. The proof is similar to that of Theorem 3.3.1. For simplicity of notation, we identify the variable x_{ba} with the element (b, a).

First, we prove that $V_i \subset \operatorname{Sing} V$, for all $1 \leqslant i \leqslant l$. Let $x \in V_i$ for some $1 \leqslant i \leqslant l$. Let $\mathcal{J}$ be the Jacobian matrix associated to the variety $V \subset \mathbb{A}(L)$, evaluated at x. Then the rows of $\mathcal{J}$ are indexed by t_j-minors in $L(j)$, $1 \leqslant j \leqslant l$, and the columns are indexed by the elements $\alpha \in L$. The (M, α)-th entry in $\mathcal{J}$ is equal to $\pm(\det M')(x)$, where M' is the matrix obtained from M by deleting the row and the column containing α, if α appears in M, and 0 otherwise.

We distinguish two cases.

(I) $s_i \in \{b_1, \ldots, b_h\}$

Let $s_i = b_j$, for some $1 \leqslant j \leqslant h$. It is easily seen that

$$\omega_j \in \mathcal{C}_{\mathbf{s},\mathbf{t}}(L)$$

(since $s_{i+1} - s_i > t_i - t_{i+1}$ and $a_j \geqslant t_i$). Now consider the one-sided ladder L' obtained from L by deleting the element ω_j, i.e. the one-sided ladder defined by the outside corners

$$\omega_1 = (b_1, a_1), \ldots, \omega_{j-1} = (b_{j-1}, a_{j-1}), \omega_{j-} = (b_j, a_j - 1),$$
$$\omega_{j+} = (b_j + 1, a_j), \omega_{j+1} = (b_{j+1}, a_{j+1}), \ldots, \omega_l = (b_l, a_l),$$

where ω_{j-} is present only if $a_j - 1 > a_{j-1}$, and ω_{j+} is present only if $b_j + 1 < b_{j+1}$.

Since $x \in V_i$, a row of $\mathfrak{J}$ indexed by a t_i-minor involving $\omega_j = x_{b_j a_j}$ is 0. Also, the column of $\mathfrak{J}$ indexed by ω_j is 0. Let $\mathfrak{J}'$ be the matrix obtained from $\mathfrak{J}$ by deleting the column indexed by ω_j and the rows indexed by t_i-minors containing ω_j. Then

$$\operatorname{rank} \mathfrak{J} = \operatorname{rank} \mathfrak{J}',$$

since $\mathfrak{J}'$ is obtained from $\mathfrak{J}$ by deleting rows and columns consisting of zero entries. Let $x' = (x_\alpha)_{\alpha \in L'}$. Then $x' \in D_{\mathbf{s},\mathbf{t}}(L')$, and $\mathfrak{J}'$ is the Jacobian matrix associated to the variety $D_{\mathbf{s},\mathbf{t}}(L') \subset \mathbb{A}(L')$, evaluated at x'. Thus

$$\operatorname{rank} \mathfrak{J}' \leqslant \operatorname{codim}_{\mathbb{A}(L')} D_{\mathbf{s},\mathbf{t}}(L').$$

Now, using Proposition 4.2.1 we obtain

$$\begin{aligned}\operatorname{codim}_{\mathbb{A}(L')} D_{\mathbf{s},\mathbf{t}}(L') &= |\mathcal{C}_{\mathbf{s},\mathbf{t}}(L')| = |\mathcal{C}_{\mathbf{s},\mathbf{t}}(L) \setminus \{\omega_j\}| \\ &< |\mathcal{C}_{\mathbf{s},\mathbf{t}}(L)| = \operatorname{codim}_{\mathbb{A}(L)} D_{\mathbf{s},\mathbf{t}}(L).\end{aligned}$$

Hence $\operatorname{rank} \mathfrak{J}' < \operatorname{codim}_{\mathbb{A}(L)} V$, which implies $\operatorname{rank} \mathfrak{J} < \operatorname{codim}_{\mathbb{A}(L)} V$, i.e. $x \in \operatorname{Sing} V$.

(II) $s_i \notin \{b_1, \ldots, b_h\}$

We have $i > 1$ and $t_{i-1} > t_i$. Let $k = i^*$, i.e. k is the largest integer such that $b_k < s_i$. Define $\mathbf{s}' = (s_1, \ldots, s_{i-1}, \widehat{s_i}, s_{i+1}, \ldots, s_l)$, $\mathbf{t}' = (t_1, \ldots, t_{i-1}, \widehat{t_i}, t_{i+1}, \ldots, t_l)$. Let $\mathcal{C} = \mathcal{C}_{\mathbf{s},\mathbf{t}}(L)$, $\mathcal{C}' = \mathcal{C}_{\mathbf{s}',\mathbf{t}'}(L)$, and

$$\mathcal{C} = \bigcup_{j \in \{1,\ldots,l\}} \mathcal{C}_j, \quad \mathcal{C}' = \bigcup_{j \in \{1,\ldots,l\} \setminus \{i\}} \mathcal{C}'_j,$$

as defined in §4.2. Then $\mathcal{C}_j = \mathcal{C}'_j$ for $j \notin \{i-1, i\}$, and

$$\begin{aligned}|\mathcal{C}| - |\mathcal{C}'| =& |\mathcal{C}_{i-1}| + |\mathcal{C}_i| - |\mathcal{C}'_{i-1}| = \\ & \big[(s_i - s_{i-1}) - (t_{i-1} - t_i)\big]\big[a_k - (t_{i-1} - 1)\big] + \\ & \big[(s_{i+1} - s_i) - (t_i - t_{i+1})\big]\big[a_k - (t_i - 1)\big] - \\ & \big[(s_{i+1} - s_{i-1}) - (t_{i-1} - t_{i+1})\big]\big[a_k - (t_{i-1} - 1)\big] = \\ & \big[(s_{i+1} - s_i) - (t_i - t_{i+1})\big](t_{i-1} - t_i) > 0\end{aligned}$$

(here $s_{i+1} = m + 1$, $t_{i+1} = 1$, if $i = l$). Therefore

$$|\mathcal{C}_{\mathbf{s}',\mathbf{t}'}(L)| < |\mathcal{C}_{\mathbf{s},\mathbf{t}}(L)|.$$

Since $x \in V_i$, a row indexed by a t_i-minor contained in $L(i)$ is 0. Let $\mathfrak{J}'$ be the matrix obtained from $\mathfrak{J}$ by deleting the rows indexed by t_i-minors contained in $L(i)$. Then

$$\operatorname{rank} \mathfrak{J} = \operatorname{rank} \mathfrak{J}'.$$

Now, $x \in D_{\mathbf{s}',\mathbf{t}'}(L)$, and $\mathcal{J}'$ is the Jacobian matrix associated to the variety $D_{\mathbf{s}',\mathbf{t}'}(L) \subset \mathbb{A}(L)$, evaluated at x. Thus

$$\operatorname{rank} \mathcal{J}' \leqslant \operatorname{codim}_{\mathbb{A}(L)} D_{\mathbf{s}',\mathbf{t}'}(L).$$

Now, using Proposition 4.2.1 we obtain

$$\operatorname{codim}_{\mathbb{A}(L)} D_{\mathbf{s}',\mathbf{t}'}(L) = |\mathcal{C}_{\mathbf{s}',\mathbf{t}'}(L)| < |\mathcal{C}_{\mathbf{s},\mathbf{t}}(L)| = \operatorname{codim}_{\mathbb{A}(L)} D_{\mathbf{s},\mathbf{t}}(L).$$

Hence $\operatorname{rank} \mathcal{J}' < \operatorname{codim}_{\mathbb{A}(L)} V$, which implies $\operatorname{rank} \mathcal{J} < \operatorname{codim}_{\mathbb{A}(L)} V$, i.e., $x \in \operatorname{Sing} V$.

Now we prove that $\operatorname{Sing} V \subset \cup_{i=1}^{l} V_i$. Let $\mathcal{C} = \mathcal{C}_{\mathbf{s},\mathbf{t}}(L)$, $\mathcal{C} = \bigcup_{i=1}^{l} \mathcal{C}_i$, as defined in §4.2.

We introduce a total order on the set of minors of L of size r, with $r \geqslant 1$ fixed, as follows: $[i_1, \dots, i_r | j_1, \dots, j_r] < [i'_1, \dots, i'_r | j'_1, \dots, j'_r]$ if there exists $1 \leqslant k \leqslant r$ such that

$$\text{either } i_1 = i'_1, \dots, i_{k-1} = i'_{k-1}, i_k < i'_k,$$
$$\text{or } i_1 = i'_1, \dots, i_r = i'_r, j_1 = j'_1, \dots, j_{k-1} = j'_{k-1}, j_k < j'_k$$

(this is simply the lexicographic order on $\{i_1, \dots, i_r, j_1, \dots, j_r\}$). Let $x \in V \setminus \cup_{i=1}^{l} V_i$. For each $1 \leqslant i \leqslant l$, let M_i be the largest $(t_i - 1)$-minor in $L(i)$ such that $(\det M_i)(x) \neq 0$. Let $\mathcal{T}_l$ be the set of elements in L_l not in the rows or the columns given by the rows and the columns of M_l. Clearly, $|\mathcal{T}_l| = |\mathcal{C}_l|$. By (decreasing) induction on i, suppose that, for some i, $1 < i \leqslant l$, the sets $\mathcal{T}_i, \dots, \mathcal{T}_l$ have been constructed, such that

$(1)_i$ $\mathcal{T}_j \subset L(j)$, $i \leqslant j \leqslant l$,

$(2)_i$ the sets $\mathcal{T}_i, \dots, \mathcal{T}_l$ are pairwise disjoint,

$(3)_i$ $|\mathcal{T}_j| = |\mathcal{C}_j|$, $i \leqslant j \leqslant l$,

$(4)_i$ $\mathcal{T}_j$ contains no elements appearing in the rows or in the columns of L given by the rows and the columns of M_j, $i \leqslant j \leqslant l$,

$(5)_i$ there exist $t_i - 1$ rows in $L(i)$ not containing any element from $\mathcal{T}_i \cup \dots \cup \mathcal{T}_l$.

We define the set $\mathcal{T}_{i-1}$ as follows. Let r be the number of the rows of M_{i-1} contained in $\tilde{L}(i-1) := L(i-1) \setminus L(i)$. We distinguish two cases.

(I) $t_{i-1} - t_i \geqslant r$

In this case $\mathcal{T}_{i-1}$ is obtained from $\tilde{L}(i-1)$ by deleting the rows given by the rows of M_{i-1}, and $t_{i-1} - t_i - r$ other rows, followed by the deletion of the $t_{i-1} - 1$ columns given by the columns of M_{i-1}. Then properties $(1)_{i-1} - (4)_{i-1}$ are obvious; the $t_{i-1} - t_i$ rows of $\tilde{L}(i-1)$ which were deleted while defining $\mathcal{T}_{i-1}$, and the $t_i - 1$ rows of $L(i)$ in $(5)_i$, intersected with $L(i-1)$, give $t_{i-1} - 1$ rows of $L(i-1)$ not containing any elements in $\mathcal{T}_{i-1} \cup \mathcal{T}_i \cup \dots \cup \mathcal{T}_l$, so that we have $(5)_{i-1}$.

(II) $t_{i-1} - t_i < r$

In this case $\mathcal{T}_{i-1}$ is obtained from $\tilde{L}(i-1)$ by deleting the r rows given by the rows of M_{i-1}, then adding $r - t_{i-1} + t_i$ rows from the $t_i - 1$ rows of $L(i)$ in $(5)_i$ which are not rows of M_{i-1}, intersected with $L(i-1)$ (this is possible, since there are $t_{i-1} - 1 - r$ rows of M_{i-1} in $L(i)$, and hence at least $(t_i - 1) - (t_{i-1} - 1 - r) = r - t_{i-1} + t_i$

rows from the $t_i - 1$ rows of $L(i)$ in $(5)_i$ are not rows of M_{i-1}), followed by the deletion of the $t_{i-1} - 1$ columns given by the columns of M_{i-1}. Again, the properties $(1)_{i-1} - (4)_{i-1}$ are obvious; the r rows of M_{i-1} which were deleted from $\tilde{L}(i-1)$, and the $(t_i - 1) - (r - t_{i-1} + t_i)$ rows from the $t_i - 1$ rows in $(5)_i$ which were not used while defining $\mathcal{T}_{i-1}$, intersected with $L(i-1)$, give $t_{i-1} - 1$ rows of $L(i-1)$ not containing any elements in $\mathcal{T}_{i-1} \cup \mathcal{T}_i \cup \cdots \cup \mathcal{T}_l$, so that we have $(5)_{i-1}$.

Thus, using induction, we obtain the disjoint sets $\mathcal{T}_j \subset L(j)$, $1 \leqslant j \leqslant l$, such that $|\mathcal{T}_j| = |\mathcal{C}_j|$, and $\mathcal{T}_j$ contains no elements in the rows or columns of L given by the rows and columns of M_j.

For $\tau \in \mathcal{T}_i \subset \mathcal{T}$, $1 \leqslant i \leqslant l$, let M^τ be the t_i-minor obtained from M_i by adding the row and the column containing τ. Obviously, $M^\tau \neq M^{\tau'}$ for $\tau, \tau' \in \mathcal{T}$, with $\tau \neq \tau'$.

We now take a total order on $\mathcal{T}$, namely $(b, a) > (b', a')$ if either $b > b'$, or $b = b'$ and $a > a'$.

Let us fix $\tau \in \mathcal{T}$, say $\tau \in \mathcal{T}_i$ for some i, $1 \leqslant i \leqslant l$. Then the (M^τ, τ)-th entry in $\mathcal{J}$ is equal to $\pm(\det M_i)(x)$, so it is nonzero. Let now $\sigma \in \mathcal{T}$, $\sigma < \tau$. If σ is not an entry of M^τ, then the (M^τ, σ)-th entry of $\mathcal{J}$ is equal to 0. Assume now that σ is the (r, s)-th entry of M^τ. Then the (M^τ, σ)-th entry of $\mathcal{J}$ is equal to $\pm(\det M')(x)$, where M' is the $(t_i - 1) \times (t_i - 1)$ matrix obtained from M^τ by deleting the r-th row and the s-th column. Let $\tau = (b, a)$, $\sigma = (b', a')$. If $b' < b$, then the indices of the first $r - 1$ rows of M' and M_i are the same, while the index of the r-th row of M' is $> b'$, which is the index of the r-th row of M_i. Thus, $M' > M_i$, and by the maximality of M_i, we obtain $(\det M')(x) = 0$. If $b' = b$, then $a' < a$. The indices of all the rows and those of the first $s - 1$ columns are the same, while the index of the s-th column in M' is $> a'$, which is the index of the s-th column of M_i. Thus $M' > M_i$, and the maximality of M_i implies that $(\det M')(x) = 0$. Thus, for $\sigma < \tau$, the (M^τ, σ)-th entry in $\mathcal{J}$ is 0.

Let $\mathcal{J}'$ be the submatrix of $\mathcal{J}$ given by the rows indexed by M^τ's and the columns indexed by τ's, with $\tau \in \mathcal{T}$. We suppose that both rows and columns of $\mathcal{J}'$ are indexed by the elements in $\mathcal{T}$, and we arrange them increasingly, with respect to the total order on $\mathcal{T}$ defined above. Then $\mathcal{J}'$ is upper triangular, and all the diagonal entries are nonzero. Thus $\det \mathcal{J}' \neq 0$, and this implies that

$$\operatorname{rank} \mathcal{J}' = |\mathcal{T}| = |\mathcal{C}| = \operatorname{codim}_{\mathbb{A}(L)} D_{\mathbf{s},\mathbf{t}}(L).$$

Consequently $\operatorname{rank} \mathcal{J} = \operatorname{codim}_{\mathbb{A}(L)} V$, i.e. $x \notin \operatorname{Sing} V$. □

4.4. The irreducible components of $\operatorname{Sing} D_{\mathbf{s},\mathbf{t}}(L)$

We preserve the notations of the previous section.

Let us fix $j \in \{1, \ldots, l\}$, and let $Z_j = V_j \times \mathbb{A}(H \setminus L)$. We shall now define $\theta_j \in W_Q^{\min}$ such that the variety Z_j identifies with the opposite cell in the Schubert variety $X(\theta_j)$ in G/Q.

Note that $w^{(a_r)}(a_r - t_j + 1) = s_j - 1$, and $s_j - 1$ is the end of a block of consecutive integers in $w^{(a_r)}$, where $r = j^*$ is the largest integer such that $b_r \leqslant s_j$. Also, the beginning of this block is $\geqslant 2$ (if the block started with 1, we would have $a_r - t_j + 1 = s_j - 1 \geqslant b_r - 1 \geqslant a_r$, which is not possible, since $t_j \geqslant 2$). Let $u_j + 1$ be the beginning of this block, where $u_j \geqslant 1$. Then it is easily seen that if $s_j - 1$ is the end of a block in $w^{(a_i)}$, $1 \leqslant i \leqslant h$, then the beginning of the block is $u_j + 1$. For each i, $1 \leqslant i \leqslant h$, such that $u_j \notin w^{(a_i)}$, let v_i be the smallest entry in $w^{(a_i)}$ which is bigger than $s_j - 1$. Note that $v_i = w^{(a_i)}(a_k - t_j + 2)$, where $k \in \{1, \dots, i\}$ is the largest such that $b_k \leqslant s_j$.

Define ${\theta_j}^{(a_i)}$, $1 \leqslant i \leqslant h$, as follows.

If $s_j - 1 \notin w^{(a_i)}$ (which is equivalent to $j > 1$, $t_{j-1} = t_j$ and $i < r$), let $\theta_j^{(a_i)} = w^{(a_i)} \setminus \{v_i\} \cup \{s_j - 1\}$.

If $s_j - 1 \in w^{(a_i)}$ and $u_j \notin w^{(a_i)}$, then $\theta_j^{(a_i)} = w^{(a_i)} \setminus \{v_i\} \cup \{u_j\}$.

If $s_j - 1$ and $u_j \in w^{(a_i)}$, then $\theta_j^{(a_i)} = w^{(a_i)}$ (note that in this case $i > r$).

Note that θ_j is well defined as an element in $W_Q^{\min}$, and $\theta_j \leqslant w$.

REMARK 4.4.1. An equivalent description of θ_j is the following. Let $t_{i_k} < t_j \leqslant t_{i_{k-1}}$.

(I) If $j \notin \{i_1, \dots, i_m\}$ (i.e. $j > 1$ and $t_{j-1} = t_j$), then

for $i < r$, $\theta_j^{(a_i)} = w_j^{(a_i)} \setminus \{e_{i_k}\} \cup \{s_j - 1\}$;

for $i = r$, $\theta_j^{(a_r)} = w_j^{(a_r)} \setminus \{e_{i_k}\} \cup \{u_j\}$, where u_j is the largest entry in $\{1, \dots, s_j - 1\} \setminus w^{(a_r)}$;

for $i > r$ and $u_j \in w^{(a_i)}$, $\theta_j^{(a_i)} = w_j^{(a_i)}$;

for $i > r$ and $u_j \notin w^{(a_i)}$, $\theta_j^{(a_i)} = w_j^{(a_i)} \setminus \{v_i\} \cup \{u_j\}$, where v_i is the smallest entry in $w^{(a_i)} \setminus \theta_j^{(a_{i-1})}$.

(II) If $j \in \{i_1, \dots, i_m\}$, (i.e. $t_{j-1} > t_j$ if $j > 1$), then

for $i \leqslant r$, $\theta_j^{(a_i)} = w_j^{(a_i)} \setminus \{e_{i_k}\} \cup \{u_j\}$, where u_j is the largest entry in

$$\{1, \dots, s_j - 1\} \setminus w^{(a_r)};$$

for $i > r$ and $u_j \in w^{(a_i)}$, $\theta_j^{(a_i)} = w_j^{(a_i)}$;

for $i > r$ and $u_j \notin w^{(a_i)}$, $\theta_j^{(a_i)} = w_j^{(a_i)} \setminus \{v_i\} \cup \{u_j\}$, where v_i is the smallest entry in $w^{(a_i)} \setminus \theta_j^{(a_{i-1})}$.

THEOREM 4.4.2. *The subvariety $Z_j \subset Z$ identifies with the opposite cell in $X(\theta_j)$, i.e. $Z_j = X(\theta_j) \cap O^-$ (scheme theoretically).*

PROOF. Let $f = \det M$, M being either a t_i-minor contained in $L(i)$, $i \in \{1, \dots, h\} \setminus \{j\}$, or a $(t_j - 1)$-minor contained in $L(j)$ be a generator of $I(Z_j)$. In the former case we have $f \in I(Z)$, and Theorem 4.0.5 implies that $f \in I(X(w) \cap O^-) \subset I(X(\theta_j) \cap O^-)$. In the latter case, M is contained in H_k, where $k \in \{1, \dots, h\}$ is the largest such that $b_k \leqslant s_j$. By Lemma 1.11.1, f can be written in the form

$f = \sum g_\phi p_\phi|_{O^-}$, with $\phi \in W^{a_k}$ such that

$$\{\phi(1), \dots, \phi(a_k)\} \cap \{a_k + 1, \dots, n\} = \{r_1, \dots, r_{t_j - 1}\},$$

and $g_\phi \in k[H]$ (here $r_1, \dots, r_{t_j-1}$ are the row indices of M). In particular we have $\phi(a_k - t_j + 2) = r_1$. Since M is contained in $L(j)$, we deduce that $r_1 \geqslant s_j$, and hence $\phi(a_k - t_j + 2) \geqslant s_j$. We have $\theta_j{}^{(a_k)}(a_k - t_j + 2) = s_j - 1$, and hence $\phi(a_k - t_j + 2) > \theta_j{}^{(a_k)}(a_k - t_j + 2)$. This shows that $\phi \not\leqslant \theta_j{}^{(a_k)}$, and therefore $p_\phi \in I(X(\theta) \cap O^-)$. Thus $f \in I(X(\theta) \cap O^-)$.

Let now $g = p_\tau|_{O^-}$, with $\tau \in W^{a_i}$ for some i, $1 \leqslant i \leqslant h$, such that $\tau \not\leqslant \theta^{(a_i)}$, be a generator of the ideal $I(X(\theta_j) \cap O^-)$. Since $\theta_j{}^{(a_i)}$ consists of several blocks of consecutive integers ending with $s_m - 1$ at the $(a_k - t_m + 1)$-th place, for some $m \in \{1, \dots, l\} \setminus \{j\}$, where $k \in \{1, \dots, i\}$ is the largest such that $b_k \leqslant s_m$, a possible block ending with $s_j - 1$ at the $(a_k - t_j + 2)$-th place, where $k \in \{1, \dots, i\}$ is the largest such that $b_k \leqslant s_j$, and a last block ending with n at the a_i-th place, it follows that either $\tau(a_k - t_m + 1) \geqslant s_m$, for some $m \neq j$, where $k \in \{1, \dots, i\}$ is the largest such that $s_m \geqslant b_k$, or $\tau(a_k - t_j + 2) \geqslant s_j$, where $k \in \{1, \dots, i\}$ is the largest such that $s_j \geqslant b_k$. In the first case we have $\tau \not\leqslant w$, and hence

$$p_\tau|_{O^-} \in I(X(w) \cap O^-) = I(Z) \subset I(Z_j).$$

Suppose now that $\tau(a_k - t_j + 2) \geqslant s_j$, $k \in \{1, \dots, i\}$ being the largest such that $s_j \geqslant b_k$. Using Lemma 1.11.2, we deduce that $p_\tau|_{O^-}$ belongs to the ideal of $k[H]$ generated by $(t_j - 1)$-minors with row indices $\geqslant s_j$, and column indices $\leqslant a_k$. Thus $p_\tau|_{O^-}$ belongs to the ideal generated by $(t_j - 1)$-minors contained in $L(j)$, which implies that $g \in I(Z_j)$. □

THEOREM 4.4.3. *The irreducible components of* $\operatorname{Sing} D_{\mathbf{s},\mathbf{t}}(L)$ *are precisely the* V_j*'s,* $1 \leqslant j \leqslant l$.

PROOF. In view of Theorem 4.4.2, we obtain that V_j, $1 \leqslant j \leqslant l$, is irreducible, and the required result follows from Theorem 4.3.1. □

Let $X(w^{\max})$ (resp. $X(\theta_j^{\max})$, $1 \leqslant j \leqslant l$) be the pull-back in $SL(n)/B$ of $X(w)$ (resp. $X(\theta_j)$, $1 \leqslant j \leqslant l$) under the canonical projection $\pi : SL(n)/B \to SL(n)/Q$. Then using Theorems 4.3.1, 4.0.5 and 4.4.2, we obtain

THEOREM 4.4.4. *The irreducible components of* $\operatorname{Sing} X(w^{max})$ *are precisely* $X(\theta_j^{max})$, $1 \leqslant j \leqslant l$.

4.5. Validity of the conjecture of Chapter 8, §16.1 for $\operatorname{Sing} X(w^{\max})$

Let Q be as above. Let $\pi : G/B \to G/Q$ be the canonical projection. Let $\tau \in W_Q^{\min}$. We have $\pi^{-1}(X_Q(\tau)) = X_B(\tau^{\max})$, where $\tau^{\max}$, as a permutation, is given by $\tau^{(a_1)}$ arranged in descending order, followed by $\tau^{(a_2)} \setminus \tau^{(a_1)}$ arranged in descending order, etc.. We shall refer to the set $\tau^{(a_i)} \setminus \tau^{(a_{i-1})}$, $1 \leqslant i \leqslant l + 1$, arranged in descending order, as the i-th block in $\tau^{\max}$ (here, $\tau^{(a_0)} = \emptyset$, and $\tau^{(a_{l+1})}$ is the set $\{1, \dots, n\} \setminus \tau^{(a_l)}$ arranged in descending order).

REMARK 4.5.1. Set $b_{h+1} - 1 = n - t_l + 1$. All of the entries in the i-th block in $w^{\max}$ are $\leqslant b_i - 1$, $2 \leqslant i \leqslant h+1$. In particular, for $1 \leqslant j \leqslant l$, s_j occurs after $s_j - 1$ in $w^{\max}$ (in view of Lemma 4.0.3).

For $\tau \in W$, let P_τ, Q_τ be as in §3.5.

LEMMA 4.5.2. *We have*

(1) $Q_{w^{max}} = Q$.

(2) *Let* $I_{w^{max}} = \{\epsilon_i - \epsilon_{i+1} \mid i = s_j - 1, 1 \leqslant j \leqslant l\}$. *Then* $S_{P_{w^{max}}} = S \setminus I_{w^{max}}$.

The assertions are clear from the description of $w^{\max}$ in view of Remark 4.5.1.

LEMMA 4.5.3. *Let* $P = P_{w^{max}}$, $Q = Q_{w^{max}}$. *Then* $\overline{B\theta_j^{max}B}$ *is P-Q stable.*

PROOF. The Q-stability of $\overline{B\theta_j^{\max}B}$ on the right is obvious. Regarding the P-stability of $\overline{B\theta_j^{\max}B}$ on the left, let x denote either e_{i_k} or v_i, where $i > j$, $u_j \notin w^{(a_i)}$ (notations being as in Section 3.4). Then $x - 1$ occurs after x in $w^{\max}$. It is clear from the definition of $\theta_j^{\max}$ that $x - 1$ occurs after x in $\theta_j^{\max}$ also. For any other entry $y \neq x$, $s_j - 1$, if $y - 1$ occurs after y in $w^{\max}$, then it does so in $\theta_j^{\max}$ also. The result now follows from this. □

LEMMA 4.5.4. *Fix* j, $1 \leqslant j \leqslant h$. *Let* C *be a block of consecutive integers in* $w^{(a_j)}$ *ending with* $s_k - 1$ *at the* $(a_j - t_k + 1)$*-th place (for some* k*) and beginning with* x_k. *Let the block preceding* C *end with* $s_i - 1$ *for some* i. *Suppose* $k^* \leqslant j$. *Then for* $\alpha = \epsilon_y - \epsilon_{y+1}$, *where* $y \in [\![s_i, x_k]\!]$, *the rank* 1 *parabolic subgroup* P_α *is contained in* $P(= P_{w^{max}})$.

PROOF. The result follows (in view of Lemma 4.5.2) from the fact that $[\![s_i, x_k]\!]$ does not contain $s_t - 1$ for any t, $1 \leqslant t \leqslant l$. □

LEMMA 4.5.5. *Fix* j, $1 \leqslant j \leqslant l$. *Let* $j^* = r$. *Then* θ_j^{max} *is the unique maximal element of the set* $\{\tau \in W \mid \tau \leqslant w^{max}, \tau^{(a_r)}(a_r - t_j + 2) \leqslant s_j - 1\}$.

The proof is clear from the definition of θ_j.

Let Type I and Type II bad occurrences be defined as in §3.5.

LEMMA 4.5.6. *A bad occurrence in* w^{max} *has to be of Type I.*

PROOF. If possible, let c, d, a, b, where $a < b < c < d$, occur in the i-th, j-th, k-th, p-th blocks respectively in $w^{\max}$. Now $c < d$ implies that $i < j$. Hence $j > 1$, and $d \leqslant b_j - 1$ (cf. Remark 4.5.1), and this implies that $b < d \leqslant b_j - 1 \leqslant b_k - 1$. But then a cannot appear before b (by definition of $w^{\max}$). □

REMARK 4.5.7. Of course, there are several Type I bad occurrences. For example, take j, $1 \leqslant j \leqslant l$. Let $j^* = r$. With notations as in Lemma 4.0.3, let $d = e_{i_k}$. We have (cf. Lemma 4.0.3) $d > s_j$. Also, in view of Remark 4.5.1, s_j is not an entry in $w^{(a_i)}$, $i \leqslant r$, and s_j appears after $s_j - 1$ in $w^{\max}$. From the definition of $w^{\max}$, it is clear that u_j appears after s_j in $w^{\max}$ (notations being as in the beginning of this section). Take $d = e_{i_k}, b = s_j - 1, c = s_j, a = u_j$.

LEMMA 4.5.8. *Let d, b, c, a be a Type I bad occurrence in w^{max}. Then*
(1) $d \in I_r$, for some r, $1 \leqslant r \leqslant m+1$.
(2) $a, c \notin I_r$, for any r, $1 \leqslant r \leqslant m+1$.

PROOF. Let d, b, c, a belong to the i-th, j-th, k-th, p-th blocks respectively in $w^{\max}$. Assertion (2) is immediate, since $p, k > 1$. Note that assertion (1) is equivalent to the assertion that $i = 1$. If $j = 1$, then $i = 1$, and (1) follows (cf. Lemma 4.0.2). Let then $j > 1$. This implies $b \leqslant b_j - 1 < c$. Suppose $i > 1$. Then we would obtain that $d \leqslant b_i - 1 \leqslant b_j - 1 < c$, which is not possible. Hence $i = 1$, and (1) follows. □

REMARK 4.5.9. With notations as in Lemma 4.5.8, we have in fact $d \in I_r$ for some $r \geqslant 2$. This is clear if $j \geqslant 2$ (since $b \leqslant b_j - 1 < c < d$). If $j = 1$, then we have $b_1 - 1 < c < d$. Thus we get that $r \geqslant 2$.

PROPOSITION 4.5.10. *The maximal elements of $F_{w^{max}}$ are precisely the θ_j^{max}, for $1 \leqslant i \leqslant l$ (here $F_{w^{max}}$ is as in Ch8, §16.1).*

PROOF. Let us denote j^* by r. Then with d, b, c, a as in Remark 4.5.7, we have that b, a, d, c occur in that order in $\theta_j^{\max}$. Let us denote $\theta_j^{\max}$ by τ'. Let w' (resp. τ) be the element of S_n obtained from τ' (resp. w) by replacing b, a, d, c (resp. d, b, c, a) respectively by d, b, c, a (resp. b, a, d, c). Then clearly $\tau \leqslant \tau'$, and $w' \leqslant w$. Further, $\overline{B\theta_j^{\max} B}$ is P-Q stable (cf. Lemma 4.5.3). Thus $\theta_j^{\max} \in F_{w^{\max}}$.

Let now $\tau' \in F_{w^{\max}}$. Let b, a, d, c be a bad occurrence in τ'. Further, let b, a, d, c appear in the p-th, q-th, r-th, s-th blocks respectively in τ' (note that $\tau' \in W_Q^{\max}$). Let $b_q = s_z$ for some z, $1 \leqslant z \leqslant l$. If $a \leqslant b_q - 1$, and $d > b_q - 1$, as in the proof of Proposition 3.5.8, we obtain $\tau'^{(a_q)}(a_q - t_z + 2) \leqslant b_q - 1 (= s_z - 1)$. This implies $\tau' \leqslant \theta_z^{\max}$ (note that $z^* = q$).

We now distinguish the following two cases

Case 1: $d \leqslant b_q - 1$

Let $d \in I_k (= [\![e_{i_k}, s_{i_k} - 1]\!])$ for some $k \geqslant 2$ (cf. Remark 4.5.9). Let $j = i_k^*$. We first observe that $j \leqslant q$. For, if $i_k = i_k^* (= j)$, then $j \leqslant q$ (since $d \leqslant b_q - 1$). If $i_k > i_k^*$, then again in view of Lemma 4.0.3, we have $s_{i_k^*} < d \leqslant b_q - 1$, and hence $b_j - 1 < b_q - 1$ (note that $s_{i_k^*} = b_j$). Hence we get $j < q$. Thus in either case we have $j \leqslant q$.

We further divide this case into the following two subcases.

Subcase 1 (a) $j < i_k$

Now, I_k appears in $w^{(a_j)}$ as a block of consecutive integers (cf. Remark 4.0.1), and $s_{i_k} - 1$ appears at the $(a_j - t_{i_k} + 1)$-th place. Let the block in $w^{(a_j)}$ preceding this block end with $s_i - 1$ at the $(a_u - t_i + 1)$-th place, for some u and i. Then $u = j$ necessarily (since $j < i_k$), and hence $i^* = u = j$. Now, in view of Lemmas 4.5.2 and 4.5.4 for $\alpha = \epsilon_y - \epsilon_{y+1}$, where $y \in [\![s_i, d-1]\!]$, the rank 1 parabolic subgroup P_α is contained in $P(= P_{w^{\max}})$. This, together with the fact that $d \notin \tau'^{(a_j)}$, implies that $[\![s_i, d]\!] \cap \tau'^{(a_j)} \neq \emptyset$ (in view of the P-stability on the left of $X(\tau')$(cf. Lemma

4.5.2)). Hence we obtain that $\tau'^{(a_j)}(a_j - t_i + 2) \leqslant s_i - 1$, where $i^* = j$. This implies $\tau' \leqslant \theta_i^{\max}$ (cf. Lemma 4.5.5).

Subcase 1 (b) $\quad j = i_k$

Note that $j > 1$ (cf. Remark 4.5.9). Consider $w^{(a_{j-1})}$. Now I_k appears in $w^{(a_{j-1})}$ as a block (cf. Remark 4.0.1, since $i_k^* > j-1$), and d belongs to this block. Further, $s_{i_k} - 1$ appears at the $(a_{j-1} - t_{i_k} + 1)$-th place. Let the block in $w^{(a_{j-1})}$ preceding this block end with $s_i - 1$ at the $(a_{j-1} - t_i + 1)$-th place for some i. Then $i^* = j-1$, necessarily (since $j = i_k$). Further, for $\alpha = \epsilon_y - \epsilon_{y+1}$, where $y \in [\![s_i, d-1]\!]$, the rank 1 parabolic subgroup P_α is contained in P (in view of Lemma 4.5.2, since $[\![s_i, d-1]\!]$ does not contain $s_t - 1$ for any t, $1 \leqslant t \leqslant l$). Now, the fact that $d \notin \tau'^{(a_q)}$ implies that $\tau'^{(a_{j-1})} \cap [\![s_i, d]\!] = \emptyset$ (in view of P-stability on the left of $X(\tau')$). Hence we obtain $\tau'^{(a_{j-1})}(a_{j-1} - t_i + 2) \leqslant s_i - 1$, where $i^* = j-1$. This implies $\tau' \leqslant \theta_i^{\max}$ (cf. Lemma 4.5.5).

Case 2: $\quad a > b_q - 1$

Let d, b, c, a appear in the i-th, j-th, k-th, x-th blocks respectively in $w^{\max}$, where $i \leqslant j < k \leqslant x$. Let u be the smallest such that $a \leqslant s_u - 1$. We have $q \leqslant u^*$ (since $q > u^*$ would imply $a \leqslant s_u - 1 < b_q - 1$, which is not true).

CLAIM. $x > u^*$.

If $j \geqslant 2$, then we have $b_q - 1 < a < b \leqslant b_j - 1$ (cf. Remark 4.5.1). Hence we obtain $u^* \leqslant j$ from which the Claim follows (since $x > j$).

If $j = 1$, let $b \in I_v$ for some $v \geqslant 2$ (cf. Lemma 4.0.2; note that $b_q - 1 < a < b$ implies $b > b_1 - 1$). We have $b_q - 1 < a < b \leqslant s_{i_v} - 1$. Hence we obtain $s_u - 1 \leqslant s_{i_v} - 1$, and $u^* \leqslant i_v^*$. Now, we have $b_k - 1 \geqslant c > s_{i_v} - 1 \geqslant s_{i_v^*} - 1$ (by the definition of $w^{\max}$). This implies $c \notin w^{(a_{i_v^*})}$, and hence $k > i_v^* \geqslant u^*$. The Claim now follows from this (since $x \geqslant k$). Thus we obtain $q \leqslant u^* < x$. Now the fact that $a \in \tau'^{(a_q)}$ implies $a \in \tau'^{(a_{u^*})}$. This, together with the P-stability on the left of $X(\tau')$, implies that $[\![a, s_u - 1]\!] \subset \tau'^{(a_{u^*})}$ (note that $s_j - 1 \notin [\![a, s_u - 1]\!]$, for any $j \neq u$, and hence for $\alpha = \epsilon_y - \epsilon_{y+1}$, where $y \in [\![a, s_u - 2]\!]$, the rank 1 parabolic subgroup P_α is contained in P). From this, we obtain $\tau'^{(a_{u^*})}(a_{u^*} - t_u + 2) \leqslant s_u - 1$ (since $\tau'^{(a_{u^*})} \leqslant w^{(a_{u^*})}$, and $a \notin w^{(a_{u^*})}$ (note that $x > u^*$)). This implies $\tau' \leqslant \theta_u^{\max}$ (cf. Lemma 4.5.5). □

THEOREM 4.5.11. *The Conjecture of Chapter 8, §16.1 holds for* Sing $X(w^{max})$.

PROOF. The result follows from Theorem 4.4.4 and Proposition 4.5.10. □

REMARK 4.5.12. Ladder determinantal varieties were first introduced by Abyankar (cf. [**4**]).

REMARK 4.5.13. A similar identification as in Theorem 3.0.3 (for the case $t_1 = \cdots = t_l$) has also been obtained by Mulay (cf. [**123**]).

REMARK 4.5.14. The w's as above arising from ladder determinantal varieties are covexillary in the sense of Lascoux (cf. [**112**]), namely, they avoid the pattern $c, d.a, b$.

REMARK 4.5.15. In [**62**], the theory of Schubert varieties and the theory of ladder determinantal varieties are complementing each other. To be more precise, geometric properties such as normality, Cohen-Macaulayness, etc., for ladder determinantal varieties are concluded by relating these varieties to Schubert varieties. The components of singular loci of Schubert varieties are determined by first determining them for ladder determinantal varieties, and then using the above mentioned relationship between ladder determinantal varieties and Schubert varieties.

5. Quiver varieties

Fix an h-tuple of non-negative integers $\mathbf{n} = (n_1, \ldots, n_h)$ and a list of vector spaces $V_1, \ldots, V_h$ over an arbitrary field k with respective dimensions $n_1, \ldots, n_h$. Define Z, the *variety of quiver representations* (of dimension $\mathbf{n}$, of the equioriented quiver of type A_h) to be the affine space of all $(h-1)$-tuples of linear maps $(f_1, \ldots, f_{h-1})$:

$$V_1 \xrightarrow{f_1} V_2 \xrightarrow{f_2} \cdots \xrightarrow{f_{h-2}} V_{h-1} \xrightarrow{f_{h-1}} V_h \,.$$

If we endow each V_i with a basis, we get $V_i \cong k^{n_i}$ and

$$Z \cong M(n_2 \times n_1) \times \cdots \times M(n_h \times n_{h-1}),$$

where $M(l \times m)$ denotes the affine space of matrices over k with l rows and m columns. The group

$$G_{\mathbf{n}} = GL(n_1) \times \cdots \times GL(n_h)$$

acts on Z by

$$(g_1, g_2, \cdots, g_h) \cdot (f_1, f_2, \cdots, f_{h-1}) = (g_2 f_1 g_1^{-1}, g_3 f_2 g_2^{-1}, \cdots, g_h f_{h-1} g_{h-1}^{-1}).$$

Now, let $\mathbf{r} = (r_{ij})_{1 \le i \le j \le h}$ be an array of non-negative integers with $r_{ii} = n_i$, and define $r_{ij} = 0$ for any indices other than $1 \le i \le j \le h$. Define the set

$$Z^\circ(\mathbf{r}) = \left\{ (f_1, \cdots, f_{h-1}) \in Z \mid \forall i<j, \ \operatorname{rank}(f_{j-1} \cdots f_i : V_i \to V_j) = r_{ij} \right\}.$$

(This set might be empty for a bad choice of $\mathbf{r}$.)

PROPOSITION 5.0.1. *The $G_{\mathbf{n}}$-orbits of Z are exactly the sets $Z^\circ(\mathbf{r})$ for $\mathbf{r} = (r_{ij})$ with*

$$r_{ij} - r_{i,j+1} - r_{i-1,j} + r_{i-1,j+1} \ge 0, \quad \forall\, 1 \le i < j \le h.$$

PROOF. This is a standard result of algebraic quiver theory [**12**], [**9**]. Since this theory is not well known among geometers, we recall it here.

Consider the abelian category $\mathcal{R}$ of quiver representations defined as follows. An object of $\mathcal{R}$ is a sequence of linear maps $(V_1 \xrightarrow{f_1} \cdots \xrightarrow{f_{h-1}} V_h)$, where the V_i are *any* vector spaces of *arbitrary* dimension. A morphism of $\mathcal{R}$ from the object $(V_1 \xrightarrow{f_1} \cdots \xrightarrow{f_{h-1}} V_h)$ to the object $(V_1' \xrightarrow{f_1'} \cdots \xrightarrow{f_{h-1}'} V_h')$ is defined to be an h-tuple of linear maps $(\phi_i : V_i \longrightarrow V_i')$ such that each of the following squares commutes:

$$\begin{array}{ccc} V_i & \xrightarrow{f_i} & V_{i+1} \\ \phi_i \downarrow & & \downarrow \phi_{i+1} \\ V_i' & \xrightarrow{f_i'} & V_{i+1}' \end{array}$$

Direct sum of objects is defined componentwise, and it is known by the Krull-Schmidt Theorem [**9**] that any object $R \in \mathcal{R}$ can be written uniquely as a direct sum of indecomposable objects. By elementary linear algebra, these indecomposables are seen to be

$$R_{ij} = (0 \to \cdots \to 0 \to \underset{V_i}{k} \overset{\sim}{\to} \cdots \overset{\sim}{\to} \underset{V_j}{k} \to 0 \to \cdots \to 0)$$

for $1 \leq i \leq j \leq h$ (corresponding to the positive roots of the root system A_h). That is, there are unique multiplicities $m_{ij} \in \mathbb{Z}^+$ with

$$R \cong \bigoplus_{1 \leq i \leq j \leq h} m_{ij} R_{ij}.$$

Our variety Z consists of representations with fixed (V_i) and all possible (f_i). Two points of Z are in the same $G_{\mathbf{n}}$-orbit exactly if they are isomorphic as objects in $\mathcal{R}$. So the orbits correspond to arrays $(m_{ij})_{1 \leq i \leq j \leq h}$ with $m_{ij} \in \mathbb{Z}^+$ and $n_i = \sum_{k \leq i \leq l} m_{kl}$.

We can compute the rank numbers $\mathbf{r} = (r_{ij})$ from the multiplicities $\mathbf{m} = (m_{ij})$:

$$r_{ij} = \sum_{k \leq i \leq j \leq l} m_{kl},$$

and conversely

$$m_{ij} = r_{ij} - r_{i,j+1} - r_{i-1,j} + r_{i-1,j+1}$$

Hence the arrays (r_{ij}) with the stated conditions classify the $G_{\mathbf{n}}$-orbits on Z. □

We define the *quiver variety* as the algebraic set

$$Z(\mathbf{r}) = \{(f_1, \cdots, f_{h-1}) \in Z \mid \forall i, j, \operatorname{rank}(f_{j-1} \cdots f_i : V_i \to V_j) \leq r_{ij}\}.$$

It will follow from Zelevinsky's theorem (cf. [**165**]) that $Z(\mathbf{r})$ is an irreducible variety and is the Zariski closure of $Z^\circ(\mathbf{r})$.

5.1. Schubert varieties

Given $\mathbf{n} = (n_1, \cdots, n_h)$, for $1 \leq i \leq h$ let

$$a_i = n_1 + n_2 + \cdots + n_i, \qquad a_0 = 0, \qquad \text{and} \qquad n = n_1 + \cdots + n_h .$$

For positive integers $i \leq j$, we shall frequently use the notations

$$[\![i, j]\!] = \{i, i+1, \ldots, j\}, \qquad [\![i]\!] = [\![1, i]\!], \qquad [\![0]\!] = \{\} .$$

Let $k^n \cong V_1 \oplus \cdots \oplus V_h$ have basis $e_1, \ldots, e_n$ compatible with the V_i. Consider its general linear group $GL(n)$, the subgroup B of upper-triangular matrices, and the parabolic subgroup Q of block upper-triangular matrices:

$$Q = \{(a_{ij}) \in GL(n) \mid a_{ij} = 0 \text{ whenever } j \leq a_k < i \text{ for some } k\} .$$

A *partial flag of type* $(a_1 < a_2 < \cdots < a_h = n)$ (or simply a *flag*) is a sequence of subspaces $U_\bullet = (U_1 \subset U_2 \subset \cdots \subset U_h = k^n)$ with $\dim U_i = a_i$. Let $E_i = V_1 \oplus \cdots \oplus V_i = \langle e_1, \ldots, e_{a_i}\rangle$, and $E'_i = V_{i+1} \oplus \cdots \oplus V_h = \langle e_{a_i+1}, \ldots, e_n\rangle$, so that $E_i \oplus E'_i = k^n$. Denoting by Fl the set of all flags $U_\bullet$ as above,we have, Fl has a transitive $GL(n)$-action induced from k^n, and $Q = \mathrm{Stab}_{GL(n)}(E_\bullet)$, so we may identify $\mathrm{Fl} \cong GL(n)/Q,\ gE_\bullet \leftrightarrow gP$. The *Schubert varieties* are indexed by *flags of subsets* of $[\![n]\!]$, of the form

$$\tau = (\tau_1 \subset \tau_2 \subset \cdots \subset \tau_h = [\![n]\!]), \qquad \#\tau_i = a_i .$$

(A permutation $w : [\![n]\!] \to [\![n]\!]$ corresponds to the subset-flag with $\tau_i = w[\![a_i]\!] = \{w(1), w(2), \ldots, w(a_i)\}$. This gives a one-to-one correspondence between cosets of the symmetric group $W = S_n$ modulo the Young subgroup $W_{\mathbf{n}} = S_{n_1} \times \cdots \times S_{n_h}$, and subset-flags.)

Given such a τ, let $E_i(\tau) = \langle e_j \mid j \in \tau_i\rangle$ be a coordinate subspace of k^n, and $E_\bullet(\tau) = (E_1(\tau) \subset E_2(\tau) \subset \cdots) \in \mathrm{Fl}$. Then we may identify the *Schubert cell* $X^\circ(\tau)$ as

$$\begin{aligned} X^\circ(\tau) &= B \cdot E_\bullet(\tau) \\ &= \left\{ (U_1 \subset U_2 \subset \cdots) \in \mathrm{Fl} \;\middle|\; \begin{array}{c} \dim U_i \cap k^j = \#\, \tau_i \cap [\![j]\!] \\ 1 \leq i \leq h,\ 1 \leq j \leq n \end{array} \right\} \end{aligned}$$

and the *Schubert variety* $X(\tau)$ may be identified as

$$\begin{aligned} X(\tau) &= \overline{X^\circ(\tau)} \\ &= \left\{ (U_1 \subset U_2 \subset \cdots) \in \mathrm{Fl} \;\middle|\; \begin{array}{c} \dim U_i \cap k^j \geq \#\, \tau_i \cap [\![j]\!] \\ 1 \leq i \leq h,\ 1 \leq j \leq n \end{array} \right\} \end{aligned}$$

where $k^j = \langle e_1, \ldots, e_j\rangle \subset k^n$.

The *opposite cell* $\mathbf{O} \subset G/Q$ gets identified with the set of flags in general position with respect to the spaces $E'_1 \supset \cdots \supset E'_{h-1}$:

$$\mathbf{O} = \{(U_1 \subset U_2 \subset \cdots) \in \mathrm{Fl} \mid U_i \cap E'_i = 0\}.$$

We denote $Y(\tau) = X(\tau) \cap \mathbf{O}$, the *opposite cell* of $X(\tau)$.

5.2. The bijection ζ

We define a special subset-flag $\tau^{\max} = (\tau_1^{\max} \subset \cdots \subset \tau_h^{\max} = [\![n]\!])$ corresponding to $\mathbf{n} = (n_1, \ldots, n_h)$. We want each $\tau_i^{\max}$ to contain numbers as large as possible given the constraints $[\![a_{j-1}]\!] \subset \tau_j^{\max}$ for all j. Namely, we define $\tau_i^{\max}$ recursively by

$$\tau_h^{\max} = [\![n]\!]; \quad \tau_i^{\max} = [\![a_{i-1}]\!] \cup \{\text{largest } n_i \text{ elements of } \tau_{i+1}^{\max}\}.$$

Furthermore, given $\mathbf{r} = (r_{ij})_{1\le i\le j\le h}$ indexing a quiver variety, define a subset-flag $\tau^{\mathbf{r}}$ to contain numbers as large as possible given the constraints

$$\#\, \tau_i^{\mathbf{r}} \cap [\![a_j]\!] = \begin{cases} a_i - r_{i,j+1} & \text{for } i \le j \\ a_j & \text{for } i > j \end{cases}$$

Namely,

$$\tau_i^{\mathbf{r}} = \{\underbrace{1\ldots a_{i-1}}_{a_{i-1}}\ \underbrace{\ldots\ldots\ldots a_i}_{r_{ii}-r_{i,i+1}}\ \underbrace{\ldots\ldots\ldots a_{i+1}}_{r_{i,i+1}-r_{i,i+2}}\ \underbrace{\ldots\ldots\ldots a_{i+2}}_{r_{i,i+2}-r_{i,i+3}}\ \cdots\ \underbrace{\ldots\ldots\ldots n}_{r_{i,h}}\}$$

where we use the visual notation

$$\underbrace{\cdots\cdots a}_{b} = [\![a-b+1, a]\!].$$

Recall that $a_j = a_{j-1} + n_j$ and $0 \le r_{ij} - r_{i,j+1} \le n_j$, so that each $\tau_i^{\mathbf{r}}$ is an increasing list of integers. Also $r_{ij} - r_{i,j+1} \le r_{i+1,j} - r_{i+1,j+1}$, so that $\tau_i^{\mathbf{r}} \subset \tau_{i+1}^{\mathbf{r}}$. Thus $\tau^{\mathbf{r}}$ is indeed a subset-flag. See §5.3 for examples.

Now define the map

$$\begin{aligned} \zeta :\qquad & Z \longrightarrow \mathrm{Fl} \\ (f_1, \ldots, f_{h-1}) & \longmapsto (U_1 \subset U_2 \subset \cdots) \end{aligned}$$

where

$$U_i = \{(v_1, \ldots, v_h) \in V_1 \oplus \cdots \oplus V_h = k^n \mid \forall\, j > i,\ v_{j+1} = f_j(v_j)\}.$$

In terms of coordinates, if we identify the linear maps $(f_1, \ldots, f_{h-1})$ with the matrices $(A_1, \ldots, A_{h-1})$, and identify $\mathrm{Fl} \cong GL(n)/Q$, we have

$$\zeta(A_1, \ldots, A_{h-1}) = \begin{pmatrix} I_1 & 0 & 0 & 0 & \cdots \\ A_1 & I_2 & 0 & 0 & \cdots \\ A_2A_1 & A_2 & I_3 & 0 & \cdots \\ A_3A_2A_1 & A_3A_2 & A_3 & I_4 & \cdots \\ \vdots & \vdots & \vdots & \vdots & \end{pmatrix} \mod Q$$

where I_i is an identity matrix of size n_i.

THEOREM 5.2.1. *(Zelevinsky* **[165]***)*
(i) ζ is a bijection of Z onto its image $Y(\tau^{max})$: $\quad \zeta : Z \tilde{\to} Y(\tau^{max})$.
Also,

$$(*)\qquad Y(\tau^{max}) = \{(U_1 \subset U_2 \subset \cdots) \mid \forall\, i,\ E_{i-1} \subset U_i,\ U_i \cap E_i' = 0\}.$$

(ii) ζ restricts to a bijection from $Z(\mathbf{r})$ onto $Y(\tau^{\mathbf{r}})$: $\quad \zeta : Z(\mathbf{r}) \tilde{\to} Y(\tau^{\mathbf{r}})$.
Also,

$$(**)\quad Y(\tau^{\mathbf{r}}) = \left\{ (U_1 \subset U_2 \subset \cdots) \;\middle|\; \begin{array}{c} \forall\, i \le j, \quad \dim U_i \cap E_j \ge a_i - r_{i,j+1}, \\ E_{i-1} \subset U_i, \quad U_i \cap E_i' = 0 \end{array} \right\}.$$

PROOF. Obviously ζ is injective. To prove (i), we first show equation $(*)$. The inclusion $\subset$ is clear. For the inclusion $\supset$, consider a flag $U_\bullet$ with $E_{i-1} \subset U_i$ for all i. Since acting by B does not change $\dim U_i \cap k^j$, we may suppose $U_\bullet = E_\bullet(\mu)$ for some $\mu = (\mu_1 \subset \cdots \subset \mu_h = [\![n]\!])$ with $[\![a_{i-1}]\!] \subset \mu_i$ for all i. By definition $\#\tau_i^{\max} \cap [\![j]\!]$ is as small as possible given $[\![a_{i-1}]\!] \subset \tau_i^{\max}$, so

$$\dim U_i \cap k^j = \#\mu_i \cap [\![j]\!] \geq \#\tau_i^{\max} \cap [\![j]\!],$$

which shows $(*)$.

Now we show that $\zeta(Z)$ is equal to the right hand side of $(*)$. The inclusion $\zeta(Z) \subset \mathrm{RHS}(*)$ is clear, so we show $\zeta(Z) \supset \mathrm{RHS}(*)$. Each U_i is transverse to E_i', so U_i is the graph of a linear map

$$(f_{i,i+1}, \ldots, f_{i,h}) : E_i \longrightarrow E_i' = V_{i+1} \oplus \cdots \oplus V_h.$$

Since $E_{i-1} \subset U_i$, we have $f_{ij}(E_{i-1}) = 0$, and we may consider

$$f_{ij} : V_i \cong E_i/E_{i-1} \to V_j.$$

Any element of U_j can be written $(v_1, \ldots, v_j, f_{j,j+1}(v_j), \ldots)$. Let $i < j$. Any $(v_1, \ldots, v_h) \in U_i$ is also an element of U_j, so $v_{j+1} = f_{j,j+1}(v_j)$. Taking $f_i = f_{i,i+1}$, we find $U_\bullet = \zeta(f_1, \ldots, f_{h-1})$.

The proof of (ii) is similar. Equation $(**)$ follows just as before. Now consider a flag $U_\bullet = \zeta(f_1 \ldots f_{h-1}) \in \zeta(Z) = Y(\tau^{\max})$. Then

$$\begin{aligned} \dim U_i \cap E_j &= \dim E_{i-1} + \dim \mathrm{Ker}(f_j f_{j-1} \cdots f_i) \\ &= \dim E_{i-1} + \dim V_i - \mathrm{rank}(f_j f_{j-1} \cdots f_i) \\ &= a_i - \mathrm{rank}(f_j f_{j-1} \cdots f_i). \end{aligned}$$

Hence $U_\bullet \in \zeta(Z(\mathbf{r})) \iff U_\bullet \in \mathrm{RHS}(**) = Y(\tau^{\mathbf{r}})$. □

COROLLARY 5.2.2. *For each* $\mathbf{r}$, $Z^\circ(\mathbf{r})$ *is an open dense* $G_{\mathbf{n}}$*-orbit in* $Z(r)$.

PROOF. Arguing as in the proof of (ii) above, we find that $\zeta(Z^\circ(\mathbf{r})) \supset Y^\circ(\tau^{\mathbf{r}})$. But it is known that $Y^\circ(\tau^{\mathbf{r}})$ is Zariski open and dense in $Y(\tau^{\mathbf{r}})$. □

5.3. Examples

EXAMPLE 5.3.1. A small generic case. Let $h = 4$, $\mathbf{n} = (2, 3, 2, 2)$,

$$\mathbf{r} = \begin{vmatrix} 2 & 2 & 0 & 0 \\ & 3 & 1 & 1 \\ & & 2 & 2 \\ & & & 2 \end{vmatrix} \qquad \mathbf{m} = \begin{vmatrix} 0 & 2 & 0 & 0 \\ & 0 & 0 & 1 \\ & & 0 & 1 \\ & & & 0 \end{vmatrix}$$

where r_{ij} and m_{ij} are written in the usual matrix positions. Note that r_{ij} is obtained by summing the entries in $\mathbf{m}$ weakly above and to the right of m_{ij}.

Then we get $(a_1, a_2, a_3, a_4) = (2, 5, 7, 9)$, $n = 9$, and

$$\tau^{\max} = (\{89\} \subset \{12589\} \subset \{1234589\} \subset [\![9]\!]),$$

$$\tau^{\mathbf{r}} = (\{45\} \subset \{12459\} \subset \{1234589\} \subset [\![9]\!]),$$

which correspond to the cosets in $W/W_{\mathbf{n}}$

$$w^{\max} = 89|125|34|67, \qquad w^{\mathbf{r}} = 45|129|38|67.$$

(The minimal-length representatives of these cosets are the permutations as written.) The partial flag variety is $\mathrm{Fl} = \{U_1 \subset U_2 \subset U_3 \subset k^9 \mid \dim U_i = a_i\}$, and the corresponding Schubert varieties are:

$$X(\tau^{\max}) = \left\{ U_\bullet \,\middle|\, \begin{matrix} k^2 \subset U_2 \\ k^5 \subset U_3 \end{matrix} \right\}, \quad X(\tau^{\mathbf{r}}) = \left\{ U_\bullet \,\middle|\, \begin{matrix} U_1 \subset k^5 \subset U_3,\ k^2 \subset U_2 \\ \dim U_2 \cap k^5 \geq 4 \end{matrix} \right\}.$$

The opposite cells $Y(\tau)$ are defined by the extra conditions $U_i \cap E'_i = 0$.

EXAMPLE 5.3.2. *Fulton's universal degeneracy schemes* **[52]**.

Given $m > 0$, let Z be the affine space associated to the quiver data

$$h = 2m, \qquad \mathbf{n} = (1, 2, \ldots, m, m, \ldots, 2, 1).$$

For each $w \in S_{m+1}$, Fulton defines a "degeneracy scheme" $\Omega_w = Z(\mathbf{r})$ as follows. Denote $\bar{i} = 2m + 1 - i$, and define $\mathbf{r} = \mathbf{r}(w) = (r_{ij})$ and $\mathbf{m} = (m_{ij})$ by:

$$\begin{aligned} r_{ij} &= r_{\bar{j}\bar{i}} = i \\ r_{i\bar{j}} &= \#\,[\![i]\!] \cap w[\![j]\!] \end{aligned} \qquad m_{ij} = \begin{cases} 1, & (i,j) = (w(k), \bar{k}),\ \exists k \leq m \\ 0, & \text{otherwise} \end{cases}$$

for $1 \leq i, j \leq m$. The associated Schubert varieties $Y(\tau^{\mathbf{r}})$ are given by

$$\begin{aligned} \tau^{\mathbf{r}} &= (\tau^{\mathbf{r}}_1 \subset \cdots \subset \tau^{\mathbf{r}}_{\bar{1}}) \\ \tau^{\mathbf{r}}_i &= [\![a_{i-1}]\!] \cup \{a_{\overline{w^{-1}(1)}}, a_{\overline{w^{-1}(2)}}, \ldots, a_{\overline{w^{-1}(i)}}\}, \\ \tau^{\mathbf{r}}_{\bar{i}} &= [\![a_{\bar{i}} - 1]\!] \cup \{a_{\bar{1}}, a_{\bar{2}}, \ldots, a_{\bar{m}}\} \end{aligned}$$

or by cosets $\widetilde{w} = \widetilde{w}_1 | \cdots | \widetilde{w}_{\bar{1}} \in W/W_{\mathbf{n}}$

$$\begin{aligned} \widetilde{w}_i &= [\![a_{i-2}+1, a_{i-1}]\!] \cup \{a_{\overline{w^{-1}(i)}}\} \\ \widetilde{w}_{\bar{m}} &= [\![a_{m-1}+1, a_m - 1]\!] \cup \{a_{\overline{w^{-1}(m+1)}}\} \\ \widetilde{w}_{\bar{j}} &= [\![a_{\bar{j}-2}+1, a_{\bar{j}-1}]\!] \end{aligned}$$

for $1 \leq i \leq m$, $1 \leq j \leq m-1$. Furthermore $\tau^{\max} = \tau^{\mathbf{r}(w)}$ and $\widetilde{w}^{\max} = \widetilde{w}^{\mathbf{r}(w)}$ for $w = e \in S_{m+1}$, the identity permutation.

EXAMPLE 5.3.3. For a given h and $\mathbf{n}$, the *variety of complexes* is defined as the union $\mathcal{C} = \cup_{\mathbf{r}} Z(\mathbf{r})$ over all $\mathbf{r} = (r_{ij})$ with $r_{i,i+2} = 0$ for all i. The subvarieties $Z(\mathbf{r})$ correspond to the multiplicity matrices $\mathbf{m} = (m_{ij})$ with $m_{ij} = 0$ for all $i + 2 \leq j$, and $m_{ii} + m_{i-1,i} + m_{i,i+1} = n_i$ for all i. Musili-Seshadri **[130]** find the irreducible components (maximal subvarieties) of $\mathcal{C}$, and show that each component is isomorphic to the opposite cell of a suitable Schubert variety.

EXAMPLE 5.3.4. The classical *determinantal variety* of $k \times l$ matrices of rank $\leq t$ is $\mathcal{D} = Z(r)$ for $\mathbf{r} = \begin{pmatrix} l & m \\ 0 & k \end{pmatrix}$, and $\mathbf{m} = \begin{pmatrix} l-m & m \\ 0 & k-m \end{pmatrix}$ where $m = \min(t+1, k, l)$. Also $n = k + l$,

$$\tau^{\max} = (\llbracket k+1, k+l \rrbracket \subset \llbracket n \rrbracket),$$
$$X(\tau^{\max}) = \mathrm{Fl} = \mathrm{Gr}(l, k^n),$$
$$\tau^{\mathbf{r}} = (\llbracket m+1, l \rrbracket \cup \llbracket k+l-m+1, k+l \rrbracket \subset \llbracket n \rrbracket)$$
$$X(\tau^{\mathbf{r}}) \cong \{U \in \mathrm{Gr}(l, k^n) \mid U \cap k^l = l - m\},$$
$$\mathcal{D} = Z(\mathbf{r}) \cong Y(\tau^{\mathbf{r}}) = \{U \in \mathrm{Gr}(l, k^n) \mid U \cap k^l = l - m,\ U \cap E' = 0\},$$

where $E' = \langle e_{l+1}, e_{l+2}, \ldots, e_n \rangle$.

5.4. The main theorem

In this section we prove the scheme-theoretic version of Zelevinsky's bijection. From now on, we assume our field k is infinite. Recall (cf. Chapter 8, §6.1.1)

PROPOSITION 5.4.1. *Let $\tau = (\tau_1 \subset \cdots \subset \tau_h = \llbracket n \rrbracket)$ be a subset-flag and $Y(\tau)$ the intersection of the Schubert variety $X(\tau)$ with the opposite cell $\mathbf{O}$. Then the (reduced) vanishing ideal $\mathfrak{I}(\tau) \subset k[\mathbf{O}]$ of $Y(\tau) \subset \mathbf{O}$ is generated by those Plücker coordinates p_σ which are incomparable with one of the p_{τ_i}:*

$$\mathfrak{I}(\tau) = \langle p_\sigma \mid \exists i,\ \#\sigma = a_i,\ \sigma \not\leq \tau_i \rangle.$$

Denote a generic element of the quiver space $Z = M(n_2 \times n_1) \times \cdots \times M(n_h \times n_{h-1})$ by $(A_1, \ldots, A_{h-1})$, so that the coordinate ring of Z is the polynomial ring in the entries of all the matrices A_i. Let $\mathbf{r} = (r_{ij})$ index the quiver variety

$$Z(\mathbf{r}) = \{(A_1, \ldots, A_{h-1}) \mid \operatorname{rank} A_{j-1} \cdots A_i \leq r_{ij}\}.$$

Let $\mathfrak{J}(\mathbf{r}) \subset k[Z]$ be the ideal generated by the determinantal conditions implied by the definition of $Z(\mathbf{r})$:

$$\mathfrak{J}(\mathbf{r}) = \left\langle \det(A_{j-1} A_{j-2} \cdots A_i)_{\lambda \times \mu} \;\middle|\; \begin{array}{c} j > i,\ \lambda \subset \llbracket n_j \rrbracket,\ \mu \subset \llbracket n_i \rrbracket \\ \#\lambda = \#\mu = r_{ij} + 1 \end{array} \right\rangle.$$

Clearly $\mathfrak{J}(\mathbf{r})$ defines $Z(\mathbf{r})$ set-theoretically.

THEOREM 5.4.2. *$\mathfrak{J}(\mathbf{r})$ is a prime ideal and is the vanishing ideal of $Z(\mathbf{r}) \subset Z$. There are isomorphisms of reduced schemes*

$$Z(\mathbf{r}) = \mathit{Spec}(k[Z] \,/\, \mathfrak{J}(\mathbf{r})) \cong \mathit{Spec}(k[\mathbf{O}] \,/\, \mathfrak{I}(\tau^{\mathbf{r}})) = Y(\tau^{\mathbf{r}}),$$

where $\mathfrak{I}(\tau^{\mathbf{r}})$ denotes the ideal defining $Y(\tau^{\mathbf{r}})$ as a closed subvariety of $\mathbf{O}$ (the opposite big cell in G/Q). That is, the quiver scheme $Z(\mathbf{r})$ defined by $\mathfrak{J}(\mathbf{r})$ is isomorphic to the reduced variety $Y(\tau^{\mathbf{r}})$, the opposite cell of a Schubert variety.

COROLLARY 5.4.3. *For a ring R, consider the polynomial ring $R[\mathbf{O}] = R[a_{kj}]$ and the ideal $\mathfrak{J}(\mathbf{r})_R \subset R[\mathbf{O}]$ generated by the same determinants as above. Define the scheme $Z(\mathbf{r})_R = \mathit{Spec}(R[\mathbf{O}]/\mathfrak{J}(\mathbf{r})_R)$*

(1) *If k is an arbitrary field, then $Z(\mathbf{r})_k$ is reduced, irreducible, Cohen-Macaulay, and normal, and has rational singularities.*
(2) *If R is a noetherian ring, then $Z(\mathbf{r})_R$ is reduced (resp. irreducible, Cohen-Macaulay, normal) exactly when Spec$R$ is reduced (resp. irreducible, Cohen-Macaulay, normal).*

PROOF. (1) Let $\overline{k}$ be the algebraic closure of k. Then the desired properties of $Z(\mathbf{r})_{\overline{k}}$ follow from the corresponding properties of Schubert varieties. But this implies these properties for $Z(\mathbf{r})_k$ as well, since $Z(\mathbf{r})_k \to Z(\mathbf{r})_{\overline{k}}$ is a faithfully flat morphism (see [**121**], §21.E)

(2) By [**122**], the morphism $Z(\mathbf{r})_{\mathbb{Z}} \to$ Spec$\mathbb{Z}$ is faithfully flat and hence so is $Z(\mathbf{r})_R \to$ Spec$R$ (see [**121**], §3.C). Now the fibers of the latter morphism are reduced, Cohen-Macaulay, and normal by (1); hence the corresponding properties hold for the total space $Z(\mathbf{r})_R$ exactly when they hold for the base (cf. [**121**], §21.E). Finally, $Z(\mathbf{r})_R \to$ Spec$R$ is a closed surjective morphism with irreducible fibers of the same dimension, and hence the total space is irreducible exactly when the base is irreducible (see for example [**152**], §I.6.3) □

Proof of Theorem 5.4.2. The map of §5.2, $\zeta : Z \tilde{\to} Y(\tau^{\max}) \subset \mathbf{O}$ is an algebraic isomorphism onto its image, since it is injective on points and on tangent vectors. (This is clear from the coordinate definition of ζ.) Thus the restriction homomorphism $\zeta^* : k[\mathbf{O}] \to k[Z]$ is surjective.

Now, ζ maps $Z(\mathbf{r})$ isomorphically onto $Y(\tau^{\mathbf{r}})$, so we have

$$\begin{aligned} Z(\mathbf{r}) = \mathrm{Spec}(k[Z]\,/\,\widetilde{\mathfrak{Z}}(\mathbf{r})) &\cong \mathrm{Spec}\big(k[\mathbf{O}]\,/\,(\zeta^*)^{-1}\widetilde{\mathfrak{Z}}(\mathbf{r})\big) \\ &= \mathrm{Spec}\big(k[\mathbf{O}]\,/\,\mathfrak{I}(\tau^{\mathbf{r}})\big) = Y(\tau^{\mathbf{r}}). \end{aligned}$$

where $\widetilde{\mathfrak{Z}}(\mathbf{r}) \subset k[Z]$ denotes the (reduced) vanishing ideal of $Y(\tau^{\mathbf{r}})$.

We must show $\mathfrak{Z}(\mathbf{r}) = \widetilde{\mathfrak{Z}}(\mathbf{r})$. Since clearly $\mathfrak{Z}(\mathbf{r}) \subset \widetilde{\mathfrak{Z}}(\mathbf{r})$, we are left with the other inclusion, which is equivalent to

$$(\zeta^*)^{-1}\mathfrak{Z}(\mathbf{r}) \supset (\zeta^*)^{-1}\widetilde{\mathfrak{Z}}(\mathbf{r}) = \mathfrak{I}(\tau^{\mathbf{r}}).$$

Let A be the generic matrix of $\mathbf{O}$. We define ideals $\mathfrak{I}_0, \mathfrak{I}_1, \mathfrak{I}_2 \subset k[\mathbf{O}]$ generated by certain minors of A:

$$\begin{aligned} \mathfrak{I}_0 &= (\zeta^*)^{-1}\mathfrak{Z}(\mathbf{r}) \\ &= (\zeta^*)^{-1}\left\langle \det(A_{j-1}A_{j-2}\cdots A_i)_{\lambda\times\mu} \;\middle|\; \begin{array}{c} j>i,\ \lambda\subset [\![n_j]\!],\ \mu\subset[\![n_i]\!] \\ \#\lambda = \#\mu = r_{ij}+1 \end{array} \right\rangle . \end{aligned}$$

$$\mathfrak{I}_1 = \left\langle \det A_{\lambda\times\mu} \;\middle|\; \begin{array}{c} i\le j,\ \lambda\subset [\![a_j+1,n]\!],\ \mu\subset[\![a_i]\!] \\ \#\lambda = \#\mu = r_{ij}+1 \end{array} \right\rangle$$

$$\mathfrak{I}_2 = \mathfrak{I}(\tau^{\mathbf{r}}) = \left\langle \det A_{\sigma\times[\![a_i]\!]} \;\middle|\; \begin{array}{c} 1\le i\le h-1,\ \sigma\subset [\![n]\!] \\ \#\sigma = a_i,\ \sigma\not\le\tau_i^{\mathbf{r}} \end{array} \right\rangle$$

To finish the proof of Theorem 5.4.2, it suffices to show

$$\mathfrak{I}_0 \supset \mathfrak{I}_1 \supset \mathfrak{I}_2 .$$

LEMMA 5.4.4. *Let $X = (x_{ij})$ and $Y = (y_{kl})$ be matrices of variables x_{ij}, y_{kl} generating a polynomial ring. Let $\mathfrak{J}_X$ (resp. $\mathfrak{J}_Y$) be the ideal generated by all $r+1$-minors of X (resp. Y). Then $\mathfrak{J}_X$ and $\mathfrak{J}_Y$ both contain all $r+1$-minors of the product XY.*

PROOF.

$$\det(XY)_{\lambda\times\mu} = \sum_{\nu} \det X_{\lambda\times\nu} \det Y_{\nu\times\mu}.$$

□

LEMMA 5.4.5. *Let $(A_1, \ldots, A_{h-1})$ be a generic element of Z, and for $j > i$ let $\mathfrak{J}_{ji}$ be the ideal generated by all $r+1$-minors of the $n_j \times n_i$ product matrix $A_{j-1} \cdots A_i$. Then $\mathfrak{J}_{ji}$ contains all $r+1$-minors of the $(n - a_{j-1}) \times a_i$ matrix*

$$\widetilde{A}^{(ji)} = \begin{pmatrix} A_{j-1} \cdots A_1 & A_{j-1} \cdots A_2 & \cdots & A_{j-1} \cdots A_i \\ A_j \cdots A_1 & A_j \cdots A_2 & \cdots & A_j \cdots A_i \\ \vdots & \vdots & & \vdots \\ A_{h-1} \cdots A_1 & A_{h-1} \cdots A_2 & \cdots & A_{h-1} \cdots A_i \end{pmatrix}$$

PROOF. Note that we can factor the matrix

$$\widetilde{A}^{(ij)} = \begin{pmatrix} I_{j-1} \\ A_j \\ \vdots \\ A_{h-1} \cdots A_j \end{pmatrix} \cdot A_{j-1} \cdots A_i \cdot (A_{i-1} \cdots A_1,\ A_{i-1} \cdots A_2,\ \cdots,\ A_{i-1},\ I_i).$$

Now apply Lemma 5.4.4 twice. □

LEMMA 5.4.6. $\quad \mathfrak{I}_0 \supset \mathfrak{I}_1 .$

PROOF. For generic elements $A \in \mathbf{O}$ and $(A_1, \ldots, A_{h-1}) \in Z$, we have by definition $\zeta^*(f(A)) = f(\zeta(A_1, \ldots, A_{h-1}))$ for any polynomial f in the matrix entries. Now let $\lambda \subset [\![a_{j-1}+1, n]\!]$, $\mu \subset [\![a_i]\!]$, $\#\lambda = \#\mu = r_{ij} + 1$, and consider a generator $\det A_{\lambda\times\mu}$ of $\mathfrak{I}_1$. Then

$$\zeta^*(\det A_{\lambda\times\mu}) = \det \zeta(A_1, \ldots, A_{h-1})_{\lambda\times\mu} = \det \widetilde{A}^{(ji)}_{\lambda'\times\mu}$$

where $\lambda' \subset [\![n - a_{j-1}]\!]$ is a translate of λ. By Lemma 5.4.5, $\det \widetilde{A}^{(ji)}_{\lambda'\times\mu} \in \mathfrak{J}(\mathbf{r})$, so $\mathfrak{I}_1 = \langle \det A_{\lambda\times\mu} \rangle \subset (\zeta^*)^{-1}\mathfrak{J}(\mathbf{r}) = \mathfrak{I}_0$. □

LEMMA 5.4.7. *Let A be a generic element of $\mathbf{O}$. Let $1 \le t \le a_i$, $1 \le s \le n$, and $\sigma = \{\sigma(1) < \sigma(2) < \cdots < \sigma(a_i)\} \subset [\![n]\!]$ with $\sigma(a_i - t + 1) \ge s$. Then $p_\sigma(A)$ belongs to the ideal of $k[\mathbf{O}]$ generated by t-minors of A with row indices $\ge s$ and column indices $\le a_i$.*

PROOF. Choose $\sigma' \subset [\![s,n]\!] \cap \sigma$ with $\#\sigma' = t$, and let $\sigma'' = \sigma \backslash \sigma'$. Then the Laplace expansion of $p_\sigma(A)$ with respect to the rows σ', σ'', gives

$$p_\sigma(A) = \det A_{\sigma \times [\![a_i]\!]} = \sum_{\lambda' \cup \lambda'' = [\![a_i]\!]} \pm \det A_{\sigma' \times \lambda'} \det A_{\sigma'' \times \lambda''},$$

where the sum is over all partitions of the interval $[\![a_i]\!]$. The first factor of each term in the sum is of the form required. □

LEMMA 5.4.8. $\mathcal{I}_1 \supset \mathcal{I}_2$.

PROOF. Let $\sigma \subset [\![n]\!]$ with $\#\sigma = a_i$, $\sigma \not\leq \tau_i^{\mathbf{r}}$ for some i, $1 \leq i \leq h-1$. Now, $\tau_i^{\mathbf{r}}$ has the largest possible entries such that

$$\tau_i^{\mathbf{r}}(a_i - r_{i,j+1}) \leq a_j, \qquad \forall j \geq i,$$

so $\sigma \not\leq \tau_i^{\mathbf{r}}$ must violate this condition for some j:

$$\sigma(a_i - r_{i,j+1}) \geq a_j + 1, \qquad \exists j \geq i.$$

Hence by Lemma 5.4.7, $p_\sigma(A)$ is in $\mathcal{I}_1$. □

Theorem 5.4.2 is therefore proved.

6. Variety of complexes

Let us fix the positive integers $h \geqslant 2, n_1, \ldots, n_{h+1}$, and let $\mathbf{n} = (n_1, \ldots, n_{h+1})$. Let $W_1, \ldots, W_{h+1}$ be finite dimensional vector spaces over a field k with $\dim W_1 = n_1, \ldots, \dim W_{h+1} = n_{h+1}$. Let Z be the affine space of all h-tuples of linear maps

$$(f_1, \ldots, f_h) : W_1 \xrightarrow{f_1} W_2 \xrightarrow{f_2} \cdots \xrightarrow{f_{h-1}} W_h \xrightarrow{f_h} W_{h+1}.$$

The group $G_{\mathbf{n}} = GL(n_1) \times \cdots \times GL(n_{h+1})$ acts on Z by

$$(g_1, g_2, \cdots, g_{h+1}) \cdot (f_1, f_2, \cdots, f_h) = (g_2 f_1 g_1^{-1}, g_3 f_2 g_2^{-1}, \cdots, g_{h+1} f_h g_h^{-1}).$$

The subvariety $\mathcal{C}$ of Z defined by

$$\mathcal{C} = \{(f_1, f_2, \cdots, f_h) \in Z \mid f_{i+1} f_i = 0 \text{ for } 1 \leqslant i \leqslant h-1\}$$

is called the *variety of complexes*.

Let $\mathcal{K}_{\mathbf{n}}$ denote the set of all $\mathbf{k} = (k_1, \ldots, k_h) \in \mathbb{Z}_+^h$, such that $k_i + k_{i+1} \leqslant n_{i+1}$ for $0 \leqslant i \leqslant h$ (here $k_0 = k_{h+1} = 0$). For $\mathbf{k} = (k_1, \ldots, k_h) \in \mathcal{K}_{\mathbf{n}}$, let

$$V(\mathbf{k}) = \{(f_1, \ldots, f_h) \in \mathcal{C} \mid \operatorname{rank}(f_i) \leqslant k_i \text{ for } 1 \leqslant i \leqslant h\}.$$

Let X_i be the $n_{i+1} \times n_i$ matrix of indeterminates, for $1 \leqslant i \leqslant h$, and let $k[X_1, \ldots, X_h]$ be the polynomial ring over the indeterminates in $X_1 \cup \cdots \cup X_h$. For $\mathbf{k} \in \mathcal{K}_{\mathbf{n}}$, let $I(\mathbf{k})$ be the ideal in $k[X_1, \ldots, X_h]$ generated by the entries of $X_{i+1}X_i$, for $1 \leqslant i \leqslant h-1$, and the $k_i + 1$ minors of X_i, for $1 \leqslant i \leqslant h$.

It is easily seen that $\mathcal{C} = \bigcup_{\mathbf{k} \in \mathcal{K}_{\mathbf{n}}} V(\mathbf{k})$.

As a special case of Theorem 5.4.2, we have,

THEOREM 6.0.1. *For* $\mathbf{k} \in \mathcal{K}_{\mathbf{n}}$, *the variety* $V(\mathbf{k})$ *is isomorphic to the opposite cell in a Schubert variety in* $SL(n)/Q$, *for* $n = n_1 + \cdots + n_{h+1}$, $Q = \bigcap_{i=1}^{h} P_{a_i}$. *Further,* $I(\mathbf{k})$ *is the defining ideal of* $V(\mathbf{k})$ *(as a closed subscheme of* Z*).*

Define the following partial order on the set $\mathcal{K}_{\mathbf{n}}$: for $\mathbf{k} = (k_1, k_2, \cdots, k_h)$, $\mathbf{k}' = (k_1', k_2', \cdots, k_h')$ in $\mathcal{K}_{\mathbf{n}}$, $\mathbf{k} \geqslant \mathbf{k}' \iff k_i \geqslant k_i'$, $1 \leqslant i \leqslant h$. Clearly, $V(\mathbf{k}') \subset V(\mathbf{k})$ if and only if $\mathbf{k}' \leqslant \mathbf{k}$. Hence we have the following

THEOREM 6.0.2. *The irreducible components of* $\mathcal{C}$ *are its subvarieties of the form* $V(\mathbf{k})$, *with* $\mathbf{k}$ *a maximal element in* $\mathcal{K}_{\mathbf{n}}$ *(with respect to the partial order above).*

As a consequence of Theorem 5.4.2, we have,

THEOREM 6.0.3. *Let* $\mathbf{k} \in \mathcal{K}_{\mathbf{n}}$.

(1) $\dim V(\mathbf{k}) = \sum_{1 \leqslant i \leqslant h+1} (n_i - k_i)(k_{i-1} + k_i)$, *where* $k_0 = k_{h+1} = 0$.

(2) $\operatorname{codim}_Z V(\mathbf{k}) = \sum_{i=1}^{h} (n_{i+1} - k_i)(n_i - k_i) + \sum_{i=1}^{h-1} k_i k_{i+1}$.

Considering some fixed basis in each W_i, we have the identifications $W_i \cong k^{n_i}$ and $Z \cong M(n_2 \times n_1) \times \cdots \times M(n_{h+1} \times n_h)$, where $M(l \times m)$ denotes the affine space of matrices over k with l rows and m columns. Then $\mathcal{C}$ can be identified with the set of all points $(A_1, \cdots, A_h)$ in $M(n_2 \times n_1) \times \cdots \times M(n_{h+1} \times n_h)$ such that $A_2 A_1 = 0, \ldots, A_h A_{h-1} = 0$. For $\mathbf{k} \in \mathcal{K}_{\mathbf{n}}$, the variety $V(\mathbf{k})$ is identified with the set of all points $(A_1, \cdots, A_h)$ in $M(n_2 \times n_1) \times \cdots \times M(n_{h+1} \times n_h)$ such that $A_2 A_1 = 0, \ldots, A_h A_{h-1} = 0$, $\operatorname{rank} A_1 \leqslant k_1, \ldots, \operatorname{rank} A_h \leqslant k_h$. Note that $V(\mathbf{k})$ is the closure of the $G_{\mathbf{n}}$-orbit in $M(n_2 \times n_1) \times \cdots \times M(n_{h+1} \times n_h)$ through a point $(A_1, \ldots, A_h)$ such that $\operatorname{rank} A_1 = k_1, \ldots, \operatorname{rank} A_h = k_h$, and due to the action of $G_{\mathbf{n}}$, we may assume, without loss of generality, that

$$A_i = \begin{pmatrix} 0 & 0 \\ M_i & 0 \end{pmatrix}, \qquad \text{for } 1 \leqslant i \leqslant h,$$

where M_i is the $k_i \times k_i$ identity matrix.

6.1. Singular locus of $V(k_1, \ldots, k_h)$

In this section we determine the singular points of $V(\mathbf{k})$, using the Jacobian criterion.

Let $\mathbf{k} = (k_1, \ldots, k_h) \in \mathcal{K}_{\mathbf{n}}$, and $V = V(k_1, k_2, \ldots, k_h)$. For $1 \leqslant i \leqslant h$, let X_i denote the $n_{i+1} \times n_i$ matrix of variables, and $\mathcal{M}_i$ the set of all $k_i + 1$ minors of X_i. Then V is the set of zeroes of the polynomials in $\mathcal{M}_1, \ldots, \mathcal{M}_h, X_2 X_1, \ldots, X_h X_{h-1}$. Let $x = (A_1, \ldots, A_h)$ be a point in V, A_i being an $n_{i+1} \times n_i$ matrix of rank at most k_i, and let J_x be the Jacobian matrix of V evaluated at x. The rows of J_x are indexed by the polynomials in the sets $\mathcal{M}_1, \ldots, \mathcal{M}_h, X_2 X_1, \ldots, X_h X_{h-1}$, while the columns of J are indexed by the entries in $X_1, \ldots, X_h$. The (M, α)-th entry of J_x, where M is a minor in $\mathcal{M}_1 \cup \cdots \cup \mathcal{M}_h$ and α is an entry in $X_1 \cup \cdots \cup X_h$, is equal to $\pm$ the minor obtained from M by deleting the row and the column of α, if α appears in M, and 0 otherwise. Let $f \in X_{i+1} X_i$ be the product of the r-th row of X_{i+1} and the s-th column of X_i, for some $1 \leqslant i \leqslant h-1$, $1 \leqslant r \leqslant n_{i+2}$, $1 \leqslant s \leqslant n_i$, and let α be an entry in $X_1 \cup \cdots \cup X_h$; then the (f, α)-th entry of J_x is equal to the (j, s)-th entry of

X_i if α is the (r, j)-th entry of X_i for some $1 \leqslant j \leqslant n_i$, the (r, l)-th entry of X_{i+1} if α is the (l, s)-th of X_i for some $1 \leqslant l \leqslant n_i$, and 0 otherwise.

The next lemma is obvious from the fact that the closure of the orbit through a point $x = (A_1, \ldots, A_h)$ in V such that $\operatorname{rank} A_i = k_i$ for all $1 \leqslant i \leqslant h$, is V itself. We include the proof of this result for establishing notations that will be used in the sequel, and for a better understanding of the Jacobian matrix of V.

LEMMA 6.1.1. *Let $x = (A_1, \ldots, A_h)$ be a point in V such that* $\operatorname{rank} A_i = k_i$ *for all $1 \leqslant i \leqslant h$. Then x is a smooth point of V.*

PROOF. Without loss of generality, we may assume that

$$A_i = \begin{pmatrix} 0 & 0 \\ M_i & 0 \end{pmatrix}, \qquad \text{for } 1 \leqslant i \leqslant h,$$

where M_i is the $k_i \times k_i$ identity matrix.

Let us denote J_x by just J.

Next we describe all the nonzero rows of J.

For $1 \leqslant i \leqslant h$, a row of J indexed by a minor $M \in \mathcal{M}_i$ of X_i is nonzero if and only if M contains the identity block M_i. The only nonzero element in such a row is equal to ± 1, placed in the column indexed by the upper right corner entry of M. When computing the rank of J, we may assume that such a ± 1 entry of J_x is actually equal to 1. Let $\mathcal{R}_i$ be the set of all minors in $\mathcal{M}_i$ containing M_i. Let Y_i be the set of all right upper corners of minors in $\mathcal{R}_i$, i.e. the set of all entries of X_i not contained in any of the rows or columns of M_i. Note that Y_i consists of the entries of the $(n_{i+1} - k_i) \times (n_i - k_i)$ right upper block of X_i, and hence its cardinality is $(n_{i+1} - k_i)(n_i - k_i)$. Also, $|\mathcal{R}_i| = (n_{i+1} - k_i)(n_i - k_i)$.

Let now f be an entry in $X_{i+1}X_i$ for some $1 \leqslant i \leqslant h - 1$. In Fig. 1, the blocks X_i, X_{i+1}, $X_{i+1}X_i$ are displayed as shown below:

X_i	
$X_{i+1}X_i$	X_{i+1}

Let f be the product of the r-th row of X_{i+1} and the s-th column of X_i, $1 \leqslant r \leqslant n_{i+2}$, $1 \leqslant s \leqslant n_i$. Then the row of J indexed by f is nonzero if and only if either the r-th row of X_{i+1} is nonzero (i.e. it is one of the last k_{i+1} rows of X_{i+1}), or the s-th column of X_i is nonzero (i.e. it is one of the first k_i columns of X_i). The set of all such f's in $X_{i+1}X_i$ can be written as $B_{i,i+1} \cup L_{i,i+1}$, where $B_{i,i+1}$ denotes the set of all entries in the last k_{i+1} rows of $X_{i+1}X_i$, and $L_{i,i+1}$ denotes the set of all entries in the first k_i columns of $X_{i+1}X_i$.

Now we find a maximal set of linearly independent rows of J among the set of rows described above.

The rows of J indexed by minors in $\mathcal{R}_1, \ldots, \mathcal{R}_h$ are linearly independent, since their nonzero entries are in different columns (these columns are indexed by the entries

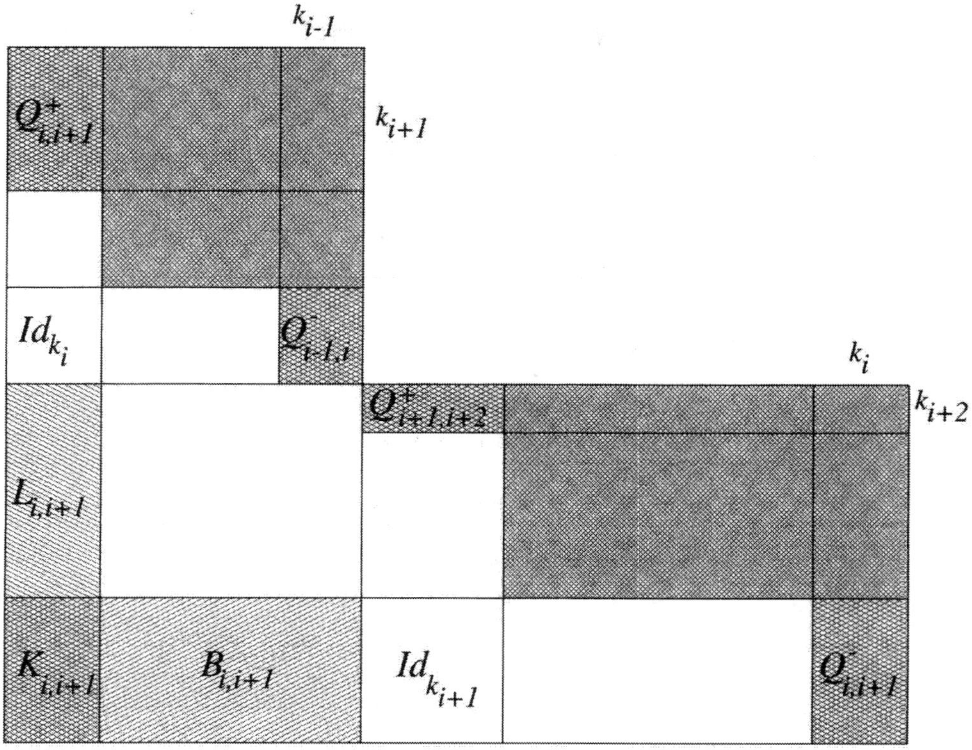

Figure 1

in $Y_1, \dots, Y_h$, and each of them contains precisely one nonzero entry in the rows indexed by $\mathcal{R}_1, \dots, \mathcal{R}_h$).

For $1 \leqslant i \leqslant h-1$, let $K_{i,i+1} = B_{i,i+1} \cap L_{i,i+1}$ (note that $K_{i,i+1}$ is the $k_{i+1} \times k_i$ left lower block of $X_{i+1}X_i$). A row of J indexed by $f \in K_{i,i+1}$ contains precisely two nonzero entries (equal to 1), none of which is in the same column with a nonzero entry of another nonzero row of J. The set of nonzero entries in the rows indexed by all $f \in K_{i,i+1}$ are in the columns indexed by the entries in the left upper block of X_i of size $k_{i+1} \times k_i$, which we denote by $Q^+_{i,i+1}$, and the right lower block of X_{i+1} of size $k_{i+1} \times k_i$, which we denote by $Q^-_{i,i+1}$.

A row of J indexed by $f \in (B_{i,i+1} \cup L_{i,i+1}) \setminus K_{i,i+1}$ contains precisely one nonzero entry (equal to 1), in a column indexed either by an element in the first k_{i+1} rows of Y_i or by an element in the last k_i columns of Y_{i+1}. Therefore these rows are identical to some of the rows indexed by minors in $\mathcal{R}_1, \dots, \mathcal{R}_h$.

Consequently, the rows of J indexed by minors in $\mathcal{R}_1, \ldots, \mathcal{R}_h$ and entries of $K_{1,2}, \ldots, K_{h-1,h}$ give a maximal set of linearly independent rows of J. Therefore

$$\begin{aligned}\operatorname{rank} J &= |\mathcal{R}_1| + \cdots + |\mathcal{R}_h| + |K_{1,2}| + \cdots + |K_{h-1,h}| \\ &= (n_1 - k_1)(n_2 - k_2) + \cdots + (n_h - k_h)(n_{h+1} - k_h) \\ &\quad + k_1 k_2 + \cdots + k_{h-1} k_h \\ &= \operatorname{codim} V.\end{aligned}$$

This shows that x is a smooth point of V. □

Notation. For the rest of this section,

$$J, Y_1, \ldots, Y_h, Q^+_{i,i+1}, Q^-_{i,i+1}, B_{i,i+1}, L_{i,i+1}, K_{i,i+1}, 1 \leqslant i \leqslant h-1$$

will denote the Jacobian matrix, respectively the sets, associated, as in Lemma 6.1.1, to the point $(A_1, \ldots, A_h)$ in V with

$$A_i = \begin{pmatrix} 0 & 0 \\ M_i & 0 \end{pmatrix}, \qquad \text{for } 1 \leqslant i \leqslant h,$$

where M_i is the $k_i \times k_i$ identity matrix.

REMARK 6.1.2. Note that the nonzero columns of J are the columns indexed by the entries in $Y_1, \ldots, Y_h, Q^+_{1,2}, Q^-_{1,2}, \ldots, Q^+_{h-1,h}, Q^-_{h-1,h}$. For $1 \leqslant j \leqslant h-1$, each column indexed by an entry in $Q^+_{j,j+1}$ is identical to a column indexed by the corresponding entry in $Q^-_{j,j+1}$. A maximal set of linearly independent columns of J consists of all distinct nonzero columns of J. We have

$$\operatorname{codim} V = |Y_1| + \cdots + |Y_h| + |Q^\pm_{1,2}| + \cdots + |Q^\pm_{h-1,h}|$$

($|Q^\pm_{j,j+1}|$ denotes the cardinality of each of $Q^+_{j,j+1}$ and $Q^-_{j,j+1}$). A nonzero column whose index is not included in the set of indices of the columns in a maximal set Σ of linearly independent columns of J is indexed by an entry in $Q^+_{j,j+1}$ or $Q^-_{j,j+1}$, for some j, and it is identical to a column whose index belongs to $Q^-_{j,j+1}$, respectively $Q^+_{j,j+1}$, and is among the indices of columns in Σ.

REMARK 6.1.3. For $1 \leqslant j \leqslant h$, a maximal set of linearly independent columns of J indexed by entries in $X_1, \ldots, X_{j-1}, X_{j+1}, \ldots, X_h$ consists of all distinct nonzero such columns. The cardinality of such a maximal set is $\operatorname{codim} V - |Y_j|$. The set of indices of the columns in such a maximal set must include $Q^+_{j-1,j}$ and $Q^-_{j,j+1}$.

REMARK 6.1.4. For $1 \leqslant j \leqslant h-1$, a maximal set of linearly independent columns of J indexed by $X_1, \ldots, X_{j-1}, X_{j+2}, \ldots, X_h$ consists of all distinct nonzero such columns. The cardinality of such a maximal set is

$$\operatorname{codim} V - (|Y_j| + |Y_{j+1}| + |Q^\pm_{j,j+1}|).$$

The set of indices of the columns in such a maximal set must include $Q^+_{j-1,j}$ and $Q^-_{j+1,j+2}$.

LEMMA 6.1.5. *Let* $x = (A_1, \dots, A_h)$ *be a point of* V *such that* $\operatorname{rank} A_j < k_j$ *for some* j, $1 \leqslant j \leqslant h$. *If* x *is smooth, then either* $k_{j-1} + k_j = n_j$, *or* $k_j + k_{j+1} = n_{j+1}$ *(here* $k_0 = k_{h+1} = 0$*).*

PROOF. Let $t_i = \operatorname{rank} A_i$, for $1 \leqslant i \leqslant h$, with $t_j < k_j$. Without loss of generality, we may assume that

$$A_i = \begin{pmatrix} 0 & 0 \\ N_i & 0 \end{pmatrix}, \qquad \text{for } 1 \leqslant i \leqslant h,$$

where N_i is the $t_i \times t_i$ identity matrix. As noticed in Remark 6.1.3, there are precisely $\operatorname{codim} V - |Y_j|$ linearly independent columns of J indexed by entries in $X_1, \dots, X_{j-1}, X_{j+1}, \dots, X_h$. Since the nonzero columns of J_x are among the nonzero columns of J, the set T of linearly independent columns of J_x indexed by entries in $X_1, \dots, X_{j-1}, X_{j+1}, \dots, X_h$ has cardinality at most $\operatorname{codim} V - |Y_j|$ (note that T consists of all distinct nonzero columns of J_x). Note that a column of J indexed by an entry in $X_1, \dots, X_{j-1}, X_{j+1}, \dots, X_h$ is either identical to some column indexed by an entry in X_j, or it is linearly independent of the columns indexed by entries in X_j. Since x is smooth, $\operatorname{rank} J_x = \operatorname{codim} V$, so there exists a set S of cardinality at least $|Y_j|$ of linearly independent columns of J_x indexed by entries in X_j such that $S \cup T$ is a set of linearly independent columns of J_x. The only nonzero columns indexed by entries in X_j are actually indexed by entries in the first k_{j+1} rows of X_j and the last k_{j-1} columns of X_j (since $\operatorname{rank} A_j < k_j$). The columns indexed by entries in $Q^-_{j-1,j} \cup Q^+_{j,j+1}$ cannot be in S, since they are in T (see Remark 6.1.3), and $S \cup T$ is a set of linearly independent columns of J_x. Thus S consists of columns indexed by entries in the first k_{j+1} rows of Y_j and the last k_{j-1} columns of Y_j. Since the cardinality of S must be at least $|Y_j|$, we deduce that the set of entries in the first k_{j+1} rows and the last k_{j-1} columns of Y_j must actually contain the set of entries in Y_j. This implies that either $k_{j-1} + k_j = n_j$, or $k_j + k_{j+1} = n_{j+1}$. □

LEMMA 6.1.6. *Let* $x = (A_1, \dots, A_h)$ *be a point of* V *such that* $\operatorname{rank} A_j < k_j$ *for some* j, $1 \leqslant j \leqslant h$, *and* $\operatorname{rank} A_i = k_i$ *for* $1 \leqslant i \leqslant h$, $i \neq j$. *Then* x *is smooth if and only if either* $k_{j-1} + k_j = n_j$, *or* $k_j + k_{j+1} = n_{j+1}$ *(here* $k_0 = k_{h+1} = 0$*).*

PROOF. In view of Lemma 6.1.5, we only have to show that if $k_{j-1} + k_j = n_j$ or $k_j + k_{j+1} = n_{j+1}$, then x is smooth. It is easily seen that the columns of J_x that are indexed by the elements in the disjoint sets

$$Y_1, \dots, Y_{j-1}, Y_{j+1}, \dots, Y_h, Q^-_{1,2}, \dots, Q^-_{j-2,j-1}, Q^+_{j+1,j+2}, \dots, Q^+_{h-1,h}, \Gamma_j,$$

where Γ_j is the set of entries in the first k_{j+1} rows and the last k_{j-1} columns of X_j, are linearly independent. Note that Y_j is contained in Γ_j (since $k_{j-1} + k_j = n_j$ or $k_j + k_{j+1} = n_{j+1}$), and the cardinality of Γ_j is precisely $|Y_j| + |Q^-_{j-1,j}| + |Q^+_{j,j+1}|$. Since the number of these columns is $\operatorname{codim} V$, we conclude that $\operatorname{rank} J_x = \operatorname{codim} V$, and hence x is smooth. □

REMARK 6.1.7. An alternate proof of Lemma 6.1.6 can be obtained as follows. First note that if $k_{j-1} + k_j = n_j$, then $|L_{j-1,j} \setminus K_{j-1,j}| = |\mathcal{R}_j|$ and if $k_j + k_{j+1} = n_{j+1}$, then $|B_{j,j+1} \setminus K_{j,j+1}| = |\mathcal{R}_j|$. Let

$$\Delta_j = \begin{cases} L_{j-1,j} \setminus K_{j-1,j}, & \text{if } k_{j-1} + k_j = n_j, \\ B_{j,j+1} \setminus K_{j,j+1}, & \text{if } k_j + k_{j+1} = n_{j+1}, \\ & \\ L_{j-1,j} \setminus K_{j-1,j} \text{ or} & \text{if } k_{j-1} + k_j = n_j \quad \text{and} \\ \quad B_{j,j+1} \setminus K_{j,j+1}, & k_j + k_{j+1} = n_{j+1} \end{cases}$$

We have $|\Delta_j| = |\mathcal{R}_j|$, and the rows of J indexed by elements in $\mathcal{R}_j$ are identical to the rows of J indexed by Δ_j. It is easily seen that the rows of J_x indexed by the $\operatorname{codim} V$ elements in the sets

$$\mathcal{R}_1, \ldots, \mathcal{R}_{j-1}, \Delta_j, \mathcal{R}_{j+1}, \ldots, \mathcal{R}_h, K_{1,2}, \ldots, K_{h-1,h}$$

are linearly independent, and hence x is smooth.

LEMMA 6.1.8. *Let $x = (A_1, \ldots, A_h)$ be a point of V such that if* $\operatorname{rank} A_j < k_j$ *for some $1 \leqslant j \leqslant h$, then*

(1) *either $1 \leqslant j-1 \leqslant h$ and* $\operatorname{rank} A_{j-1} = k_{j-1}$, *or $1 \leqslant j+1 \leqslant h$ and* $\operatorname{rank} A_{j+1} = k_{j+1}$
(2) *either $k_{j-1} + k_j = n_j$, or $k_j + k_{j+1} = n_{j+1}$ (here $k_0 = k_{h+1} = 0$)*

Then x is a smooth point of V.

PROOF. Let Δ_j be as in Remark 6.1.7. Then the rows of J_x indexed by the $\operatorname{codim} V$ elements in the sets

$$\begin{aligned} &\mathcal{R}_i, \text{ with } i \text{ such that } \operatorname{rank} A_i = k_i, \\ &\Delta_j, \text{ with } j \text{ such that } \operatorname{rank} A_j < k_j, \\ &K_{1,2}, \ldots, K_{h-1,h} \end{aligned}$$

are linearly independent. Therefore x is a smooth point of V.

□

REMARK 6.1.9. An alternate proof of Lemma 6.1.8 can be obtained as follows. For $1 \leqslant j \leqslant h$ with $\operatorname{rank} A_j < k_j$, let Γ_j be as in the proof of Lemma 6.1.6. It is easily seen that the columns of J_x that are indexed by the elements in the disjoint sets

$$\begin{aligned} &Y_i, \text{ with } i \text{ such that } \operatorname{rank} A_i = k_i, \\ &Q^+_{i,i+1}, \text{ with } i \text{ such that } \operatorname{rank} A_i = k_i, \operatorname{rank} A_{i+1} = t_{i+1}, \\ &\Gamma_j, \text{ with } i \text{ such that } \operatorname{rank} A_j < k_j, \end{aligned}$$

are linearly independent. Since the number of these columns is $\operatorname{codim} V$, we conclude that $\operatorname{rank} J_x = \operatorname{codim} V$, and hence x is smooth.

LEMMA 6.1.10. *Let $x = (A_1, \ldots, A_h)$ be a point of V such that* $\operatorname{rank} A_j < k_j$ *and* $\operatorname{rank} A_{j+1} < k_{j+1}$ *for some j, $1 \leqslant j \leqslant h-1$. Then x is a singular point of V.*

PROOF. Let $\operatorname{rank} A_i = t_i$, with $t_j < k_j, t_{j+1} < k_{j+1}$. Without loss of generality, we may assume that

$$A_i = \begin{pmatrix} 0 & 0 \\ N_i & 0 \end{pmatrix}, \qquad \text{for } 1 \leqslant i \leqslant h,$$

where N_i is the $t_i \times t_i$ identity matrix.

The nonzero columns of J_x are among the nonzero columns of J. Let T be the indices of a maximal set of linearly independent columns of J_x indexed by entries in $X_1, \ldots, X_{j-1}, X_{j+2}, \ldots, X_h$. Note that a column of J indexed by an entry in $X_1, \ldots, X_{j-1}, X_{j+2}, \ldots, X_h$ is either identical to some column indexed by an entry in $X_j \cup X_{j+1}$, or it is linearly independent of the columns indexed by entries in $X_j \cup X_{j+1}$. Thus there exists a set $S \subset X_j \cup X_{j+1}$ such that the columns of J_x indexed by elements of $S \cup T$ give a maximal set of linearly independent columns of J_x, and we have $\operatorname{rank} J_x = |S| + |T|$.

As noticed in Remark 6.1.4, there are precisely

$$\operatorname{codim} V - (|Y_j| + |Y_{j+1}| + |Q^{\pm}_{j,j+1}|)$$

linearly independent columns of J indexed by entries in $X_1, \ldots, X_{j-1}, X_{j+2}, \ldots, X_h$. Therefore $|T| \leqslant \operatorname{codim} V - (|Y_j| + |Y_{j+1}| + |Q^{\pm}_{j,j+1}|)$.

Now we show that $|S| < |Y_j| + |Y_{j+1}| + |Q^{\pm}_{j,j+1}|$. Each column indexed by an element of S contains precisely one nonzero entry, in a row indexed by an entry of X_jX_{j-1}, $X_{j+1}X_j$ or $X_{j+2}X_{j+1}$ (since $\operatorname{rank} A_j < k_j$ and $\operatorname{rank} A_{j+1} < k_{j+1}$). The columns indexed by entries in $Q^-_{j-1,j}$ and $Q^+_{j+1,j+2}$ are identical to columns in T (see Remark 6.1.4). Therefore $S \cap Q^-_{j-1,j} = \emptyset$, $S \cap Q^+_{j+1,j+2} = \emptyset$, and

$$S \subset Y_j \cup Y_{j+1} \cup Q^+_{j,j+1} \cup Q^-_{j,j+1}.$$

The set of columns indexed by entries in $Q^+_{j,j+1}$ is the same as the set of columns indexed by entries in $Q^-_{j,j+1}$, so we may assume that $S \cap Q^-_{j,j+1} = \emptyset$. Thus we have $|S| = |S \cap Y_j| + |S \cap Y_{j+1}| + |S \cap Q^+_{j,j+1}|$. Obviously, $|S \cap Y_j| \leqslant |Y_j|$, and $|S \cap Y_{j+1}| \leqslant |Y_{j+1}|$ (note that one can have equalities here). On the other hand, we have $|S \cap Q^+_{j,j+1}| \leqslant t_j t_{j+1} < k_j k_{j+1} = |Q^{\pm}_{j,j+1}|$. Consequently, $|S| < |Y_j| + |Y_{j+1}| + |Q^{\pm}_{j,j+1}|$.

Thus $\operatorname{rank} J_x < \operatorname{codim} V$, which shows that x is a singular point of V. □

Next we describe the singular locus of V.

For $1 \leqslant i \leqslant h$, let V_i denote the set of all points $(A_1, \ldots, A_h)$ in V such that $\operatorname{rank} A_i < k_i$, i.e.

$$V_i = V(k_1, \ldots, k_{i-1}, k_i - 1, k_{i+1}, \ldots, k_h).$$

For $1 \leqslant j \leqslant h-1$, let $V_{j,j+1}$ denote the set of all points $(A_1, \ldots, A_h)$ in V such that $\operatorname{rank} A_j < k_j$ and $\operatorname{rank} A_{j+1} < k_{j+1}$, i.e.

$$V_{j,j+1} = V(k_1, \ldots, k_{j-1}, k_j - 1, k_{j+1} - 1, k_{j+2}, \ldots, k_h).$$

Note that $V_{j,j+1} = V_j \cap V_{j+1}$.

THEOREM 6.1.11. *The irreducible components of* Sing V *are the subvarieties of* V *of the form* V_i, *with* $i \in \Omega$, *and* $V_{j,j+1}$, *with* $j \notin \Omega$, $j+1 \notin \Omega$, *where* Ω *is the set of all* $1 \leqslant i \leqslant h$ *such that* $k_{i-1} + k_i < n_i$ *and* $k_i + k_{i+1} < n_{i+1}$.

PROOF. Note that Sing V is a union of $G_{\mathbf{n}}$-orbit closures, i.e. subvarieties of V of the form $V(\mathbf{r})$, with $\mathbf{r} \in \mathcal{K}_{\mathbf{n}}$.

Next we determine for which $\mathbf{r} = (r_1, \ldots, r_h) \in \mathcal{K}_{\mathbf{n}}$, with $\mathbf{r} \leqslant \mathbf{k}$, the subvariety $V(\mathbf{r})$ is singular. In order to check that $V(r_1, \ldots, r_h)$ is singular, it is enough to check that the point $x = (A_1, \ldots, A_h)$ with rank $A_i = r_i$ is singular.

If there is no i with $r_i < k_i$, by Lemma 6.1.1, $V(\mathbf{r})$ is smooth.

If there is precisely one index i with $r_i < k_i$, by Lemma 6.1.6, the subvariety $V(\mathbf{r}) = V_i$ is singular if and only if $i \in \Omega$.

Next, we analyze the subvarieties $V(\mathbf{r})$ with $r_i < k_i$ for more than one index i.

If there is some j with $r_j < k_j$ and $r_{j+1} < k_{j+1}$, then $V(\mathbf{r}) \subset V_{j,j+1}$. By Lemma 6.1.10, $V_{j,j+1}$ is singular, and therefore $V(\mathbf{r})$ is singular.

If for each j, we have either $r_j = k_j$ or $r_{j+1} = k_{j+1}$, then by Lemma 6.1.8, $V(\mathbf{r})$ is singular only if $i \in \Omega$ for some i with $r_i < k_i$. But in this case $V(\mathbf{r}) \subset V_i$, and since V_i is singular, $V(\mathbf{r})$ is also singular.

Thus we have determined all the subvarieties of V of the form $V(\mathbf{r})$ with $\mathbf{r} \in \mathcal{K}_{\mathbf{n}}$, $\mathbf{r} \leqslant \mathbf{k}$ that are contained in Sing V. By taking the maximal such subvarieties (with respect to set inclusion), we obtain the required description for the irreducible components of Sing V. □

As a consequence, we obtain the following result, that describes the singular loci of the irreducible components of the variety of complexes:

THEOREM 6.1.12. *(cf.* **[60]***) If* V *is an irreducible component of the variety of complexes* $\mathcal{C}$, *then the irreducible components of* Sing V *are* $V_{1,2}, \ldots, V_{h-1,h}$.

PROOF. In this case the set Ω is empty (by Theorem 6.0.2), and the result follows immediately from Theorem 6.1.11. □

7. An application

In this section, we take $h = 2$. Let $V(k_1, k_2)$ be an irreducible component of $\mathcal{C}$, and let $w \in W^Q$ be the element such that $V(k_1, k_2) \cong Y_Q(w)$. In this section, using Theorem 6.1.11, we deduce results on Sing$X(w^{\max})$, $w^{\max}$ being the maximal representative in W of the coset wW_Q (note that $X(w^{\max})$ is the pull back of $X(w)$ under the canonical projection $G/B \to G/Q$).

Let $a_1 = n_1$, $a_2 = n_1 + n_2$, $a_3 = n_1 + n_2 + n_3 (= n)$, $G = SL(n)$, and $Q = P_{a_1} \cap P_{a_2}$. We have (cf. Theorem 5.4.2)

$$V(k_1, k_2) \cong Y_Q(w),$$
$$V(k_1 - 1, k_2 - 1) \cong Y_Q(\theta),$$

where $w, \theta \in W^Q$ are given by

$$w^{(a_1)} = (\underbrace{\ldots a_1}_{a_1 - k_1}\ \underbrace{\ldots . a_2}_{k_1}),\ w^{(a_2)} = (\underbrace{1 \ldots a_1}_{a_1}\ \underbrace{\ldots . a_2}_{n_2 - k_2}\ \underbrace{\ldots . a_3}_{k_2}),$$

$$\theta^{(a_1)} = (\underbrace{\ldots a_1}_{a_1 - k_1 + 1}\ \underbrace{\ldots . a_2}_{k_1 - 1}),\ \theta^{(a_2)} = (\underbrace{1 \ldots a_1}_{a_1}\ \underbrace{\ldots . a_2}_{n_2 - k_2 + 1}\ \underbrace{\ldots . a_3}_{k_2 - 1}).$$

LEMMA 7.0.1. *$V(k_1, k_2)$ being an irreducible component of V, we have, either $k_1 + k_2 = n_2$ or $k_1 = n_1,\ k_2 = n_3$.*

PROOF. Note that (k_1, k_2) is a maximal element of $A_{\mathbf{n}}$. We divide the proof into the following two cases.

Case 1. $n_2 \geqslant n_1 + n_3$.

In this case, there is a unique maximal choice for k_1, k_2, namely $k_1 = n_1,\ k_2 = n_3$.

Case 2. $n_2 < n_1 + n_3$.

If $n_2 \geqslant n_1$, then $(k_1, k_2) = (n_1 - i, n_2 - n_1 + i),\ i = 0, 1, \cdots, n_1$ are all the possible maximal choices for (k_1, k_2), and for all such choices, we have, $k_1 + k_2 = n_2$.

Similarly, if $n_2 \geqslant n_3$, then $(k_1, k_2) = (n_2 - n_3 + i, n_3 - i),\ i = 0, 1, \cdots, n_3$ are all the possible maximal choices for (k_1, k_2), and for all such choices, we have, $k_1 + k_2 = n_2$.

If $n_2 < n_1,\ n_2 < n_3$, then $(k_1, k_2) = (i, n_2 - i),\ i = 0, 1, \cdots, n_2$ are all the possible maximal choices for (k_1, k_2), and for all such choices, we have, $k_1 + k_2 = n_2$. □

REMARK 7.1.1. From the proof of Lemma 7.0.1, we have, if $n_2 \geqslant n_1 + n_3$, then $k_1 = n_1,\ k_2 = n_3$; if $n_2 < n_1 + n_3$, then $k_1 + k_2 = n_2$.

7.1. We have the following two cases (cf. proof of Lemma 7.0.1):

Case1. $n_2 \geqslant n_1 + n_3$.

In this case, we have $k_1 = n_1,\ k_2 = n_3$ (cf. Remark 7.1.1). Hence

$$w^{(a_1)} = ([\![x, a_2]\!]),\ w^{(a_2)} = ([\![1, a_1]\!], [\![x', a_2]\!], [\![y, a_3]\!]),$$

$$\theta^{(a_1)} = ([\![n_1]\!], [\![x+1, a_2]\!]),\ \theta^{(a_2)} = ([\![1, a_1]\!], [\![x'-1, a_2]\!], [\![y+1, a_3]\!]),$$

where $x = a_2 + 1 - k_1\ (= n_2 + 1),\ x' = a_2 + 1 - (n_2 - k_2)\ (= n_1 + n_3 + 1),\ y = a_3 + 1 - k_2\ (= a_2 + 1)$.

Case 2. $n_2 < n_1 + n_3$.

In this case, we have, $k_1 + k_2 = n_2$ (cf. Remark 7.1.1). Hence

$$w^{(a_1)} = ([\![p, a_1]\!], [\![x, a_2]\!]), \qquad w^{(a_2)} = ([\![1, a_1]\!], [\![x, a_2]\!], [\![y, a_3]\!]),$$

$$\theta^{(a_1)} = ([\![p-1, a_1]\!], [\![x+1, a_2]\!]), \qquad \theta^{(a_2)} = ([\![1, a_1]\!], [\![x-1, a_2]\!], [\![y+1, a_3]\!]),$$

where $p = a_1 + 1 - (n_1 - k_1),\ x = a_2 + 1 - k_1\ (= n_1 + k_2 + 1),\ y = a_3 + 1 - k_2$.

REMARK 7.1.2. Note that θ is the unique maximal element τ in W^Q such that $\tau \leqslant w$ and $\tau^{(a_i)}(a_i - k_i + 1) \leqslant a_i,\ i = 1, 2$ (here, $\tau^{(a_i)}(t)$ denotes the entry at the t-th place in $\tau^{(a_i)}$).

7.2. Validity of the conjecture of Chapter 8, §16.1 for Sing $X(w^{\max})$

Let Q be as above. As above, we shall denote an element τ in W^Q as $\tau = (\tau^{(a_1)}, \tau^{(a_2)})$. Let $\tau^{\max}$ denote the maximal representative in W of the coset τW_Q. Note that $X(\tau^{\max})$ is the inverse image of $X_Q(\theta))$ under the canonical projection $G/B \to G/Q$; also, $\tau^{\max}$ as a permutation is given by writing the entries in $\tau^{(a_1)}$ in descending order, followed by the entries in $\tau^{(a_2)} \setminus \tau^{(a_1)}$ in descending order, followed by the entries in $\{1, \cdots, n\} \setminus \tau^{(a_2)}$ in descending order. In the sequel, we shall refer to these as the I, II, III blocks respectively of $\tau^{\max}$. Also, in the sequel, for an interval $[\![a, b]\!]$, $[\![a, b]\!]_-$ shall denote the entries of $[\![a, b]\!]$ written in descending order. Thus, in Case 1 of 7.1, we have (with notations as in 7.1),
I block in $w^{\max} = [\![n_2 + 1, a_2]\!]_-$, I block in $\theta^{\max} = ([\![n_2 + 2, a_2]\!]_-, [\![n_1]\!])$,
II block in $w^{\max} = ([\![y, a_3]\!]_-, [\![x', n_2]\!]_-, [\![1, a_1]\!]_-)$, II block in

$$\theta^{\max} = ([\![y + 1, a_3]\!]_-, [\![x' - 1, n_2 + 1]\!]_-, [\![1, a_1 - 1]\!]_-),$$

III block in $w^{\max} = [\![a_1 + 1, x' - 1]\!]_-$, III block in $\theta^{\max} = ([\![a_2 + 1]\!], [\![a_1 + 1, x' - 2]\!]_-)$.
In Case 2 of 7.1, we have (with notations as in 7.1),
I block in $w^{\max} = ([\![x, a_2]\!]_-, [\![p, a_1]\!]_-)$, I block in $\theta^{\max} = ([\![x+1, a_2]\!]_-, [\![p-1, a_1]\!]_-)$,
II block in $w^{\max} = ([\![y, a_3]\!]_-, [\![1, p - 1]\!]_-)$, II block in

$$\theta^{\max} = ([\![y + 1, a_3]\!]_-, [\![x - 1, x]\!]_-, [\![1, p - 2]\!]_-),$$

III block in $w^{\max} = ([\![a_2 + 1, y - 1]\!]_-, [\![a_1 + 1, x - 1]\!]_-)$, III block in

$$\theta^{\max} = ([\![a_2 + 1, y]\!]_-, [\![a_1 + 1, x - 2]\!]_-).$$

For $\tau \in W$, we shall denote the associated Schubert variety by $X(\tau)$.

THEOREM 7.2.1. *Sing* $X(w^{max}) = X(\theta^{max})$
PROOF. The result follows from Theorems 5.4.2, 6.1.11. □

We now show that the Conjecture of Chapter 8, §16.1 holds for $X := X(w^{\max})$. Let bad patterns of Type I and Type II be defined as in §3.5.

PROPOSITION 7.2.2. *A bad pattern in w^{max} has to be of Type II.*
PROOF. If possible, assume that there is a Type I bad pattern d, b, c, a in $w^{\max}$, where $a < b < c < d$. We have, d belongs to the I or II block in $w^{\max}$.
(1) Let d belong to the I block in $w^{\max}$. This implies that $b \in$ I or II block.

Let b belong to the I block in which case we have c does not belong to the I block. In Case 1 of 7.1, we have, $[\![b, d]\!] \subset$ I block in $w^{\max}$, and this contradicts the facts that $b < c < d$ and c does not belong to the I block. Hence our assumption is wrong and the result follows. In Case 2 of 7.1, we have, $d \geqslant x$, $p \leqslant b \leqslant a_1$. Hence we obtain $a \leqslant p$, and hence a, c should belong to the II block in $w^{\max}$. This in turn implies that $c \in [\![y, a_3]\!]$, which is not possible since $c < d$, and $d < y$. The result follows from this.

Let now b belong to the II block in $w^{\max}$ in which case we have c, a should belong to the III block. In Case 1 of 7.1, we have, $[\![a, c]\!]$ is contained in the III block which is not possible, since $a < b < c$ and b does not belong to the III block. In Case 2 of 7.1, we have, $b \in [\![1, p-1]\!]$, since $b < d < y$. But now, $a \geqslant a_1 + 1\,(> p-1)$ (since a belongs to the III block), which contradicts the fact that $b > a$.

(2) Let d belong to the II block in $w^{\max}$. This implies that b belongs to the II block, and c, a belong to the III block. In Case 1 of 7.1, we have, $[\![a, c]\!] \subset$ III block, which is not possible, since $a < b < c$ and b does not belong to the III block. In Case 2 of 7.1, we have, $b \in [\![1, p-1]\!]$ necessarily (since $b < c < y$) in which case, we obtain $a > b$ (since $a \geqslant a_1 + 1 > p - 1$), which is not possible. □

PROPOSITION 7.2.3. *There do exist Type II bad patterns in w^{max}. Further for any Type II bad pattern c, d, a, b in w^{max}, where $a < b < c < d$, we have, $c \in [\![x, a_2]\!]$, $d \in [\![y, a_3]\!]$, $a \in [\![1, u]\!]$, $b \in [\![a_1 + 1, v]\!]$, where $x =$ or $> n_2 + 1$ according as $n_1 =$ or $> k_1$, $y =$ or $> a_2 + 1$ according as $n_3 =$ or $> k_2$, $u = a_1$ (respectively $p - 1$) in Case 1 (respectively Case 2) of 7.1, $v = (x' - 1)$ (respectively $x - 1$) in Case 1 (respectively Case 2) of 7.1.*

PROOF. We have a Type II pattern in $w^{\max}$, namely, $c = x,\ d = y,\ a = u,\ b = v$ in $w^{\max}$ (here, x, y, u, v are as in Proposition 7.2.3).

Let now c, d, a, b be a Type II pattern in $w^{\max}$. We have c belongs to the I block, d, a belong to the II block, and b belongs to the III block (necessarily). We distinguish the following two cases:

Case 1. $n_2 \geqslant n_1 + n_3$.

In this case we have $n_1 = k_1$, and clearly $c \in [\![n_2 + 1, a_2]\!]$. This implies

$$d \in [\![y, a_3]\!], \quad a \in [\![1, a_1]\!], \quad b \in [\![a_1 + 1, x' - 1]\!].$$

Case 2. $n_2 < n_1 + n_3$.

Again clearly $c \in [\![x, a_2]\!]$. This implies

$$d \in [\![y, a_3]\!], \quad a \in [\![1, p-1]\!], \quad b \in [\![a_1 + 1, x - 1]\!].$$

□

COROLLARY 7.2.4. *With notations as in Proposition 7.2.3, we have, w^{max} is non-covexillary.*

PROOF. This follows (in view of Proposition 7.2.3) from the definition of a covexillary permutation; recall (cf. **[112]**) that a permutation $(a_1, \cdots, a_n)$ is covexillary if it avoids the pattern c, d, a, b, where $a < b < c < d$. □

LEMMA 7.2.5. (1) $Q_{w^{max}} = P_{w^{max}} = Q$.

(2) *$\overline{B\theta^{max}B}$ is Q-Q is stable , i.e., stable for multiplication on the left and right by Q.*

PROOF. The assertions follow from the definition of $w^{\max}$ and $\theta^{\max}$ □

THEOREM 7.2.6. *The Conjecture of Chapter 8, §16.1 holds for* Sing $X(w^{max})$.

PROOF. We first observe that $\theta^{\max} \in F_{w^{\max}}$ ($F_{w^{\max}}$ being as in Ch8, §16.1); for, corresponding to the Type II pattern $c = x,\ d = y,\ a = u,\ b = v$ in $w^{\max}$ (here,

x, y, u, v are as in Proposition 7.2.3), we have the bad pattern a, c, b, d in $\theta^{\max}$. Let us denote $\theta^{\max}$ by ϕ. Let w_1 (resp. τ) be the element of S_n obtained from ϕ (resp. $w^{\max}$) by replacing a, c, b, d (resp. c, d, a, b) respectively by c, d, a, b (resp. a, c, b, d). Then clearly $\tau \leqslant \phi$, and $w_1 \leqslant w$. Further, $\overline{B\theta^{\max}B}$ is Q-Q stable (cf. Lemma 7.2.5). Thus $\theta^{\max} \in F_{w^{\max}}$.

Let now $\tau_1 \in F_{w^{\max}}$.

We have an occurrence in τ_1 of the form (a, c, b, d), $a < b < c < d$, corresponding to the occurrence (c, d, a, b) in $w^{\max}$ (cf. Proposition 7.2.2). The fact that $\tau_1 \in F_{w^{\max}}$ in particular implies that $\tau_1 \in W_Q^{\max}$, and this in turn implies that a belongs to the I block, c, b belong to the II block, and d belongs to the III block in τ_1, i.e., $a \in \tau_1^{(a_1)}$, $b, c \in \tau_1^{(a_2)}$. Let w_1 be the element of S_n obtained from τ_1 by replacing a, c, b, d respectively by c, d, a, b. We have $w_1 \leqslant w^{\max}$ (by definition of $F_{w^{\max}}$).

Let i (respectively j) be such that $w_1^{(a_1)}(i) \leqslant a_1$, $w_1^{(a_1)}(i+1) > a_1$, $w_1^{(a_1)}(j) = c$. Note that $i \geqslant a_1 - k_1$ (since $w_1 \leqslant w^{\max}$ and $w^{(a_1)}(a_1 - k_1) \leqslant a_1$), $j \geqslant i + 1$ (since $c \geqslant x\,(= a_2 + 1 - k_1) > n_1\,(= a_1)$). Now, $\tau_1^{(a_1)}$ is obtained from $w_1^{(a_1)}$ by replacing c by a, where note that $a \leqslant a_1$ (cf. Proposition 7.2.3). From this we obtain $\tau_1^{(a_1)}(i+1) = w_1^{(a_1)}(i) \leqslant a_1$ and hence

$$\tau_1^{(a_1)}(a_1 - k_1 + 1) \leqslant a_1 \tag{1}$$

(note that $a_1 - k_1 + 1 \leqslant i + 1$). Let k (respectively l) be such that

$$w_1^{(a_2)}(k) \leqslant a_2, \quad w_1^{(a_2)}(k+1) > a_2, \quad w_1^{(a_2)}(l) = d.$$

Note that $k \geqslant a_2 - k_2$ (since $w_1 \leqslant w^{\max}$ and $w^{(a_2)}(a_2 - k_2) \leqslant a_2$), $l \geqslant k + 1$ (since $d \geqslant y > a_2$ (cf. Proposition 7.2.3)). Now, $\tau_1^{(a_2)}$ is obtained from $w_1^{(a_2)}$ by replacing d by b, where note that $b \leqslant a_2$. From this we obtain $\tau_1^{(a_2)}(k+1) = w_1^{(a_2)}(k) \leqslant a_2$ and hence

$$\tau_1^{(a_2)}(a_2 - k_2 + 1) \leqslant a_2 \tag{2}$$

(note that $a_2 - k_2 + 1 \leqslant k + 1$). Now (1) and (2) together with Remark 7.1.2 imply that $\tau_1 \leqslant \theta^{\max}$. The required result follows from this. □

Example: Let $\mathbf{n} = (2, 3, 2)$. We have $n = 7$, $Q = P_2 \cap P_5$. Further, $V(1,2)$, $V(2,1)$ are the two irreducible components of V. Let w_1, w_2 be the elements of W^Q such that $V(1,2) \cong Y_Q(w_1)$, $V(2,1) \cong Y_Q(w_2)$. We have,

$$w_1^{(2)} = (25), \qquad w_1^{(5)} = (12\ 5\ 67),$$
$$w_2^{(2)} = (45), \qquad w_2^{(5)} = (12\ 45\ 7).$$

Let θ_1, θ_2 be the elements of W^Q such that $V(0,1) \cong Y_Q(\theta_1)$, $V(1,0) \cong Y_Q(\theta_2)$. We have,

(1) $$\theta_1^{(2)} = (12), \qquad \theta_1^{(5)} = (12\ 45\ 7),$$

(2) $$\theta_2^{(2)} = (25), \qquad \theta_2^{(5)} = (12\ 345).$$

Further, we have,

(3) $$w_1^{\max} = (5276143), \qquad \theta_1^{\max} = (2175463),$$

(4) $$w_2^{\max} = (5472163), \qquad \theta_2^{\max} = (5243176).$$

Note that w_1 (resp. w_2) is non-covexillary with $5, 6, 1, 4$ (resp. $4, 7, 2, 3$) as a Type II occurrence. Note also that the occurrence $1, 5, 4, 6$ in θ_1 (resp. $2, 4, 3, 7$ in θ_2) corresponds to the occurrence $5, 6, 1, 4$ in w_1 (resp. $4, 7, 2, 3$ in w_1).

CHAPTER 13

The cohomology and homology of the flag variety

In this chapter, we present results on $H^*(Fl(n))$, $H_*(Fl(n))$, the cohomology and homology of the flag variety $Fl(n)$. Our treatment here follows closely that of [**54**]. After reviewing $H^*(X)$, $H_*(X)$, Poincaré duality, Chern classes in §1-3, we describe an algebraic $\mathbb{Z}$-basis for $H^*(Fl(n))$ in §4. In §5 we give a presentation for the $\mathbb{Z}$-algebra $H^*(Fl(n))$. In §6, we describe a geometric basis for $H_*(Fl(n))$. Schubert polynomials are discussed in §7. These results are spelled out for the Grassmannian in §8 and the relationship between Littlewood-Richardson coefficients and the structure coefficients in $H^*(G_{d,m})$ is brought out.

1. $H^*(X)$ and $H_*(X)$

Let X be an algebraic variety. Let H_iX and H^iX be the i-th singular homology and cohomology of X with integer coefficients (see [**153**] for the definition of H_iX, H^iX, X being a topological space). The cohomology $H^*X = \oplus H^iX$ has a ring structure,

$$\begin{aligned} H^iX \otimes H^jX &\longrightarrow H^{i+j}X \\ \alpha \otimes \beta &\longmapsto \alpha \cup \beta \end{aligned}$$

(the cup product); it is an associative skew-commutative ring with $1 \in H^0X$. The homology $H_*X = \oplus H_iX$ is a left module over H^*X defined by

$$\begin{aligned} H^iX \otimes H_jX &\longrightarrow H_{j-i}X \\ \alpha \otimes \beta &\longmapsto \alpha \cap \beta \end{aligned}$$

(the cap product).

A morphism $f : X \to Y$ induces *"pullback"* homomorphisms

$$f^* : H^iY \to H^iX$$

and *"pushforward"* homomorphisms

$$f_* : H_iX \to H_iY.$$

Both f^* and f_* are functorial: $(g \circ f)_* = g_* \circ f_*$ and $(g \circ f)^* = f^* \circ g^*$, where g is a morphism $g : Y \to Z$. The pullback f^* is a homomorphism of rings, while the pushforward satisfies the *"projection formula"*

$$f_*(f^*(\alpha) \cap \beta) = \alpha \cap f_*(\beta),\ \alpha \in H^iY,\ \beta \in H_jX.$$

2. Poincaré duality

Let X be a smooth projective complex variety of dimension N. Then $H_{2N}(X) \cong \mathbb{Z}$ (since X is an oriented $2N$-dimensional real manifold), with a generator denoted $[X]$ and called the *fundamental class* of X. The map $H^iX \to H_{2N-i}X, \alpha \mapsto \alpha \cap [X]$ is an isomorphism (see **[153]** for details). The above isomorphism is referred to as the Poincaré duality map. By means of this isomorphism, we will identify the homology and cohomology groups of X.

3. Chern classes

Given a vector bundle E of rank r on a variety X, consider the corresponding projective bundle $\pi : \mathbb{P}(E) \to X$ whose fiber over a point in X is the projective space of lines through origin in E. Let $L = \mathcal{O}_{\mathbb{P}(E)}(1)$. Then L determines an element of $H^2(\mathbb{P}(E))$, namely the class of a hyperplane which is the zero set of a nonzero vector in E^* (here we work with real dimensions). Let us denote this class by ξ. Now, on any open set U of X over which E is trivial, ξ restricts to the class of a hyperplane in $\mathbb{P}(E|_U) = U \times \mathbb{P}^{r-1}$. From this it follows (by a Mayer-Vietoris argument) that $H^*(\mathbb{P}(E))$ is free over $H^*(X)$ with basis $\{1, \xi, \xi^2, \dots, \xi^{r-1}\}$. Hence there exist unique elements $a_1, \dots, a_r$, $a_i \in H^{2i}(X)$ such that $\xi^r + \pi^*(a_1)\xi^{r-1} + \cdots + \pi^*(a_{r-1})\xi + \pi^*(a_r) = 0$ in $H^{2r}(\mathbb{P}(E))$, i.e. $H^*(\mathbb{P}(E)) \cong H^*(X)[T]/(T^r + a_1T^{r-1} + a_{r-1}T + a_r)$, T mapping to ξ.

DEFINITION 3.0.1. The class $a_i \in H^{2i}(X)$ is defined to be the i-th *Chern class* of E, and is denoted $c_i(E)$, $1 \leqslant i \leqslant r$. We define $c_0(E)$ to be 1 and $c_i(E) = 0$ for $i < 0$ or $i > r$.

4. A $\mathbb{Z}$-basis for $H^*(Fl(n))$

Now take $X = Fl(n)(= \{\text{complete flags } F_\bullet : 0 = V_0 \subset V_1 \subset \cdots \subset V_n = \mathbb{C}^n, \dim V_i = i\})$. Let E_X be the trivial bundle of rank n over X. There is a universal filtration of submodules $0 = U_0 \subset U_1 \subset \cdots \subset U_n = E_X = E$, say, rank $U_i = i$. Over a point in X (corresponding to a flag $F_\bullet$) the fibers of these bundles are simply the vector spaces occurring in $F_\bullet$. Let $L_i = U_i/U_{i-1}$, $x_i = c_1(L_i)$.

THEOREM 4.0.1. *The set* $\{x_1^{i_1} x_2^{i_2} \dots x_n^{i_n}, 0 \leqslant i_j \leqslant n - j\}$ *is a* $\mathbb{Z}$*-basis for* $H^*(X)$.

PROOF. $X(= Fl(n))$ may be constructed as a sequence of projective bundles as follows: first we have $\mathbb{P}(E) \to \{\text{point}\}$, with $\mathcal{O}_{\mathbb{P}(E)}(1) = U_1 = L_1$, $c_1(L_1) = x_1$. Hence $H^*(\mathbb{P}(E))$ has a $\mathbb{Z}$-basis given by $\{1, x_1, x_1^2, \dots, x_1^{n-1}\}$. On $\mathbb{P}(E)$, we have the bundle E/U_1 of rank $n - 1$. Hence considering $\mathbb{P}(E/U_1) \to \mathbb{P}(E)$, we have

$$\mathcal{O}_{\mathbb{P}(E/U_1)}(1) = U_2/U_1 = L_2, \qquad c_1(L_2) = x_2,$$

and $H^*(\mathbb{P}(E/U_1))$ has a $H^*(\mathbb{P}(E))$-basis consisting of $\{1, x_2, \dots, x_2^{n-2}\}$; hence $H^*(\mathbb{P}(E/U_1))$ has a $\mathbb{Z}$-basis consisting of $\{x_1^{i_1} x_2^{i_2}, 0 \leqslant i_1 \leqslant n-1, 0 \leqslant i_2 \leqslant n-2\}$.

Continuing thus, we obtain a $\mathbb{Z}$-basis for $H^*\big(Fl(n)\big)$ consisting of $\{x_1^{i_1}x_2^{i_2}\ldots x_n^{i_n}, 0 \leqslant i_j \leqslant n-j\}$. □

COROLLARY 4.0.2. *$H^*(Fl(n))$ is free over $\mathbb{Z}$ of rank $n!$.*

5. A presentation for the $\mathbb{Z}$-algebra $H^*(Fl(n))$

We first recall

Whitney's Formula

Let $0 \to E' \to E \to E'' \to 0$ be an exact sequence of vector bundles on an algebraic variety X. Then $c_k(E) = \sum_{i+j=k} c_i(E') \cdot c_j(E'')$.

Now take $X = Fl(n)$ and $E = E_X$, the trivial bundle of rank n over X. From above, we have, E has a filtration with line bundles L_i as quotients. Hence it follows from Whitney formula that $c_i(E)$ is the i-th elementary symmetric polynomial in $c_1(L_i)(= x_i)$, $1 \leqslant i \leqslant n$. But $c_i(E) = 0$ for all i, since E is the trivial bundle. Thus we obtain $e_i(x_1, \ldots, x_n) = 0$, where $e_i(X_1, \ldots, X_n) = \sum_{i_1 < \cdots < i_j} X_{i_1} \ldots X_{i_j}$, the i-th elementary symmetric polynomial.

Let

$$R(n) = \mathbb{Z}[X_1, \ldots, X_n]/\langle e_i(X_1, \ldots, X_n),\ i = 1, \ldots, n\rangle,$$

where $\mathbb{Z}[X_1, \ldots, X_n]$ is the polynomial algebra over $\mathbb{Z}$ in n variables.

THEOREM 5.0.1. *The map*

$$f : \mathbb{Z}[X_1, \ldots, X_n] \longrightarrow H^*(Fl(n))$$
$$X_i \longmapsto x_i$$

defines an isomorphism of $R(n)$ onto $H^*(Fl(n))$.

PROOF. From above, f induces a surjective homomorphism

$$f_n : R(n) \longrightarrow H^*(Fl(n)).$$

In order to show that f_n is an isomorphism, we shall show that the images of $X_1^{i_1} \ldots X_n^{i_n}$, $0 \leqslant i_j \leqslant n-j$ in $R(n)$ span $R(n)$. Clearly they are linearly independent in $R(n)$, since their images under f_n are linearly independent in $H^*(Fl(n))$ (cf. Theorem 4.0.1). Let us denote by y_i the image of X_i in $R(n)$. We have (in $R(n)$)

$$\prod_{i=1}^{n}(1 - y_i t) = \sum_{k=0}^{n}(-1)^k e_k(y_1, \ldots, y_n)t^k = 1.$$

Hence for $1 \leqslant p \leqslant n$ we have (in $R(n)$)

$$(*) \qquad \prod_{i=1}^{p} \frac{1}{(1 - y_i t)} = \prod_{i=p+1}^{n} (1 - y_i t).$$

The left hand side of $(*)$ equals $\sum_{i\geqslant 0} h_i(y_1,\dots,y_p)t^i$, where $h_i(y_1,\dots,y_p)$ is the i-th complete symmetric polynomial in $y_1,\dots,y_p$ (i.e., the sum of monomials in $y_1,\dots,y_p$ of degree i). Comparing the coefficients of t^i in $(*)$, we get $h_i(y_1,\dots,y_p)=0, i>n-p$. In particular, using $h_{n-p+1}(y_1,\dots,y_p)(=y_p^{n-p+1}+\dots)=0$, $1\leqslant p\leqslant n$, we get $y_p^{n-p+1}=\sum a_I \underline{y}^I$, where $\underline{y}^I=y_1^{i_1}\dots y_p^{i_p}$, with $i_p\leqslant n-p$. From this, it follows (by decreasing induction on p) that

$$\{y_1^{i_1}\dots y_m^{i_n},\ 0\leqslant i_j\leqslant n-j\}$$

spans $R(n)$. The required result now follows. □

COROLLARY 5.0.2. (A presentation for $H^*(Fl(n))$) *The $\mathbb{Z}$-algebra $H^*(Fl(n))$ has a set of algebra generators $\{x_i,\ 1\leqslant i\leqslant n\}$ and the ideal of relations is generated by $\{e_i(x_1,\dots,x_n),\ 1\leqslant i\leqslant n\}$.*

6. The homology $H_*(Fl(n))$

For $w\in W(=S_n)$, we shall denote by $[X_w]$ the class determined by the Schubert variety X_w in $H_{2l(w)}(Fl(n))$ (recall that $\dim X(w)=l(w)$). For $w\in W$, let Y_w be the closure of the B^--orbit $B^-\cdot e_w$, where B^- is the Borel subgroup opposite to B.

LEMMA 6.0.1. *Let $u,v\in S_n$. If $X_w\cap Y_v\neq\emptyset$, then $l(v)\leqslant l(u)$, with equality if and only if $u=v$. The varieties X_w and Y_w meet transversally at e_w.*

PROOF. Now X_u, Y_v being stable under B, B^- respectively, $Z:=X_u\cap Y_v$ is T-stable, where T is the maximal torus given by $T=B\cap B^-$. Hence Z must contain a T-fixed point e_w, for some $w\in S_m$. Now e_w belongs to X_u implies $l(w)\leqslant l(u)$; similarly e_w is in Y_v implies that $l(w)\geqslant l(v)$. We obtain $l(v)\leqslant l(w)\leqslant l(u)$, and hence $l(v)=l(u)$ is equivalent to $v=u=w$. The first part of the statement now follows.

Now $X_w=B\cdot e_w\dot{\bigcup}_{u<w}X_u$, $Y_w=B^-\cdot e_w\dot{\bigcup}_{v>w}Y_v$. Hence from the first part we obtain $Y_w\cap X_u=\emptyset$, $u<w$, $X_w\cap Y_v=\emptyset$, $v>w$. This implies $X_w\cap Y_w=B\cdot e_w\cap B^-\cdot e_w$. Clearly, $B\cdot e_w$ and $B^-\cdot e_w$ intersect transversally at e_w. □

The pairing $\langle\,,\,\rangle$ on $H_*(Fl(n))$

There exists a bilinear map

$$\langle\,,\,\rangle : H_{2d}\big(Fl(n)\big)\times H_{2N-2d}\big(Fl(n)\big)\longrightarrow H_0\big(Fl(n)\big)(\cong\mathbb{Z}),$$

where $N=\binom{n}{2}(=\dim Fl(n))$. This has the property that for a pair of classes $[E]$, $[F]$ of two closed subvarieties of complementary dimension, $\langle[E],[F]\rangle=0$ or 1, according as $E\cap F=\emptyset$ or E and F meet transversally at a point. This together with Lemma 6.0.1 implies

$$(*)\qquad \big\langle[X_u],[Y_v]\big\rangle=\delta_{uv},$$

where u, v vary over permutations of given length d. Hence it follows that the classes $[X_w]$, $w \in S_n$ are linearly independent, and there are $n!$ of these classes. Hence we obtain that $\{[X_w], w \in S_n\}$ is a $\mathbb{Q}$-basis for $H_*(Fl(n)) \otimes \mathbb{Q}$ (note that in view of Corollary 4.0.2 and Poincaré duality, we have, $H_*(Fl(n))$ is free over $\mathbb{Z}$ of rank $n!$). But if a homology class is written as a rational linear combination of the classes $[X_w]$, then (*) implies that all these coefficients are in fact integers. Hence we obtain

THEOREM 6.0.2. (A geometric basis for $H_*(Fl(n))$) *The classes*

$$\{[X_u], l(u) = d\}$$

form a $\mathbb{Z}$-basis for $H_{2d}(Fl(n))$, and the classes

$$\{[Y_v], l(v) = d\}$$

form a dual basis for $H_{2N-2d}(Fl(n))$.

REMARK 6.0.3. We have in fact $[Y_w] = [X_{w_0 w}]$, for $w \in S_m$, w_0 being the permutation in S_n of largest length, namely $w_0 = (n\,n-1\ldots 21)$ (see **[54]** for a proof).

7. Schubert polynomials

Let $w \in S_n$ and let σ_w be the class in $H^{2l(w)}(Fl(n))$ corresponding to $[Y_w]$ in $H_{2N-l(w)}(Fl(n))$ under Poincaré duality; in particular, we have $[\sigma_{s_i}] = x_1 + \cdots + x_i$, $1 \leqslant i \leqslant n-1$. We shall now define certain canonical lifts in $\mathbb{Z}[X_1, \ldots, X_n]$ for the Schubert classes σ_w, $w \in S_n$, and these lifts will be called *Schubert polynomials* (cf. **[112]**).

7.1. The difference operators ∂_i on $\mathbb{Z}[X_1, \ldots, X_n]$

For $P \in \mathbb{Z}[X_1, \ldots, X_n]$ define $\partial_i(P) = \dfrac{P - s_i(P)}{X_i - X_{i+1}}$, $1 \leqslant i \leqslant n-1$, where $s_i(P)$ is the polynomial obtained by switching the variables X_i and X_{i+1}.

REMARK 7.1.1. (1) Note that $P - s_i(P)$ is divisible by $X_i - X_{i+1}$.

(2) If P is homogeneous of degree d, then $\partial_i(P)$ is homogeneous of degree $d-1$.

(3) $\partial_i(P)$ is symmetric in X_i and X_{i+1}.

(4) $\partial_i(P) = 0 \iff s_i(P) = P$, i.e., if and only if P is symmetric in X_i and X_{i+1}. Hence $\partial_i(\partial_i(P)) = 0$, for any P.

(5) $\partial_i(PQ) = \partial_i(P)Q + s_i(P)\partial_i(Q)$.

The polynomial $\mathfrak{S}_w$

Let $w \in S_m$; let $w_0 = (n\,n-1\ldots 21)$. Set

$$\mathfrak{S}_{w_0} = X_1^{n-1}X_2^{n-2}\ldots X_{n-2}^2 X_{n-1}.$$

Write $w = w_0 s_{i_1} s_{i_2} \ldots s_{i_r}$, where $l(w_0 s_{i_1} s_{i_2} \ldots s_{i_p}) = l(w_0) - p$, $1 \leqslant p \leqslant r$, and define

$$\mathfrak{S}_w := \partial_{i_1} \circ \cdots \circ \partial_{i_r}(\mathfrak{S}_{w_0}).$$

REMARK 7.1.2. $\mathfrak{S}_w$ is independent of the decomposition chosen.

Examples

1. $\mathfrak{S}_{\text{id}} = 1$.
2. $\mathfrak{S}_{s_i} = X_1 + X_2 + \cdots + X_i$.
3. In S_3, $\mathfrak{S}_{w_0} = X_1^2 X_2$, $\mathfrak{S}_{s_1 s_2} = X_1 X_2$, $\mathfrak{S}_{s_2 s_1} = X_1^2$.

PROPOSITION 7.1.3. *Let $w \in S_m$. We have*

(1) $\partial_i(\mathfrak{S}_w) = \mathfrak{S}_{ws_i}$, *if $w > ws_i$.*
(2) $\partial_i(\mathfrak{S}_w) = 0$, *if $w < ws_i$.*
(3) $\mathfrak{S}_w = \sum_I a_I X^I$, *where $X^I = x_1^{i_1} \dots x_n^{i_n}$ and the sum runs over $I = (i_1, \dots, i_n)$ with $0 \leqslant i_j \leqslant n - j$ for all j, and $a_I \in \mathbb{Z}_+$.*

See [54] for a proof.

Let $S_\infty = \cup S_m$ (here, for $k > m$, we identify S_m with the subgroup of S_k leaving $m+1, m+2, \cdots, k$ fixed).

We have, for $w \in S_\infty$, $\mathfrak{S}_w \in \mathbb{Z}[X_1, \dots, X_n]$ if and only if $w(i) < w(i+1)$, $i \geqslant n$.

THEOREM 7.1.4. (1) *$\{\mathfrak{S}_w \mid w \in S_\infty, w(i) < w(i+1), i \geqslant n\}$ is a $\mathbb{Z}$-basis for $\mathbb{Z}[X_1, X_2, \dots, X_n]$.*

(2) *Under the isomorphism*

$$f : \mathbb{Z}[X_1, X_2, \dots, X_n]/\langle e_i, i = 1, \dots, n\rangle \longrightarrow H^*(Fl(n)),$$

we have $f(\mathfrak{S}_w) = \sigma_w$.

See [54] for a proof.

REMARK 7.1.5. In view of Theorem 7.1.4, multiplication of two classes $\sigma_{w_1} \cdot \sigma_{w_2}$ is known by knowing $\mathfrak{S}_{w_1} \cdot \mathfrak{S}_{w_2}$ as a $\mathbb{Z}$-linear combination of the $\mathfrak{S}_w$'s.

8. Results for the Grassmannian

8.1. Relationship between Schur polynomials and Schubert polynomials

Fix d, $1 \leqslant d \leqslant n-1$. Then the canonical projection

$$\begin{aligned} Fl(n) \quad &\longrightarrow G_{d,n} \\ F_. = (V_0 \subset V_1 \subset \cdots \subset V_n = \mathbb{C}^n) &\longmapsto V_d \end{aligned}$$

identifies $H^*(G_{d,n})$ as a subring of $H^*(Fl(n))$. Under this identification, for $\underline{i} \in I_{d,n}$, $\sigma_{\underline{i}}$ gets identified with the class (in $\mathbb{Z}[X_1, \dots, X_n]/\langle e_i(X_1, \dots, X_n), i = 1, \dots, n\rangle$) of the Schubert polynomial $\mathfrak{S}_w$, where w is the permutation with the entries of $\underline{i}$ in the first d places, and the rest of the entries appearing in ascending order (such a permutation is called a *Grassmannian permutation*).

Now $\underline{i}$ determines a Young diagram $\lambda : (\lambda_1 \geqslant \dots \geqslant \lambda_d)$, where $\lambda_t = i_{d+1-t} - (d+1-t)$. We have (cf. [112]), the Schubert polynomial $\mathfrak{S}_w$ is simply the Schur polynomial $S_\lambda(X_2, \dots, X_d)$ (cf. Chapter 5).

8.2. Schubert classes and Littlewood-Richardson coefficients

Let $G = SL_n(\mathbb{C})$. Recall (cf. Chapter 5) that the finite dimensional irreducible representations of $SL_n(\mathbb{C})$ are indexed by Young diagrams with columns of length at most $n-1$, two Young diagrams $\lambda = (\lambda_1 \geqslant \lambda_2 \geqslant \cdots \geqslant \lambda_{n-1} \geqslant 0)$, $\mu = (\mu_1 \geqslant \mu_2 \geqslant \cdots \geqslant \mu_{n-1} \geqslant 0)$ being identified if $\lambda_i - \mu_i$ is the same for all i. Let λ, μ be two Young diagrams. Let V^λ, V^μ denote the irreducible G-modules with highest weight λ, μ respectively. Let

$$V^\lambda \otimes V^\mu = \bigoplus c^\nu_{\lambda\mu} V^\nu.$$

The integers $c^\nu_{\lambda\mu}$ are called the *Littlewood-Richardson coefficients*. From Chapter 5, we have, for $g \in G$,

$$\chi_{V^\lambda}(g) = S_\lambda(x_1, \ldots, x_m),$$

where $\chi_{V^\lambda}(g) : G \to k^*$ is the character map sending g to the trace of g, and S_λ is the Schur polynomial (here, $x_1, \ldots, x_m$ are the eigen values of g on V^λ). Hence we obtain

$$S_\lambda \cdot S_\mu = \sum c^\nu_{\lambda\mu} S_\nu.$$

On the other hand S_λ, S_μ, S_ν coincide with the respective Schubert polynomials $\mathfrak{S}_\tau$, $\mathfrak{S}_\phi$, $\mathfrak{S}_\theta$, where τ, ϕ, θ are the Grassmannian permutations determined by λ, μ, ν respectively. From our discussion above, we obtain that

$$[\sigma_\tau] \cdot [\sigma_\phi] = \sum c^\nu_{\lambda\mu} [\sigma_\theta].$$

Thus the coefficients appearing in the product of two Schubert classes in $H^*(Fl(n))$ are simply the Littlewood-Richardson coefficients.

REMARK 8.2.1. The above result was first proved by Lesieur (cf. [**113**]). The proof in [**113**] is carried out by showing that the problems of the determination of the Littlewood-Richardson coefficients and the structure coefficients in $H^*(G_{d,n})$ are both controlled by the same algebra of symmetric functions.

Index

List of symbols

Bibliography

[1] S. ABEASIS, *Codimension 1 orbits and semi-invariants for the representations of an oriented graph of type A_n*, Trans. Amer. Math. Soc. 282 (1984), 463–485.

[2] S. ABEASIS, A. DEL FRA, *Degenerations for the representations of a quiver of type A_m*, J. of Alg., 93 (1985), 376-412.

[3] S. ABEASIS, A. DEL FRA, H. KRAFT, *The geometry of representations of A_m*, Math. Ann., 256 (1981), 401–418.

[4] S.S. ABHYANKAR, *Enumerative combinatorics of Young tableaux*, Monographs and Textbooks in Pure and Applied Mathematics, 115, Marcel Dekker, Inc., New York, 1988.

[5] M. AIGNER, *Combinatorial Theory*, Springer-Verlag, New York, 1979.

[6] H.H. ANDERSON, *Schubert varieties and Demazure's character formula*, Inv. Math., 79 (1985), 611-618.

[7] H.H. ANDERSEN, J.C. JANTZEN, AND W. SOERGEL, *Representations of quantum groups at a p-th root of unity and semisimple groups in characteristic p: Independence of p*, Astèrisque 220 (1994), 1-320.

[8] A. ARABIA, *Cohomologie T-quivariante de G/B pour un groupe G de Kac-Moody*, C.R. Acad. Sci. Paris Sr. I Math. 302 (1986), no. 17, 631–634.

[9] M. AUSLANDER, I. REITEN, S. SMALO, *Representation theory of Artin algebras*, Cambridge University Press, 1995.

[10] A. BEILINSON AND J. BERNSTEIN, *Localization of $\mathfrak{g}$-modules*, C. R. Acad.Sci. Paris Ser. I Math, 292 (1981), 15–18.

[11] I. BERNSTEIN, I. GELFAND, AND S. GELFAND, *Structure of representations generated by highest weight vectors*, Funct. Anal. and Appl., 5 (1971), 1–8.

[12] J. BERNSTEIN, I. GELFAND, V. PONOMAREV *Coxeter functors and Gabriel's theorem*, Russ. Math. Surveys, 28 (1973), 17-32.

[13] ———, *Schubert Cells and Cohomology of the Spaces G/P*, Russian Math. Surveys, 28 (1973), 1–26.

[14] S. BILLEY, *Kostant polynomials and the cohomology ring for G/B*, Proc. Nat. Acad. Sci. U.S.A., 1 (1997), 29–32.

[15] S. C. BILLEY, *Pattern avoidance and rational smoothness of schubert varieties*, to appear in Adv. in Math.

[16] S. BILLEY AND V. LAKSHMIBAI, *On the singular locus of a Schubert variety*, to be published by Birkhaüser.

[17] A. BJÖRNER, *Shellabel and Cohen-Macaulay partially ordered sets*, Trans. A.M.S., 260 (1980), 159-183.

[18] A. BJÖRNER AND M. WACHS, *Bruhat order of classical groups and shellability*, Adv. Math., 43 (1982), 87-100.

[19] A. BJÖRNER AND M. WACHS, *On lexicographically shellable posets*, Trans. A.M.S., 277 (1983).

[20] B.D. BOE, *Kazhdan-Lusztig polynomials for Hermitian symmetric spaces*, Trans. Amer. Math. Soc., 309, (1988), 279-294.

[21] M. BÓNA, *The permutation classes equinumerous to the smooth class*, Elec. J. Combinatorics, 5 (1998).

[22] A. BOREL, *Linear algebraic groups*, GTM, 126, Second edition, Springer-Verlog, New York, 1991.

[23] A. BOREL, *Intersection Cohomology*, Birkhauser, 1984.

[24] A. BOREL, *Représentations linéaires et espaces homogènes Kähleriens des groupes simples compacts* (1954), Coll. Papers I, 392-396, Springer-Verlog 1983.

[25] R. BOTT AND H. SAMELSON, *Application of the theory of Morse to symmetric spaces*, Amer. J. Math., 80 (1958), 964-1029.

[26] R. BOTT *Homogeneous vector bundles*, Ann. Math., series 2, 66 (1957), 203-248.

[27] N. Bourbaki, *Groupes et Algèbres de Lie*, Chapitres 4, 5 et 6, Hermann, Paris, 1968.

[28] F. BRENTI, *Combinatorial expansions of Kazhdan-Lusztig polynomials*, J. London Math. Soc., 55 (1997), 448–472.

[29] ———, *Kazhdan-Lusztig polynomials and r-polynomials from a combinatorial point of view*, Discrete Math, (to appear).

[30] M. BRION, *Equivariant cohomology and equivariant intersection theory*, Representation Theories and Algebraic Geometry, NATO ASI series, Series C: Mathematical and Physical Sciences, vol. 514, 1-38, Kluwer Academic Publishers.

[31] M. BRION AND P. POLO, *Generic singularities of certain Schubert varieties*, preprint (1998).

[32] J.-L. BRYLINSKI AND M. KASHIWARA, *Kazhdan-Lusztig conjectures and holonomic systems*, Invent. Math., 64 (1981), 387–410.

[33] J. B. CARRELL, *The Bruhat graph of a Coxeter group, a conjecture of Deodhar, and rational smoothness of Schubert varieties*, Proceedings of Symposia in Pure Math., 56 (1994), 53–61.

[34] J.CARRELL, *On the smooth points of a Schubert variety*, CMS Conference Proceedings, vol 16, 15-24, Proceedings of the coference on " Representations of Groups: Lie, Algebraic, Finite,and Quantum", Banff, Alberta, June 1994.

[35] J.CARRELL AND R. KUTTLER, *On the Smooth Points of T-stable varieties in G/B and the Peterson Map*, preprint (1999)

[36] C. CHEVALLEY, *Classification de groups de Lie algébriques*, Séminaire, 1956-58, Secrétariat mathématique, vol II, rue Pierrie-Curie, Paris, 1958.

[37] C. CHEVALLEY, *Sur les décompositions cellulaires des spaces G/B*, Proc. Symp. Prue. Math., 56 (1994), Part I, 1-25.

[38] D. Cox, J. Little and D. O'Shea, Ideals, Varieties and Algorithms, Springer-Verlag, New York, 1992.

[39] C. DE CONCINI, D. EISENBUD AND C. PROCESI, *Hodge Algebras*, Astérisque 91 (1982).

[40] C. DE CONCINI AND C. PROCESI, *A characteristic-free approach to invariant Theory*, Adv. Math., 21 (1976), 330-354.

[41] C. DECONCINI, E. STRICKLAND, *On the variety of complexes*, Adv. in Math., 41 (1981), 57-77.

[42] M. DEMAZURE, *Désingularisation des variétés de Schubert généralisées*, Ann. Sci. E.N.S., 7 (1974), 53-88

[43] V. DEODHAR, *Local Poincaré duality and non-singularity of schubert varieties*, Comm. Algebra, 13 (1985), 1379–1388.

[44] V. DEODHAR, *A combinatorial settting for questions in Kazhdan-Lusztig theory*, Geom. Dedicata, 36 (1990).

[45] V. DEODHAR, *A brief survey of Kazhdan-Lusztig theory and related topics*, Proceedings of Symposia in Pure Math, 56 (1994).

[46] P. DOUBILET, G.C. ROTA AND J. STEIN, *On the foundations of combinatorial Theory IX*, Studies in Applied Math., 53 (1974), 185-216.

[47] J. EGON AND M. HOCHSTER, *Cohen-Macaulay rings, invariant theory and the generic perfection of determinantal loci*, Amer. J. Math., 93 (1971), 1020-1058.

[48] C. EHRESMANN, *Sur la topologie des certaines espaces homogenes*, Ann. Math., 35 (1934), 396-443.

[49] D. EISENBUD, *Commutative algebra with a view toward Algebraic Geometry*, Springer-Verlag, GTM, 150.

[50] D. EISENBUD, *Introduction to algebras with straightening laws*, Proceedings of the third Oklahoma conference on Ring Theory and Algebra, edited by B. Mac Donald, Marcel-Dekker (1980).

[51] D. EISENBUD AND B. STURMFELS, *Binomial Ideals*, Duke Math. Journal 1, 84 (1996), 1-45.

[52] W. FULTON, *Universal Schubert polynomials*, Duke Math. J., 96 (1999), no:3, 575-594.

[53] W. FULTON, *Introduction to Toric Varieties*, Annals of Math. Studies 131, Princeton U. P., Princeton N. J., 1993.

[54] W. FULTON, *Young Tableaux; with applications to representation theory and geometry*, London Mathematical Society Student Texts, vol 35, Cambridge University Press, New York, 1997.

[55] W. FULTON AND J. HARRIS, *Representation Theory*, a first course, Grad. Texts in Math., 129, Springer-Verlog.

[56] P. GABRIEL, V. DLAB (eds.), Proc. ICRA II: Ottawa 1979, Springer LNM 831.

[57] V. GASHAROV, Factoring the Poincaré polynomials for the Bruhat order on s_n, to appear in J. Comb. Theory.

[58] V. GASHAROV, *Sufficiency of Lakshmibai-Sandhya singularity conditions for Schubert varieties*, to appear in Compositio Mathematica.

[59] D. GLASSBRENNER AND K. E. SMITH, *Singularities of certain ladder determinantal varieties*, Journal of Pure and Applied Algebra 100 (1995), 59-75.

[60] N. GONCIULEA, *Singular loci of varieties of complexes II*, to appear in J. Alg.

[61] N. GONCIULEA AND V. LAKSHMIBAI, *Degenerations of Flag and Schubert varieties to Toric varieties*, Transformation Groups, vol 1, no:3 (1996), 215-248.

[62] N. GONCIULEA AND V. LAKSHMIBAI, *Singular loci of Schubert varieties and Ladder determinantal varieties*, to appear in J. Alg.

[63] M. GORESKY AND R. MACPHERSON, *Intersection homology-II*, Invent. Math., 72 (1983), 77–129.

[64] A. GROTHENDIECK, *Local cohomology*, Lecture Notes in Mathematics 41, Springer-Verlag (1966).

[65] A. GROTHENDIECK, J.A. DIEUDONNE - EGA (Eléments de Géométrie Algébrique), Chapter IV, Third part, I.H.E.S. 32 (1967).

[66] H. HANSEN, *On cycles in flag manifolds*, Math. Scand. 33 (1973), 269-274.

[67] R. HARTSHORNE, Algebraic Geometry, Springer-Verlog, Graduate Texts in Mathematics 52.

[68] J. HERZOG AND N.V. TRUNG, *Gröbner bases and multiplicity of determinantal and pfaffian ideals*, Adv. Math., 96 (1992), 1-37.

[69] T. Hibi, *Distributive lattices, affine semigroup rings, and algebras with straightening laws*, Commutative Algebra and Combinatorics, Advanced Studies in Pure Math. 11 (1987), 93-109.

[70] F. HIRZEBRUSCH, *Topological methods in Algebraic Geometry*, Spriger-Verlag, 1978.

[71] W.V.D. HODGE, *Some enumerative results in the theory of forms*, Proc. Camb. Phil. Soc. 39 (1943), 22-30.

[72] W.V.D. HODGE AND D. PEDOE, *Methods of Algebraic Geometry, Vol.II*, Cambridge University Press, (1952).

[73] JAMES E. HUMPHREYS, *Linear Algebraic Groups*, volume 21 of *Graduate texts in mathematics*. Springer-Verlag, New York, 1975.

[74] JAMES E. HUMPHREYS. *Reflection groups and Coxeter groups*. Cambridge University Press, 1990.

[75] JAMES E. HUMPHREYS, *Introduction to Lie Algebras and Representation Theory*, GTM 9. Springer-Verlag, New York, 1972.

[76] C. HUNEKE AND V. LAKSHMIBAI, *A characterization of Kempf varieties and the geometric consequences*, J. Alg., 94 (1985), 52-105.

[77] C. HUNEKE AND V. LAKSHMIBAI, *Degeneracy of Schubert varities*, Contemporary Math, 139 (1992), 181-235.

[78] J.C. JANTZEN, *Representations of algebraic groups*, Academic Press, 1987.

[79] M. KASHIWARA AND T. TANISAKI, *Kazhdan-Lusztig conjecture for affine Lie algebras with a negative level*, RIMS 983, (1994).

[80] D. KAZHDAN AND G. LUSZTIG, *Representations of Coxeter groups and Hecke algebras*, Inv. Math., 53 (1979), 165–184.

[81] D. KAZHDAN AND G. LUSZTIG, *Schubert varieties and Poincaré duality*, Proc. Symp. Pure. Math., A.M.S., 36 (1980), 185–203.

[82] D. KAZHDAN AND G. LUSZTIG, *Affine algebras and quantum groups*, Duke Math. Journal, 1991.

[83] G. KEMPF, *Linear systems on homogeneous spaces*, Ann. Math., 103 (1976), 557-591.

[84] G. KEMPF ET AL, *Toroidal Embeddings*, Lecture notes in Mathematics 339, Springer-Verlag, 1973.

[85] G. KEMPF AND A. RAMANATHAN, *Multicones over Schubert varieties*, Inv. Math., 87 (1987), 353-363.

[86] B. KOSTANT AND S. KUMAR, *The Nil Hecke Ring and Cohomology of G/P for a Kac-Moody Group G^**, Advances in Math., 62 (1986), 187–237.

[87] S. KUMAR, *The nil Hecke ring and singularities of Schubert varieties*, Inv. Math.,123, 1996, 471-506.

[88] S. KUMAR AND P. LITTELMANN, *Frobenius splitting in characteristic zero and the quantum Frobenius map*, preprint 1999.

[89] V. LAKSHMIBAI, *Kempf varieties*, J. Ind. Math. Soc., Vol. 40 (1976), 299-349.

[90] V. LAKSHMIBAI, *Geometry of G/P - VI*, J. Alg., 108 (1987), 355-402.

[91] V. LAKSHMIBAI, *Bases for quantum Demazure modules*, CMS Conference Proceedings, 16, 199-216, Proceedings of the conference on " Representations of Groups: Lie, Algebraic, Finite, and Quantum", Banff, Alberta, June 1994.

[92] V. LAKSHMIBAI, *Tangent spaces to Schubert Varieties*, Math. Research Letters 2 (1995), 473-477.

[93] V. LAKSHMIBAI, *On Tangent Spaces to Schubert Varieties-I*, to appear in J. Algebra.

[94] V. LAKSHMIBAI, *On Tangent Spaces to Schubert Varieties -II*, to appear in J. Algebra.

[95] V. LAKSHMIBAI, *Singular Loci of Varieties of Complexes*, to appear in J. Alg. Combinatorics.

[96] V. LAKSHMIBAI, *Kempf varieties*, J. Ind. Math. Soc., 40 (1976), 299-349.

[97] V. LAKSHMIBAI, P. LITTELMANN AND C.S. SESHADRI, *Schubert varieties*, to be published by Birkhaüser.

[98] V. LAKSHMIBAI AND P. MAGYAR, *Degeneracy schemes, Schubert varieties and Quiver varieties*, International Math. Research Notices, no:12, 1998, 627-640.

[99] V. LAKSHMIBAI, C. MUSILI AND C.S. SESHADRI, *Cohomology of line bundles on G/B*, Ann. E.N.S., t. 7 (1974), 89-137.

[100] V. LAKSHMIBAI, C. MUSILI, AND C. S. SESHADRI, *Geometry of $G/P - III$*, Proc. Ind. Acad. Sci., 87A (1978), 93-177.

[101] V. LAKSHMIBAI, C. MUSILI AND C.S. SESHADRI, *Geometry of G/P IV*, Proc. Ind. Acad. Sci., 88A (1979), 279-362.

[102] V. LAKSHMIBAI, C. MUSILI, AND C. S. SESHADRI, *Geometry of G/P*, Bull. Amer. Math. Soc., (1979), 432-435.

[103] V. LAKSHMIBAI AND B. SANDHYA, *Criterion for smoothness of Schubert varieties in $SL(n)/B$*, Proc. Indian Acad. Sci. (Math. Sci.), **100**, (1990), 45-52.

[104] V. LAKSHMIBAI AND C.S. SESHADRI *Geometry of G/P - II*, Proc. Ind. Acad. Sci., 87A (1978), 1-54.

[105] V. LAKSHMIBAI AND C.S. SESHADRI, *Singular locus of a Schubert variety*, Bull. A.M.S., 11 (1984), 363-366.

[106] V. LAKSHMIBAI AND C.S. SESHADRI, *Geometry of G/P - V*, J. Alg., 100 (1986), 462-557.

[107] V. LAKSHMIBAI AND C. S. SESHADRI, *Standard monomial theory and Schubert varieties*-a survey, Proccedings of the Conference on "Algebraic Groups and Applications," 1991, 279-322, Manoj Prakashan.

[108] V. LAKSHMIBAI AND J. WEYMAN, *Multiplicities of points on a Schubert variety in a minuscule G/P*, Advances in Math., 84 (1990), 179-208.

[109] A. LASCOUX, *Foncteurs de Schur et Grassmannienne*, Thèse, Université de Paris, VII, 1977.

[110] A. LASCOUX, *Polynômes de Kazhdan-Lusztig pour les variétés de Schubert vexillaires*, C. R. Acad. Sci. Paris Série I,, 321, (1995), 511-516.

[111] A. LASCOUX AND M.-P. SCHÜTZENBERGER, *Polynômes de Kazhdan and Lusztig pour les grassmanniennes*, Astérisque, 87–88 (1981), p.249–266. Young tableaux and Schur functions in algebra and geometry (Toruń, 1980).

[112] A. LASCOUX AND M.P. SCHUTZENBERGER, *Schubert polynomials and Littlewood-Richardson rule*, Letters in Math. Phys., 10 (1985), 111-124.

[113] L.LESIEUR, *Les problemes d'intersection sur une variété de Grassmann*, C.R. Acad. Sci. 225 (1947), 916-917.

[114] P. LITTELMANN, *A Littelwood-Richardson rule for symmetrizable Kac-Moody algebras*, Inv. Math.,116 (1994), 329-346.

[115] P. LITTELMANN, *Contracting modules and Standard Monomial Theory for symmetrizable Kac-Moody algebras*, J. Amer. Math. Soc. II (1998), no. 3, 551-567.

[116] P. LITTELMANN, *The path model, the quantum Frobenius map and Standard Monomial Theory*, to appear in Algebraic groups and their representations, eds., R. Carter and J. Saxl Kluwer Academic Publishers.

[117] G. LUSZTIG, *Green functions and singularities of unipotent classes*, Adv. Math., 42 (1981), 169-178.

[118] G. LUSZTIG, *Modular representations and quantum groups*, Contemporary Mathematics 82, A.J.M., (1989), 59-77.

[119] I. G. MACDONALD, *The Poincaré series of a coxeter group*, Math. Ann., 199 (1972), 161–174.

[120] J. MATIJEVIC, Three local conditions on a graded ring, Trans. A.M.S., 205 (1975), 275-284.

[121] H. MATSUMURA, *Commutative Algebra*, 2d ed, Math. Lecture Note Ser. 56, Benjamin/Cummings, Reading, Massachusettes., 1980.

[122] V.B. MEHTA AND A. RAMANATHAN, *Frobenius splitting and cohomology vanishing for Schubert varieties*, Ann. Math., 122 (1985), 27-40.

[123] S. B. MULAY, *Determinantal loci and the flag variety*, Adv. Math. 74 (1989), 1-30.

[124] D. MUMFORD, *The red book of varieties and schemes*, Lecture Notes in Mathematics 1358, Springer-Verlog.

[125] D. MUMFORD, *Abelian varieties*, Oxford University Press, Bombay, 1970.

[126] D. MUMFORD, *Complex projective varieties*, Grundlehren der math. Wissenschaften 221, Springer-Verlog, 1976.

[127] D. MUMFORD AND J. FOGARTY, *Geometric Invariant Theory*, Second Edition, Ergebnisse der Math. und ihrer Grenzgebiete 34, A series of modern surveys in Mathematics, Springer-Verlog.

[128] C. MUSILI *Postulation formula for Schubert varieties*. J. Indian. Math. Soc., 36 (1972), 143-171.

[129] C. MUSILI1, *Some properties of Schubert varieties*, J. Indian Math. Society, 38 (1974) 131-145.

[130] C. MUSILI AND C.S. SESHADRI, *Schubert varieties and the variety of complexes*, Arithmetic and Geometry, Progress in Math. 36,(1983), 329-359.

[131] M.S. NARASIMHAN AND S. RAMANAN, *Moduli of vector bundles on a compact Riemann surface*, Ann. Math. 89 (1969), 14-51.

[132] P.E. NEWSTEAD, *Introduction to moduli problems and orbit spaces*, Tata Institute Lecture Notes published by Springer-Verlog, 1978.

[133] P. POLO, *On Zariski tangent spaces of Schubert varieties, and a proof of a conjecture of Deodhar*, Indag. Math., 5 (1994), 483-493.

[134] R. PROCTOR, *Classical Bruhat orders are lexicographic shellable*, J. Alg., 77 (1982), 104-126.

[135] S. RAMANAN AND A. RAMANATHAN, *Projective normality of Flag varieties and Schubert varieties*, Inv. Math., 79 (1985), 217-224.

[136] A. RAMANATHAN, *Schubert varieties are arithmetically Cohen-Macualay*, Inv. Math., 80 (1985), 283-294.

[137] A. RAMANATHAN, *Equations defining Schubert varieties and Frobenius splitting of diagonals*, Publ. Math. I.H.E.S., 65 (1987), 61-90.

[138] G. REISNER, *Cohen-Macaulay quotients of polynomial rings*, Adv. Math., 21 (1976).

[139] J. ROSENTHAL, *An explicit formula for the multiplicity of points on a classical Schubert variety*, preprint (1998).

[140] J. ROSENTHAL AND A. ZELEVINSKY, *Multiplicities of points on Schubert varieties in Grassmannians*, preprint (1999).

[141] P. SANKARAN AND P. VANCHINATHAN, *Small resolutions of Schubert varieties in symplectic and orthogonal Grassmannians* , Publ. RIMS, 30 (1994), 443-458.

[142] P. SANKARAN AND P. VANCHINATHAN, *Small resolutions of Schubert varieties and Kazhdan-Lusztig polynomials*, Publ. RIMS, 31 (1995). 465-480.

[143] M. SCHLESSINGER, *Rigidity of quotient singularities*,Inv. Math., 14 (1971), 17-26.

[144] H. SCHUBERT, *Kalkül der abzählenden geometrie*, Teubner, Leipzig, 1879, reprinted, Springer-Verlog 1979.

[145] I. SCHUR, *Über eine klasse von matrizen die sich einen gegebenen matrix zuordnen lassen*, Dissertation (1901), Berlin.

[146] I. SCHUR, *Über die darstellurg der symmetrischen und der alternierenden gruppe durch gebrochene lineare substitutionen*, Crelle's Journal, 139 (1911), 155-250.

[147] J.P. SERRE, *Faisceaux algèbriques cohérentsés*, Ann. Math., 61 (1955), 197-278.

[148] J.P. SERRE, *Algèbre locale multiplicités*, Lecture Notes in Mathematics, 11, Springer-Verlog (1965).

[149] C.S. SESHADRI, *Geometry of G/P-I*, C.P. Ramanujam: A Tribute, 207 (Springer-Verlag), published by Tata Institute, Bombay (1978), 207-239.

[150] C.S. SESHADRI, *Line bundles on Schubert varieties*, Bombay colloquium on vector bundles, (1984).

[151] C.S. SESHADRI, *Introduction to the theory of standard monomials*, Brandeis Lecture Notes 4, (1985).

[152] I. SHAFAREVICH, *Basic Algebraic Geometry*, Springer-Verlag, Berlin (1977).

[153] E.H. SPANIER, *Algebraic Topology*, published by McGraw-Hill, 1966.

[154] T. A. SPRINGER, *Quelques applications de la cohomologie d'intersection*, Astrisque, 92-93 (1982), pp. 249–273. Bourbaki Seminar, Col 1981/1982.

[155] Z. STANKOVA, *Forbidden subsequences*, Discrete Math., 132 (1994), 291–316.

[156] B. STURMFELS, *Gröbner bases and convex poytopes*, AMS University Lecture Series, 8, Providence, RI., (1995).

[157] P. SVANES, *Coherent cohomology on Schubert subschemes of flag schemes*, Adv. Math., 14 (1974), 369-453.

[158] A. VAN DEN HOMBERGH, *Note on a paper by Bernstein, Gelfand, Gelfand on Verma modules*, Proc. Konin. Neder. Aka.Amsterddam Ser. A, 77 (1974), 352–356.

[159] D.-N. VERMA, *Structure of certain induced representations of complex semisimple lie algebras*, Bull. Amer. Math. Soc., 74 (1968), 160–166.

[160] P. WEBB (ed.), *Representations of Algebras: Durham 1985*, Lond. Math. Soc. Lect. Note Ser. 116, Cambridge Univ. Press. (1986).

[161] A. WEIL *On algebraic groups and homogeneous spaces*, Amer. J. Math., 77 (1955), 493-512.

[162] H. WEYL, *The classical groups*, Princeton University Press, 1946.

[163] A. YOUNG *Quantitative substitutional analysis I-IX*, Proc. London. Math. Soc., 1901-1952.

[164] A. ZELEVINSKY, *Small resolutions of singularities of Schubert varieties*, Funt. Anal. Appl., 17, (1983), 142-144.

[165] A. ZELEVINSKY, *Two remarks on graded nilpotent classes*, Uspekhi Math. Nauk., 40 (1985), no. 1 (241), 199-200.

Tatsuo Suwa

Indices of Vector Fields and Residues of Singular Holomorphic Foliations

Vector fields arise naturally in many branches of mathematics. An interesting problem in geometry is to study the relation between the structure of the space and the property of vector fields that can exist on it. A typical example is the classical Poincaré-Hopf theorem, which relates the local indices of a vector field and the Euler-Poincaré characteristic of the manifold.

In this book, the invariant subvarieties are treated systematically and generalized to the ones for vector fields and holomorphic foliations on singular varieties.

The author tried to make the book as self-contained as possible. In fact, the preparatory material here could serve as an introduction to complex analytic geometry and the techniques explained should be effective as well for other problems related to characteristic classes.

Basic material. Further topics. Poincaré-Hopf indices and the Baum-Bott residues of holomorphic vector fields. Indices and residues of vector fields on singular varieties. Singular foliations on complex surfaces. Singular holomorphic foliations.

Le professeur Tatsuo Suwa appartient au département de Mathématiques de l'Université de Hokkaido, à Sapporo.

ISBN 2 7056 6361 4. Collection *Actualités mathématiques*. 240 x 175, 222 p. 210 F

Mutsuo Oka

Non-Degenerate Complete Intersection Singularity

Singularities which are encountered in many branches of mathematics are usually given by explicit polynomal equations. When these polynomials have non-degenerate Newton boundaries, it turns out that the geometry of the singularity is quite simple. In fact the combinatorial data for the Newton boundary is enough to understand the apparently complicated geometry of singularity. On the other hand, such a class of singularity is large enough to cover most of important singularities.

In this book is introduced an elementary method to study this class of singularities from the combinatorial data of the Newton boundary which is called *toric modifications.* The reader is not requested to have any knowledge of algebraic geometry or differential geometry. For those who are familiar with linear algebra, calculus and elementary topology, there will be no difficulty in understanding the main idea of this method.

This book is self-contained. Most of the assertions and proofs given here are very explicit. The reader will enjoy constructing explicitly the resolution of singularities and he will find it easy to compute Milnor numbers and other invariants using this resolution.

Le professeur Mutsuo Oka enseigne à l'Université de Tokyo.

The author takes care to present ideas in a form which can be easily generalised.
Mathematical Reviews

ISBN 2 7056 6343 6. Collection *Actualités mathématiques.* 240 x 175, 224 p. 220 F

Marie-Hélène Schwartz

Classes de Chern des ensembles analytiques

Les classes de Chern, classes de cohomologie représentant des caractères fondamentaux du fibré tangent, figurent parmi les principaux invariants des variétés analytiques complexes. Le présent travail de Marie-Hélène Schwartz traite de la généralisation de cette idée au cas d'ensembles analytiques complexes avec singularités. Le problème est considérablement plus difficile et pour le traiter, elle utilise une décomposition, appelée stratification de Whitney, d'un espace analytique complexe en sous-variétés analytiques qui satisfont certaines propriétés d'incidence et elle invente plusieurs techniques très intéressantes qui lui permettent de définir les classes de Chern au moyen de l'indice de champs de repères qui satisfont certaines conditions par rapport aux strates de la stratification de Whitney.

Ce travail reprend des idées que l'auteur avaient présentées en 1964. Il se trouve qu'après qu'elle ait donné sa première définition des classes de Chern des ensembles complexes singuliers, MacPherson a donné en 1974 une définition plus synthétique et moins technique, mais l'expérience a appris que la définition donnée par Marie-Hélène Schwartz était très utile à la fois d'un point de vue conceptuel et d'un point de vue pratique. Les techniques qu'elle introduit dans ce texte ont une portée plus grande que la simple définition des classes de Chern et aident à comprendre la structure des ensembles analytiques singuliers.

ISBN 2 7056 6393 2. Collection *Actualités mathématiques*. 240 x 175, 224 p. 210 F

Imprimé en France par
LIGHTNING SOURCE
www.lightningsource.fr

IMPRIM'VERT®
Votre imprimeur agit pour l'environnement

Printed by Amazon Italia Logistica S.r.l.
Torrazza Piemonte (TO), Italy

10638369R00190